WITHDRAWN

AF412764

This item must be returned by 10 a.m. on the last date shown below. A fine will be due if it is not returned on time.

DATE OF RETURN

BORROWABLE ONLY OVERNIGHT OR FOR THE WEEKEND

3 0 JAN 1996

1 2 APR 1999

UNIVERSITY COLLEGE LONDON

Gower Street London WC1E 6BT

LF4B   ucl.B0459.6.7.90

Spon's Mechanical and Electrical Services Price Book

Spon's Mechanical and Electrical Services Price Book

Edited by

DAVIS LANGDON & EVEREST
Chartered Quantity Surveyors

1996

Twenty seventh edition

London · Glasgow · Weinheim · New York · Tokyo · Melbourne · Madras

Published by E & FN Spon, an imprint of Chapman & Hall, 2–6 Boundary Row, London SE1 8HN

Chapman & Hall, 2–6 Boundary Row, London SE1 8HN, UK

Blackie Academic & Professional, Wester Cleddens Road, Bishopbriggs, Glasgow G64 2NZ, UK

Chapman & Hall GmbH, Pappelallee 3, 69469 Weinham, Germany

Chapman & Hall USA, 115 Fifth Avenue, New York NY10003, USA

Chapman & Hall Japan, ITP-Japan, Kyowa Building, 3F, 2–2–1 Hirakawacho, Chiyoda-ku, Tokyo 102, Japan

Chapman & Hall Australia, 102 Dodds Street, South Melbourne, Victoria 3205, Australia

Chapman & Hall India, R. Seshadri, 32 Second Main Road, CIT East, Madras 600 035, India

First edition 1968
Twenty seventh edition 1995
© 1995 E & FN Spon Ltd

Printed in Great Britain by
T. J. Press (Padstow) Ltd, Padstow, Cornwall

ISBN 0 419 20890 9
ISSN 0305-4543

Apart from any fair dealing for the purposes of research or private study, or criticism or review, as permitted under the UK Copyright Designs and Patents Act, 1988, this publication may not be reproduced, stored, or transmitted, in any form or by any means, without the prior permission in writing of the publishers, or in the case of reprographic reproduction only in accordance with the terms of the licences issued by the Copyright Licensing Agency in the UK, or in accordance with the terms of licences issued by the appropriate Reproduction Rights Organization outside the UK. Enquiries concerning reproduction outside the terms stated here should be sent to the publishers at the London address printed on this page.

The publisher makes no representation, express or implied, with regard to the accuracy of the information contained in this book and cannot accept any legal responsibility or liability for any errors or omissions that may be made.

A catalogue record for this book is available from the British Library.

♾ Printed on permanent acid-free text paper, manufactured in accordance with ANSI-/NISO Z39.48-1992 and ANSI/NISO Z39.48-1984 (Permanence of Paper).

UNIVERSITY
COLLEGE LONDON
LIBRARY

Preface

The twenty seventh edition of *Spon's Mechanical and Electrical Services Price Book* has a revised presentation to previous editions although the organisation of information is as the last edition. All material and labour pricing data has been developed and stored by electronic means to allow future updating and distribution. The order of the book reflects the order of estimating, first outline costs, followed by more detailed costs, then unit rate items. The major sections of the book, Rates of Wages and Working Rules and Materials Costs/Measured Work Prices, are sub-divided into Mechanical Installations and Electrical Installations.

Before referring to prices or other information in the book, readers are advised to study the `Directions' which precede each section of the Materials Costs/Measured Work Prices. As before, no allowance has been made in any of the sections for Value Added Tax.

Approximate Estimating sections give a broad scope of outline costs for various building types and all-in rates for certain elements and selected specialist activities; these are followed by elemental analysis and then by a quantified analysis of a specific building type.

The prime purpose of the Materials Costs/Measured Work Prices part is to provide average prices for mechanical and electrical and other engineering services enabling a Bill of Quantities to be priced, giving a reasonably accurate indication of the likely project cost. Supplementary information is included which will enable readers to make adjustments to suit their own requirements. It cannot be emphasised too strongly that it is not intended that prices should be used in the preparation of an actual tender without adjustment for the circumstances of the particular project such as productivity, locality, project size and current market conditions. Adjustments should be made to standard rates for time, location, local conditions, site constraints and any other factor likely to affect the costs of a specific scheme. Readers are referred to the build up of the gang rates, where allowances are included for supervision, labour related insurances etc., and to the pages which introduce each section, where the percentage allowances for overheads and profit are defined. However, it should be noted that other than these items, no other allowances have been included for specific preliminaries. In addition Materials Costs/Measured Work Prices for the electrical services have been printed on different coloured paper to distinguish them from the mechanical services section.

Readers are reminded of the service available in *Spon's Price Book Update*, where details of significant changes to the published information are given. The printed update is circulated free of charge every 3 months, until the publication of next year's *Price Book*, to those readers who have requested it. In order to receive this the coloured card bound in with this volume should be completed and returned.

As with previous editions the Editors invite the views of readers, critical or otherwise, which might usefully be considered when preparing future editions of this work.

While every effort is made to ensure the accuracy of the information given in this publication, neither the Editors nor Publishers in any way accept liability for loss of any kind resulting from the use made by any person of such information.

In conclusion, the Editors record their appreciation of the indispensable assistance received from many individuals and organisations in compiling this book.

DAVIS LANGDON & EVEREST
Chartered Quantity Surveyors
Princes House
39 Kingsway
London WC2B 6TP

Contents

Mechanical Installations *(continued)*

Electrical Installations

Electrical Installations *(continued)*

PART FOUR: DAYWORK

PART FIVE: FEES FOR PROFESSIONAL SERVICES

Construction Quality and Quality Standards

The European perspective

George Atkinson, Construction Consultant, UK

✦ A practical and accessible guide to the changes taking place in the Single Market

✦ Authoritative explanations of the making and use of EU codes, regulations, specifications and standards

Quality is a vital issue to be addressed by all construction professionals working in Europe today.

Construction Quality and Quality Standards gives clear, concise guidance to the making and use of codes, regulations and technical specifications in Europe.

The author presents a general overview to construction before going on to specific issues relating to quality and the relevant British Standards. This is an invaluable source of information on international organizations throughout Europe as well as French, English and German translations of commonly used terms and definitions in the codes, regulations and standards.

Construction Quality and Quality Standards: the European perspective will provide a unique reference source for all construction professionals.

Contents: Construction quality and its achievement. Building well. The role of the client. Quality in traditional construction. The route to faultless building. Achievement of quality through management. Fitness for intended use. The single European market. Construction products directive. European standards and technical specifications. The procurement and other directives. Achievement of construction quality in six countries. Some unresolved issues. Conclusion. Organisations, terms and definitions. International and European Organisations. National Organisations in Member States of the European Union and European Free Trade Association. Terms and definitions, relating to codes, regulations, standards, certification and testing. Reference sources. Index.

246x189mm 352 pages 8 line illus
July 1995 Hardback 0-419-18490-2 £55.00

For further information and to order please contact: The Marketing Dept.,
E & F N Spon, 2-6 Boundary Row, London SE1 8HN
Tel: 0171 865 0066 Fax: 0171 522 9623

E & F N Spon

An imprint of Chapman & Hall

Acknowledgements

The editors wish to record their appreciation of the assistance given by many individuals and organisations in the compilation of this edition.

Manufacturers, Distributors and Sub-Contractors who have contributed this year include:-

ABB-Wylex Sales Ltd
Wylex Works
Wythenshaw
Manchester M22 4RA
Cabling Systems,
Switchgear & Distribution
Tel: 0161 998 5454
Fax: 0161 945 1587

A E L
Acoustics & Environmetrics Ltd
7 Berkeley Court
Manor Park
Runcorn
Cheshire WA7 1TQ
Energy Recovery Systems
Tel: 01928 579068
Fax: 01928 579523

Allaway Acoustics Ltd
Old Police Station
1 Queens Road
Hertford
Herts SG14 1EN
Air Distribution Equipment
Tel: 01992 550825
Fax: 01992 554982

Angus Fire Armour Ltd
Thame Park Road
Thame
Oxon OX9 3RT
Fire Fighting & Detection Equipment
Tel: 01844 214545
Fax: 01844 213511

Appleby Ltd
Forge Road
Willenhall
West Midlands
Cabling Systems
Tel: 01922 602111
Fax: 01922 601113

APV-Vent Axia Industrial Division
Fleming Way
Crawley
West Sussex RH10 2NN
Fans
Tel: 01293 526062
Fax: 01293 560257

S G Baldwin
Loughborough Road
Bunny
Notts NG11 6QN
Concrete Cable Protection
Tel: 0115 9211261
Fax: 0115 9848456

BICC Cables Ltd
Energy Cables Division
P O Box 1
Warrington Road
Prescot
Merseyside L34 5SZ
Cables
Tel: 0151 430 8080
Fax: 0151 430 3434

BICC Pyrotenax Ltd
P O Box 20
Prescot
Merseyside L34 5GB
MI Cables
Tel: 0151 430 4000
Fax: 0151 430 4004

Biddle Air Systems
St Mary's Road
Nuneaton
Warwickshire CV11 5AV
Air Distribution Equipment
Tel: 01203 384233
Fax: 01203 373621

Bill Switchgear Ltd
Aston Lane
Perry Barr
Birmingham B20 3BT
Switchgear & CCT Protection
Tel: 0121 356 6001
Fax: 0121 356 1962

Braithwaite Eng Ltd
Neptune Works
Uskway
Newport
Gwent
NP9 2UY
Liquid Storage Systems
Tel: 01633 262141
Fax: 01633 250631

S Brannan & Sons Ltd
Leconfield Industrial Estate
Cleator Moor
Cumbria
CA25 5QE
Thermometers
Tel: 01946 810413
Fax: 01946 813694

Brights of London Ltd
Westgate Business Park
Westgate Carr Road
Pickering
N Yorks YO18 8LX
Clock Systems
Tel: 0181 393 2045 or 01751 74333
Fax: 0181 393 2441

British Fittings Ltd
Stanton House
PO Box 297
Holyhead Road
Birmingham B21 0AH
Pipes, Fittings, Fixings & Sundries
Tel: 0121 553 6666
Fax: 0121 525 0952

British Steel Plc
Tubes & Pipes
Corby Works
PO Box 101
Corby
Northants NN17 5UA
Pipes, Fittings, Fixings & Sundries
Tel: 01536 402121
Fax: 01536 404111

Broadcrown Ltd
Alliance Works
Stone
Staffordshire ST15 8BA
Standy Engine Power
Tel: 01785 817513
Fax: 01785 812474

Donald Brown (Brownall) Ltd
Stretford Road
Manchester M16 9AR
Lab. Tap
Tel: 0161 872 6941
Fax: 0161 848 7681

Brown & Tawse Plc
Bromley by Bow
London E3 3JQ
Flanges, Pipe Fittings, Fixings & Sundries
Tel: 0181 980 4466
Fax: 0181 981 6836

Brush Transformers Ltd
P O Box 20
Loughborough
Leics LE11 1HN
Switchgear & Distribution
Tel: 01509 612822
Fax: 01509 612825

British Transformers Ltd
PO Box 20
Loughborough
Leics LE11 1HN
Power Transformers
Tel: 01509 612822
Fax: 01509 610550

BSS (UK) Ltd
Fleet House
Lee Circle
Leicester LE1 3QQ
**Flanges, Pipes, Fittings, Fixings & Sundries,
Expansion Joints**
Tel: 0116 2623232
Fax: 0116 2531343

BSS (Wirral)
Rossmore Road East
Rossmore Road Industrial Estate
Ellesmere Port
South Wirral L65 3DD
Tel: 0151 355 8281
Fax: 0151 375 2001

Caradon MK Electric Ltd
Shrubbery Road
Edmonton
London N9 0PB
**Cabling Systems, Switchgear &
Distribution, Wiring, Fittings
& Accessories**
Tel: 0181 803 3355
Fax: 0181 807 5733

Caradon Stelrad Ideal Boilers
PO Box 103
National Avenue
Kingston-upon-Hall
North Humerside
HU5 4JN
Boilers/Heating Products
Tel: 01482 492251
Fax: 01482 448858

Caradon Stelrad Ideal Radiators
Marriot Road
Swinton
Mexborough S64 8EN
Heating Products
Tel: 01709 578950
Fax: 01709 589410

Caradon Terrain Ltd
Aylesford
Kent M20 7PJ
Drainage Systems
Tel: 01622 717811
Fax: 01622 716920

CMP Products Ltd
Glasshouse Street
St Peters
Newcastle Upon Tyne NE6 1BE
Cabling Systems, Wiring, Fittings & Accessories
Tel: 0191 265 7411
Fax: 0191 265 0581

Conex Sanbra Ltd
Whitehall Road
Tipton
West Midlands DY4 7JU
Copper Pipes, Fittings, Fixings & Sundries
Tel: 0121 557 2831
Fax: 0121 520 8778

Constant Power Services Ltd
Unit 8A, The Cam Centre
Wilbury Way
Hitchin
Herts SG4 0TW
UPS Systems
Tel: 01462 422955
Fax: 01462 422754

Coley Instruments
Burnside Industrial Estate
Kilsyth
Glasgow
G65 9JX
Thermometers
Tel: 01236 826040
Fax: 01236 824090

Crabtree Electrical Industries Ltd
Lincoln Works
Walsall WS1 2DN
**Switchgear & Distribution,
Wiring, Fittings & Accessories**
Tel: 01922 721202
Fax: 01922 721321

Crane Fluid Systems
Trafford House
Chester Road
Stretford
Manchester M32 0RS
Valves, Fittings, Actuators
Tel: 0161 872 6026
Fax: 0161 872 3840

Cumbria Heating Components
Block Two
Brymau Three Industrial Estate
River Lane
Saltney
Chester CH4 8RH
Domestic & Industrial Heating Equipment
Tel: 01244 671877
Fax: 01244 671305

Cutler Hammer Limited
Mill Street
Ottery St Mary
Devon
EX11 1AG
Cable Management Systems
Tel: 01404 812131
Fax: 01404 815471

Davis Group Limited
Rookwood Way
Havenhill
Suffolk CB9 8PB
Cable Management System
Tel: 01440 704411
Fax: 01440 702822

Delta Crompton Cables Ltd
Millmarsh Lane
Brimsdown
Enfield
Middlesex EN3 7QD
Rubber and Plastic Cables
Tel: 0181 804 2468
Fax: 0181 443 2281

Dimplex Heating Limited
Marketing Dept
Millbrook
Southampton S09 2DP
Electrical Heating
Tel: 01703 777117
Fax: 01703 771096

Bryan Donkin Company Ltd
Derby Road
Chesterfield
S40 2EB
Valves
Tel: 01246 273153
Fax: 01246 235273

Dorman-Smith Switchgear Ltd
Blackpool Road
Preston PR2 2DQ
Switchgear & Distribution
Tel: 01772 728271
Fax: 01772 726276

Drainage Systems Ltd
Cary Avenue
St Marys Cray
Orpington
Kent B45 3RH
C.I. Pipes
Tel: 01689 890900

Dunham Bush
Brittania House
Newton Road
Hoylake
Merseyside L47 3DG
Air Distribution, Heating
Tel: 0151 632 3393
Fax: 0151 632 4453

Durapipe Glynwed Plastics Ltd
Walsall Road
Norton Canes
Cannock
Staffs WS11 3NS
Thermoplastic Piping
Tel: 01543 279909
Fax: 01543 279450

Engineering Appliances Ltd
Unit 11
Sunbury Cross Ind Est
Brooklands Close
Sunbury On Thames
TW16 7DX
Expansion Joints
Tel: 01932 788889
Fax: 01932 761263

Euro-Diesel (UK) Ltd
Strato House
Somerford Court
Somerford Road
Circencester
Gloustershire GL7 1TW
UPS Systems
Tel: 01285 640879
Fax: 01285 640879

Farnell Electrical Services
Edinburgh Way
Harlow
Essex
CM20 2DF
Cables
Tel: 01279 441144
Fax: 01279 441687

Furse and Company Limited
Wilford Road
Nottingham NG2 1EB
Lightning Protection
Tel: 0115 9863471
Fax: 0115 9860071

GEC-Alsthom Installation Equipment
East Lancashire Road
Liverpool L10 5HB
Switchgear & Distribution
Tel: 0151 525 8371
Fax: 0151 523 7007

Gent Ltd
140 Waterside Road
Hamilton Industrial Park
Leicester LE5 1TN
Fire Detection Systems
Tel: 0116 2462000
Fax: 0116 2462300

Gwynedd Foundaries
Sinclair Works
PO box 3
Ketley
Telford
Salop TF1 4AD
C.I. Pipe and Drainage
Tel: 01952 252308
Fax: 01952 243760

Graham
10 Hartford Way
Chester CH1 4NT
Plumbers Merchants
Tel: 01244 373886
Fax: 01244 328924

Grundfos Pumps Ltd
Gawsnorth Court
Risley Road
Risley, Warrington
Cheshire WA3 6NJ
Pumps, Circulation
Tel: 01925 813300
Fax: 01925 830014

Harvey Fabrication Ltd
Hancock Road
London
E3 3DA
Cisterns, Tanks & Cylinders
Tel: 0181 981 7811
Fax: 0181 981 7815

Hattersley, Newman, Hender Ltd
Burscough Road
Ormskirk
Lancashire L39 2XG
Valves
Tel: 01695 577199
Fax: 01695 578775

Hawker Siddeley Switchgear Ltd
P O Box 19
Falcon Works
Loughborough
Leicestershire LE11 1HL
Switchgear & Distribution
Tel: 01509 611311
Fax: 01509 610404

Heap and Partners
Brittania House
Newton Road
Hoylake
Wirral
Merseyside L47 3DG
Fluid Control Engineer
Tel: 0151 632 3393
Fax: 0151 632 4453

Hepworth Building Products
Hazle Head
Stocksbridge
Sheffield
S30 5HG
Clay and Plastic Drainage
Tel: 01226 763561
Fax: 01226 764827

Horne Engineering Ltd
Rankine Street
Johnstone
Scotland PA5 8BD
Valves
Tel: 01505 321455
Fax: 01505 336287

Horseley Bridge Tanks
27/29 Thornleigh Trading Estate
Blowers Green
Dudley
West Midlands DY2 8UB
Storage Tanks
Tel: 01384 459119
Fax: 01384 459117

Industrial Hangars Ltd
IHL House
Thorpe Close
Thorpe Way Industrial Estate
Banbury
Oxen OX16 8UU
Pipe Support Systems
Tel: 01295 257401
Fax: 01295 257477

IMI Yorkshire Fittings Ltd
PO Box 166
Leeds LS1 1RD
**Copper Pipes, Fittings, Fixings
& Sundries**
Tel: 0113 2706945
Fax: 0113 2705644

IMI Rycroft Ltd
Duncombe Road
Bradford
BD8 9TB
Calorifiers
Tel: 01274 490911
Fax: 01274 482565

John Millar UK Limited
Brittania House
Newton Road
Hoylake
Wirral
Merseyside L47 3DG
Flexible Joints
Tel: 0151 632 3393
Fax: 0151 632 4453

Kiddiethorn Fire Protection
1 Summers Road
Brunswick Business Park
Liverpool L3 4BL
Fire Protection
Tel: 0151 207 1243
Fax: 0151 709 6062

Kiloheat Ltd
Enterprise Way
Edenbridge
Kent TN8 6HF
Fans
Tel: 01732 866000
Fax: 01295 257477

Knowsley SK
Centre-Point
Westinghouse Road
Trafford Park
Manchester M17 1AE
Fire Fighting Equipment
Tel: 0161 872 7511
Fax: 0161 848 8508

Kopex International Ltd
189 Bath Road
Slough
Berks SL1 4AR
Wiring, Fittings & Accessories
Tel: 01753 534931
Fax: 01753 693521

Lancashire Fittings Limited
The Science Village
Claro Road
Harrogate
North Yorkshire HG1 4AF
Stainless Steel Fittings
Tel: 01423 522355
Fax: 01423 506111

Lennox Industries Ltd
PO Box 174
Westgate Interchange
Northampton
NN5 5AG
Heating Units
Tel: 01604 591159
Fax: 01604 587536

Marflex International Limited
Unit 40
Llandow Industrial Estate
Cowbridge
South Glamorgan CF7 7PB
Chimneys, Flues
Tel: 01446 775551
Fax: 01446 772468

Marley Extrusions Ltd
Dickley Lane
Lenham
Maidstone
Kent ME17 2DE
UPVC, Pipes, Fittings, Fixings & Sundries
Tel: 01622 858888
Fax: 01622 858725

Marshall Tufflex Ltd
Ponswood Industrial Estate
Hastings
East Sussex TN34 1YJ
Cabling Management Systems
Tel: 01424 427691
Fax: 01424 720670

MC Integ Limited
Integ House
Rougham Industrial Estate
Barry St Edmonds
Suffolk IP30 9ND
Tanks and Vessels
Tel: 01359 270610
Fax: 01359 270458

MEM Ltd
Whitegate
Broadway, Chadderton
Oldham OL9 9QG
Wiring, Fittings & Accessories
Tel: 0161 652 1111
Fax: 0161 626 1709

Menvier Ltd
Southam Road
Banbury
Oxon OX16 7RX
Fire Detection Systems
Tel: 01295 256363
Fax: 01295 270102

Mita (UK) Ltd
Copthorne House
The Broadway
Abergele
Clwyd LL22 7DD
Cabling Management Systems
Tel: 01745 826224
Fax: 01745 833161

Mucklow Bros Ltd
Narrow Lane
Halesowen
West Midlands B62 9PA
Flanges, Pipes, Fittings, Fixings & Sundries
Tel: 0121 423 1212
Fax: 0121 423 2020

Myson RCM
Old Wolverton Road
Milton Keynes
Buckinghamshire MK12 5PT
Skirting Radiator
Tel: 01908 321155
Fax: 01908 317387

National Vulcan Ltd
St Mary's Parsonage
Manchester M6 9AP
Valves
Tel: 0161 834 8124
Fax: 0161 839 3227

Owens-Corning Building Products (UK) Ltd
PO Box 10
Stafford Road
St Helens
Merseyside LA10 3NS
Thermal & Accoustic Insulation
Tel: 01744 24022
Fax: 01744 612007

Ottermill Ltd
Mill Street
Ottery St Mary
Devon EX11 1AG
**Cabling Systems, Switchgear &
Distribution**
Tel: 01404 812131
Fax: 01404 815471

Ozonair Limited
Quarrywood Industrial Estate
Aylesford
Maidstone
Kent ME20 7NB
Air Distribution Equipment
Tel: 01622 717861
Fax: 01622 719291

Pegler Ltd
St Catherines Avenue
Doncaster
South Yorkshire DN4 8DF
**Radiators, Heaters & Controls,
Valves**
Tel: 01302 329777
Fax: 01302 730515

Philmac PTY Ltd
Diplocks Way
Hailsham
East Sussex
BN27 3JF
Pipes
Tel: 01323 847323
Fax: 01323 844775

Pipeline Centre
Textilose Road
Trafford Park
Manchester M17 1WA
Pipes & Fittings
Tel: 0161 872 8431
Fax: 0161 872 7208

Pirelli Cables Ltd
Power Cables Division
PO Box 6
Leight Road
Eastleigh
Hants SO5 5YE
Power Cables
Tel: 01703 644522
Fax: 01703 640214

Pirelli Cables Ltd
Special Cables Division
PO Box 30
Chickenhall Lane
Eastleigh
Hants SO5 5XA
Cables
Tel: 01703 644544
Fax: 01703 649649

Plumb Centre Ltd
Boroughbridge Road
Ripon
N Yorks HG4 1SL
General
Tel: 01765 690690
Fax: 01765 690100

Plumb Centre (Saltney, Chester)
Units 1 & 2
Marley Way
Off High Street
Saltney
Chester CH4 8SX
Merchants, Plumbing, Electrical
Tel: 01244 674352
Fax: 01244 680937

Polypipe
Broomhouse Lane
Edlington
Doncaster DN12 1ES
UPVC Pipes & Fittings
Tel: 01709 770000
Fax: 01709 770001

E Poppleton & Son Limited
Conway Road
Colwyn Bay
Clwyd LI28 4AS
Ductwork
Tel: 01492 546061
Fax: 01492 544076

Potterton Myson Limited
Eastern Avenue
Team Valley Trading Estate
Gateshead
Newcastle upon Tyne NE11 0PG
Heating Products
Tel: 0191 491 7500
Fax: 0191 491 7568

Pullen Pumps Ltd
Unit 2, Mercury Park
Mecury Way (off Barton Dock Road)
Urmston
Manchester M31 2LR
Pumps, Circulators
Tel: 0161 866 9311
Fax: 0161 866 9323

Richard Vanspall Associates Ltd
2nd Floor
Drayton Bridge House
Tavistock Road
West Drayton
Middlesex UB7 7BB
Cooling Water Specialists
Tel: 01895 440546
Fax: 01895 431944

Salamandre Plc
Hunts Rise
South Marston Ind. Estate
South Marston
Swindon
Wiltshire SN3 4RE
Cabling Trunking Systems
Tel: 01793 828000
Fax: 01793 828597

Selkirk Manufacturing Ltd
Bassett House
High Street
Banstead
Surrey
SM7 2LZ
Chimney Systems, Exhausts
Tel: 01737 353388
Fax: 01737 362501

Sheffield Insulation (Industrial)
Tollgate Road
Burscough
Lancashire L40 8LD
Pipe/Duct Insulation
Tel: 01704 892552
Fax: 01704 892676

Sheffield Insulation
Textilose Road
Trafford Park
Manchester M17 1PY
Insulation General
Tel: 0161 876 4776
Fax: 0161 876 4775

Sound Attenduators Ltd
Eastgates
Colcheser
Essex
CO1 2TW
Noise Control Engineers
Tel: 01206 866911
Fax: 01206 865987

South Wales Transformers Ltd
PO Box 20
Longhborough
Leicestershire LE11 1HN
Transformers
Tel: 01509 612822
Fax: 01509 612825

Spirax-Sarco Ltd
Charlton House
Cheltenham
Glos GL53 8ER
Boiler Controls/Pipeline Products
Tel: 01242 521361
Fax: 01242 573342

Tecra Limited
Bumpers Lane
Sealand Industrial Estate
Chester CH1 4LT
Copper Pipe
Tel: 01244 377539
Fax: 01244 378644

Swifts of Scarborough Limited
Cayton Low Road
Eastfield
Scarborough
North Yorkshire
YO11 3BY
Cable Trays
Tel: 01723 583131
Fax: 01723 584625

Teddington Bellows Ltd
Teilo Works
Portardulais
Swansea
SA4 1RP
Expansion Joints
Tel: 01792 882591
Fax: 01792 885199

Thorn Lighting Ltd
Elstree Way
Borehamwood
Herts WD6 1HZ
Lighting
Tel: 0181 905 1313
Fax: 0181 905 1287

Vent-Axia Limited
Unit 2
Caledonia Way
Stretford Motorway Estate
Barton Dock Road
Manchester M32 0ZH
Ventilation/Cooling
Tel: 0161 865 8421
Fax: 0161 865 0098

Victaulic Systems Ltd
PO Box 13
46-48 Wilbury Way
Hitchin
Herts SG4 0UD
Pipe Jointing Systems
Tel: 01462 422622
Fax: 01462 422072

Walsall Conduits Ltd
Dial Lane
West Bromwich
West Midlands B70 0EB
Cabling Management Systems
Tel: 0121 557 1171
Fax: 0121 557 5631

Waterloo Air Distribution/Control
14 Parsons Road
Manor Trading Estate
South Benfleet
Essex SS7 4PT
Air Diffusers/Controls
Tel: 01268 794121
Fax: 01268 751841

Wellman Robey Ltd
Newfield Road
Oldbury
West Midlands
B69 3ET
Boilers
Tel: 0121 552 3311
Fax: 0121 552 4571

Wibe Ltd
Unit 8D, Castle Vale Ind. Estate
Maybrook Road
Minworth
Sutton Coldfield
West Midlands
B76 1AL
Cabling Support Systems
Tel: 0121 313 1010
Fax: 0121 313 1020

Weidmuller (Klippon Products) Ltd
Power Station Road
Sheerness
Kent ME12 3AB
Wiring Accessories
Tel: 01795 580999
Fax: 01795 580115

THE UK'S BEST SELLING BUILDING AND CONSTRUCTION NEWSPAPER

Keeps you up to date with events in your industry, brings you news of contracts and jobs, and provides a guide to the latest developments in construction plant, equipment, materials and products

AVAILABLE EVERY THURSDAY FROM YOUR LOCAL NEWSAGENT

CONSTRUCTION NEWS

OR CALL 01858 468888 TO SUBSCRIBE

Construction News, 2-6 Boundary Row, London SE1 8HN. Tel; 0171 410 6611 Fax; 0171 522 9646

DIRECTIONS

Prices shown are average prices on a fluctuating basis for typical buildings during the second quarter, of 1995. Unless otherwise shown, they exclude external services and professional fees.

The information in this section has been arranged to follow more closely the order in which estimates may be developed:

a)   Cost Indices and Regional Variations - gives indices and variations to be applied to estimates in general

b)   Outline Costs - gives a range of data (based on a rate per square metre) for all-in engineering costs associated with a wide variety of building functions.

   For certain specified building types these initial all-in costs are further subdivided to give a broad division between the main work elements.

c)   Elemental and All-in-Rates - given for a number of items and complete component parts i.e. boiler plant, ductwork and mains and switchgear together with some other items not covered in parts of this or other sections of the book i.e. lifts, escalators, kitchen equipment, medical gases.

d)   Elemental Costs - are shown for four building types - a speculative office block, a hotel, a data processing centre and a district general hospital. In each case, a full analysis of engineering services costs is given to show the division between all elements and their relative costs to the total building area.

   A further analysis is given to indicate the additional stage of cost information required when, for a complex building such as a hospital, prices are also required for each functional area or department.

Prices should be applied to the total floor area of all storeys of the building under consideration. The area should be measured between the external walls without deduction for internal walls and staircases/lift shafts.

Although prices are reviewed in the light of recent tenders it has only been possible to provide a range of prices for each building type. This should serve to emphasise that these can only be average prices for typical requirements and that such prices can vary widely depending upon a number of features. Rates per square metre should not therefore be used indiscriminately and each case must be assessed on its merits.

The prices do not include for incidental builder's work nor for profit and attendance by a main contractor where the work is executed as a sub-contract: they do however include for profit and overheads for the services contractor and for 2.5% cash discount for the main contractor. Capital contributions to statutory authorities and public undertakings and the cost of work carried out by them have been excluded.

COST INDICES

The following tables reflect the major changes in cost to contractors but do not necessarily reflect changes in tender levels. In addition to changes in labour and materials costs tenders are affected by other factors such as the degree of competition in the particular industry and area where the work is to be carried out, the availability of labour and the general economic situation. This has meant in recent years that, when there has been an abundance of work, tender levels have often increased at a greater rate than can be accounted for by increases in basic labour and material costs and, conversely, when there is a shortage of work this has often resulted in keener tenders. Allowances for these factors are impossible to assess on a general basis and can only be based on experience and knowledge of the particular circumstances. In compiling the tables the cost of labour has been calculated on the basis of a notional gang as set out elsewhere in the book. The proportion of labour to materials has been assumed as follows:
Mechanical Services - 30:70 / Electrical Services - 50:50 (1976 = 100)

Mechanical Services

Year	First Quarter	Second Quarter	Third Quarter	Fourth Quarter
1978	121	123	126	133
1979	138	141	150	156
1980	165	171	172	174
1981	181	185	187	190
1982	197	201	201	201
1983	204	205	208	211
1984	215	221	223	226
1985	232	236	236	236
1986	238	240	239	243
1987	244	249	251	254
1988	257	266	268	272
1989	276	284	284	286
1990	288	298	299	301
1991	312	316	317	319
1992	323	325	326	329
1993	333	334	336	338
1994	341	342	345	350
1995	357P	359F	364F	370F
1996	373F	375F	380F	387F

Electrical Services

Year	First Quarter	Second Quarter	Third Quarter	Fourth Quarter
1978	121	128	130	131
1979	143	145	147	154
1980	169	169	177	192
1981	195	197	199	200
1982	202	211	211	212
1983	214	225	225	226
1984	228	229	236	237
1985	240	240	247	250
1986	251	249	249	257
1987	268	269	271	274
1988	286	289	290	293
1989	306	306	306	308
1990	323	324	326	330
1991	350	350	350	350
1992	367	368	368	370
1993	373	377	378	378
1994	380	381	383	389
1995	398P	399F	400F	402F
1996	415F	416F	418F	419F

(P = Provisional)
(F = Forecast)

COST INDICES

Regional Variations

The table of regional variation factors have been derived from data rebased so that Outer London = 1.00. The figures have been extracted from an analysis of contract prices for schemes tendered between mid 1982 and the end of 1994. As such it is important to realise that the factors represent the average state of affairs over the twelve year period, rather than an exact indication of the regional differences at this precise point in time.

The factors include an adjustment to the general building factors to reflect the lower level of regional variations that exist for engineering sub-contract work, as compared to main contract work.

Outer London	1.00	Inner London	1.06
Kent, Surrey & Sussex	0.97	Beds, Essex & Herts	0.94
Berks, Bucks, Herts & Oxon	0.93	South West	0.90
East Midlands	0.90	West Midlands	0.90
East Anglia	0.94	Yorkshire & Humberside	0.89
North West	0.96	Northern	0.92
Scotland	0.96	Wales	0.89
Northern Ireland	0.73	Channel Islands	1.18

OUTLINE COSTS

SQUARE METRE RATES FOR BASIC SERVICES BY BUILDING TYPE
The undernoted examples indicate the range of rates within which normal engineering services would be contained for each of the more common building types.

	Square metre £
Industrial Buildings (CI/SfB 2)	
Factories	
for letting	32 to 73
owner occupation	48 to 100
Warehouses	
high bay for owner occupation	48 to 104
Administrative, public, commercial and office buildings; general (CI/SfB 3)	
Civic offices	
fully air conditioned	286 to 410
Offices for letting	
non air conditioned	136 to 205
fully air conditioned	222 to 374
Offices for owner occupation	
medium rise, air conditioned	231 to 450
high rise, air conditioned	231 to 472
Fitted out offices including air conditioning	298 to 448
Fitted out department store ditto	234 to 363
Health and welfare facilities (CI/SfB 4)	
District general hospitals	310 to 440
Private hospitals	328 to 516
Refreshment, entertainment, recreation buildings (CI/SfB 5)	
Arts and drama centres	233 to 318
Theatres, large	302 to 410
Educational, scientific and information buildings (CI/SfB 7)	
Secondary/middle schools	129 to 191
Universities	
arts buildings	168 to 211
science buildings	
physics	168 to 266
biology	236 to 279
chemistry	266 to 312
Museums	
national	570 to 712
Residential buildings (CI/SfB 8)	
Local authority schemes	
two storey houses	56 to 77
medium rise flats	76 to 100

DAVIS LANGDON & EVEREST

Authors of Spon's Price Books

Davis Langdon & Everest is an independent practice of Chartered Quantity Surveyors with over 500 staff in 19 UK offices.

We believe that cost awareness and control are responsibilities that must be accepted by all members of the team, but the ability to act on that responsibility is dependent upon the quality of the cost advice given. Our approach to cost consultancy is therefore:

*   to be positive and creative in our advice, rather than simply reactive;

*   to concentrate on value for money and value engineering rather than on superficial cost-cutting;

*   to give advice that is matched to the Client's own criteria, rather than to impose standard or traditional solutions;

*   to see cost as one component of a successful design solution, which needs to be balanced with many others, and to work as an integrated member of a design team in achieving that balance;

*   to pay attention to the life-long costs of owning and operating a facility, rather than to the initial capital cost only.

We work in small teams, under the active leadership of a Partner, to provide continuity of involvement throughout the stages of a project. At the same time, we can draw on the resources of a large firm, and upon departments able to give specialist advice on engineering services, life cycle costing, cost analyses and research.

Davis Langdon & Everest - a first point of call for anyone contemplating a construction project. **Consult the Source.**

TRUST DLE TO PROVIDE THE DEFINITIVE PRICE DATA

DAVIS LANGDON & EVEREST
CHARTERED QUANTITY SURVEYORS
CONSTRUCTION COST CONSULTANTS

LONDON
Princes House
39 Kingsway
London
WC2B 6TP
Tel : (0171) 497 9000
Fax : (0171) 497 8858

BRISTOL
St Lawrence House
29/31 Broad Street
Bristol BS1 2HF
Tel : (0117) 927 7832
Fax : (0117) 925 1350

CAMBRIDGE
36 Storey's Way
Cambridge CB3 ODT
Tel : (01223) 351258
Fax : (01223) 321002

CARDIFF
3 Raleigh Walk
Brigantine Place
Atlantic Wharf
Cardiff CF1 5LN
Tel : (01222) 471306
Fax : (01222) 471465

CHESTER
Ford Lane Farm
Lower Lane
Aldford
Chester CH3 6HP
Tel : (01244) 620222
Fax : (01244) 620303

EDINBURGH
74 Great King Street
Edinburgh
EH3 6QU
Tel : (0131) 557 5306
Fax : (0131) 557 5704

GATESHEAD
11 Regent Terrace
Gateshead
Tyne and Wear
NE8 1LU
Tel : (0191) 477 3844
Fax : (0191) 490 1742

GLASGOW
Cumbrae House
15 Carlton Court
Glasgow G5 9JP
Tel : (0141) 429 6677
Fax : (0141) 429 2255

IPSWICH
17 St Helens Street
Ipswich IP4 1HE
Tel : (01473) 253405
Fax : (01473) 231215

LEEDS
Duncan House
14 Duncan Street
Leeds
LS1 6DL
Tel : (0113) 243 2481
Fax : (0113) 2424601

LIVERPOOL
Cunard Building
Water Street
Liverpool
L3 1JR
Tel : (0151) 236 1992
Fax : (0151) 227 5401

MANCHESTER
Boulton House
Chorlton Street
Manchester M1 3HY
Tel : (0161) 228 2011
Fax : (0161) 228 6317

MILTON KEYNES
6 Bassett Court
Newport Pagnell
Buckinghamshire
MK16 OJN
Tel : (01908) 613777
Fax : (01908) 210642

NEWPORT
34 Godfrey Road
Newport
Gwent NP9 4PE
Tel : (01633) 259712
Fax : (01633) 215694

NORWICH
63 Thorpe Road
Norwich NR1 1UD
Tel : (01603) 628194
Fax : (01603) 615928

OXFORD
Avalon House
Marcham Road
Abingdon
Oxford OX14 1TZ
Tel : (01235) 555025
Fax : (01235) 554909

PLYMOUTH
3 Russell Court
St Andrews Street
Plymouth
Devon PL6 2AX
Tel : (01752) 668372
Fax : (01752) 221219

PORTSMOUTH
Kings House
4 Kings Road
Portsmouth PO5 3BQ
Tel : (01705) 815218
Fax : (01705) 827156

SOUTHAMPTON
Clifford House
New Road
Southampton
SO14 OAB
Tel : (01703) 333438
Fax : (01703) 226099

**DAVIS LANGDON
CONSULTANCY**
Princes House
39 Kingsway
London WC2B 6TP
Tel : (0171) 379 3322
Fax : (0171) 379 3030

A MEMBER OF
DAVIS LANGDON & SEAH INTERNATIONAL

OUTLINE COSTS

ELEMENTAL RATES FOR BASIC SERVICES BY BUILDING TYPE

There follows typical examples of how rates per square metre for the basic engineering services of a project may be allocated to individual elements.

	Factories	
	Speculative	*Prestige*
	£	£
Cold water services	2.60	2.60
Hot water services	1.92	1.92
Heating and ventilation	19.85	44.10
Fire protection	1.92	1.92
Mains and switchgear	9.25	9.48
Power	4.85	6.10
Lighting	8.01	9.14
Luminaires	9.74	11.85
Fire protection	2.48	4.28
Communications	0.00	1.25
Total approximate cost of above services	60.62	92.64

		Offices Owner	
	Speculative	*occupied*	*Prestige*
	£	£	£
Cold water services	6.65	7.56	8.34
Hot water services	5.08	5.97	6.65
Heating and ventilation	59.11	0.00	0.00
Air conditioning	0.00	194.48	224.15
Fire protection	5.18	5.18	5.18
Mains and switchgear	13.54	16.81	20.20
Power	27.07	30.80	32.37
Lighting	8.23	9.70	11.05
Luminaires	21.99	26.97	35.87
Emergency lighting	4.40	6.20	7.79
Telephone wireways	0.79	1.02	1.12
Clock	0.00	0.00	0.45
Public address and television	0.00	0.00	1.69
Fire alarms	6.32	6.32	7.56
Security	0.00	3.72	6.09
Lightning protection	0.79	0.79	0.79
External lighting	0.23	1.02	1.69
Total approximate cost of above services	159.38	316.54	370.99

Approximate Estimating

OUTLINE COSTS

ELEMENTAL RATES FOR BASIC SERVICES BY BUILDING TYPE *continued*

	Non-teaching hospitals £	Secondary schools £	University arts & administration buildings £	Local authority houses £	Local authority flats £
Cold water services	27.30	12.40	20.20	7.56	10.04
Hot water services	20.30	11.63	13.42	6.88	8.80
Heating	59.56	65.54	62.16	28.54	39.03
Ventilation	53.69	12.07	0.00	0.00	2.71
Fire protection	2.71	2.14	2.71	0.00	0.00
Ancillary services	37.56	4.40	0.00	2.26	2.26
Mains and switchgear	21.66	7.90	11.85	2.48	3.83
Power	27.86	11.85	16.47	6.65	7.79
Lighting	18.95	11.05	12.52	7.56	8.80
Luminaires	25.83	14.77	18.39	3.16	3.96
Fire protection	11.05	7.66	7.45	0.00	0.79
Communications	22.56	5.08	4.74	1.35	1.69
Ancillary services	2.71	0.00	7.80	0.00	0.00
Total approximate cost of above services	331.74	166.49	177.71	66.44	89.70

Notes

The above prices do not include any public utility charges that may be made.

Prices for local authority flats and houses are based upon tenders for large schemes and on dwellings accommodating 4 persons.

Cold and hot water services exclude the cost of sanitary fittings.

Heating in the case of non-teaching hospitals excludes the heat source. The price for heating to Local Authority flats assumes individual gas fired boilers serving panel radiators.

Fire protection includes extinguishers, hose reels, wet and dry risers and foam inlets. Where required sprinklers would cost an additional £11.41 to £19.87 per m² of area protected.

Ancillary services includes, where applicable, natural gas, compressed air, treated water, refrigeration, laboratory services, special piped services, steam, and condense and medical vacuum installations.

The prices do not allow for such items as kitchen and laundry equipment, hoists and balers.

In the case of non-teaching hospitals, mains and switchgear includes stand-by facilities.

Fire protection includes alarms and detectors.

Communications, includes where applicable, burglar and security alarms, clocks, telephone conduit, television and radio aerials and outlets, public address and call systems.

Ancillary services include common services trunking and cable trays.

ELEMENTAL AND ALL-IN-RATES

ELEMENTAL RATES FOR ALTERNATIVE MECHANICAL SERVICES SOLUTIONS FOR OFFICES
The undernoted examples, when considered within the context of office accommodation, indicate the range of rates for alternative design solutions for each of the basic mechanical services elements.

	Square metre £
Sanitary and disposal installation	
Normal services for low rise building (up to 1,000 m²)	3 to 10
Normal services for low rise building (over 1,000 m²)	8 to 10
Normal services for medium rise building (up to 1,000 m²)	8 to 12
Normal services for medium rise building (over 1,000 m²)	10 to 16
Normal services for high rise building (over 1,000 m²)	10 to 16
Hot and cold water installation	
Services to speculative offices (up to 1,000 m²)	4 to 12
Services to speculative offices (medium to high rise)	7 to 16
Services to owner occupied offices (up to 1,000 m²)	7 to 14
Services to owner occupied offices (medium to high rise)	10 to 19
Heating installation	
LPHW radiator system to speculative offices (up to 1,000 m²)	39 to 52
LPHW radiator system to speculative offices (over 1,000 m²)	43 to 61
LPHW radiator system to owner occupied offices (up to 1,000 m²)	43 to 61
LPHW radiator system to owner occupied offices (over 1,000 m²)	47 to 64
LPHW convector system to speculative offices (up to 1,000 m²)	35 to 55
LPHW convector system to speculative offices (over 1,000 m²)	47 to 64
LPHW convector system to owner occupied (up to 1,000 m²)	48 to 61
LPHW convector system to owner occupied (over 1,000 m²)	53 to 71
Ventilated heating system	53 to 69
Ventilation and air conditioning	
Ventilation to internal WC's only	2 to 9
Warm air heating and ventilation to offices	96 to 125
Comfort cooling, fan coil, speculative offices	130 to 176
Comfort cooling, fan coil, owner occupied offices	137 to 193
Comfort cooling, variable air volume, speculative offices	146 to 221
Comfort cooling, variable air volume, owner occupied offices	155 to 243
Full air conditioning, fan coil, speculative offices	146 to 207
Full air conditioning, fan coil, owner occupied offices	153 to 233
Full air conditioning, variable air volume, speculative offices	168 to 242
Full air conditioning, variable air volume, owner occupied offices	170 to 259
Special services	
Firefighting installation - hose reels and dry risers	3 to 10
Single level sprinkler installation	10 to 19
Double level sprinkler installation	21 to 26
External services	
Incoming mains services (gas, water)	2 to 9

ELEMENTAL AND ALL-IN-RATES

ELEMENTAL RATES FOR AIR CONDITIONING FOR OFFICES

Approximate costs of air conditioning installations for two specimen office blocks with floor areas of 6,000 m² and 15,000 m² in four and eight storeys respectively. The types of installation selected are the variable air volume and the induction system.

The costs allow for all plant and equipment, distribution ductwork, pipework for heating, chilled and cooling water, automatic controls, fire protection systems and all associated electrical work.

	Floor area	
	6,000 m²	*15,000 m²*
	£	*£*
Variable air volume system, per m²	219.00	176.00
Induction system, per m²	194.00	148.00

Air conditioning installations vary considerably according to the type of plant selected, the ancillary services chosen, the methods of heating and cooling, the sophistication of automatic controls, the requirements for fire protection, the type of fuel available for heating and many other considerations. No two buildings will have precisely the same requirements.

The following information has been compiled to indicate the average cost of a number of different design solutions. Brief specification notes are provided to enable the user to make his own adjustments to the cost of individual elements to take into account his own design criteria. It has been assumed that the building is double glazed, has better than average insulation and a window to wall ratio not exceeding 50%.

ELEMENTAL AND ALL-IN-RATES

ELEMENTAL RATES FOR AIR CONDITIONING FOR OFFICES *continued*

VARIABLE AIR VOLUME SYSTEM

Elements	Office block of 6,000 m²		Office block of 15,000 m²		Spec'n Ref.
	Cost of element £	Cost of element per m² of floor area £	Cost of element £	Cost of element per m² of floor area £	
Boilers					
Plant and instruments	32,820.00	5.47	57,170.00	3.81	A
Flue	11,060.00	1.84	17,150.00	1.14	B
Water treatment	11,060.00	1.84	21,780.00	1.45	C
Gas installation	5,070.00	0.85	5,100.00	0.34	D
Space heating					
Distribution pipework	31,950.00	5.33	64,270.00	4.28	E
Convectors and/or radiators	6,240.00	1.04	15,300.00	1.02	F
Heating to batteries	62,910.00	10.49	113,560.00	7.57	H
Chilled water to batteries	47,590.00	7.93	87,140.00	5.81	J
Condenser cooling water					
Distribution pipework	22,370.00	3.73	38,630.00	2.58	K
Cooling plant					
Chillers	113,340.00	18.89	191,270.00	12.75	L
Cooling towers	12,670.00	2.11	19,310.00	1.29	M
Automatic controls	99,000.00	16.50	155,890.00	10.39	N
Ductwork					
Supply	481,240.00	80.21	1,095,870.00	73.06	P
Extract	101,660.00	16.94	229,900.00	15.33	P
Air conditioning plant					
Heating batteries	32,820.00	5.47	57,630.00	3.84	Q
Humidifiers & cooling batteries	58,400.00	9.73	101,040.00	6.74	Q
Fans and filters	89,610.00	14.94	153,110.00	10.21	R
Sound attenuation	43,510.00	7.25	106,610.00	7.11	S
Fire protection	25,400.00	4.23	46,200.00	3.08	T
Electrical work in connection	29,480.00	4.91	55,930.00	3.73	V
Totals	1,318,200.00	219.70	2,632,860.00	175.53	
	say	220.00	say	176.00	

ELEMENTAL AND ALL-IN-RATES

ELEMENTAL RATES FOR AIR CONDITIONING FOR OFFICES *continued*

INDUCTION SYSTEM

Elements	Office block of 6,000 m²		Office block of 15,000 m²		Spec'n Ref.
	Cost of element £	Cost of element per m² of floor area £	Cost of element £	Cost of element per m² of floor area £	
Boilers					
Plant and instruments	32,820.00	5.47	57,170.00	3.81	A
Flue	11,060.00	1.84	17,150.00	1.14	B
Water treatment	11,060.00	1.84	21,780.00	1.45	C
Gas installation	5,070.00	0.85	5,100.00	0.34	D
Space heating					
Distribution pipework	31,950.00	5.33	64,270.00	4.28	E
Convectors and/or radiators	6,240.00	1.04	15,300.00	1.02	F
Induction units	117,850.00	19.64	315,490.00	21.03	G
Heating to batteries	129,970.00	21.66	215,990.00	14.40	H
Chilled water					
Distribution pipework	167,050.00	27.84	309,150.00	20.61	J
Condenser cooling water					
Distribution pipework	22,310.00	3.72	37,080.00	2.47	K
Cooling plant					
Chillers	88,000.00	14.67	153,260.00	10.22	L
Cooling towers	11,490.00	1.92	17,300.00	1.15	M
Automatic controls	109,260.00	18.21	195,910.00	13.06	N
Ductwork					
Supply	141,150.00	23.53	254,620.00	16.97	P
Extract	71,010.00	11.84	151,870.00	10.12	P
Air conditioning plant					
Heating batteries	24,660.00	4.11	44,960.00	3.00	Q
Humidifiers & cooling batteries	42,700.00	7.12	65,660.00	4.38	Q
Fans and filters	61,060.00	10.18	111,700.00	7.45	R
Sound attenuation	32,820.00	5.47	84,820.00	5.65	S
Fire protection	22,430.00	3.74	40,020.00	2.67	T
Electrical work in connection	29,170.00	4.86	49,440.00	3.30	V
Totals	1,169,130.00	194.88	2,228,040.00	148.52	
	say	195.00	say	149.00	

ELEMENTAL AND ALL-IN-RATES

ELEMENTAL RATES FOR AIR CONDITIONING FOR OFFICES *continued*

Brief Specification Notes

Ref.

Boilers

A Plant and instruments: Three gas-fired boilers each of approximately 250 and 580 kW capacity for the two buildings respectively; together with burners, pumps, direct-mounted instruments, feed and expansion tanks. Normal standby facilities are included.

B Flue: Mild steel insulated in boiler house, internal lining to vertical builders' stack.

C **Water treatment**: Chemical dosage equipment.

D **Gas installation**: Pipework internal to building, meter, solenoid valves.

Space heating

E Distribution pipework: pipework from boilers to terminal equipment, all valves, fittings and supports, insulation.

F Convector and/or radiators: Panel radiators, or natural convectors in circulation areas and staircases.

G **Induction units**: High velocity units suitable for four-pipe system utilizing ducted fresh air.

H **Heating to air heater batteries**: Distribution pipework to batteries, valves, fittings and supports, insulation.

J **Chilled water to batteries and induction units**: Distribution pipework, valves, fittings and supports, insulation.

K **Condenser cooling water:** Distribution pipework, valves, fittings and supports, insulation.

Cooling plant

L Chillers: Centrifugal chiller units of approximately 190 and 470 tons total capacity for the two buildings respectively, including mountings and supports, insulation and pumps. Normal standby facilities are included.

M Cooling towers: Forced or induced draught fans, roof-mounted cooling towers with supports.

N **Automatic controls**: Pneumatic controls including motorised valves, all thermostats, control panels, actuators, interconnecting wiring and tubing.

Ductwork

P Supply and extract: Galvanised mild steel ductwork, fittings and supports, terminal units (for V.A.V system), dampers, grilles and diffusers, insulation.

Air conditioning plant

Q Heating and cooling batteries: Humidifiers, batteries and casing and connections.

R Fans and filters: Centrifugal and axial flow fans with casings and connections, and automatic roll type filters.

S Sound attenuation: Silencers and duct lining (short lengths only).

T **Fire protection**: Heat detectors, smoke detectors, gas detectors, control panel, interconnecting wiring (excluding other fire protection services not directly associated with air conditioning installation).

V **Electrical work in connection**: Electrical supplies to control panels and mechanical plant, mechanical services distribution board.

ELEMENTAL AND ALL-IN-RATES

AIR CONDITIONING DESIGN LOADS

Recommended air conditioning loads for various applications:

Computer rooms	10 m² of floor area per ton
Restaurant	16 m² of floor area per ton
Banks (main area)	22 m² of floor area per ton
Large Office Buildings (exterior zone)	25 m² of floor area per ton
Supermarkets	30 m² of floor area per ton
Large Office Block (interior zone)	32 m² of floor area per ton
Small Office Block (interior zone)	35 m² of floor area per ton

ALL IN RATES FOR PRICING MECHANICAL APPROXIMATE QUANTITIES

COLD WATER

Light gauge copper tube to B.S. 2871 part 1 table X with joints as described including allowance for waste, fittings and brackets assuming average runs.

	Cost per metre £
Capillary Joints	
15mm	17.57
22mm	16.23
28mm	19.85
35mm	22.87
42mm	33.08
54mm	34.89
Compression Joints	
15mm	16.59
22mm	15.65
28mm	19.85
35mm	25.45
42mm	36.39
54mm	40.30
Bronze Welded Joints	
76mm	121.49

HOT WATER

Light gauge copper tube to B.S. 2871 part 1 table X with joints as described including allowance for waste, fittings and brackets assuming average runs.

	Cost per metre £
Capillary Joints	
15mm	17.57
22mm	13.06
28mm	17.44
35mm	19.72
42mm	28.88
54mm	33.08
Compression Joints	
15mm	16.37
22mm	12.22
28mm	16.37
35mm	23.63
42mm	34.82
54mm	38.61
Bronze Welded Joints	
76mm	110.79

ELEMENTAL AND ALL-IN-RATES

ALL IN RATES FOR PRICING MECHANICAL APPROXIMATE QUANTITIES *continued*

DISTRIBUTION PIPEWORK HEATING OUTSIDE PLANT ROOM

Mild steel tube to B.S. 1387 with joints in the running length allowance for waste, fittings and brackets assuming average runs.

	Cost per metre	
	Screwed £	Welded £
Black Medium Weight		
15mm	36.87	27.84
20mm	16.11	19.07
25mm	17.92	20.81
32mm	17.62	20.81
40mm	18.47	21.29
50mm	41.75	49.74
65mm	49.39	59.48
80mm	60.92	73.25
100mm	50.66	64.23
125mm	77.33	88.53
150mm	103.87	141.27
Black Heavy Weight		
15mm	38.74	39.26
20mm	17.49	19.92
25mm	19.24	23.22
32mm	19.13	23.40
40mm	19.93	24.02
50mm	44.87	53.05
65mm	49.98	67.36
80mm	63.93	78.66
100mm	68.44	62.60
125mm	70.06	90.82
150mm	110.49	118.49

                              Approximate Estimating

ELEMENTAL AND ALL-IN-RATES

ALL IN RATES FOR PRICING MECHANICAL APPROXIMATE QUANTITIES *continued*

TANKS, CALORIFIERS, PUMPS AND VALVES
These have not been detailed in this section, for sizes and types see 'Prices for Measured Work' section.

HEAT SOURCE

BOILERS

Low pressure hot water boiler plant of welded construction comprising 2 No. boilers with instruments and boiler mountings and burners for gas combustion including delivery and commissioning, 2 No. heating circulating pumps, boiler control panel and thermostatic controls for heating circuit, automatic controls, pipework, fittings and valves with all necessary insulation, feed and expansion tank, cold feed, vent pipes and flue.

	Total cost of plant £	
Total boiler capacity 275 kW	49,290.00	- 65,350.00
Total boiler capacity 1100 kW	67,500.00	- 89,990.00
Total boiler capacity 2250 kW	92,130.00	- 116,770.00

Low pressure hot water boiler plant of welded construction comprising 2 No. boilers with instruments and boiler mounting and burners for gas combustion including delivery and commissioning, 2 No. heating circulating pumps, 2 No. hot water service circulation pumps, 2 No. calorifiers, boiler control panel and thermostatic controls for heating and hot water service circuits, automatic controls, all interconnecting pipework, fittings and valves, with all necessary insulation, feed and expansion tank, cold feed and vent pipes and flue.

	Total cost of plant £	
Total boiler capacity 275 kW	54,640.00	- 73,910.00
Total boiler capacity 1100 kW	77,140.00	- 100,700.00
Total boiler capacity 2250 kW	98,560.00	- 124,270.00

ELEMENTAL AND ALL-IN-RATES

ALL IN RATES FOR PRICING MECHANICAL APPROXIMATE QUANTITIES *continued*

BOILERS *continued*

Low pressure hot water boiler plant of welded construction comprising 2 No. boilers with instruments and boiler mountings and burners for 35 second oil including delivery and commissioning, 2 No. heating circulating pumps, 2 No. oil storage tanks, boiler control panel and thermostatic controls for heating and hot water service circuits, automatic controls, all interconnecting pipework, fittings and valves with all necessary insulation, feed and expansion tank, cold feed and vent pipes and flue.

	Total cost of plant £		
Total boiler capacity 275 kW	52,500.00	-	70,710.00
Total boiler capacity 1100 kW	70,710.00	-	100,700.00
Total boiler capacity 2250 kW	98,560.00	-	128,550.00

Low pressure hot water boiler plant of welded construction comprising 2 No. boilers with instruments and boiler mountings and burners for 35 second oil including delivery and commissioning, 2 No. heating circulation pumps, 2 No. hot water service circulating pumps, 2 No. calorifiers, 2 No. oil storage tanks, boiler control panel and thermostatic controls for heating and hot water service circuits, automatic controls, all interconnecting pipework, fittings and valves to feed and expansion tank, cold feed and vent pipes and flue.

	Total cost of plant £		
Total boiler capacity 275 kW	60,000.00	-	83,560.00
Total boiler capacity 1100 kW	82,490.00	-	109,270.00
Total boiler capacity 2250 kW	106,060.00	-	137,130.00

ELEMENTAL AND ALL-IN-RATES

ALL IN RATES FOR PRICING MECHANICAL APPROXIMATE QUANTITIES *continued*

SPACE HEATING AND AIR TREATMENT

DOMESTIC CENTRAL HEATING

Solid fuel central heating installation comprising either open fire room heater or boiler, fuel storage, pump, small or microbore distribution pipework, steel radiators, valves, expansion tank, room thermostat, and all insulation:

	£
Heating for 3 rooms comprising 2 radiators plus hot water service	1,720.00 - 1,980.00
Heating for 4 rooms comprising 3 radiators plus hot water service	1,950.00 - 2,350.00
Heating for 5 rooms comprising 4 radiators plus hot water service	2,350.00 - 2,650.00
Heating for 6 rooms comprising 5 radiators plus hot water service	2,530.00 - 2,770.00
Heating for 7 rooms comprising 6 radiators plus hot water service	2,770.00 - 3,050.00
Heating for 7 rooms comprising 7 radiators plus hot water service	2,990.00 - 3,420.00

Gas fired central heating installation comprising boiler, pump, small or microbore distribution pipework, steel radiators, valves, expansion tank, room thermostat and all insulation:

	£
Heating for 3 rooms comprising 3 radiators plus hot water service	1,930.00 - 2,560.00
Heating for 4 rooms comprising 4 radiators plus hot water service	2,520.00 - 2,800.00
Heating for 5 rooms comprising 5 radiators plus hot water service	2,800.00 - 2,980.00
Heating for 6 rooms comprising 6 radiators plus hot water service	2,980.00 - 3,150.00
Heating for 7 rooms comprising 7 radiators plus hot water service	3,150.00 - 3,830.00

Oil fired central heating installation comprising oil storage tank and supports, pump, room thermostats, small or microbore distribution pipework, steel radiators, valves and insulation together with hot water cylinder for hot water supply:

	£
Heating for 7 rooms comprising 7 radiators and boiler plus hot water service	3,390.00 - 4,060.00
Heating for 10 rooms comprising 10 radiators and boiler plus hot water service	3,720.00 - 4,270.00

ELEMENTAL AND ALL-IN-RATES

ALL IN RATES FOR PRICING MECHANICAL APPROXIMATE QUANTITIES *continued*

RADIATORS

Pressed steel panel type radiator prime finish including brackets and fixing, taking down once for painting by others (cost per m² of heating surface)		305mm high £	432mm high £	584mm high £	740mm high £
Single panel	m²	65.67	47.25	48.10	47.57
Double panel	m²	48.74	36.64	36.10	66.21

For floor or wall mounted individual convector and unit heaters see 'Materials Costs/Measured Work Prices' section.

CHILLED AND COOLING WATER

	Cost per ton of cooling capacity £
Chilled water installation comprising two refrigeration compressors, cooling towers, pumps, pipework valves and fittings, controls, thermostatic controls, insulation, anti-vibration mountings and starters	1,180.00 - 2,370.00

DUCTWORK

To calculate the weight of ductwork multiply the duct length (measured on the centre line overall fittings) by their respective girths and then apply the appropriate sheeting weights to the superficial areas without making any allowance for the additional sheeting in joints, seams, welts, waste, etc. The rates below allow for ductwork and for all other labour and material in fabrication fittings, supports and jointing to equipment, stop and capped ends, elbows, bends, diminishing and transition pieces, regular and reducing couplings, branch diffuser and 'snap on' grille connections, ties, 'Ys', crossover spigots, etc., turning vanes, regulating dampers, access doors and openings, handholes, test holes and covers, blanking plates, flanges, stiffeners, tie rods and all supports and brackets fixed to concrete or brickwork.

	Per tonne £
Rectangular low velocity galvanised mild steel ductwork in accordance with table 5 of the HVCA DW 142 (metric).	8,570.00
Rectangular high velocity galvanised mild steel ductwork in accordance with tables 7 and 8 of the HVCA DW 142 (metric)	10,070.00

PACKAGE AIR HANDLING UNITS, FANS AND ROOF EXTRACT UNITS

These have not been detailed in this section, for sizes and types see 'Material Costs/Measured Work Prices' section.

ELEMENTAL AND ALL-IN-RATES

ALL IN RATES FOR PRICING MECHANICAL APPROXIMATE QUANTITIES *continued*

PROTECTIVE INSTALLATIONS

SPRINKLER INSTALLATION

Sprinkler installation including sprinkler head and all associated pipework, valve sets, booster pumps and water storage:

	£
Price per sprinkler head	145.00

Recommended maximum area coverage per sprinkler head:
Extra light hazard, 21 m² of floor area
Ordinary hazard, 12 m² of floor area
Extra high hazard, 9 m² of floor area

HOSE REELS, WET AND DRY RISERS

100mm dry riser main including dry riser breeches horizontal inlet with 2 No. 64mm male instantaneous coupling and landing valve. Complete with padlock and leather strap.

	£
Price per landing	1,060.00

Hose reel pedestal mounted type with 37 metres of hose including approximately 15 metres of pipework:

	£
Price per hose reel	611.00

THERMAL INSULATION

Foil-faced pre-formed mineral fibre glass fibre rigid sections, 25mm thick, fixed with adhesive and aluminium bands at 450mm intervals all joints and ends vapour sealed, including all pipe fittings, flanges and valves.

Nominal size	*Cost per metre* £
15mm	8.14
20mm	8.67
25mm	9.21
32mm	9.86
40mm	10.07
50mm	14.57
65mm	17.99
80mm	22.61
100mm	25.29
125mm	29.46
150mm	35.03

ELEMENTAL AND ALL-IN-RATES

ALL IN RATES FOR PRICING MECHANICAL APPROXIMATE QUANTITIES *continued*

THERMAL INSULATION *continued*

Canvas-faced pre-formed mineral fibre glass fibre rigid sections, fixed with aluminium bands at 450mm intervals with 0.5mm thick aluminium sheet secured with rivets or self tapping screws, including all pipe fittings, flanges and valves.

	Cost per metre
Nominal size	*£*
15mm	21.74
20mm	22.81
25mm	24.10
32mm	25.29
40mm	26.14
50mm	31.71
65mm	36.00
80mm	44.67
100mm	49.28
125mm	55.18
150mm	63.10

ELEMENTAL AND ALL-IN-RATES

ALL IN RATES FOR PRICING ELECTRICAL APPROXIMATE QUANTITIES

MAINS AND SWITCHGEAR

Indoor substation equipment.
The cost of an indoor substation is governed by various factors e.g. the duty it is required to perform, local electricity authority requirements, earthing requirements dependent on the soil resistivity of the ground in the vicinity of the substation and the actual location of the site etc.
An accurate cost of a proposed substation can be estimated only when full design details and all relevant information is available. However for budgetary purposes cost figures given below in respect of typical equipment utilized in a modern substation may serve as a helpful guide.

Cubicle pattern 11kV switchboards.
The cost of cubicle pattern switchgear will depend on the type of circuit breakers employed and the level of protection and metering required. As an approximate guide, the cost per section might be of the order of the following:

	Cost per Section £
Utilizing oil circuit breakers	12,750.00
Utilizing vacuum circuit breakers	13,940.00

Cubicle pattern LV switchboards.
The cost of cubicle pattern LV switchboards will depend on quality, short- circuit rating, level of metering and protection required, and the number and ratings of incoming and outgoing circuit breakers and fuse switches.
In the absence of such information an approximate cost can be determined on the basis of installed capacity. For general applications this will range between £19.28 and £25.29 per kVA.

2MVA - 2 Transformer sub-station
The total approximate cost of a sub-station can therefore be determined as follows:

	£
5 Panel HV switchboard using oil circuit breakers	64,670.00
2 x 1000 kVA ON AN type transformers at £13,400.0 each	26,800.00
LV switchboard - 2000 kVA at say £24.10 per kVA	48,200.00
HV and LV cabling, say	3,650.00
Earthing, say	3,110.00
Total approximate cost of 2 MVA sub-station	£146,430.00

ELEMENTAL AND ALL-IN-RATES

ALL IN RATES FOR PRICING ELECTRICAL APPROXIMATE QUANTITIES *continued*

MAINS AND SWITCHGEAR *continued*

Rising main busbar trunking

Four-pole mild steel sheet rising main busbar trunking with copper busbars, complete with tap-off points, fire barriers at 3.5 m intervals, one thrust block, one expansion joint assembly and one cable entry chamber with glands for incoming cables and including tap-off cable clamps, interconnections, internal wiring, shields, phase buttons, labels and external earth tape; cost of installations based on a 60 m length of rising main.

	Per Metre £
200 amp rising main	205.00
300 amp rising main	225.00
400 amp rising main	242.00
600 amp rising main	308.00

Overhead plug-in busbar trunking

Four-pole overhead plug-in busbar trunking with copper busbars enclosed within extended aluminium casing. Cost of installation based on network system of 300 metres running length including various fittings, e.g. feed units, bends, tees, etc.

	Per Metre £	PC each £
100 amp plug-in overhead busbar system fixed and connected	49.23	
20 amp T.P. & N. plug-in unit		17.20
30 amp T.P. & N. plug-in unit		43.88
60 amp T.P. & N. plug-in unit		71.79

	Per Metre £	PC each £
300 amp plug-in overhead busbar system fixed and connected	81.47	
30 amp T.P. & N. plug-in unit		111.45
60 amp T.P. & N. plug-in unit		133.90
100 amp T.P. & N. plug-in unit		146.26

Note P.C. values of plug-in units are inclusive of HRC fuse links.

STANDBY GENERATING SETS

Diesel

For ratings up to 1000 kVA see 'Prices For Measured Work' section.

For ratings over 1000 kVA, approximate installed cost … £122.00 per kVA

Gas Turbine

Approximate installed cost
50 - 1000 kVA	£243.00 per kVA
Over 1000 kVA	£194.00 per kVA

ELEMENTAL AND ALL-IN-RATES

ALL IN RATES FOR PRICING ELECTRICAL APPROXIMATE QUANTITIES *continued*

POWER
Approximate prices for wiring of power points complete, including socket outlets with plugs but excluding consumer control units.

	Per Point £
13 amp Socket outlets	
Wired in PVC insulated and PVC sheathed cable in flats and houses on a ring main circuit protected, where buried, by heavy gauge PVC conduit	48.82
As above but in commercial property	59.95
Wired in PVC insulated cable in screwed welded conduit on a ring main circuit in flats and houses	69.22
As above but in commercial property	80.75
As above but in industrial property	92.80
Wired in M.I.C.C. cable on a ring main circuit in flats and houses	70.25
As above but in commercial property	82.30
As above but in industrial property (PVC sheathed cable)	104.85

	Each £
Cooker and control units	
45 amp Circuit including control unit, wired in PVC insulated and PVC sheathed cable, protected where buried by heavy gauge PVC conduit	95.38
As above but wired in PVC insulated cable in screwed welded conduit	139.26
As above but wired in M.I.C.C. cable	156.46

Low voltage power circuits
Three phase four wire circuit feeding an individual load, wired with PVC insulated cable drawn in heavy gauge black enamelled screwed welded conduit including standard associated fittings and flexible PVC sheathed metallic conduit, not exceeding one metre long, fixed to brickwork or concrete with distance saddles in surface work, *per 10 metre run.*

Cable size mm²	*Conduit size* mm	£
1.5	20	143.58
2.5	20	146.78
4	20	156.46
6	25	191.79
10	25	217.54
16	32	246.48

Three phase four wire circuit feeding an individual load item, wired with four core M.I.C.C./PVC cable including terminations, with glands and cable loop for vibration-free connection, fixed with PVC covered clips, *per 10 metre run.*

Cable size mm²	£
1.5	136.06
2.5	168.20
4	190.76
6	217.54
10	261.41
16	344.95

ELEMENTAL AND ALL-IN-RATES

ALL IN RATES FOR PRICING ELECTRICAL APPROXIMATE QUANTITIES *continued*

POWER *continued*

Power and communications common underfloor ducting   £
225 x 25 mm three compartment underfloor steel
cable duct including fittings, laid in screed,
excluding the builder's work . . . . per metre   31.11

Power and communications common flush floor trunking
395 x 60 mm three compartment flush floor trunking,
including associated fittings, fixed to floor . . . . per metre   61.08

Flush floor outlet boxes
Three compartment flush floor outlet box complete
with twin socket outlet, twin telephone outlet
plate, blank plate for data section, power circuit
wiring only.
Fixed to underfloor ducting . . . . per box   109.28

Ditto but fixed to flush floor trunking . . . . per box   133.90

Ditto but fixed to cavity floor panel and complete
with flexible metallic conduit, not exceeding two
metres long . . . . per box   145.75

Wiring and connections only, to mechanical services.
The cost of electrical connections to mechanical services equipment will vary depending on the type of
building and complexity of the equipment but and allowance of £6.43 per m² of total floor area should be
a useful guide to allow for power and control wiring in the conduit system including local isolators and
remote push button stations.

ELEMENTAL AND ALL-IN-RATES

ALL IN RATES FOR PRICING ELECTRICAL APPROXIMATE QUANTITIES *continued*

LIGHTING

Approximate prices for wiring of lighting points complete, including accessories but excluding lighting fittings.

	Per Point £
Final circuits	
Wired in PVC insulated and PVC sheathed cable in flats and houses installed in cavities and roof space protected, where buried, by heavy gauge PVC conduit	43.98
As above but in commercial property	53.24
Wired in PVC insulated cable in screwed welded conduit in flats and houses	91.67
As above but in commercial property	113.20
As above but in industrial property	129.06
Wired in M.I.C.C. cable in flats and houses	77.45
As above but in commercial property	92.80
As above but in industrial property (PVC sheathed cable)	107.12

ELEMENTAL AND ALL-IN-RATES

ALL IN RATES FOR PRICING ELECTRICAL APPROXIMATE QUANTITIES *continued*

EMERGENCY LIGHTING

Self-contained, non maintained 3 hour duration fitting wired with 1.5 mm² single core PVC insulated cables drawn in 20 mm heavy gauge black enamelled screwed welded conduit.

		£
tungsten	*per fitting*	180.04
fluorescent	*per fitting*	208.88

ELECTRIC HEATING
Electric underfloor heating by heating cable
The approximate cost of solid embedded underfloor heating including distribution fuseboard, time switches and thermostats is £22.50 per m² of floor area based on 145 watts per m².

Storage heaters
Storage heater with built in thermostat, temperature controller and safety cut-out.

Capacity kW		Supply and installation cost of storage heating including wiring. Unit £
1.7	P.C. £145.75	316.00
2.55	P.C. £186.33	366.37
3.4	P.C. £225.06	414.58

Note
Cost of wiring is based on 10 metre run of circuit including a 25 amp double pole switch with metal box, share of distribution fuse board and 4 mm² PVC cable drawn into 20 mm conduit.
For wiring with PVC insulated and PVC sheathed cable the above costs may be reduced by £64.28 in each case.

CLOCK SYSTEMS
Clock circuits
D.C. clock circuit wired with 2 x 1.5 mm² single core PVC insulated cable drawn in 20 mm heavy gauge black enamelled screwed welded conduit including standard associated fittings, clock connector and share of the battery unit fixed to brickwork or concrete with spacer bar saddles in surface work.
Cost based on an average circuit run of 30 metres per clock point (excluding provisions for clocks).

	£
per clock point	362.15

		£
Crystal controlled master clock	P.C.	604.20
300 mm Wall mounted slave clock	P.C.	67.47

A.C. mains clock circuit wired as above but cost based on an average circuit run of 4 m per clock point.

	£
per clock point	51.40

		£
300 mm Wall mounted mains clock	P. C.	54.69

ELEMENTAL AND ALL-IN-RATES

ALL IN RATES FOR PRICING ELECTRICAL APPROXIMATE QUANTITIES *continued*

FIRE ALARMS

Fire alarm circuits
Fire alarm system wired in single core PVC insulated cables drawn into black enamelled conduit.

	Average cost per point
	£
Bell	133.90
Break glass contact	122.16
Heat detector	117.94
Smoke detector	175.72

For costs of zone control/indicator panels, battery chargers and batteries, see 'Prices for Measured Work' section.

EXTERNAL LIGHTING

Estate road lighting
Post type road lighting lantern 80 watt MBF/U complete with 4.5 m high column with hinged lockable door, control gear and cut-out including all internal wiring, interconnections and earthing fed by 2.5 mm² two core PVC SWA PVC underground cable.
Approximate installed price *per metre road length* (based on 300 metres run) including time switch but excluding all trench and builder's work
    Columns erected on same side of road at 30 m intervals £22.45
    Columns erected on both sides of road at 60 m intervals
    in staggered formation £27.81

Bollard lighting
Bollard lighting fitting 50 watt MBF/U including controlgear,
all internal wiring, interconnections, earthing and 10 metres
of 2.5 mm² two core PVC SWA PVC underground cable
    Approximate installed price excluding trench work and
    builder's work £258.22 *each*

Outdoor flood lighting
Wall mounted outdoor flood light fitting complete with tungsten
halogen lamp, mounting bracket, wire guard and all internal
wiring; fixed to brickwork or concrete and connected
    Installed price 500 watt £94.25 *each*
    Installed price 1000 watt £114.64 *each*

Pedestal mounted outdoor floor light fitting complete with
1000 watt MBF/U lamp, mounting bracket, control gear,
contained in weatherproof steel box, all internal wiring,
interconnections and earthing; fixed to brickwork or
concrete and connected
    Approximate installed price excluding builder's work £769.20 *each*

ELEMENTAL AND ALL-IN-RATES

ALL IN RATES FOR PRICING SPECIALIST APPROXIMATE QUANTITIES

MEDICAL GASES

Tubing and Fittings
Light gauge copper tube to B.S.2871, Part 1,
Table X degreased and sealed with silver brazed
end feed fittings including allowance for waste,
fittings and brackets, assuming average runs

	Cost per metre
Nominal size	*£*
12mm	17.20
15mm	19.26
22mm	19.26
28mm	23.59
35mm	36.46
42mm	41.82
54mm	50.37
67mm	58.92
76mm	71.79
108mm	101.87

	Cost each
Flush mounted terminal unit, with 12 mm copper tail brazed to terminal inlet and distribution pipe	*£*
	73.95

Lockable isolating valve, with lockable
head mounted in flush mounted box

Size	
15mm	24.62
22mm	31.11
28mm	39.66
35mm	49.34
42mm	62.11
54mm	73.95

Equipment

	Total Plant Cost £
Cylinder manifold, 2 x 5 cylinders, comprising two header assemblies, bottom frame, inlet pipe and control panel, regulator, non-return and isolating valves, relief valves, spanners and spanner rack, and one spare cylinder rack with ten spare cylinders	3,680.00
2 cylinder emergency supply manifold with header and header support regulator, non-return and isolating valves, two relief valves	396.00

ELEMENTAL AND ALL-IN-RATES

ALL IN RATES FOR PRICING SPECIALIST APPROXIMATE QUANTITIES *continued*

MEDICAL GASES *continued*

Equipment *continued*

	Total Plant Cost £
Packaged compressed air unit, one duty and one stand-by compressor, two stage reciprocating piston type, duty of 1587 litres/min of oil free air to B.S.5475/HMT22 at 7 bar from each compressor, complete with horizontal air receiver, separator, silencers, air dryers, air cooled coolers and cylinders, controls and control panel, all necessary valves, gauges, alarms, motors and flexible connections	32,520.00
Packaged medical vacuum plant, one duty and one standby direct driven vacuum pumps having a duty of 1200 litres/min of free air displacement at 500 mm Hg below atmosphere, total design flow of 1480 litres/min of free air, with horizontal vacuum reservoir bacterial filters, silencers, controls and control panel and all valves, gauges, alarms, motors and meters, flexible connections	18,130.00
Liquid oxygen supply plant, with vacuum insulated evaporator (VIE), liquid capacity of 3080 litres, standby emergency oxygen twin manifold, air vaporiser, control and control panel, safety valves, isolating and non-return valves, flexible connections	12,960.00

ELEMENTAL AND ALL-IN-RATES

ALL IN RATES FOR PRICING SPECIALIST APPROXIMATE QUANTITIES *continued*

CATERING EQUIPMENT

Commercial Catering Establishment, 215m², 175 Nr covers/places, including cold storage preparation, cooking, display/servery, waste disposal, cash point, dishwashing/food disposal and cleaning facilities and local electrical/mechanical connections only. Excludes dining furniture/fitments.

ELEMENTAL COST PLAN

	Total Cost £	Cost/m² £	Cost/Cover £
Food Storage (Cold rooms/ storage racks, etc.)	14,860.00	69.12	84.91
Food Preparation/Cooking	23,960.00	111.44	136.91
Display Counters/Serving	19,180.00	89.21	109.60
Washing/Clearing/Disposal	16,360.00	76.09	93.49
Cash Tills/Counters	5,020.00	23.35	28.69
Total	79,380.00	369.21	453.60

ELEMENTAL AND ALL-IN-RATES

ALL IN RATES FOR PRICING SPECIALIST APPROXIMATE QUANTITIES *continued*

LIFT INSTALLATIONS

The cost of lift installations will vary depending upon a variety of circumstances. The following prices assume a floor to floor height of 3.5 metres and standard finishes to cars, doors and gates.

Passenger lifts

Electrically operated A.C. driven 8 or 10 person lift
serving 6 levels with directional collective controls
and a speed of 1 metre per second .... £61,590.00 to £74,140.00
    Add to above for
        Bottom motor room .... £5,120.00 to £7,400.00
        Extra levels served .... £2,360.00 to £3,650.00
        Increase speed of travel from 1 to 1.60 metres
        per second .... £2,360.00 to £4,920.00
        Increase speed of travel from 1 to 2.50 metres
        per second .... £32,140.00 to £36,960.00
        Enhanced finish to car .... £3,650.00 to £9,860.00
Electrically operated A.C. driven 21 person lift
serving 6 levels with directional collective controls
and a speed of 1.60 metres per second .... £98,829.00 to £123,530.00
    Add to above for
        Bottom motor room .... £6,108.00 to £12,220.00
        Extra levels served .... £3,646.00 to £4,930.00
        Increased speed of travel from 1.60 to 2.50
        metres per second .... £36,967.00 to £43,060.00
        Enhanced finish to car .... £6,108.00 to £12,220.00
For any floor bypassed add 45% of the cost of an
an extra level served

Goods lifts

Electrically operated two speed general purpose
goods lift serving 5 levels, to take 500 kg load,
inter manually operated shutters and automatic
push button control and a speed of 0.25 metres per
second .... £43,281.00 to £61,820.00
    Add to above for
        Extra levels served .... £2,359.00 to £3,650.00
        Increased capacity up to 2000 kg .... £12,216.00 to £18,640.00
        Increase speed of travel from 0.25 to 0.50
        metres per second .... £6,108.00 to £12,220.00
Oil hydraulic operated goods lift 5 levels, to take
500 kg load with manually operated shutter and
automatic push button control and a speed of 0.25
metres per second .... £49,200.00 to £60,030.00
    Add to above for
        Increased capacity up to 1000 kg .... £3,646.00 to £6,110.00
        Increased capacity up to 1500 kg .... £6,108.00 to £12,220.00
        Increased capacity up to 2000 kg .... £12,216.00 to £18,530.00

ELEMENTAL AND ALL-IN-RATES

ALL IN RATES FOR PRICING SPECIALIST APPROXIMATE QUANTITIES *continued*

ESCALATOR INSTALLATIONS

30°Pitch escalator with a rise of 3 to 5
metres and enamelled sheet steel or glass
balustrades

600mm step width	£49,290.00 to	£61,820.00
800mm step width	£53,140.00 to	£65,570.00
1000mm step width	£55,600.00 to	£68,030.00

Add to above for

Stainless steel balustrades	£4,930.00 to	£7,400.00
Glass balustrades with balustrade lighting	£3,650.00 to	£6,220.00

Building Regulations Explained

1995 Revision

5th Edition

John Stephenson

(from the 3rd and 4th editions)

"...more than a guide in the usual sense, it sets the Regulations in context"

- AJ Focus

Thoroughly revised and updated, the latest edition of Building Regulations Explained incorporates the most recent amendments to the Regulations which came into effect during 1995. It starts with the development of building control and then describes procedural aspects before dealing with all 14 Approved Documents in detail, showing their relevance and method of operation, with the help of numerous tables and diagrams.

An excellent reference source you can't afford to be without!

August 1995: 297x210: 512pp, 627 line illus
Hardback: 0-419-19690-0: £49.50

For further information and to order please contact: The Marketing Dept.,
E & F N Spon, 2-6 Boundary Row, London SE1 8HN
Tel: 0171 865 0066 Fax: 0171 522 9623

E & F N Spon

An imprint of Chapman & Hall

Approximate Estimating

ELEMENTAL COSTS

OFFICE BUILDING

Speculative ten storey city office block with a gross floor area of 6,585 m², upon which this cost analysis is based. Fully air conditioned with heating and cooling, and 3 ten person passenger lifts.

Cost Summary

El. Ref.	Element	Total Cost £	Cost / m² £
5D	Water Installations		
	Cold water services	56,331.00	8.55
	Hot water services	36,682.00	5.57
	Chilled water	200,095.00	30.39
5E	Heating and Heat Source	265,199.00	40.27
5G	Ventilating Systems		
	Air conditioning	619,954.00	94.15
	Automatic controls	131,966.00	20.04
5H	Electrical Installation		
	Electrical source and power supplies	109,719.00	16.66
	To mechanical installation	104,603.00	15.89
	General lighting	83,254.00	12.64
	Lighting fittings	151,890.00	23.07
5J	Lift and Conveyor Installation		
	Lifts	239,709.00	36.40
5K	Protective Installation		
	Sprinkler installation	139,893.00	21.24
	Fire-fighting installation	27,301.00	4.15
	Lighting protection	6,494.00	0.99
5L	Communication Installation		
	Telephones	2,458.00	0.37
	Fire alarms and security	40,508.00	6.15
	Summary total	2,216,056.00	336.53

ELEMENTAL COSTS

HOTEL BUILDING

The Hotel has a gross floor area of 10,153 m², upon which the cost analysis is based, and associated car parking area of 4,149 m².

The construction is on seven levels, having three half area floors containing service yard and goods entrance, reception areas and public areas, with the remaining Hotel on four full floors.

The private and associated public car parking is constructed on the half area levels.

The Hotel provides a total of 150 bedrooms, comprising 116 double/twin bedded rooms, 24 single/syndicate rooms, 6 interconnecting rooms and 4 suites. All of the bedrooms have en-suite bathrooms and are fitted to a high standard but are not air conditioned.

The public areas of the Hotel include a restaurant with 150 covers, a lounge bar and cocktail bar, all fully air conditioned. The major facilities also include a Banqueting Hall and Foyer to seat 300 and two large Conference / Promotion rooms to cater for medium sized meetings. The main Hall and one of the meeting rooms is capable of being subdivided into two smaller units if required. Two smaller Conference rooms are provided as part of a suite which allows the option of incorporating a number of related syndicate rooms. The Banqueting and Conference facilities, like the public areas are fully air conditioned.

Five lifts are provided, two main passenger lifts serving all floors, two passenger lifts serving lower and car park floors and a service/goods lift.

Cost Summary

El. Ref.	Element	Total Cost £	Cost / m² £
5D	Water Installations		
	Hot and cold water services	371,514.00	36.59
5E	Heating and Heat Source	374,739.00	36.91
	Carried forward	746,253.00	73.50

ELEMENTAL COSTS

HOTEL BUILDING (Cont'd)

Cost Summary

El. Ref.	Element	Total Cost £	Cost / m² £
	Brought forward	746,253.00	73.50
5G	Ventilating Systems		
	Car park ventilation	15,078.00	1.49
	Bathroom ventilation	74,418.00	7.33
	Air conditioning	623,396.00	61.40
5H	Electrical Installation		
	Electrical source and power supplies	281,338.00	27.71
	To mechanical installation	39,929.00	3.93
	General lighting	428,252.00	42.18
	Emergency lighting	51,603.00	5.08
	Car park lighting	114,334.00	11.26
	External lighting	32,682.00	3.22
5J	Lift and Conveyor Installation		
	Lifts	280,652.00	27.64
5K	Protective Installation		
	Fire-fighting installation	44,357.00	4.37
	Lighting protection	11,399.00	1.12
5L	Communication Installation		
	Telephones	5,841.00	0.58
	Fire alarms	177,424.00	17.48
5M	Special Services		
	Whirlpool, sauna and steam room	79,115.00	7.79
	Kitchen equipment	230,845.00	22.74
	Summary total	3,236,916.00	318.82

ELEMENTAL COSTS

DATA PROCESSING CENTRE

Commercial computer centre comprising three distinct areas - energy centre, main computer hall / air handling corridors, and a two storey office complex with atrium, over a total floor area of 5,300 m², upon which the cost analysis is based. The plant room for the Office Complex being located with the main computer hall / A.H.C. areas. The external area of the building site including access roads, car parks and footpaths is 13,525 m².

Piping systems have been designed and installed to permit the installation of additional plant items (eg. PCU's, Dri-Coolers, Chillers), to cope with the increased computer load without the existing plant having to be shut down. The infra structure has been ultimately designed for 750 watts/m² although the internal process cooling plant has been sized to handle 400 watts/m² (with an allowance of 25% spare capacity).

All of the electrical systems are designed to avoid system failure. A single circuit or component failure will not effect more equipment than the redundancy facilities allow, and will not therefore impair the operation of the Centre.

As computer loads increase additional plant items (eg. additional generator set and UPS set) can be added into the system without the existing system having to be shut-down.

Common services, trunking, trays and ladder racks, also have inbuilt allowances of 25% spare capacity.

Cost Summary

El. Ref.	Element	Total Cost £	Cost / m² £
5D	Water Installations		
	Mains supply	8,912.00	1.68
	Cold water services	28,213.00	5.32
	Hot water services	8,117.00	1.53
	Steam and condensate	9,996.00	1.89
	Carried forward	55,238.00	10.42

ELEMENTAL COSTS

DATA PROCESSING CENTRE (Cont'd)

Cost Summary

El. Ref.	Element	Total Cost £	Cost / m² £
	Brought forward	55,238.00	10.42
5E	Heat Source		
	Condenser water installation	297,411.00	56.12
	Chilled water installation	407,829.00	76.95
	Pressurisation installation	18,956.00	3.58
	L.P. hot water installation	59,108.00	11.15
	Water treatment system	6,873.00	1.30
5F	Space Heating and Air Treatment		
	Water and/or steam	81,847.00	15.44
	Heating with cooling air (treated locally)	349,723.00	65.99
5G	Ventilating Systems		
	Mechanical extract system	60,916.00	11.49
	Halon extract system	44,315.00	8.36
5H	Electrical Installation		
	Electrical source and power supplies	1,791,788.00	338.07
	Electric power supplies	644,401.00	121.59
	Electric lighting and fittings	249,020.00	46.98
5J	Lift and Conveyor Installation	37,285.00	7.03
5K	Protective Installation		
	Halon extinguishing system	62,520.00	11.80
	Lighting protection	17,586.00	3.32
5L	Communication Installation		
	Warning installation (security and fire)	461,376.00	87.05
	Visual and audio installation	22,319.00	4.21
	Telephone installation	9,170.00	1.73
5M	Special Services		
	Energy management system	472,383.00	89.13
	Summary total	5,150,064.00	971.71

ELEMENTAL COSTS

DISTRICT GENERAL HOSPITAL

The Hospital is based on the `Nucleus' design concept and comprises four `cruciform' templates on two /three levels arranged on both sides of a communication corridor (Street).

Adjoining the templates is a Service and Energy Centre consisting of Linen Distribution, Kitchen, Switch Rooms and Generators, Boiler Hall and Incinerator / Waste Handling Facility.

The total floor area of the Hospital is 15,520 m², upon which the cost analysis is based.

Cost Summary

El. Ref.	Element	Total Cost £	Cost / m² £
5B	Services Equipment		
	(Mechanical)		
	Cold room installation	120,084.00	7.74
	Services equipment	211,860.00	13.65
	(Electrical)		
	Equipment - all areas	28,294.00	1.82
5B	Water Installation		
	(Mechanical)		
	Boosted mains cold water	104,194.00	6.71
	Potable water storage	28,726.00	1.85
	Non potable water storage	22,047.00	1.42
	Utility water storage	19,200.00	1.24
	Boilerhouse wash down	3,144.00	0.20
	Incinerator wash down	3,592.00	0.23
	Incinerator bin wash	2,995.00	0.19
	Emergency shower	357.00	0.02
	Kitchen potable cold water	16,440.00	1.06
	Potable tank cold water	185,198.00	11.93
	Kitchen D.H.W.S. (cold feed)	4,095.00	0.26
	Kitchen D.H.W.S	54,205.00	3.49
	Non potable cold storage	32,983.00	2.13
	D.H.W.S	406,567.00	26.20
	Carried forward	1,243,981.00	80.14

ELEMENTAL COSTS

DISTRICT GENERAL HOSPITAL (Cont'd)

Cost Summary

El. Ref.	Element	Total Cost £	Cost / m² £
	Brought forward	1,243,981.00	80.14
5B	Water Installation (cont'd)		
	Hydrotherapy cold water	5,016.00	0.32
	Patient hoist	1,195.00	0.08
	Cascade fountains	12,063.00	0.78
	Incoming water	3,016.00	0.19
5E	Heat Source		
	(Mechanical)		
	Soft water	32,628.00	2.10
	Boiler feed water	40,247.00	2.59
	INT blowdown (cold feed)	8,555.00	0.55
	INT blowdown	29,677.00	1.91
	CONT blowdown	6,594.00	0.42
	Boiler sight glass/aux. drains	2,355.00	0.15
	Incinerator boosted cold water	1,629.00	0.10
	Heavy fuel oil	155,072.00	9.99
	High pressure steam	249,526.00	16.08
	Flash steam	2,765.00	0.18
	HP steam (temp)	55,175.00	3.56
	Existing steam	117.00	0.01
	HP condensate	8,292.00	0.53
	Pumped condensate	15,552.00	1.00
	HP condensate (temp)	14,668.00	0.95
	Pumped condensate (temp)	44,835.00	2.89
	Existing condensate	454.00	0.03
	MPHW cold feed	3,476.00	0.22
	MPHW heating	123,602.00	7.96
	LPHW to tank farm	8,268.00	0.53
	LPHW common	54,278.00	3.50
	LPHW 12 hour	29,655.00	1.91
	LPHW 24 hour	28,304.00	1.82
	LPHW CT	8,106.00	0.52
	LPHW kitchen	6,109.00	0.39
	LPHW hydrotherapy pool	4,968.00	0.32
	Chilled water installation	186,523.00	12.02
	Heat recovery	15,940.00	1.03
	Incinerator installation	164,243.00	10.58
	Connections to equipment	6,032.00	0.39
	Carried forward	2,572,916.00	165.74

ELEMENTAL COSTS

DISTRICT GENERAL HOSPITAL (Cont'd)

Cost Summary

El. Ref.	Element	Total Cost £	Cost / m² £
	Brought forward	2,572,916.00	165.74
5F	**Space Heating**		
	(Mechanical)		
	Low pressure steam	3,323.00	0.21
	Humidifier steam	19,741.00	1.27
	Humidifier condensate	10,445.00	0.67
	MPHW heating	144,332.00	9.30
	LPHW to tank farm	19,761.00	1.27
	LPHW 12 hour	143,052.00	9.22
	LPHW 24 hour	115,432.00	7.44
	LPHW CT	64,162.00	4.13
	LPHW kitchen	26,017.00	1.68
	LPHW hydrotherapy pool	2,432.00	0.16
	Chilled water (hydrotherapy pool)	5,336.00	0.34
	Low velocity supply vent	361,111.00	23.27
	Low velocity supply (kitchen)	48,589.00	3.13
	Hydrotherapy pool supply	18,708.00	1.21
	Plant actuation	12,063.00	0.78
	Acoustic work	9,048.00	0.58
	Fire dampers	96,510.00	6.22
5G	**Ventilating Systems**		
	(Mechanical)		
	Clean extract	116,091.00	7.48
	Dirty extract	120,369.00	7.76
	Fume cupboard extract	7,117.00	0.46
	Fume canopy extract	6,559.00	0.42
	Clean extract (kitchen)	15,684.00	1.01
	Dirty special/extract (kitchen)	9,300.00	0.60
	Special extract	3,130.00	0.20
	Hydrotherapy pool extract	9,340.00	0.60
	Waste handling dirty extract	1,039.00	0.07
	Carried forward	3,961,607.00	255.22

ELEMENTAL COSTS

DISTRICT GENERAL HOSPITAL (Cont'd)

Cost Summary

El. Ref.	Element	Total Cost £	Cost / m² £
	Brought forward	3,961,607.00	255.22
5H	Electrical Installation		
	(Electrical)		
	High voltage installation	110,703.00	7.13
	Standby generator	268,845.00	17.32
	Sub-main distribution	326,086.00	21.01
	General power	135,934.00	8.76
	Lighting installation	272,824.00	17.58
	Luminaires	222,933.00	14.36
	Cable trunking/tray	65,794.00	4.24
	Earthing	37,878.00	2.44
	Canopy lighting	12,063.00	0.78
	Emergency lighting plant room and roof	12,063.00	0.78
	Emergency lighting hydrotherapy pool	9,048.00	0.58
	Additional lighting to artwork	12,063.00	0.78
	Additional fire barriers for electrical work	12,063.00	0.78
	Additional connections	9,048.00	0.58
	Cable supports future use	2,412.00	0.16
	Services to external signs	12,063.00	0.78
	Additional switchgear	12,063.00	0.78
5I	Gas Installation		
	(Mechanical) Natural Gas Incoming (Attendance) (See 6C)		
	Interruptible natural gas	24,007.00	1.55
	Uninterruptable natural gas	18,226.00	1.17
5K	Protective Installations		
	(Mechanical)		
	Fire hose reels	57,444.00	3.70
	Charged riser	28,836.00	1.86
	Hydrant mains (see also 6C)	4,859.00	0.31
	Smoke input ventilation	25,120.00	1.62
	Smoke extract ventilation	32,471.00	2.09
	(Electrical)		
	Lightning protection	16,550.00	1.07
	Carried forward	5,703,003.00	367.43

ELEMENTAL COSTS

DISTRICT GENERAL HOSPITAL (Cont'd)

Cost Summary

El. Ref.	Element	Total Cost £	Cost / m² £
	Brought forward	5,703,003.00	367.43
5L	Communications Installations		
	(Electrical)		
	Telephone installation	37,253.00	2.40
	Intercommunication system	24,525.00	1.58
	Radio distribution system	30,224.00	1.95
	Television system	8,995.00	0.58
	Nurse call system	139,906.00	9.01
	Fire alarm system	299,282.00	19.28
	Intruder alarm system	3,220.00	0.21
	Panic and lift alarm system	2,720.00	0.18
	Call bell system	664.00	0.04
	Computer system	2,797.00	0.18
	Cable trunking/tray	81,161.00	5.23
	Additional fire alarm needs	24,128.00	1.55
5M	Special Installations		
	(Mechanical)		
	Proportional chemical dosing	1,385.00	0.09
	Shot chemical dosing	4,295.00	0.28
	Diesel fuel oil	63,996.00	4.12
	Hydrotherapy pool	53,579.00	3.45
	Medical gas alarms	10,181.00	0.66
	Oxygen installation	47,354.00	3.05
	Compressed air installation	50,868.00	3.28
	Vacuum installation	51,176.00	3.30
	Condenser installation	2,806.00	0.18
	Bin cleaning equipment	8,445.00	0.54
	Lifting beams	12,063.00	0.78
6C	External Services		
	(Mechanical)		
	Incoming water main	157.00	0.01
	Existing hydrant main connection	998.00	0.06
	Fire hydrant main	10,905.00	0.70
	Incoming natural gas	156.00	0.01
	(Electrical)		
	Main electric supply	213.00	0.01
	External lighting and luminaires	38,360.00	2.47
	Courtyard lighting	10,590.00	0.68
	Summary total	6,725,405.00	433.29

ELEMENTAL COSTS

DISTRICT GENERAL HOSPITAL (Example 2)

(The details which follow show how costs per square metre based upon the functional or departmental area of the building can fluctuate within each element, reflecting the varying demands for different services within different departments.)

For this example, an alternative Hospital (to that used for example 1) has been used. This Hospital, again designed around the `Nucleus' concept, comprises 5 Nr. two storey templates to either side of the communications corridor together with a two storey Service Centre to one end of the corridor. Auxiliary works - generators, fuel tanks and minor control rooms are sited outside the main buildings.

The total floor area is 16,970 m², upon which this cost analysis is based but it should be noted that this is also the first phase of a planned larger development. In addition, the Service Centre also caters for an adjacent existing Hospital and as a consequence major areas (Kitchens etc) are oversized in relation to the template (department) areas.

Construction Quality and Quality Standards

The European perspective

George Atkinson, Construction Consultant, UK

✦ A practical and accessible guide to the changes taking place in the Single Market

✦ Authoritative explanations of the making and use of EU codes, regulations, specifications and standards

Quality is a vital issue to be addressed by all construction professionals working in Europe today. **Construction Quality and Quality Standards** gives clear, concise guidance to the making and use of codes, regulations and technical specifications in Europe.
The author presents a general overview to construction before going on to specific issues relating to quality and the relevant British Standards. This is an invaluable source of information on international organizations throughout Europe as well as French, English and German translations of commonly used terms and definitions in the codes, regulations and standards.

246x189mm 352 pages 8 line illus
July 1995 Hardback 0-419-18490-2 £55.00

For further information and to order please contact: The Marketing Dept.,
E & F N Spon, 2-6 Boundary Row, London SE1 8HN
Tel: 0171 865 0066 Fax: 0171 522 9623

E & F N Spon
An imprint of Chapman & Hall

ELEMENTAL COSTS

DISTRICT GENERAL HOSPITAL (cont'd)

Element	Adult Acute 168 beds Area m² 3,094 Cost £	Rate m²	Children 46 beds Area m² 1,031 Cost £	Rate m²	Operating Theatres 4 theatres Area m² 855 Cost £	Rate m²
5A Sanitary Appliances	90,935	29.39	32,268	31.30	29,594	34.61
5B Services Equipment	18,918	6.11	5,672	5.50	5,618	6.57
5C Disposal Installation	45,366	14.66	15,497	15.03	14,315	16.74
5D Water Installation	58,685	18.97	24,946	24.20	29,762	34.81
5E Heat Source						
5F Space Heating/Air Treatment	108,999	35.23	36,506	35.41	113,733	133.02
5G Ventilating System	97,635	31.56	23,787	23.07	54,942	64.26
5H Electrical Installation	204,126	65.97	58,711	56.95	86,295	100.93
5I Gas Installation						
5J Lift/Conveyor Installation						
5K Protective Installations	56,758	18.34	18,913	18.34	15,685	18.35
5L Communication Installation	127,192	41.11	35,648	34.58	13,485	15.77
5M Special Installations	92,953	30.04	42,627	41.35	99,572	116.46
5N Builders Work to Services	110,130	35.59	35,392	34.33	53,820	62.95
5O Builders Profit and Attendance on Services	57,410	18.56	18,446	17.89	28,108	32.87
Totals	1,069,107	345.53	348,413	337.95	544,929	637.34

Element	Adult Day Care 15 beds Area m² 482 Cost £	Rate m²	Accident and Emergency 46 beds Area m² 787 Cost £	Rate m²	Fracture Clinic 27 sessions Area m² 222 Cost £	Rate m²
5A Sanitary Appliances	21,260	44.11	24,680	31.36	10,577	47.64
5B Services Equipment	2,963	6.15	5,310	6.75		
5C Disposal Installation	10,204	21.17	11,698	14.86	6,168	27.78
5D Water Installation	25,818	53.56	10,743	13.65	10,038	45.22
5E Heat Source						
5F Space Heating/Air Treatment	16,149	33.50	22,787	28.95	29,893	134.65
5G Ventilating System	14,561	30.21	15,272	19.41	12,743	57.40
5H Electrical Installation	30,125	62.50	37,416	47.54	27,515	123.94
5I Gas Installation						
5J Lift/Conveyor Installation						
5K Protective Installations	8,842	18.34	14,438	18.35	4,073	18.35
5L Communication Installation	8,412	17.45	23,936	30.41	12,046	54.26
5M Special Installations	10,714	22.23	34,752	44.16	8,569	38.60
5N Builders Work to Services	17,789	36.91	24,365	30.96	14,191	63.92
5O Builders Profit and Attendance on Services	9,275	19.24	12,691	16.13	7,428	33.46
Totals	176,112	365.37	238,088	302.53	143,241	645.22

Approximate Estimating

ELEMENTAL COSTS

DISTRICT GENERAL HOSPITAL (cont'd)

Element	Intensive Therapy Unit 8 beds Area m² 381		Outpatients Department 67.5 beds Area m² 649		Pharmacy (excl. manf.) 300 beds Area m² 278	
	Cost £	Rate m²	Cost £	Rate m²	Cost £	Rate m²
5A Sanitary Appliances	7,087	18.60	15,085	23.24	1,942	6.99
5B Services Equipment	2,654	6.97	30,698	47.30		
5C Disposal Installation	5,767	15.14	8,351	12.87	2,800	10.07
5D Water Installation	11,144	29.25	25,297	38.98	3,949	14.21
5E Heat Source						
5F Space Heating/Air Treatment	64,640	169.66	53,629	82.63	16,745	60.23
5G Ventilating System	37,111	97.40	17,233	26.55	11,075	39.84
5H Electrical Installation	52,968	139.02	52,077	80.24	15,299	55.03
5I Gas Installation					1,906	6.86
5J Lift/Conveyor Installation						
5K Protective Installations	6,991	18.35	11,907	18.35	5,099	18.34
5L Communication Installation	11,262	29.56	20,257	31.21	4,809	17.30
5M Special Installations	20,563	53.97	26,707	41.15	1,873	6.74
5N Builders Work to Services	25,636	67.29	30,955	47.70	8,076	29.05
5O Builders Profit and Attendance on Services	13,419	35.22	16,202	24.96	4,226	15.20
Totals	259,242	680.43	308,398	475.18	77,799	279.86

Element	X-Ray 4 - R/D rooms Area m² 565		Administration 9 points Area m² 597		Main Entrance 9 points Area m² 306	
	Cost £	Rate m²	Cost £	Rate m²	Cost £	Rate m²
5A Sanitary Appliances	8,697	15.39	2,899	4.86	5,029	16.43
5B Services Equipment	5,310	9.40			1,484	4.85
5C Disposal Installation	7,761	13.74	2,896	4.85	3,029	9.90
5D Water Installation	20,246	35.83	6,749	11.30	8,432	27.56
5E Heat Source						
5F Space Heating/Air Treatment	35,233	62.36	30,630	51.31	1,508	4.93
5G Ventilating System	15,483	27.40	12,534	20.99	467	1.53
5H Electrical Installation	40,119	71.01	38,405	64.33	7,829	25.58
5I Gas Installation						
5J Lift/Conveyor Installation						
5K Protective Installations	10,366	18.35	10,951	18.34	5,614	18.35
5L Communication Installation	16,948	30.00	3,105	5.20	1,444	4.72
5M Special Installations	14,829	26.25	1,505	2.52		
5N Builders Work to Services	21,091	37.33	13,964	23.39	4,729	15.45
5O Builders Profit and Attendance on Services	11,039	19.54	7,308	12.24	2,475	8.09
Totals	207,122	366.60	130,946	219.33	42,040	137.39

ELEMENTAL COSTS

DISTRICT GENERAL HOSPITAL (cont'd)

Element	Medical Records 16 points Area m² 204		Staff Changing 436 places Area m² 156		Med./Nursing Management 5 offices Area m² 64	
	Cost £	Rate m²	Cost £	Rate m²	Cost £	Rate m²
5A Sanitary Appliances			13,357	85.62		
5B Services Equipment						
5C Disposal Installation	559	2.74	6,224	39.90	175	2.73
5D Water Installation			16,846	107.99		
5E Heat Source						
5F Space Heating/Air Treatment	12,656	62.04	13,580	87.05	11,918	186.22
5G Ventilating System	3,712	18.20	4,973	31.88	4,893	76.45
5H Electrical Installation	14,820	72.65	8,486	54.40	25,351	396.11
5I Gas Installation						
5J Lift/Conveyor Installation						
5K Protective Installations	3,743	18.35	2,861	18.34	1,173	18.33
5L Communication Installation	3,306	16.21	2,434	15.60	4,084	63.81
5M Special Installations					1,005	15.70
5N Builders Work to Services	5,053	24.77	8,228	52.74	5,554	86.78
5O Builders Profit and Attendance on Services	2,645	12.97	4,307	27.61	2,932	45.81
Totals	46,494	227.93	81,296	521.13	57,085	891.94

Element	Education 2 rooms Area m² 33		Rehab. (part) (excl. gym pool & workshop) Area m² 537		On call Suite 3 rooms Area m² 29	
	Cost £	Rate m²	Cost £	Rate m²	Cost £	Rate m²
5A Sanitary Appliances			11,655	21.70		
5B Services Equipment			5,310	9.89		
5C Disposal Installation	90	2.73	8,853	16.49	78	2.69
5D Water Installation			23,694	44.12		
5E Heat Source						
5F Space Heating/Air Treatment	1,761	53.36	37,504	69.84	5,876	202.62
5G Ventilating System	776	23.52	24,805	46.19	3,373	116.31
5H Electrical Installation	1,337	40.52	35,489	66.09	9,415	324.66
5I Gas Installation			636	1.18		
5J Lift/Conveyor Installation						
5K Protective Installations	605	18.33	9,851	18.34	530	18.28
5L Communication Installation	1,114	33.76	13,080	24.36	2,003	69.07
5M Special Installations	1,005	30.45	2,508	4.67	3,146	108.48
5N Builders Work to Services	836	25.33	20,836	38.80	2,807	96.79
5O Builders Profit and Attendance on Services	439	13.30	10,906	20.31	1,470	50.69
Totals	7,963	241.30	205,127	381.98	28,698	989.59

Approximate Estimating

ELEMENTAL COSTS

DISTRICT GENERAL HOSPITAL (cont'd)

Element	Dental (Excl. W'shop Recovery) 16 sessions Area m² 23		E.N.T. (Part) Area m² 93		Laboratory Services (Outstation) Area m² 69	
	Cost £	Rate m²	Cost £	Rate m²	Cost £	Rate m²
5A Sanitary Appliances						
5B Services Equipment						
5C Disposal Installation	164	7.13	253	2.72	486	7.04
5D Water Installation						
5E Heat Source						
5F Space Heating/Air Treatment	1,508	65.57	11,317	121.69	6,696	97.04
5G Ventilating System	467	20.30	3,503	37.67	4,428	64.17
5H Electrical Installation	1,567	68.13	11,743	126.27	6,118	88.67
5I Gas Installation	1,783	77.52				
5J Lift/Conveyor Installation						
5K Protective Installations	422	18.35	1,707	18.35	1,266	18.35
5L Communication Installation			1,925	20.70	960	13.91
5M Special Installations	2,005	87.17	1,005	10.81	14,949	216.65
5N Builders Work to Services	949	41.26	3,767	40.51	4,086	59.22
5O Builders Profit and Attendance on Services	495	21.52	1,972	21.20	2,140	31.01
Totals	9,360	406.95	37,192	399.92	41,129	596.06

Element	Mortuary/P.M. 18 places 2 P.M. rooms Area m² 198		Dining 600 meals Area m² 460		Kitchen 1200 meals Area m² 726	
	Cost £	Rate m²	Cost £	Rate m²	Cost £	Rate m²
5A Sanitary Appliances	28,093	141.88			16,995	23.41
5B Services Equipment	1,086	5.48	30,661	66.65	226,777	312.37
5C Disposal Installation	12,816	64.73	1,261	2.74	5,725	7.89
5D Water Installation	16,834	85.02	14,013	30.46	127,736	175.94
5E Heat Source						
5F Space Heating/Air Treatment	67,226	339.53	50,372	109.50	82,487	113.62
5G Ventilating System	18,505	93.46	14,599	31.74	24,073	33.16
5H Electrical Installation	12,587	63.57	19,259	41.87	59,352	81.75
5I Gas Installation			4,020	8.74	10,958	15.09
5J Lift/Conveyor Installation						
5K Protective Installations	3,632	18.34	8,439	18.35	13,319	18.35
5L Communication Installation	3,246	16.39	3,650	7.93	11,359	15.65
5M Special Installations	57,851	292.18			40,770	56.16
5N Builders Work to Services	25,667	129.63	18,363	39.92	75,494	103.99
5O Builders Profit and Attendance on Services	13,423	67.79	9,581	20.83	39,457	54.35
Totals	260,966	1318.00	174,218	378.73	734,502	1011.73

ELEMENTAL COSTS

DISTRICT GENERAL HOSPITAL (cont'd)

Element	H.D.S.U. 4 theatres 300 beds Area m² 569 Cost £	Rate m²	Stores Area m² 749 Cost £	Rate m²	Linen Store at source 300 beds Area m² 96 Cost £	Rate m²
5A Sanitary Appliances	2,011	3.53	443	0.59		
5B Services Equipment	256,497	450.79	2,255	3.01		
5C Disposal Installation	2,022	3.55	2,244	3.00	263	2.74
5D Water Installation	138,754	243.86	1,794	2.40		
5E Heat Source						
5F Space Heating/Air Treatment	61,314	107.76	35,818	47.82	10,233	106.59
5G Ventilating System	17,958	31.56	9,638	12.87	2,752	28.67
5H Electrical Installation	12,856	22.59	32,554	43.46	5,426	56.52
5I Gas Installation			3,288	4.39		
5J Lift/Conveyor Installation						
5K Protective Installations	10,439	18.35	13,741	18.35	1,761	18.34
5L Communication Installation	3,248	5.71	7,301	9.75	1,216	12.67
5M Special Installations	3,340	5.87	21,341	28.49		
5N Builders Work to Services	61,964	108.90	18,356	24.51	2,882	30.02
5O Builders Profit and Attendance on Services	32,388	56.92	9,607	12.83	1,508	15.71
Totals	602,791	1059.39	158,380	211.47	26,041	271.26

Element	Works Dept. 300 beds Area m² 425 Cost £	Rate m²	Telephone Exchange 400 extensions Area m² 96 Cost £	Rate m²	Auxiliary Buildings Area m² 399 Cost £	Rate m²
5A Sanitary Appliances	4,530	10.66				
5B Services Equipment	14,838	34.91				
5C Disposal Installation	4,962	11.68	263	2.74	734	1.84
5D Water Installation	12,860	30.26				
5E Heat Source						
5F Space Heating/Air Treatment	51,169	120.40	15,350	159.90		
5G Ventilating System	13,767	32.39	4,129	43.01		
5H Electrical Installation	43,406	102.13	5,426	56.52	34,155	85.60
5I Gas Installation						
5J Lift/Conveyor Installation						
5K Protective Installations	7,796	18.34	1,761	18.34	7,319	18.34
5L Communication Installation	4,461	10.50	406	4.23	7,230	18.12
5M Special Installations						
5N Builders Work to Services	19,676	46.30	3,517	36.64	7,893	19.78
5O Builders Profit and Attendance on Services	10,299	24.23	1,842	19.19	4,131	10.35
Totals	187,764	441.80	32,694	340.57	61,462	154.03

ELEMENTAL COSTS

DISTRICT GENERAL HOSPITAL (cont'd)

Element	Communications Area m² 2,432 Cost £	Rate m²	Stores Area m² 365 Cost £	Rate m²	TOTAL Area m² 16,970 Cost £	Rate m²
5A Sanitary Appliances					327,138	19.28
5B Services Equipment					616,050	36.30
5C Disposal Installation	21,237	8.73	4,969	13.61	207,230	12.21
5D Water Installation			126,822	347.46	715,161	42.14
5E Heat Source			427,969	1172.52	427,969	25.22
5F Space Heating/Air Treatment	67,171	27.62	220,581	604.33	1,294,988	76.31
5G Ventilating System	29,020	11.93	38,459	105.37	536,672	31.62
5H Electrical Installation	55,390	22.78	184,520	505.53	1,230,142	72.49
5I Gas Installation			8,602	23.57	31,191	1.84
5J Lift/Conveyor Installation	174,565	71.78			174,565	10.29
5K Protective Installations	44,612	18.34	6,696	18.35	311,308	18.34
5L Communication Installation	4,264	1.75	4,337	11.88	358,168	21.11
5M Special Installations	104,230	42.86			607,819	35.82
5N Builders Work to Services	67,003	27.55	71,858	196.87	784,923	46.25
5O Builders Profit and Attendance on Services	35,069	14.42	38,215	104.70	410,855	24.21
Totals	602,561	247.76	1,133,028	3104.19	8,034,179	473.43

Spon's Gulf States Construction Price Book
(Volume 1)

and

Spon's Gulf States Construction Price Book
(Volume 2)

Hunter Fleming Associates Ltd

✦ only sources of detailed price data on this very wealthy and active construction market

These two volumes are the first, fully researched and comprehensive price books for the Gulf States in English. They will be an essential reference for any construction professional involved in the Middle East market.

Contents for Volume One:
Acknowledgements. Preface. Introduction to the Gulf. Principles of measurement (international). Form of contract FIDIC. Lebanon. Oman. United Arab Emirates. Index.

October 1995: 210x297: ap512pp, 4 line illus
Hardback: 0-419-21120-9: c.£125.00

Contents for Volume Two:
Acknowledgements. Preface. Introduction to the Gulf. Principles of measurement (international). Form of contract (FIDIC). Bahrain. Qatar. Saudi Arabia. Index.

October 1995: 210x297: ap512pp
Hardback: 0-419-21130-6: c.£125.00

For further information and to order please contact: The Marketing Dept., E & F N Spon, 2-6 Boundary Row, London SE1 8HN
Tel: 0171 865 0066 Fax: 0171 522 9623

E & F N Spon
An imprint of Chapman & Hall

Mechanical Installations
RATES OF WAGES AND WORKING RULES

RATES OF WAGES

HEATING, VENTILATING, AIR CONDITIONING, PIPING AND
DOMESTIC ENGINEERING INDUSTRY

Extracts from National Agreement made between:

Heating and Ventilating   and   Manufacturing Science Finance
  Contractor's Association   Union (MSF)
ESCA House,   79 Camden Road,
34 Palace Court,   London NW1 9ES
Bayswater,   Telephone: 0171-267 4422
London W2 4JG
Telephone: 0171-229 2488

WAGE RATES, ALLOWANCES AND OTHER PROVISIONS

Hourly rates of wages
All districts of the United Kingdom

Main grades	% of Fitters rate	From 1 March 1993 p/hr	From 5 September 1994 p/hr
Pipework Foreman	135	667	682
Ductwork Foreman	130	642	657
Chargehand	125	618	631
Advanced Fitter	110	543	556
Fitter	100	494	505
Improver	95	469	480
Assistant	90	445	455
Mate over 18	80	395	404
Junior Mate 17-18	50	247	253
Junior Mate up to 17	35	173	177

Note: Ductwork Erection Operatives are
entitled to the same rates and allowances as
the parallel Fitter grades shown (Note 1)

Craft Apprentice	From 1 March 1993 p/hr	From 5 September 1994 p/hr
Year 1 - if taken on as an apprentice	203	207
Year 2	247	253
Year 3	325	328
Year 4	395	404

RATES OF WAGES

Junior Ductwork Erectors (Probationary)
Trainee Rates of Pay

Age at entry	From 1 March 1993 p/hr	From 5 September 1994 p/hr
17	203	207
18	247	253
19	321	328
20	395	404

Junior Ductwork Erectors (Year of Training)

	From 1 March 1993			From 5 September 1994		
	1 yr p/h	2 yr p/hr	3 yr p/hr	1 yr p/hr	2 yr p/hr	3 yr p/hr
17	247	321	395	253	328	404
18	321	395	423	328	404	432
19	395	399	444	404	408	454
20	395	420	445	404	429	455

Welding supplements	From 1 March 1993 p/hr	From 5 September 1994 p/hr
gas/arc	49	51
gas or arc	25	25

Payable to Fitters and Advanced Fitters
qualified in accordance with Clause 8f and
becomes part of the normal hourly rate

Daily travelling allowance
C: Craftsmen including Improvers
M&A: Assistants, Mates, Junior Mates
  and Craft Apprentices
Direct distance from centre to
job in miles

		From 1 March 1993		From 5 September 1994	
		C	M&A	C	M&A
Over	Not exceeding	p/hr	p/hr	p/hr	p/hr
0	2	116	100	119	102
2	5	236	204	241	208
5	10	321	270	328	276
10	15	394	336	403	343
15	20	511	437	522	447
20	25	629	538	643	550
25	30	739	635	755	649
30	35	820	717	838	733
35	40	902	775	922	792
40	45	986	846	1008	865
45	50	1069	916	1093	936

Weekly holiday credit welfare stamp (Note 3)	From 5 October 1992				
	£	£	£	£	£
	a	b	c	d	e
Weekly Holiday Credit	29.29	26.95	24.60	21.09	15.22
Weekly Welfare Premium	2.70	2.70	2.70	2.70	2.70
Total	31.99	29.65	27.30	23.79	17.92

RATES OF WAGES

	From *1 March 1993*	From *5 September 1994*
Daily abnormal conditions money p per day	237	242
Exposed work at heights money Over 125 ft, p per day	95	97
Swings, cradles and ladders money p per hour	24	25
Lodging allowance £ per day	18.10	(tba)

Notes

1   The grades of Ductwork Erection Operatives as defined in Clause 6g of the National Agreement are:
    Ductwork Chargehand Erector
    Ductwork Advanced Erector
    Ductwork Erector

2   From 7 January 1991 the percentage of the Fitter's rate upon which the rates for a Foreman (other than Ductwork Foreman) is calculated will increase from 130 to 135 per cent.

3   The grades of Operatives covered by the range of credit values and entitled to the different rates of sickness and accident benefit are as follows:

a	b	c	d	e
Foreman Chargehand*	Advanced Fitter qualified gas/arc Advanced Fitter qualified gas or arc Advanced Fitter* Fitter qualified gas/arc	Fitter qualified gas or arc Fitter* Improver 4th year Apprentice	Assistant Mate 3rd Year Apprentice	1st and 2nd Year Apprentices and Junior Mates aged 17/18 and 16/17

* Ductwork Erection Operatives are entitled to the same credit values and rates of sickness and accident benefit as the parallel Fitter grade shown.

4   Payment of death benefit is subject to Inland Revenue requirements which currently provide that it may not exceed four times annual earnings of the deceased, subject to a minimum of £5,000.

RATES OF WAGES

Authorised rates of wages agreed by the Joint Industry Board for the Plumbing Mechanical Engineering Services Industry in England and Wales

The Joint Industry Board for Plumbing Mechanical
Engineering Services in England and Wales,
Brook House, Brook Street,
St Neots, Huntingdon, Cambs. PE19 2HW   Telephone: 01480 476925

WAGE RATES, ALLOWANCES AND OTHER PROVISIONS

EFFECTIVE FROM 4 APRIL 1994

All districts of the United Kingdom

	Hourly rate £
Technical plumber	6.20
Advanced plumber	5.49
Trained plumber	4.96
Apprentice	
1st year of training	1.55
2nd year of training	2.38
3rd year of training	2.94
4th year of training	3.63
Adult Trainees	
1st 6 months of employment	3.94
2nd 6 months of employment	4.20
3rd 6 months of employment	4.38

Overtime

Overtime premium rates shall be paid after 39 hours are worked. The working week shall be 37.5 hours.

Daily travelling allowance plus return fares
All daily travel allowances are to be paid at the daily rate as follows:

	Miles									
Over	2	5	10	15	20	25	30	35	40	45
Not exceeding	5	10	15	20	25	30	35	40	45	50
	£	£	£	£	£	£	£	£	£	£
Technical plumber	2.43	3.21	4.03	4.82	5.64	6.42	7.22	7.52	7.75	7.98
Advanced plumber	2.11	2.81	3.52	4.22	4.92	5.65	6.34	6.65	6.85	7.05
Trained plumber	1.94	2.55	3.21	3.83	4.49	5.12	5.76	6.06	6.24	6.42
Apprentice										
1st year	0.57	0.77	0.95	1.16	1.35	1.53	1.72	1.82	1.87	1.92
2nd year	0.86	1.15	1.43	1.71	1.99	2.27	2.56	2.71	2.79	2.87
3rd year	1.06	1.40	1.74	2.09	2.45	2.79	3.16	3.36	3.46	3.56
4th year	1.35	1.76	2.23	2.66	3.13	3.57	4.02	4.27	4.40	4.53

For all distances over 35 miles operatives to be paid lodging allowance in accordance with the rules. The travel allowances set out are to be paid when public transport is used.

Mileage allowance (see new working rule 8.4.3(b))   0.26 per mile

RATES OF WAGES

Abnormal conditions . . . . £2.00 per day

Lodging allowance (Note 1)
i)　London - within the London Orbital Motorway i.e. M25 . . . . £20.91 per night
ii)　Standard - all areas other than London . . . . £16.73 per night

Responsibility money . . . . £0.33 per hour

Plumbers welding supplement
Possession of Gas or Arc Certificate . . . . £0.24 per hour
Possession of Gas and Arc Certificate . . . . £0.43 per hour

Tool allowance . . . . £2.24 per week

Weekly holiday credit/sick benefit stamp
Effective from 5 April 1993　　　　　　　　　　　　　　　　　　　　　£

Technical Plumber 21.01
Advanced Plumber 18.76
Trained Plumber, Ex Skillcentre Trainee and
Apprentice in last year of training 17.23
Working Principals and Plumbers over 65 . . . . 11.20
Apprentice Plumber 1st year of training 4.74
Apprentice Plumber 2nd to 3rd year of training 7.75
Ancillary employee 13.80

Note 1 Important - Taxation Treatment

Please note that by way of concession from the Inland Revenue the Lodging Allowance of £16.73 is payable without the deduction of income tax. The subsistence allowance is subject to Schedule E Income Tax through PAYE System.

WORKING RULES

Extracts from National Working Rules (HVCA)

HOURS OF WORK (CLAUSE 3)

(a)   The normal working week shall consist of 38 hours to be worked in five days from Monday to Friday inclusive. The length of each normal working day shall be determined by the Employer but shall not be less than six hours or more than eight hours unless otherwise agreed between the Employer and the Operative concerned.

(b)   The Employer and the Operative concerned may agree to extend the working hours to more than 38 hours per week for particular jobs, provided that overtime shall be paid in accordance with Clause 9. (Attention is also drawn to the provisions for containing overtime - see Clause 9a).

MEAL AND TEA BREAKS (CLAUSE 4)

(a)   The normal break for lunch shall be one hour except when such a break would make it impossible for the normal working day to be worked, in which case the break may be reduced to not less than half an hour.

(b)   An Operative directed to start work before his normal starting time or to continue work after his normal finishing time shall be entitled to a quarter of an hour meal interval with pay at the appropriate overtime rate for each two hours of working (or part thereof exceeding one hour) in excess of the normal working day, which on Saturdays and Sundays shall mean eight hours. Where an Operative is entitled to a morning and/or evening meal interval under this clause, the meal interval shall replace the morning and/or afternoon tea break referred to in 4(c).

(c)   A tea break shall, subject to 4(b), be allowed in the morning and in the afternoon without loss of pay, provided that Operatives co-operate with the Employer in minimising the interruption to production. To this end the duration of the tea break shall be limited to the time necessary to drink tea and the tea shall be drunk at the Operatives workplace wherever possible.

GUARANTEED WEEK (CLAUSE 5)

(a)   Subject to the provisions of this clause an Operative who has been continuously employed by the same Employer for not less than two weeks is guaranteed wages equivalent to his inclusive hourly normal time earnings for 38 hours in any normal working week; provided that during working hours he is capable of, available for and willing to perform satisfactorily the work associated with his usual occupation, or reasonable alternative work if his usual work is not available.

(b)   In the case of a week in which holidays recognised by agreement, custom or practice occur, the guaranteed week shall be reduced for each day of holiday by the normal working day as determined in Clause 3(a).

(c)   In the event of a dislocation of production as a result of industrial action the guarantee shall be automatically suspended. In the event of such dislocation being caused by Operatives working under other Agreements and the Operatives covered by this Agreement not being Parties to the dislocation, the Employers shall, in accordance with Clause 5(d), endeavour to provide other work or if not able to do so will provide for the return of the Operatives to the shop or office from which they were sent. The Operatives will receive instructions as soon as is practicable as to proceedings to other work or return to shop.

(d)   The basis upon which the Employer shall endeavour to provide alternative work as required in Clause 5c shall be as follows:

    (i)   where possible the Employer shall try to organise work on each job so as to provide a normal day's work for five days, Monday to Friday.

    (ii)   where this is not possible on any particular job, the Employer shall endeavour to arrange to transfer Operatives to other sites to make up working hours to a normal day's work for five days, Monday to Friday.

    (iii)   where an Employer finds it impossible to provide a normal day's work for five days, Monday to Friday, he should rearrange the working hours in agreement with the Operatives concerned so that normal time earnings for 38 hours in the normal working week can be earned but in less than five days

    (iv)   where it is not possible to provide Operatives with a minimum of 38 hours during the week, rather than resort to dismissals a reduced working week may be agreed.

WORKING RULES

GUARANTEED WEEK (CLAUSE 5) (CONTD)

(e)   In the event of dislocation of production as a result of civil commotion, the guarantee shall be automatically suspended at the termination of the pay week after the dislocation first occurs and the Operative may be required by the Employer to register as an unemployed person.

GRADES AND SKILLS OF OPERATIVES (CLAUSE 6)

(a)   Operatives covered by this Agreement shall be graded in accordance with the definitions in Clauses 6(f) and 6(g).

(f)   The definitions of the grades (other than ductwork grades in Clause 6(g)) shall be:

Junior Mate
A Junior Mate is a Mate under 18 years of age.

Mate
A Mate must be at least 18 years of age.

Assistant
An assistant must have the following qualifications:
(i)   Have worked in the trade at least three years
(ii)   Be at least 20 years of age
(iii)   Be capable of performing semi-skilled tasks including support work for craftsmen without constant direct supervision
(iv)   Have had at least six months' continuous employment prior to appointment as an assistant with the Employer so appointing him.

With due consideration of the requirements of firms specialising in repetitive or prefabricated work the ratio of Assistants to craftsmen shall not normally exceed one to two employed in each company as a whole (not a branch office or a job by job basis).

Improver
An Improver must have either Qualification A or Qualification B.
Qualification A:
(i)   Have worked in the trade for at least five years
(ii)   Have been employed with his current employer for not less than one year
(iii)   Be at least 24 years of age
(iv)   Conform with such other requirements that may be laid down from time to time by the Heating Ventilating and Domestic Engineers' National Joint Industrial Council (NJIC) and have received the prior approval of the NJIC. Applications for up-grading shall be made on the form approved by the NJIC, who shall keep a register of Operatives up-graded under this Clause.

Qualification B:
(i)   Be at least 24 years of age
(ii)   Have satisfactorily completed the Department of Employment Adult Training Course in Heating and Ventilating Fitting Craft Practice as approved by the NJIC.

Craft Apprentice
A Craft Apprentice shall be undertaking an approved course of apprenticeship by duly executed Agreements of Service in the form prescribed by the NJIC.

Fitter
A Fitter must have one of the following qualifications:
(i)   Have successfully completed an apprenticeship approved by the NJIC, and have passed the practical examination of an appropriate City and Guilds of London Institute basic craft course which has been recognised by the NJIC and approved by the Parties

or
(ii)   Have successfully completed an improvership of one year under Qualification A or 18 months under Qualification B in the reference to Improvers or
(iii)   Be already employed as a craftsman in the industry on 24 February 1969.

A Fitter who is qualified in accordance with Clause 8(f) shall receive the appropriate welding supplement.

WORKING RULES

Advanced Fitter

An Advanced Fitter must have one of the qualifications of a Fitter and in addition:

    (i)   Must have had at least five years' service in the industry as a craftsman or since successful completion of an approved basic craft course as defined in the Fitter grade

    (ii)   Must have technical knowledge and skill beyond that of a Fitter including <u>either</u> successful completion of an appropriate City and Guilds of London Institute advanced craft course which has been recognised by the NJIC and approved by the Parties <u>or</u> competence, both practical and theoretical, in at least one of the following:
         - Commissioning and testing of simple systems
         - Layout and installation of plant and associated pipework
         - fault diagnosis
         - installation of large bore pipework over six inches
         - installation of high pressure steam and hot water pipework

    (iii)   Must have general competence and organising ability beyond that of a Fitter so that he is able without detailed supervision to set out jobs from drawings and specifications, requisition sundry materials, work in an efficient and economical manner and liaise effectively with other trades.

An Advanced Fitter who is qualified in accordance with Clause 8(f) shall receive the appropriate welding supplement.

Chargehand

A craftsman who is designated by the Employer as a Chargehand to carry out Chargehand duties in the course of working with the tools of the trade, shall receive remuneration dependent on the character of the charge as provided below:

    (i)   Where the Chargehand duties require the Chargehand to take sole responsibility for smaller contracts with an average labour force including three other craftsmen and/or of a gang of that size on larger contracts, the Chargehand shall receive the appropriate rate for the Chargehand grade and his rate shall include for any welding skill.

    (ii)   Where the charge is of a lesser character than that detailed in (i) above, the Chargehand shall receive a remuneration which shall be agreed between the Chargehand and the Employer.

A Chargehand may be appointed on a temporary or short term basis to cover peak periods of working or to allow a craftsman to gain experience in that role and shall be paid in accordance with (i) or (ii) above, as appropriate.

Foreman

A craftsman who satisfies the qualifications of an Advanced Fitter may be designated by the employer as a Foreman and shall receive the rate for the Foreman grade, provided he is competent to perform all the duties listed below (or the vast majority of them as appropriate to and in accordance with the requirements of the Employer):

    (i)   Assign tasks to the Chargehands and other Operatives under his direct control

    (ii)   Redeploy Chargehands or Operatives under his direct control, in order to achieve the optimum productivity including on-site batch production and fabrication

    (iii)   Decide methods to be used for individual operations and instruct Operatives accordingly

    (iv)   Ensure variation work does not proceed without authority from the office

    (v)   Maintain site contract control procedure

    (vi)   Requisition and progress supply of necessary equipment and materials to Operatives when required

    (vii)   Ensure that Operatives take all reasonable steps to safeguard, maintain and generally take care of Employer's tools and materials

    (viii)   Maintain day to day liaison and programme of work with main contractor and other sub-contractors

    (ix)   Inspect and review progress of work of sub-contractors

    (x)   Monitor progress of main contractor, in order that agreed programme is met

    (xi)   Measure and record progress of work

    (xii)   Inspect the work of Operatives for quality, progress and satisfactory completion

    (xiii)   Check weekly progress against programme and identify deviations therefrom

    (xiv)   Verify bookings on time and job cards and despatch them promptly to the office

    (xv)   Notify office of impending delays likely to affect progress or give rise to a claim

    (xvi)   Establish reasons for delays to work and notify office

    (xvii)   Provide information for cost variation investigations when necessary

    (xviii)   Forecast labour requirements

WORKING RULES

(xix)	Ensure company instructions and standards of discipline, workmanship and safety are maintained on site
(xx)	Ensure that the conditions of the National Agreement and any other conditions of employment are complied with
(xxi)	Supervise training of Craft Apprentices assigned to his control
(xxii)	Take overall charge of all his Employer's labour on site and act where necessary as the Employer's site agent
(xxiii)	Evolve and/or agree order of work within overall programme and control its progress
(xxiv)	Decide or agree locations of site office, site stores, site workshop and other work stations and adjust same to suit site progress and changing conditions
(xxv)	Ensure compliance of all work, whether executed by own Operatives or sub-contractors with drawings and specifications
(xxvi)	Organise, supervise and record such tests (e.g. hydraulic) and/or inspections as are required during progress of contract
(xxvii)	Requisition or otherwise procure such attendances and facilities as are required of the main contractors and/or of other sub-contractors
(xxviii)	Attend site meetings (if so required by the Employer)
(xxix)	Ensure that safe methods of work are adopted by Operatives under his direct control
(xxx)	Ensure clearance of rubbish as specified
(xxxi)	Arrange and supervise testing on completion, including compliance with specifications, snagging and operational handing over as directed and final site clearance
(xxxii)	Such other duties as are reasonably required by the Employer

(g)   The definitions of the ductwork grades shall be:

Junior Ductwork Trainee
A Junior Ductwork Trainee shall undertake the approved in-company scheme of training as set out in Clause 23.

Ductwork Erector
A Ductwork Erector shall be at least 20 years of age; or shall have trained in accordance with the training agreement set out in Clause 23; or shall have undergone training in the ductwork industry for not less than four years; shall be able, without constant direct supervision, to carry out the installation of ductwork and have the skill to perform the following duties (or such others as may from time to time be agreed upon between the Association and the Union):

(i)	Set out duct runs and plant equipment from drawings to datum lines
(ii)	Make simple supports on site and fix supports of all types
(iii)	Fix items of plant within the duct runs, together with grilles, diffusers, terminal units etc., and make final connections to ductwork
(iv)	Position and assemble sections of air handling units, complete with fans and associated equipment
(v)	Operate all machines and equipment used in ductwork erection other than those used for manual welding (Note - It is agreed that the spot welding of steel sheet shall be included in the list of the Ductwork Erector's skills)
(vi)	Carry out on-site minor modifications and repairs
(vii)	Measure and specify any simple make up pieces of ductwork
(viii)	Blank off and prepare ductwork runs for pressure testing of high-velocity high-pressure ductwork
(ix)	Fit thermostats, temperature probes and control equipment
(x)	Sling and hoist loads of various shapes and sizes, provided adequate training has been given.
(xi)	Rig portable and temporary working platforms to conform with safety regulations
(xii)	Assemble and erect light steel structures and associated sheet metal cladding, including insulation
(xiii)	Have knowledge (by instructions and training) of and comply with all site safety regulations
(xiv)	Be responsible for grades junior to the Ductwork Erector who directly assist him in carrying out any of the above operations
(xv)	Undertake any training in relation to his own grading and co-operate in the training of the other Operatives junior to him
(xvi)	Carry out any other operations incidental to the foregoing

WORKING RULES

Ductwork Advanced Erector

A Ductwork Advanced Erector shall be at least 22 years of age and shall have had at least two years' service as a Ductwork Erector with his Employer at the time of upgrading to a Ductwork Advanced Erector. He shall have the qualifications of a Ductwork Erector as set out above and in addition the technical knowledge and skill including competence, both theoretical and practical, in the following duties (or others as may from time to time be agreed between the Association and the Union):

(i)	Appreciation of a contract situation and of his Employer's obligations thereunder
(ii)	General understanding of the operation of ducted air systems and equipment
(iii)	Installation of the ductwork system in sequence with other services
(iv)	Measuring and specifying any modifications of ductwork as required
(v)	Assigning tasks to Operatives under his control and deploying labour on site as necessary
(vi)	Pressure testing of high-velocity high-pressure ductwork systems to the requirements of any relevant HVCA specification
(vii)	Installation fault diagnosis in the ductwork system
(viii)	Recording, measuring and documenting variations from contract.

He shall have general competence and organising ability beyond that of a Ductwork Erector as defined herein, so that he is able without detailed supervision to set out jobs from drawings and specifications; to requisition sundry materials; to work in an efficient and economical manner; and to liaise effectively with other trades.

A Ductwork Advanced Erector may be designated a 'Temporary Chargehand', for example, to cater for peak periods of work volume, high labour force and unusual circumstances. Where such appointment is made it shall be for a minimum period of two weeks and the grade rate shall apply during the period which he is designated as Temporary Chargehand, and at the end of the period he shall revert to his previous grade.

Ductwork Chargehand Erector

A craftsman who has the qualifications of a Ductwork Advanced Erector may be designated by the Employer as a Ductwork Chargehand Erector to carry out Chargehand duties in the course of working with the tools of the trade, and shall receive the rate for the grade, if he is competent to perform all or most of the duties listed below in accordance with the requirements of the Employer:

(i)	Assign tasks to the Ductwork Advanced Erectors and other Operatives under his control
(ii)	Deploy Operatives under his control so as to achieve the optimum productivity including off-site assembly and minor modifications
(iii)	Decide method to be used for individual operations and instruct Operatives accordingly
(iv)	Requisition and progress supply of necessary equipment and materials to Operatives as required
(v)	Ensure that Operatives take all reasonable steps to safeguard, maintain and generally take care of the Employer's tools and materials
(vi)	Inspect the work of Operatives for quality, progress and satisfactory completion
(vii)	Ensure that the Employer's instructions and standards of discipline, workmanship and safety are maintained on site
(viii)	Ensure that variation work does not proceed without authority from the Employer
(ix)	Verify bookings on time and job cards and despatch them promptly to the Employer
(x)	Notify the Employer of any delay likely to affect progress or give rise to additional costs
(xi)	Maintain such records of installation progress and variations as may be required by the Employer
(xii)	Forecast labour requirements on the job in hand
(xiii)	Supervise craft training of Operatives assigned to his control
(xiv)	Organise, supervise and record such tests (for example, leakage testing) and/or inspections as are required during the progress of the contract
(xv)	Requisition or otherwise procure such attendances, services or facilities as are required from the main contractor and/or other sub-contractors
(xvi)	Ensure that safe methods of work are adopted by Operatives under his control
(xvii)	Ensure clearance of rubbish and scrap for which the Employer is responsible
(xviii)	Ensure compliance with specifications, carry out any necessary remedial work and hand over as required by the contract and clear site of Employer's equipment
(xix)	Carry out any other operations incidental to the above.

It is accepted that a Ductwork Chargehand Erector does not necessarily carry this grading with him to a new employer. It is understood that the Ductwork Chargehand Erector's rate of pay shall be dependent on the character of the charge, for example, where his duties require him to take sole responsibility for a small contract or a section or sections of a large contract, and according to the number of Operatives.

WORKING RULES

(h)		Erection, Alteration and Dismantling of Scaffolding and Mobile Towers
Operatives shall, where properly trained or supervised, undertake the erection, alteration and dismantling of mobile towers and easy-fix scaffolding as part of their normal work. The Employer shall ensure that such supervision is undertaken by Operatives who are properly instructed in the necessary working and safety procedures.

BALANCE OF GANGS (CLAUSE 7)

The balance of gangs as between Craftsmen, Assistants, Mates and Craft Apprentices shall be on the basis that:

(i)	Assistants may do semi-skilled tasks including support work for craftsmen without constant direct supervision and shall perform the manual work of fetching and carrying, receiving and checking materials as required by the Employer.

(ii)	Support work for skilled men may be done by the skilled men themselves or by Assistants, Mates or Craft Apprentices in order to secure the maximum utilisation of labour and the optimum economic production; thus one Mate can be used to do the support work for two or more Craftsmen or conversely two or more Mates may work with one Craftsman.

(iii)	Mates shall not be confined to the manual work of fetching and carrying. They shall within their capacity, carry out semi-skilled tasks; one object being to improve productivity and the other being to permit those who wish to do so to qualify for consideration for regrading as Assistants.

(iv)	In order to provide Craft Apprentices with appropriate practical experience and to permit them to make the fullest possible contribution to production, they shall be permitted to work with the tools with the maximum of supervision necessary but always on work which is under the control of a recognised Craftsman. In the case of welding, this shall mean that Craft Apprentices shall not weld until they have completed the appropriate City and Guilds welding course. Craft Apprentices shall not be employed solely and continuously on heavy labouring work.

WAGES AND ALLOWANCES (CLAUSE 8)

(f)		*Welding Supplement*
A Fitter or an Advanced Fitter who holds one or both current certificates of Competency issued by the Heating, Ventilating and Domestic Engineer's National Joint Industrial Council, in oxy-acetylene welding and/or metal arc welding and who is competent in such welding to the standard(s) required by such certificate(s) shall receive an hourly welding supplement for one welding skill or for both welding skills as appropriate, which shall be the following percentage of the Fitter's hourly rate:
One welding skill - 5%
Two welding skills - 10%

(g)		*Merit Money*
Payment of merit money to an Operative may be made at the option of the Employer for mobility, loyalty, long service etc., and for special skill over and above that detailed in definition of the Operative's grade.

(h)		*Payment of Wages*
(i)	Unless otherwise agreed between the Operative and the Employer, the pay week shall end at midnight on Friday and wages shall be paid on the following Thursday.

(ii)	The Employer at his discretion may pay each Operative to the nearest £1 upwards each week carrying the credit forward, deducting it from the next wage payment which is again paid to the nearest £1 upwards.

(iii)	Where any packets cannot be calculated on time sheets, the Employer shall make an assessed payment for the days worked. The pay packet shall be corrected the following week.

(i)		*Junior Mate*
Junior Mates shall be paid the same rates as Craft Apprentices. They shall be subject to the conditions of employment set out in this Agreement.

WORKING RULES

(j)   *Abnormal Conditions*
Operatives engaged on exceptionally dirty work, or work under abnormal conditions, of such a character as to be equally onerous, shall receive an allowance extra per day or part of a day. The determination of the conditions to which this allowance shall apply shall be agreed between the Employer and the Operative concerned in each case. The allowance shall be agreed from time to time by the Association and the Union and shall be enumerated in an Appendix to this Agreement. It shall be in addition to any part payment for exposed work at heights made under Clause 8(k).

(k)   *Exposed Work at Heights*
An Operative working in exposed conditions at heights over 125 feet on an unclad building having no other form of protection from weather conditions shall be paid an allowance extra per day or part of a day. The allowance shall be agreed from time to time by the Association and the Union and shall be enumerated in an Appendix to this Agreement.

(l)   *Swings, Cradles, Ladders*
An Operative working in swings or cradles shall be paid an allowance extra per hour for the time actually worked in those conditions. An Operative working on ladders shall be paid an allowance extra per hour for the time actually worked at a height of 20 feet and an additional allowance per hour for each additional 10 feet. The height shall be measured from the nearest fixed flooring or fixed scaffolding to the actual work. The allowances shall be agreed from time to time by the Association and the Union and shall be enumerated in an Appendix to this Agreement.

(m)   *Target Incentive Schemes*
Where it has been agreed between the Employer and the majority of his workforce, that target incentive schemes shall be operated in connection with works on which they are or are to be employed, such schemes shall be operated in accordance with general principles established by the Association and the Union for their operation, which are set out in Appendix B.

OVERTIME (CLAUSE 9)

(a)   It is accepted by the parties that overtime must be contained. To this end, except in cases of urgency or emergency, actual working hours should not exceed:
     (i)   An average of 45 per week in the case of travelling men who should only work on Saturdays and/or Sundays in cases of urgency or emergency
     (ii)   An average of 55 per week in the case of lodging men whose work on Saturdays and/or Sundays should be reasonably contained
(c)   The difference between the normal hourly rate and the overtime rate shall be known as the 'premium' payment.
(d)   For the purpose of calculating overtime, and regardless of the length of the normal working day as determined under Clause 3, time worked on Monday to Friday inclusive shall be paid for as follows:
     (i)   First eight hours worked after normal starting time - normal hourly rate.
     (ii)   Thereafter, until four hours after the normal finishing time - time-and-a-half.
     (iii)   Thereafter, until normal starting time next morning - double time.
     (iv)   If time is lost through the fault of the Operative the time lost shall be added to the normal starting time and the resultant time shall be used for the purposes of calculating overtime payable at time-and-a-half.
     (v)   An Operative directed to start work before the normal starting time shall be paid the appropriate overtime rates for all hours worked before the normal starting time, but if through the action of the Operative the normal working day is not worked, ordinary hourly rates shall be paid for all hours worked.
     (vi)   The calculation of overtime for any day shall not be affected by any hours of absence arising from,
     -certified sickness
     -absence with the concurrence of the Employer
     -absence for which the Operative can produce evidence to the satisfaction of the Employer that his absence was due to causes beyond his control.
     (vii)   An Operative called back to work at any time between the period commencing two hours after the normal finishing time and until two hours before normal starting time shall be paid such overtime rates as would apply had work been continuous from normal finishing time and shall be paid a minimum of two hours at the appropriate rate.

WORKING RULES

(e)　Time worked on Saturday and Sunday shall be paid for as follows:
    (i)　Saturday - first five hours, time and a half; after the first five hours, double time but if time is lost through the fault of the Operative the double time rate shall not apply until time lost has been made up.
    (ii)　Sunday - double time for all hours worked until starting time on Monday Morning.

PAYMENT FOR HOLIDAYS WORKED (CLAUSE 10)

(a)　This clause applies to all recognised holidays as defined in Clause 18, and in Scotland - three days of the Winter Holiday period as defined in Clause 10(c).

(b)　An Operative who works on any of the days in Clause 10(a) shall be paid a minimum of two hours at the appropriate rate. In addition an Operative shall be granted a day's holiday with pay for each holiday day worked as provided in Clause 18(c).

(c)　Time worked on such days shall be paid as follows:

In England and Wales
New Year's Day, Good Friday, Easter Monday, May Bank Holiday, Spring Bank Holiday, Late Summer Bank Holiday, Christmas Day, Boxing Day:
    - Double time for all hours worked.
In Scotland
Three consecutive days of the Winter Holiday period including New Year's Day and the one or two holiday days which immediately follow it (if any), Spring Bank Holiday, Friday before Spring Bank Holiday, May Bank Holiday, Autumn Holiday (one day): Boxing Day.
    - Double time for all hours worked.
Christmas Day and the one day of recognised holiday to be agreed locally.
The normal working day as determined in Clause 3(a), time and a half,　thereafter double time.
Friday before Autumn Holiday.
The normal working day as determined in Clause 3(a), normal hourly rates; thereafter overtime rates in accordance with Clause 9.

In Northern Ireland
The eight days of recognised holidays agreed by the Northern Ireland Branches of the Association and Union. Double time for all hours worked.

(d)　The general conditions of the Agreement shall apply to men called back to work on these holidays.

NIGHT SHIFTS AND NIGHT WORK (CLAUSE 11)

(a)　For an Operative who works for at least five consecutive nights:
    (i)　The basic rate, called the night shift rate, shall be one and a third times the normal rate.
    (ii)　Overtime rates and conditions shall be as for normal working days (provided in clause 9) but the basic rate shall be the night shift.
(b)　An Operative who works for less than five nights and does not work during the day shall be paid at overtime rates as if the normal day had already been worked.

CONTINUOUS SHIFT WORK (CLAUSE 12)

Where jobs have to be continuously operated the work shall be carried out in two or three shifts of eight hours each according to requirements. The Operatives concerned shall be paid time and a third in cases where a six day shift is worked and time and a half in cases where a seven day shift is worked, overtime and night shift rates being compounded in these rates. Arrangements shall be made to change the shifts worked by each Operative.

PROVISION OF TOOLS (CLAUSE 13)

(a)　The Operative shall provide a rule and spirit level. Other tools shall be provided by the Employer but the Operative shall take all reasonable steps to safeguard, maintain and generally take care of the Employer's tools.

(b)　The Operative shall co-operate in the implementation of reasonable procedures properly designed to prevent loss of or damage to tools.

WORKING RULES

ALLOWANCES TO OPERATIVES WHO TRAVEL DAILY (CLAUSE 15)

(a)   Except where his centre is the job, an Operative who is required by the Employer to travel daily up to 50 miles to the job shall be paid fares and travelling time as stated in (i) and (ii) below:

      (i)   Return daily travelling fares from his centre to the job. Where cheap daily or period fares or other cheap travel arrangements by public transport are available the Employer may pay fares on that basis. Where, however, a change in such travel arrangements results from a change in the working arrangements the Employer must pay the Operative for any additional cost. The Employer at his option may provide suitable conveyance for the Operative to and from the job in which case fares shall not be paid.

      (ii)   Allowances for travelling time, provided that the normal hours are worked on the job. The allowances for travelling time shall be agreed from time to time by the Association and the Union and shall be enumerated in an Appendix to this Agreement. When a reasonably direct journey is not possible, a claim for special consideration may be made by the Operative and in case of dispute the matter shall be referred to the Chief Officials of the parties, whose decision shall be final.

(b)   Except where his centre is the job, payment to the Operative of allowances for travelling time and fares for journeys beyond fifty miles daily from his centre to the job will be for agreement between the Employer and the Operative concerned.

ALLOWANCES TO OPERATIVES WHO LODGE (CLAUSE 16)

(a)   Where an Operative is sent to a job to which it is impracticable to travel daily and where the Operative lodges away from his place of residence he shall (except if he is engaged at the job or if his centre is the job) be paid the items in (i) to (v) below where appropriate:

      (i)   A nightly lodging allowance including the night of the day of return and when on week-end leaves in accordance with Clause 17(a). The nightly lodging allowance shall be agreed from time to time by the Association and the Union and shall be enumerated in an Appendix to this Agreement. The lodging allowance shall not be paid when an Operative is absent from work without the concurrence of the Employer, nor when suitable lodging is arranged by the Employer at no expense to the Operative, nor during the annual holidays defined in Clause 20 including the week of Winter Holiday. The Operative shall provide the Employer with a statement signed by himself to the effect that he is in lodgings for the period of payment of lodging allowance under this clause. Without such evidence, the Employer shall deduct tax on lodging allowance paid.

      (ii)   Any V.A.T. charged on the cost of lodgings, subject to the provision by the Operative of a valid tax invoice on which the Employer can claim input credit.

      (iii)   When suitable lodgings are not available within two miles from the job, daily return fares from lodgings to job. The Employer at his option may provide suitable conveyance for the Operative between the lodgings and the job, in which case the fares shall not be paid.

      (iv)   Time spent in travelling to and from the centre at the commencement and completion of the job at the normal hourly rate but when an excessive number of hours of travelling is necessarily incurred, a claim for special consideration may be made by the Operative to the Employer or by the Employer to the Operative and in case of dispute the matter shall be referred to the Chief Officials of the parties, whose decision shall be final.

      (v)   Fares between his centre and the job at the commencement and the completion of the job. Return fares shall be used when available.

      (vi)   Week-end leaves in accordance with Clause 17(a).

(b)   An Operative whose employment is terminated in accordance with Clause 2a during the course of a job, shall be entitled to travelling time and a single fare for the journey from the job to his centre. This condition shall not apply to an Operative who is discharged for misconduct or who leaves the job without the concurrence of his Employer.

WORKING RULES

WEEK-END LEAVES (CLAUSE 17)

(a)   An Operative who is in receipt of lodging allowance in accordance with Clause 16 shall be allowed a week-end leave every two weeks. Such Operative shall be entitled to return to his respective centre for the recognised holidays prescribed in Clause 18 and to facilitate this, the nearest normal week-end leave shall, where necessary, be deferred or brought forward to coincide with the holiday.

(b)   Unless the Employer and the Operative agree otherwise the week-end leave shall be from normal finishing time on Friday to normal starting time on Monday.

(c)   An Operative shall not normally be required to start his return journey before 6.00a.m. on the appropriate day of return to the job but shall, where the return journey makes it impossible to commence work at the normal starting time, agree with his Employer the working arrangements for the day.

(d)   Weekend return fares shall be paid for weekend leaves. If an Operative does not elect to return to his centre a single fare from the job to his centre shall be paid.

(e)   The following travelling time arrangements shall apply to an operative on weekend leave for journeys to and from his centre:

    (i)   Where the job is up to 150 miles from his centre, he shall travel in his own time from the job to his centre, but travelling time from his centre to the job shall be paid at the normal hourly rate.

    (ii)   Where the job is 150 miles or more from his centre, he shall be paid four hours at the normal hourly rate from the job to his centre, and travelling time from his centre to the job shall be paid at the normal hourly rate.

If an Operative elects to stay at the job travelling time shall not be paid.

(f)   When a reasonably direct journey is not possible or when an excessive number of hours travelling is necessarily incurred on jobs more than 150 miles from an Operative's centre, a claim for special consideration in respect of travelling time may be made by the Operative to the Employer or by the Employer to the Operative and in case of dispute the matter shall be referred to the Chief Officials of the parties, whose decision shall be final.

(g)   An Operative on week-end leaves (including holidays, provided under Clause 18), shall be paid the nightly lodging allowance, provided that the leave is within this Agreement or is agreed with the Employer.

RECOGNISED HOLIDAYS (CLAUSE 18)

(a)   The following days have been designated as recognised holidays and shall be paid in accordance with the H & V Holiday & Welfare Scheme. The pay shall consist of the appropriate holiday credits standing to the credit of the Operative

If any of these days comes within the annual holidays as provided in Clause 20, mutual arrangements shall be made to substitute some other day for the day or days included.

In England and Wales
New Year's Day; Good Friday; Easter Monday; May Bank Holiday; Spring Bank Holiday; Late Summer Bank Holiday; Christmas Day; Boxing Day.

In Scotland
Spring Holiday; Friday before Spring Holiday; May Bank Holiday; Autumn Holiday (two days); Christmas Day; Boxing Day; plus one other day to be agreed locally.

In Northern Ireland
Note: The days when recognised holidays are to be taken in Northern Ireland are subject to discussion between the Northern Ireland Branches of the Association and the Union.

(b)   Any Operative who has insufficient credits on his holiday card to pay for the three days of recognised holiday included in the winter holiday period, because he entered the industry after the commencement of the appropriate stamping period, shall be entitled to three days pay at the normal hourly rate for eight hours. The Employer shall be responsible for paying the difference between this sum and the value of any holiday credits that may have been accrued in the appropriate stamping period.

WORKING RULES

(c)   Operatives who work on a recognised holiday as set out in Clause 18(a) shall be paid in accordance with Clause 10(c) and shall be entitled to a day's holiday in lieu, at a mutually agreed time, payment for which shall be the sum of the appropriate holiday credits for that day of recognised holiday.

(d)   The general conditions of the Agreement shall apply to Operatives called back to work on these holidays.

H. & V. HOLIDAY AND WELFARE SCHEME (CLAUSE 19)

(a)   All Operatives shall be provided with a card in accordance with the H. & V. Holiday and Welfare Scheme. The rules of the Scheme which are incorporated into and form part of this Agreement are set out in a separate Supplement. The card of each Operative shall, subject to the rules of the Scheme, be stamped on a weekly basis by the Employer first employing him in any week in an accounting period. The stamp shall cover:

    (i)   A weekly credit in respect of annual recognised holidays (the value of the credit shall be agreed from time to time between the Association and the Union and shall be enumerated in an Appendix to this Agreement).

    (ii)   A weekly premium in respect of welfare benefits (the value of the premium shall be determined from time to time by the Association and shall be enumerated in an Appendix to this Agreement).

(b)   Variation or Amendment: Clauses 19, 20, 21 and 22 of this Agreement may be varied or amended by agreement of the Parties but any variation or amendment shall, subject to the rules of the H. & V. Holiday and Welfare Scheme, only become operative at the beginning of a new accounting period. Notice of any proposed variation must be given in writing to each of the other Parties at least six months prior to the commencement of any accounting period.

(c)   Termination: Either of the Parties to this Agreement may terminate Clauses 19, 20, 21 or 22 at the end of any accounting period by giving notice in writing to the other Party at least 12 months before the end of the accounting period. In the event of termination of the `Annual and Recognised Holidays Provision' the Parties agree to provide the holiday facilities and holiday payments until such time as the rights acquired by the Operatives in respect of holiday credits under this section have been met.

ANNUAL HOLIDAYS (CLAUSE 20)

All Operatives shall have annual holidays with pay in accordance with the H. & V. Holiday and Welfare Scheme. The pay shall consist of the appropriate annual holiday credits standing to the credit of the Operative.

Annual holidays shall consist of:

    (i)   Four days of Spring Holiday

    (ii)   Two weeks of Summer Holiday

    (iii)   Seven days of Winter Holiday

ANNUAL AND RECOGNISED HOLIDAY CREDITS - PAYMENT (CLAUSE 21)

The sum standing to the credit of each Operative, being the sum of the weekly credits less any administrative charge approved by the Parties to the Agreement, shall be paid to the Operative on taking his annual and/or recognised holidays by the Employer in accordance with the H. & V. Holiday and Welfare Scheme.

WELFARE BENEFITS - PAYMENT (CLAUSE 22)

(a)   All Operatives shall be entitled to sickness and accident benefit and other welfare benefits in accordance with the H. & V. Holiday and Welfare Scheme.

Electrical Installations
RATES OF WAGES AND WORKING RULES

RATES OF WAGES

ELECTRICAL CONTRACTING INDUSTRY

Extracts from National Working Rules determined by:

The Joint Industry Board for the Electrical Contracting Industry
Kingswood House
47/51 Sidcup Hill, Sidcup, Kent, DA14 6HP
Telephone 0181-302 0031

Operatives in possession of XVth Edition Grade Cards

From and including 3rd December 1994 the JIB rates of wages will be as set out below;

Electrician, Instrument Mechanic, Instrument Pipefitter (or equivalent specialist grade)	Approved Electrician, Approved Instrument Mechanic, Approved Instrument Pipefitter (or equivalent specialist grade)	Technician (or equivalent specialist grade)	Labourer
National Standard Rates			
£	£	£	£
5.72	6.20	7.18	4.46
London Rates			
6.12	6.60	7.58	4.86

On Shore Agreement

Operatives engaged upon On Shore Work in connection with oil and gas exploration from the seabed, as laid down in section 5.3 of the JIB Handbook.

Electrician, Instrument Mechanic, Instrument Pipefitter (or equivalent specialist grade) £8.08

Approved Electrician, Approved Instrument Mechanic, Approved Instrument Pipefitter, (or equivalent specialist grade) £8.55

Technician, (or equivalent specialist grade) £9.53

Labourer £6.81

RATES OF WAGES

1983 Joint Industry Board Apprentice Training Scheme
Apprentice rates effective from and including 12 August 1995

Junior Apprentice . . . . Training allowance of £56.25 per week
Senior Apprentice (stage 1) . . . . £2.52 per hour
Senior Apprentice (stage 2) . . . . £3.72 per hour

Senior Apprentices qualifying for London Weighting shall receive the amount determined, from time to time, by the Joint Industry Board. This is currently 21p per hour. Junior Apprentices will not receive London Weighting.

Apprentices who obtain a Pass with Distinction in both relevant components of the City & Guilds 236 Part I Examination (or an equivalent examination approved by the Joint Industry Board) shall be paid an additional amount of 13p per hour as a Senior Apprentice (Stage 1) until becoming a Senior Apprentice (Stage 2).

Apprentices who obtain a pass with Distinction in both relevant components of the City & Guilds 236 Part II Examination (or an equivalent examination approved by the Joint Industry Board) and an overall pass shall be paid an additional amount of 17p per hour until termination of the Training Contract.

1989 Joint Industry Board Adult Craft Training Scheme

	Adult Trainee (under 21)	*Adult Trainee (over 21)*	*Senior Graded Electrical Trainee*
From and including 3rd December 1994			
	£	£	£
National Standard Rate	3.35	4.46	5.15
London Weighting	3.75	4.86	5.55

RATES OF WAGES

Travel time and travel allowances from 3 December 1994

Operatives who are required to start and finish at the normal starting and finishing time on jobs which are up to and including 35 miles from the shop - in a straight line - shall receive both travel allowances and payment for travelling time as follows:

Distance Point to Point for journeys of	*Total daily trav. allow- ance*	*Total daily travelling time*			
		Tech	*Apvd Elec*	*Elec*	*Lab*
(a) National Standard Rate: From 3 December 1994					
Return journey of					
Up to 1 mile each way	83p	Nil	Nil	Nil	Nil
Up to 2 miles each way	95p	143p	123p	113p	89p
Up to 3 miles each way	109p	238p	205p	189p	148p
Up to 4 miles each way	121p	285p	246p	227p	177p
Up to 5 miles each way	171p	379p	327p	302p	237p
Between 5 and 10 miles each way	241p	569p	491p	454p	354p
Between 10 and 15 miles each way	343p	759p	656p	605p	472p
Between 15 and 20 miles each way	482p	854p	738p	680p	531p
Between 20 and 25 miles each way	552p	948p	819p	756p	591p
Between 25 and 35 miles each way	622p	1043p	901p	832p	649p
(b) London area: From 3 December 1994					
Return journey of					
Up to 1 mile each way	83p	Nil	Nil	Nil	Nil
Up to 2 miles each way	95p	150p	131p	122p	97p
Up to 3 miles each way	109p	251p	218p	203p	161p
Up to 4 miles each way	121p	301p	262p	243p	193p
Up to 5 miles each way	171p	401p	349p	324p	258p
Between 5 and 10 miles each way	241p	602p	524p	486p	387p
Between 10 and 15 miles each way	343p	802p	699p	648p	515p
Between 15 and 20 miles each way	482p	903p	786p	729p	580p
Between 20 and 25 miles each way	552p	1002p	873p	810p	645p
Between 25 and 35 miles each way	622p	1103p	960p	891p	708p

RATES OF WAGES

Travel time and travel allowance from 12 August 1995

Apprentices who are required to start and finish at the normal starting and finishing time on jobs which are up to and including 35 miles from the shop - in a straight line - shall receive travel allowances and payment for travelling time as follows:

Distance Point to Point for journeys of	Total daily travel allowance	Total daily travelling time	
		Senior Apprent. (Stage 1)	Senior Apprent. (Stage 2)
(a) National Standard Rate: From 12 August 1995			
Up to 1 mile each way	83p	Nil	Nil
Up to 2 miles each way	95p	63p	93p
Up to 3 miles each way	109p	105p	155p
Up to 4 miles each way	121p	126p	186p
Up to 5 miles each way	171p	168p	248p
Between 5 and 10 miles each way	241p	252p	372p
Between 10 and 15 miles each way	343p	336p	496p
Between 15 and 20 miles each way	482p	378p	558p
Between 20 and 25 miles each way	552p	420p	620p
Between 25 and 35 miles each way	622p	462p	682p
(b) London area: From 12 August 1995			
Up to 1 mile each way	83p	Nil	Nil
Up to 2 miles each way	95p	68p	98p
Up to 3 miles each way	109p	114p	164p
Up to 4 miles each way	121p	137p	197p
Up to 5 miles each way	171p	182p	262p
Between 5 and 10 miles each way	241p	273p	393p
Between 10 and 15 miles each way	343p	364p	524p
Between 15 and 20 miles each way	482p	410p	590p
Between 20 and 25 miles each way	552p	455p	655p
Between 25 and 35 miles each way	622p	501p	721p

Travelling costs for Junior Apprentices remain as laid down in the 1983 JIB Training Scheme.

RATES OF WAGES

Travel time and travel allowance from 3 December 1994

Adult trainees who are required to start and finish at the normal starting and finishing time on jobs which are up to and including 35 miles from the shop - in a straight line - shall receive both travel allowances and payment for travelling time as follows:

Distance Point to Point for journeys of	*Total daily travel allow- ance*	*Total daily travelling time*		
		**(1)*	**(2)*	**(3)*
(a) National Standard Rate: From 3 December 1994				
Up to 1 mile each way	83p	Nil	Nil	Nil
Up to 2 miles each way	95p	67p	89p	103p
Up to 3 miles each way	109p	111p	148p	171p
Up to 4 miles each way	121p	133p	177p	204p
Up to 5 miles each way	171p	177p	237p	272p
Between 5 and 10 miles each way	241p	266p	354p	408p
Between 10 and 15 miles each way	343p	354p	472p	544p
Between 15 and 20 miles each way	482p	399p	531p	612p
Between 20 and 25 miles each way	552p	443p	591p	680p
Between 25 and 35 miles each way	622p	487p	649p	748p
(b) London area: From 3 December 1994				
Up to 1 mile each way	83p	Nil	Nil	Nil
Up to 2 miles each way	95p	75p	97p	110p
Up to 3 miles each way	109p	124p	161p	184p
Up to 4 miles each way	121p	149p	193p	220p
Up to 5 miles each way	171p	199p	258p	294p
Between 5 and 10 miles each way	241p	298p	387p	441p
Between 10 and 15 miles each way	343p	397p	515p	588p
Between 15 and 20 miles each way	482p	447p	580p	661p
Between 20 and 25 miles each way	552p	497p	645p	734p
Between 25 and 35 miles each way	622p	546p	708p	808p

Note: *(1) Adult Trainee (under 21)
         *(2) Adult Trainee (over 21)
         *(3) Senior Graded Electrical Trainee

RATES OF WAGES

Country Allowance
£16.56 from and including 3rd December 1994

Lodging Allowances
£17.00 from and including 3rd December 1994

Lodgings weekend retention fee, maximum reimbursement
£17.00 from and including 3rd December 1994

Annual Holiday Lodging Allowance Retention
A maximum of £3.40 per night (£23.80 per week) from and including 3rd December 1994

Responsibility money
Approved Electricians in charge of work, who undertake supervision of other operatives, shall be paid "Responsibility Money" of not less than 2.5p and not more than 50p per hour.

From and including 4 January 1992 responsibility payments shall be enhanced by overtime and shift premiums where appropriate.

Combined JIB Benefits Stamp Value (from week commencing 4 January 1993)

	Weekly stamp value
Technician	£26.35
Approved Electrician	£23.57
Electrician	£22.20
Senior Graded Electrical Trainee	£20.60
Labourer & Adult Trainee	£18.63
Adult Trainee(Under 21)	£15.45

JIB Welfare Benefits

Sick Pay, Death Benefit, Accidental Death Benefit

(a)   The following will apply with effect from Monday 1 October 1990

   (i)   Sick Pay in addition to Statutory Sick Pay

1st three days	No payment
4th to the 14th day inclusive	£6.00 per day
3rd to the 28th week inclusive	£24.50 per week
29th to the 52nd week inclusive	£21.00 per week
Benefit ceases after 52 weeks.	

   (ii)   Death Benefit
£15,000.00 for death from any cause.

   (iii)   Accidental Death Benefits with effect from 4 January 1992
£12,500.00 for adult operatives and £6,250.00 for apprentices for death from an accident either at work, or travelling to or from work, making a total of £27,500.00 for death due to an accident at work.

WORKING RULES

INTRODUCTION

The JIB National working Rules are made under rule 80 of the Rules of the Joint Industry Board (Section 1) as the National Joint Industrial Council for the Electrical Contracting Industry.

The principal objects of the Joint Industry Board are to regulate the relations between employers and employees engaged in the industry and to provide all kinds of benefits for persons concerned with the Industry in such ways as the Joint Industry Board may think fit, for the purpose of stimulating and furthering the improvement and progress of the industry for the mutual advantage of the employers and employees engaged therein, and in particular, for the purpose aforesaid, and in the public interest, to regulate and control employment and productive capacity within the Industry and the levels of skill and proficiency, wages, and welfare benefits of persons concerned in the Industry.

"The Industry" means the Electrical Contracting Industry in all its aspects in England, Wales, Northern Ireland, the Isle of Man and Channel Islands and such other places as may from time to time be determined by the Joint Industry Board, including the design, manufacture, sale, distribution, installation, erection, maintenance, repair and renewal of all kinds of electrical installations, equipment and appliances and ancillary plant activities.

1: General

These JIB National Working Rules and Industrial Determinations supersede previous Rules and Agreements made between the constituent parties of the National Joint Industrial Council for the Electrical Contracting Industry and shall govern and control the conditions for electrical instrumentation and control engineering, data and communications transmission work, its installation, maintenance and its dismantling and other ancillary activities covered by the Joint Industry Board for the Electrical Contracting Industry ("JIB") and shall come into effect in respect of work performed on and after Monday, 2 March 1970. These Rules apply nationally and in such a manner as may be determined from time to time by the JIB National Board.

2: Grading

Graded operatives shall comply in all respects with the Grading Definitions set out in Section 4 in carrying out the work of the Industry, erect their own mobile scaffolds and use such power operated and other tools, plant, etc, as may be provided by their Employer and the standard JIB Graded Rates of Wages shall be paid. Grading shall only be valid by the possession of a Grade Card issued by the Joint Industry Board.

Nothing in these rules shall prevent the maximum flexibility in the employment of skilled operatives.

3: Working Hours

The working week shall be 37.5 hours per week worked on five days, Monday to Friday inclusive.

The normal day shall be not more than eight hours worked during any consecutive nine hours between 7.30 am and 6.30 pm. Where shifts are required which fall outside these limits, payments and conditions of work shall be determined by the Joint Industry Board.

Meal breaks, including washing time, shall be unpaid of one hour duration or lesser period at the Employers discretion and shall not be exceeded. The Employer shall declare the working days and hours (including breaks) on each job.

4: Utilisation of Working Hours

There shall be full utilisation of working hours which shall not be subject to unauthorized "breaks". Time permitted for tea breaks shall not be exceeded.

Bad timekeeping and/or unauthorized absence from the place of work, during working hours, shall be construed as Industrial Misconduct.

WORKING RULES

INTRODUCTION (contd)

4: Utilisation of Working Hours (contd)

Meeting of operatives shall not be held during working hours except by arrangement with the Job/Shop Representative and with the prior permission of the Employer or the Employers site management or the Employers representative.

5: Tools

The Employer shall provide all power-operated and expendable tools as required; operatives shall act with the greatest possible responsibility in respect of the use, maintenance and safe-keeping of tools and equipment of their Employers.

The operative shall have a kit of hand tools appropriate for carrying out efficiently the work for which he is employed; the kit shall include a lockable tool box.

The Employer shall provide, where practicable, suitable and lockable facilities for storing operatives tool-kits.

6: Wages (Graded Operatives)

(a)   The National Standard JIB Graded Rates of Wages (hereinafter called "the JIB Rates of Wages") shall be those from time to time determined by the Joint Industry Board pursuant to Rule 80 of the Rules of the Joint Industry Board.

The JIB Rates of Wages appropriate to operatives shall be such rates for their grades as the Joint Industry Board may from time to time determine to be appropriate for their grade in the place where they are working and they shall be paid no more and no less wages.

This rule does not permit the introduction of any scheme of payment by result of production bonuses, except as may be determined from time to time by the JIB National Board.

(b)   London Weighting

    (i)   Definition of the London Zone
    From 7 January 1989, the definition of the London Zone is: That area lying within and including the M25 London Orbital Motorway.
    (ii)   Application
    London Weighting as determined from time to time by the Joint Industry Board shall apply to all operatives and (at separate rates) to apprentices working on jobs in the London Zone as defined above in Rule 6 (b) (i) and for avoidance of doubt it is intended that such London Weighting shall be deemed to be part of the standard JIB Rates of Wages.

Such London Weighting shall also apply to any operative or apprentice who has been working from a London based shop in the London Zone for not less than 12 weeks and who is sent by his employer to a job out of the London Zone for a period of not more than 12 weeks or for the duration of one particular contract, whichever is the longer.
London Weighting shall apply to all paid hours including overtime and shift premium payments (National Working Rules 9 and 10), statutory holiday payments (National Working Rule 13) and travelling time payments (National Working Rule 12(b)), but not to incentive payments under JIB authorized schemes.

    (iii) The amount of London Weighting shall be
        from and including 4 January 1992:
    Graded Operative . . . . 40p per hour

WORKING RULES

(c)   Responsibility Money
Approved Electricians in charge of work, who undertake the supervision of other operatives, shall be paid "responsibility money" as determined from time to time by the JIB National Board, currently not less than 2.5p and not more than 50p per hour.

From and including **4th January,1992** responsibility payments shall be enhanced by overtime and shift premiums where appropriate.

7: Payment of Wages

Wages shall normally be paid by Credit Transfer. Alternatively, another method of payment may be adopted by mutual arrangement between Employer and Operative.

Wages shall be calculated for weekly periods and paid to the Operative within 5 normal working days of week termination, unless alternative arrangements are agreed.

Each operative shall receive an itemised written pay statement in accordance with the Employment Protection (Consolidation) Act 1978.

9: Overtime

(a)   Hours
Overtime is deprecated by the Joint Industry Board; systematic overtime in particular is to be avoided.

A Regional Joint Industry Board may, from time to time, declare permissible hours of overtime in the Region which shall not be exceeded without permission of the Regional Joint Industry Board.

Overtime will not be restricted in the case of Breakdowns or Urgent Maintenance and Repairs.

(b)   Payment
> (i)   The number of hours to be worked at normal rates in any one week (Monday to Friday) before any overtime premium is calculated shall be:-
>
> From and including 4 January 1992   38 hrs

Premium time shall be paid at time-and-a-half. All hours worked between 1 p.m. on Saturday and normal starting time on Monday shall be paid at double time. Overtime premium payments shall be calculated on the appropriate standard rate of pay.

Exceptions: For the purpose of premium payment, an Operative shall be deemed to have worked normal hours on days where, although no payment is made by the Employer, the Operative:
(a)   has lost time through certified sickness.
(b)   was on a rest period for the day following continuous working the previous night.
(c)   was absent with the Employers permission.
(ii)   Any Operative who has not worked five days (as determined in Rule 3) from Monday to Friday, taking into account the exceptions detailed above, is precluded from working the following Saturday or Sunday.

(c)   Call out
Notwithstanding the previous Clause, when an operative is called upon to return to work after his normal finishing time and before his next normal starting time he shall be paid at time-and-a-half for all the hours involved (home-to-home), subject to the guaranteed minimum payment. Between 1 pm on Saturday, and the normal starting time on Monday the appropriate premium rate shall be double time.

The guaranteed minimum payment for a single call out under this clause shall be the equivalent of 4 hours at the Operatives JIB Rates of Wages.

WORKING RULES

INTRODUCTION (contd)

10: Shiftwork

Operatives may be required to undertake shiftworking arrangements in order to meet the requirements of the job or client. Operatives may not be so required without reasonable notice.

(a)   Permanent Night Shift
      (i)   Night shift is where operatives (other than as overtime after the end of a day shift) work throughout the night for not less than three consecutive nights.
A full night shift shall consist of 37.5 hours worked on five nights, Monday night to Friday night inclusive, with unpaid breaks for meals each night, to be mutually arranged. The employer shall declare working hours including breaks on each contract.
      (ii)   Payment
Night shifts shall be paid at the rate of time and one-third for all hours worked up to 37.5 in any one week, Monday to Friday.

(b)   Double day Shift (Rotating)
      (i)   The shift week will be from Monday to Friday. Each shift shall be of 7.5 hours worked with an unpaid half hour meal break. The distribution of the hours will be subject to local requirements. Shifts will normally be on an early and late basis.
      (ii)   Payment
Rotating double-day shift working will be paid at the rate of time plus 20% for normal hours in the early shift and time plus 30% for normal hours worked in the late shift.

(c)   Three Shift Working (Rotating)
      (i)   The shift week will be from Monday to Friday. Each shift shall be of 7.5 hours duration with an unpaid half hour meal break. The distribution of the hours will be subject to local requirements. Shifts will normally be on an early, late and night shift basis.
      (ii)   Payment
Rotating Three-Shift work will be paid at the rate of time plus 20%, time plus 30% and time plus 33 1/3% for the early, late and night shifts respectively.

(d)   Three Shift Working (Seven day continuous)
      (i)   Occasional
Where continuous shift work is occasionally required to cover both weekdays and weekends, weekend working shall attract the appropriate premiums contained in Rule 9(b) above. Weekday working shall attract the premiums contained in Rule 10(c) above. Generally speaking, "occasional" shiftwork shall be defined as a shiftwork requirement for a period of four weeks or less to meet some short term or emergency exigency.
      (ii)   Rostered
Where continuous three shift working is required to cover a regular seven day working pattern the following conditions shall be observed:
            (i)   Unless the requirements for continuous three shift working is   specified in the operatives contract of employment or terms of engagement, four weeks prior notice shall be given before the introduction of a rostered three shift working system.

(d)   Three Shift Working (Seven day continuous)(contd)

            (ii)   Prior to the introduction of a rostered three shift working system the employer will discuss and agree with his employees representatives the most suitable pattern of hours to achieve the required cover.
            (iii)   Subject to the above, rostered three shift working shall not be restricted.
            (iv)   The normal shift week shall be from Monday to Sunday and will   comprise a maximum 37.5 hours in any one week for which the employees shall be paid at time plus thirty per cent.
            (v)   All hours rostered, or unrostered, in excess of 37.5 hours in any week, Monday to Sunday, shall fall within the terms of Rule 10(e) below.

WORKING RULES

(e)   Overtime on Shifts
The number of hours to be worked at the appropriate shift rates before overtime premium is calculated shall be:-

From and including 4 January 1992   38 hrs

Premium payments shall be calculated on the appropriate standard rate of pay and not on the shift rate.

(f)   Other Shift Arrangements
Detailed arrangements for any other shift system of those operating on sites covered by the JIB/NJC Treaty Arrangement will be as approved by the JIB.

12: Travelling Time and Travel Allowances, Country and Lodging Allowances

(a)   Wages and Allowances
Operatives who are required to book on and off at the Employers Shop shall be entitled to time from booking on until booking off with overtime if the time so booked exceeds the normal working day. They shall also be entitled to a travel allowance as determined from time to time.

(b)   Travelling Time and Travel Allowance from 4 January 1992
Operatives who are required to start and finish at the normal starting and finishing time on jobs which are up to and including 35 miles from the shop - in a straight line - shall receive both Travel Allowance and Payment for Travelling Time as determined.

N.B. Site Transport
Where an employer provides transport from shop to site and from site to shop or between sites, those operatives so transported shall not be entitled to Travel Allowance.

(c)   Country Allowance
Operatives sent from the Employers Shop who are required to start and finish at the normal starting and finishing time on jobs over 35 miles from the Shop and who elect to travel daily to the job instead of taking lodgings will be paid Country Allowance in lieu of travelling time and Travel Allowance. In the case of London jobs this is over and above the JIB Rates of Wages for the London Zone.

(d)   Lodging Allowance
    (i)   Operatives sent from the Employers Shop who are required to start and finish at the normal starting and finishing time on jobs over 35 miles from the Shop, who elect to lodge away from home and provide proof of lodging will be paid Lodging Allowance.
    (ii)   Travelling time and travel allowance between the lodgings and the job shall not normally be paid. Where it is proved to the Employers satisfaction that suitable lodging and accommodation is not available near the job, travelling time and travel allowances for any distance of more than 10 miles each way will be paid in accordance with the scale contained in paragraph 12(b) above on the excess distance.
    (iii)   On being sent to the job the Operative shall receive his actual fare and travelling time at ordinary rates from the Employers Shop and when he returns to the Employers Shop except that when, of his own free will, he leaves the job within one calendar month from the date of his arrival and in cases where he is dismissed by the Employer for proved bad timekeeping, improper work or similar misconduct, no return travelling time or fares shall be paid.
    (iv)   The payment of Lodging Allowance shall not be made when suitable board and lodging is arranged by the Employer at no cost to the Operative.
    (v)   The payment of Lodging Allowance shall not be made during absence from employment unless a Medical Certificate is produced for the whole of the period claimed. When an operative is sent home by the firm at their cost the payment of Lodging Allowance shall cease.

WORKING RULES

INTRODUCTION (contd)

12: Travelling Time and Travel Allowances, Country and Lodging Allowances (contd)

(d)   Lodging Allowance (contd)
  (vi)   No payment for the retention of lodgings during Annual Paid Holiday shall be made by the Employer except in cases where the Operative is required to pay a retention fee during Annual Paid Holiday when reimbursement shall be of the amount actually paid to a maximum of £3.31 per day (£23.17 per week) from and including 1 March 1993 upon production of proof of payment to the Employers satisfaction.
  (vii) Where an Operative is away from his lodgings at a weekend under Rule 12(e) but has to pay a retention fee for his lodgings, reimbursement shall be the amount actually paid, to a maximum of £16.56 from and including 1 March 1993, upon production of proof of payment to the Employers satisfaction.

(e)   Period Return Fares for Operatives who lodge:
  (i)   On jobs up to and including 100 miles from the Employers Shop, return railway fares from the Job to the Employers Shop, without travelling time, shall be paid for every two weeks.
  (ii)   On jobs over 100 miles and up to and including 250 miles from the Employers Shop, return railway fares from the Job to the Employers Shop, with 4 hours travelling time at ordinary rate time, shall be paid every four weeks.
  (iii) On jobs over 250 miles from the Employers Shop, return railway fares from the Job to the Employers Shop, with 7.5 hours travelling time at ordinary rates, shall be paid every four weeks.
  (iv)   In cases under sub-clauses (ii) and (iii) above, where the Employer, through necessity or expediency, requires his Operatives to work during the specified weekend leave period, he shall arrange that they shall have another period in substitution but this provision shall not apply under sub-clause (i) above.

          N.B. All distances shall be calculated in a straight line (point to point).

When Annual Holidays with pay are taken the period returns may be moved forward or backward from the date upon which they become due, to enable the period returns to coincide with the date of the Annual Paid Holiday.
Special consideration shall be given to Operatives where it is necessary for them to return home on compassionate grounds, e.g. domestic illness.

(f)   Locally Engaged Labour:
Where an Employer does not have a Shop within 25 miles of the Job, he can engage labour domiciled within a 25 miles radius of that Job. Operatives shall receive the JIB Rates of Wages applicable to the Zone of the Job and travelling time and Travel Allowance in accordance with Clause (b), but with the exception of "home" being substituted for "shop" in Clause (b).

Locally engaged labour, domiciled within a 25 miles radius of the Job, can be transferred to other Jobs within that radius without affecting their entitlements under this Rule. Operatives transferred to a Job outside that radius and within the Zone Rate will be entitled to Country Allowance in accordance with the Rules.

13: Statutory Holidays

(a)   Qualification
Seven and a half hours pay at the appropriate JIB Rates of Wages shall be paid for a maximum of eight Statutory Holidays per annum. In general, the following shall constitute such paid holidays:
          New Years Day; Good Friday; Easter Monday; May Day; Spring Time Bank Holiday; Late Summer Bank Holiday; Christmas Day; Boxing Day.
In areas where any of these days are not normally observed as holidays in the Electrical Contracting Industry, traditional local holidays may be substituted by mutual agreement and subject to the determination of the appropriate Regional Joint Industry Board.

WORKING RULES

When Christmas Day and/or Boxing Day or New Years Day falls on a Saturday or Sunday, the following provisions apply:

Christmas Day

When Christmas Day falls on a Saturday or a Sunday, the Tuesday next following shall be deemed to be a paid holiday.

Boxing Day or New Years Day

When Boxing Day or New Years Day falls on a Saturday or Sunday, the Monday next following shall be deemed to be a paid holiday.

In order to qualify for payment, operatives must work full time for the normal day on the working days preceding and following the holiday.

For the purpose of this Rule, an operative shall be deemed to have worked on one or both of the qualifying days when the Operative

(i)   has lost time through certified sickness.

(ii)   was on a rest period for the day following continuous working all the previous night.

(iii)   was absent with the Employers permission.

(b) Payment for working Statutory Holidays

When operatives are required to work on a Paid Holiday within the scope of this Agreement, they shall receive wages at the following rates for all hours worked:

CHRISTMAS DAY - Double time and a day or shift off in lieu for which they shall be paid wages at bare time rates for the hours constituting a normal working day. The alternative day hereunder shall be mutually agreed between the Employer and the Operatives concerned.

In respect of all other days: either

(a)   Time-and-a-half plus a day or shift off in lieu for which they shall be paid wages at bare time rates for the hours constituting a normal working day. The alternative day hereunder shall be mutually agreed between the Employer and the Operatives concerned; or

(b)   at the discretion of the Employer 2.5 times the bare time rate in which event no alternative day is to be given.

In the case of night shift workers required to work on a Statutory Holiday, the premiums mentioned above shall be calculated upon the night shift rate of time-and-a-third. Time off in lieu of Statutory Holidays shall be paid at bare time day rates.

14: Annual Holidays

Operatives shall be entitled to payment for Annual Holidays as determined from time to time under the JIB Annual Holiday with Pay Scheme, depending upon their continuity of service in the Industry.

The JIB Annual Holiday with Pay Scheme is carried out on behalf of the Joint Industry Board by the Electrical Contracting Industry Benefits Agency (ECIBA) who operate a Benefits Credit collection system. Details of the scheme are shown in Section 9.

GRADING DEFINITIONS

1.1 Technician

Qualification and Training

Must have been a Registered Apprentice and have had practical training in electrical installation work and must have obtained the City and Guilds of London Institute Electrical Installation Work Part III Course Certificate (or approved equivalent), and either:

(a)   Must be at least 27 years of age.
Must have had at least five years' experience as a Approved Electrician with "responsibility money", including a minimum of three years in a supervisory capacity in charge of electrical engineering installations of such a complexity and dimension as to require wide technical experience and organisational ability.

WORKING RULES

GRADING DEFINITIONS (contd)

1.1 Technician (contd)

Qualification and Training (contd)

or   (b)   Must have exceptional technical skill, ability and experience beyond that expected of an Approved Electrician, so that his value to the Employer would be as if he were qualified as a Technician under (a) above and, with the support of his present Employer, may be granted this grade by the Joint Industry Board

Duties

Technicians must have knowledge of the most economical and effective layout of electrical installations together with the ability to achieve a high level of productivity in the work which they control. They must also be able to apply a thorough working knowledge of the National Working Rules for the Electrical Contracting Industry, of the current IEE Regulations for Electrical Installations, of the Electricity (Factories Act) Special Regulations, 1908 and 1944, the Electricity Supply Regulations, Installations (ie Regulations 22-29 inclusive and 31), of any Regulations dealing with Consumers Installations which may be issued, relevant British Standards and Codes of Practice, and of the Construction Industry Safety Regulations.

1.2 Approved Electrician

Qualification and Training

Must have been a Registered Apprentice or undergone some equivalent method of training and have had practical training in electrical installation work.
          and
must have obtained at least the City and Guilds Electrical Installation Work Part II Course Certificate (or approved equivalent)
          and
have obtained Achievement Measurement 2 or must be able, with the application for Grading and any other relevant supporting evidence (i.e. the City and Guilds Electricians Certificate) which may be required, to satisfy the Grading Committee of his experience and suitability
          and
must have had two years experience working as an Electrician subsequent to the satisfactory completion of training and immediately prior to the application for this grade, or be 22 years of age, whichever is the sooner.

Duties

Approved Electricians must possess particular practical, productive and electrical engineering skills with adequate technical supervisory knowledge so as to be able to work on their own proficiently and carry out electrical installation work without detailed supervision in the most efficient and economical manner; be able to set out jobs from drawings and specifications and requisition the necessary installation materials. They must also have a thorough working knowledge of the National Working Rules for the Electrical Contracting Industry, of the current IEE Regulations for Electrical Installations, of the Electricity Supply Regulations, 1937, issued by the Electricity Commissioners so far as they deal with Consumers' Installations (i.e. Regulations 22-29 inclusive and 31), of any Regulations dealing with Consumers' Installations which may be issued, relevant British Standards and Codes of Practice, and of the Construction Industry Safety Regulations.

WORKING RULES

1.3. Electrician

Qualification and Training

Must have been a registered Apprentice or undergone some equivalent method of training and have had adequate training in electrical installation work
>   and

Must have completed the City and Guilds Electrical Installation Work Part II Course (or approved equivalent) and have obtained Achievement Measurement 2 or must be able, with the application for grading and any other relevant supporting evidence which may be required, to satisfy the Grading Committee of his experience and suitability.
>   and

Must be at least 21 years of age (which requirement may be waived if the applicant has obtained a pass in the City and Guilds Electrical Installation Part II Course or approved equivalent).

Duties

Must be able to carry out electrical installation work efficiently in accordance with the National Working Rules for the Electrical Contracting Industry, the current IEE Regulations for Electrical Installations, and the Construction Industry Safety Regulations;

1.4. Labourer

Labourers may be employed to assist in the installation of cables in accordance with Section 5.1 - Cable Agreement: and to do other unskilled work under supervision provided that they should not be used to re-introduce pair working. Nothing in these rules should be taken to imply that labourers must be employed where there is not sufficient unskilled work to justify their employment, nor to prevent skilled men from doing a complete electrical installation job including the unskilled elements in these circumstances. On any Site at any time there shall be employed in total no more than one Labourer to four skilled JIB Graded Operatives. This particular requirement may be reviewed in the light of the particular circumstances in respect of a particular site upon application, by either Party to the appropriate Regional Joint Industry Board.

WORKING RULES

INTRODUCTION (contd)

14: Annual Holidays

Operatives shall be entitled to payment for Annual Holidays as determined from time to time under the JIB Annual Holiday with Pay Scheme, depending upon their continuity of service in the Industry.

The JIB Annual Holiday with Pay Scheme is carried out on behalf of the Joint Industry Board by the Electrical Contracting Industry Benefits Agency (ECIBA) who operate a Benefits Credit collection system. Details of the scheme are shown in Section 9.

GRADING DEFINITIONS

1.1 Technician

Qualification and Training

Must have been a Registered Apprentice and have had practical training in electrical installation work and must have obtained the City and Guilds of London Institute Electrical Installation Work Part III Course Certificate (or approved equivalent), and either:

  (a) Must be at least 27 years of age.
     Must have had at least five years' experience as a Approved Electrician with "responsibility money", including a minimum of three years in a supervisory capacity in charge of electrical engineering installations of such a complexity and dimension as to require wide technical experience and organisational ability.

or (b) Must have exceptional technical skill, ability and experience beyond that expected of an Approved Electrician, so that his value to the Employer would be as if he were qualified as a Technician under (a) above and, with the support of his present Employer, may be granted this grade by the Joint Industry Board

Duties

Technicians must have knowledge of the most economical and effective layout of electrical installations together with the ability to achieve a high level of productivity in the work which they control. They must also be able to apply a thorough working knowledge of the National Working Rules for the Electrical Contracting Industry, of the current IEE Regulations for Electrical Installations, of the Electricity (Factories Act) Special Regulations, 1908 and 1944, the Electricity Supply Regulations, Installations (ie Regulations 22-29 inclusive and 31), of any Regulations dealing with Consumers Installations which may be issued, relevant British Standards and Codes of Practice, and of the Construction Industry Safety Regulations.

WORKING RULES

1.2 Approved Electrician

Qualification and Training

Must have been a Registered Apprentice or undergone some equivalent method of training and have had practical training in electrical installation work.

　　　and

must have obtained at least the City and Guilds Electrical Installation Work Part II Course Certificate (or approved equivalent)

　　　and

have obtained Achievement Measurement 2 or must be able, with the application for Grading and any other relevant supporting evidence (i.e. the City and Guilds Electricians Certificate) which may be required, to satisfy the Grading Committee of his experience and suitability

　　　and

must have had two years experience working as an Electrician subsequent to the satisfactory completion of training and immediately prior to the application for this grade, or be 22 years of age, whichever is the sooner.

Duties

Approved Electricians must possess particular practical, productive and electrical engineering skills with adequate technical supervisory knowledge so as to be able to work on their own proficiently and carry out electrical installation work without detailed supervision in the most efficient and economical manner; be able to set out jobs from drawings and specifications and requisition the necessary installation materials. They must also have a thorough working knowledge of the National Working Rules for the Electrical Contracting Industry, of the current IEE Regulations for Electrical Installations, of the Electricity Supply Regulations, 1937, issued by the Electricity Commissioners so far as they deal with Consumers' Installations (i.e. Regulations 22-29 inclusive and 31), of any Regulations dealing with Consumers' Installations which may be issued, relevant British Standards and Codes of Practice, and of the Construction Industry Safety Regulations.

WORKING RULES

GRADING DEFINITIONS (contd)

1.3. Electrician

Qualification and Training

Must have been a registered Apprentice or undergone some equivalent method of training and have had adequate training in electrical installation work
      and
Must have completed the City and Guilds Electrical Installation Work Part II Course (or approved equivalent) and have obtained Achievement Measurement 2 or must be able, with the application for grading and any other relevant supporting evidence which may be required, to satisfy the Grading Committee of his experience and suitability.
      and
Must be at least 21 years of age (which requirement may be waived if the applicant has obtained a pass in the City and Guilds Electrical Installation Part II Course or approved equivalent).

Duties

Must be able to carry out electrical installation work efficiently in accordance with the National Working Rules for the Electrical Contracting Industry, the current IEE Regulations for Electrical Installations, and the Construction Industry Safety Regulations;

WORKING RULES

1.4. Labourer

Labourers may be employed to assist in the installation of cables in accordance with Section 5.1 - Cable Agreement: and to do other unskilled work under supervision provided that they should not be used to re-introduce pair working. Nothing in these rules should be taken to imply that labourers must be employed where there is not sufficient unskilled work to justify their employment, nor to prevent skilled men from doing a complete electrical installation job including the unskilled elements in these circumstances. On any Site at any time there shall be employed in total no more than one Labourer to four skilled JIB Graded Operatives. This particular requirement may be reviewed in the light of the particular circumstances in respect of a particular site upon application, by either Party to the appropriate Regional Joint Industry Board.

Spon's Contractors' Handbook: Electrical Installation 1995

5th Edition

Edited by **Tweeds**, Chareterd Quantity Surveyors, UK

- covers electrical supply; power and lighting; communications; security and control systems; mechanical and electrical services; plant and tool hire

- includes SMM6/SMM7 index

- general guidance chapters on VAT and other taxation, estimating techniques and business management

This handbook is written for the electrical contractor undertaking small to medium contracts (£50-£50,000). It gives materials prices, labour times and costs and unit prices for a wide range of electrical works.

Contents: Preface. SMM6/SMM7 index. Introduction. Starting up in business. Running the business. Taxation. Estimating. **Rates for measured works**: Conduit. Cable trunking. CAble tray. Wiring, fitting and accessories. Cables. Fire and intruder detection systems. Lighting protection. Telephone and communication systems. Switchgear and distribution equipment. Alterations and repairs. Testing. Approxiamte estimating. General data. Index.

Spon's Contractors' Handbooks Series

September 1994: 216x138: 376pp
Paperback: 0-419-18600-X: £25.00

For further information please contact: The Marketing Dept,. E & F N Spon, 2-6 Boundary Row, London SE1 8HN Tel: 0171 865 0066 Fax: 0171 522 9623

E & F N Spon

An imprint of Chapman & Hall

Material Costs/
Measured Work Prices

DAVIS LANGDON & EVEREST

Authors of Spon's Price Books

Davis Langdon & Everest is an independent practice of Chartered Quantity Surveyors with over 500 staff in 19 UK offices.

We believe that cost awareness and control are responsibilities that must be accepted by all members of the team, but the ability to act on that responsibility is dependent upon the quality of the cost advice given.   Our approach to cost consultancy is therefore:

* to be positive and creative in our advice, rather than simply reactive;

* to concentrate on value for money and value engineering rather than on superficial cost-cutting;

* to give advice that is matched to the Client's own criteria, rather than to impose standard or traditional solutions;

* to see cost as one component of a successful design solution, which needs to be balanced with many others, and to work as an integrated member of a design team in achieving that balance;

* to pay attention to the life-long costs of owning and operating a facility, rather than to the initial capital cost only.

We work in small teams, under the active leadership of a Partner, to provide continuity of involvement throughout the stages of a project. At the same time, we can draw on the resources of a large firm, and upon departments able to give specialist advice on engineering services, life cycle costing, cost analyses and research.

Davis Langdon & Everest - a first point of call for anyone contemplating a construction project. **Consult the Source.**

TRUST DLE TO PROVIDE THE DEFINITIVE PRICE DATA

Mechanical Installations
MATERIAL COSTS/MEASURED WORK PRICES

DIRECTIONS

The following explanations are given for each of the column headings and letter codes. It should be noted that not only are full material costs per item declared but also the published list price, the latest published price increase (update).

Unit	Prices for each unit are given as singular (1 metre, 1 nr).
Net price	Manufacturer's latest issued material/component price list, plus nominal allowance for fixings, plus any percentage uplift advised by manufacturer, plus where appropriate waste less percentage discount. The net price also reflects the applicable quantity for the minimum purchase of each batch of material and against which discounts etc. apply.
Material cost	Net price plus percentage allowance for overheads and profit.
Labour constant	Gang norm (in manhours) for each operation.
Labour cost	Labour constant multiplied by the appropriate all-in manhour cost.(See also relevant Rates of Wages Section)
Measured work price	Material cost plus Labour cost.

MATERIAL COSTS

The Material Costs given includes for delivery to sites in the London area at March/April 1995 with an allowance for waste, overheads and profit, and represents the price paid by contractors after the deduction of all trade discounts but excludes any charges in respect of VAT.

MEASURED WORK PRICES

These prices are intended to apply to new work in the London area and include allowances for all charges, preliminary items and profit. The prices are for reasonable quantities of work and the user should make suitable adjustments if the quantities are especially small or especially large. Adjustments may also be required for locality (eg outside London) and for the market conditions (eg volume of work on hand or on offer) at the time of use.

DIRECTIONS

MECHANICAL INSTALLATIONS
The labour rate on which these prices have been based is £8.87 per man hour which is the London Standard Rate at September 1994 plus allowances for all other emoluments and expenses. To this rate has been added 25% to cover site and head office overheads and preliminary items together with a further 2½% for profit, resulting in an inclusive rate of **£11.36** per man hour. The rate of £8.87 per man hour has been calculated on a working year of 1,755.60 hours; a detailed build-up of the rate is given at the end of these Directions.

PLUMBING INSTALLATIONS
The labour rate on which these prices have been based is £8.87 per man hour which is the rate at September 1994 plus allowances for all other emoluments and expenses. To this rate has been added similar percentages as for MECHANICAL INSTALLATIONS resulting in an inclusive rate of **£11.36** per man hour.

DUCTWORK INSTALLATIONS
The labour rate on which these prices have been based is £8.89 per man hour which is the rate at September 1994. To this rate has been added similar percentages as for MECHANICAL INSTALLATIONS resulting in an inclusive rate of **£11.39** per man hour.

In calculating the 'Measured Work Prices' the following assumptions have been made:
- (a)   That the work is carried out as a sub-contract under the Standard Form of Building Contract and that such facilities as are usual would be afforded by the main contractor.
- (b)   That, unless otherwise stated, the work is being carried out in open areas at a height which would not require more than simple scaffolding.
- (c)   That the building in which the work is being carried out is no more than six storeys high.

Where these assumptions are not valid, as for example where work is carried out in ducts and similar confined spaces or in multi-storey structures when additional time is needed to get to and from upper floors, then an appropriate adjustment must be made to the prices. Such adjustment will normally be to the labour element only.

No allowance has been made in the prices for any cash discount to the main contractor.

DIRECTIONS

LABOUR RATE - MECHANICAL & PLUMBING

The following detail shows how the labour rate of £8.87 per man hour has been calculated.

The annual cost of notional eleven man gang.

	FOREMAN	ADVANCED FITTER/ WELDER GAS/ARC	ADVANCED FITTER/ WELDER GAS OR ARC	ADVANCED FITTER	FITTER	MATE	APPRENTICE	SUB-TOTALS
	1 NR	1 NR	2 NR	3 NR	2 NR	1 NR	1 NR	
Hourly Rate from 05/09/1994	6.82	6.07	5.81	5.56	5.05	4.04	3.28	
Working hours/annum	1,755.60	1,755.60	3,511.20	5,266.80	3,511.20	1,755.60	1,755.60	
x Hourly rate = £/annum	11,973.19	10,656.49	20,400.07	29,283.41	17,731.56	7,092.62	5,758.37	102,895.71
Incentive schemes @ 5%	598.66	532.82	1,020.00	1,464.17	886.58	354.63	287.92	5,144.78
Daily travel rate/day	3.28	3.28	3.28	3.28	3.28	2.76	2.76	
Days/annum	231	231	462	693	462	231	231	
£/annum	757.68	757.68	1,515.36	2,273.04	1,515.36	637.56	637.56	8,094.24
Daily travel fare	4.00	4.00	4.00	4.00	4.00	4.00	4.00	
Days/annum	226	226	452	678	452	226	182	
£/annum	904.00	904.00	1,808.00	2,712.00	1,808.00	904.00	728.00	9,768.00
Weekly holiday/welfare (nr of weeks)	52	52	104	156	104	52	52	
Stamp value	32.68	30.29	30.29	30.29	27.89	24.30	24.30	
£/annum	1,699.36	1,575.08	3,150.16	4,725.24	2,900.56	1,263.60	1,263.60	16,577.60
National insurance contributions								
Weekly gross pay EACH	278.86	249.94	239.91	230.27	210.60	169.14	139.83	
% Contributions	0.102	0.102	0.102	0.102	0.102	0.076	0.056	
£ Contributions EACH	28.44	25.49	24.47	23.49	21.48	12.85	7.83	
x Nr of men = £ contributions/week	28.44	25.49	48.94	70.47	42.96	12.85	7.83	
£ Contributions/annum	1,359.43	1,218.42	2,339.33	3,368.47	2,053.49	614.23	374.27	11,327.64
Tool and clothing allowance	75.00	75.00	150.00	225.00	150.00	65.00	65.00	805.00
Welding supplement		0.51	0.25					
Hours/annum		1,755.60	3,511.20					
£/annum		895.36	877.80					1,773.16

SUB-TOTAL	156,386.13
TRAINING (INCLUDING ANY TRADE REGISTRATIONS) - SAY 1%	1,563.86
SEVERANCE PAY AND SUNDRY COSTS - SAY 1.5%	2,369.25
EMPLOYER'S LIABILITY AND THIRD PARTY INSURANCE - SAY 2%	3,206.38
ANNUAL COST OF NOTIONAL 11 MAN GANG (10.5 MEN ACTUALLY WORKING)	163,525.62
ANNUAL COST PER PRODUCTIVE MAN (ie÷10.5)	15,573.87
ALL IN MAN HOUR (BASED ON 1755.6 HOURS)	8.87

Notes:
(1)　The following assumptions have been made in the above calculations:-
　　(a)　The working week of 37.5 hours is made up of 7.5 hours Monday to Friday.
　　(b)　The actual hours worked are five days of 8.5 hours each.
　　(c)　Five days in the year are lost through sickness or similar reasons.
　　(d)　A working year of 1755.6 hours.
(2)　National insurance contributions are those effective from March 1994.
(3)　Weekly Holiday Credit/Welfare Stamp values are those effective from 31 January 1994.

DIRECTIONS

LABOUR RATE - DUCTWORK

The following detail shows how the labour rate of £8.89 per man hour has been calculated.

The annual cost of notional seven man gang.

	CHARGEHAND ERECTOR 1 NR	ADVANCED ERECTOR 1 NR	ERECTOR 4 NR	TRAINEE 1 NR	SUB-TOTALS
Hourly Rate from 05/09/1994	6.31	5.56	5.05	4.55	
Working hours/annum	1,755.60	1,755.60	7,022.40	1,755.60	
x Hourly rate = £/annum	11,077.84	9,761.14	35,463.12	7,987.98	64,290.08
Incentive schemes @ 5%	553.89	488.06	1,773.16	399.40	3,214.51
Daily travel rate/day	3.28	3.28	3.28	2.76	
Days/annum	231	231	924	231	
£/annum	757.68	757.68	3,030.72	637.56	5,183.64
Daily travel fare	4.00	4.00	4.00	4.00	
Days/annum	226	226	904	226	
£/annum	904.00	904.00	3,616.00	904.00	6,328.00
Weekly holiday/welfare (nr of weeks)	52	52	208	52	
Stamp value	32.68	30.29	27.89	24.30	
£/annum	1,699.36	1,575.08	5,801.12	1,263.60	10,339.16
National insurance contributions					
Weekly gross pay EACH	259.19	230.27	210.60	188.81	
% Contributions	0.102	0.102	0.102	0.076	
£ Contributions EACH	26.44	23.49	21.48	14.35	
x Nr of men = £ contributions/week	26.44	23.49	85.92	14.35	
£ Contributions/annum	1,263.83	1,122.82	4,106.98	685.93	7,179.56
Tool and clothing allowance	75.00	75.00	300.00	65.00	515.00

SUB-TOTAL	97,049.95
TRAINING (INCLUDING ANY TRADE REGISTRATIONS) - SAY 1%	970.50
SEVERANCE PAY AND SUNDRY COSTS - SAY 1.5%	1,470.31
EMPLOYER'S LIABILITY AND THIRD PARTY INSURANCE - SAY 2%	1,989.82
ANNUAL COST OF NOTIONAL 7 MAN GANG (6.5 MEN ACTUALLY WORKING)	101,480.58
ANNUAL COST PER PRODUCTIVE MAN (ie÷6.5)	15,612.40
ALL IN MAN HOUR (BASED ON 1755.6 HOURS)	8.89

Notes:
(1) The following assumptions have been made in the above calculations:-
  (a) The working week of 38 hours is made up of 8 hours Monday and 7.5 hours Tuesday to Friday.
  (b) The actual hours worked are five days of 8.5 hours each.
  (c) Five days in the year are lost through sickness or similar reasons.
  (d) A working year of 1755.60 hours.
(2) National insurance contributions are those effective from March 1994.
(3) Weekly Holiday Credit/Welfare Stamp values are those effective from 31 January 1994.

S:PIPED SUPPLY SYSTEMS

Item	Net Price £	Material £	Labour hours	Labour £	Unit	Total Rate £
S41 : FUEL OIL STORAGE/DISTRIBUTION : STORAGE TANKS AND VESSELS						
Fuel Storage Tanks; mild steel; with all necessary screwed bosses; oil resistant joint rings (Includes placing in position)						
Rectangular; 3mm plate						
1130 litres capacity	105.00	115.50	7.04	78.52	nr	**194.02**
1360 litres capacity	113.00	124.30	10.00	111.50	nr	**235.80**
2730 litres capacity	171.00	188.10	22.22	247.78	nr	**435.88**
4550 litres capacity	328.00	360.80	45.45	506.82	nr	**867.62**
Fuel Storage Tanks; plastic; with all necessary screwed bosses; oil resistant joint rings (Includes placing in position)						
Premium; cylinderical; 6mm thick						
1360 litres maximum capacity	217.00	238.70	2.00	22.30	nr	**261.00**
2720 litres maximum capacity	332.00	365.20	2.00	22.30	nr	**387.50**
Standard; cylindericaL; 6mm thick						
1300 litres maximum capacity	209.00	229.90	2.00	22.30	nr	**252.20**
1300 litres maximum capacity	133.00	146.30	2.00	22.30	nr	**168.60**
1360 litres maximum capacity	148.00	162.80	2.00	22.30	nr	**185.10**
1500 litres maximum capacity	222.00	244.20	2.00	22.30	nr	**266.50**
2460 litres maximum capacity	309.00	339.90	2.00	22.30	nr	**362.20**

Keep your figures up to date, free of charge

This section, and most of the other information in this Price Book, is brought up to date every three months, until the next annual edition, in the *Price Book Update.*

The *Update* is available free to all Price Book purchasers.

To ensure you receive your copy, simply complete the reply card from the centre of the book and return it to us.

S:PIPED SUPPLY SYSTEMS

Item	Net Price £	Material £	Labour hours	Labour £	Unit	Total Rate £
S60 : FIRE HOSE REELS : HOSE REELS						
Hose Reels; automatic; first aid pedestal mounted reel; connection to 25 mm screwed joint; reel with 30.5 metres, 19 mm rubber hose; (suitable for working pressure up to 7 bar.)						
Reels						
Non-swing pattern	228.27	251.10	2.50	28.40	nr	**279.50**
Swinging pattern	296.11	325.72	2.50	28.40	nr	**354.12**
Recessed swinging pattern	328.08	360.89	2.50	28.40	nr	**389.29**

S:PIPED SUPPLY SYSTEMS

Item	Net Price £	Material £	Labour hours	Labour £	Unit	Total Rate £
S61 : DRY RISERS : INLET BOX, OUTLET BOXES AND VALVES						
Dry Rising Main; (Note: for tubing and flanged connections see other sections.)						
Bronze/gunmetal inlet beeching for pumping in with 64mm dia. instantaneous male coupling; with cap, chain and 25mm drain valve						
Single inlet with black pressure valve, screwed or flanged to steel	115.90	127.49	1.15	13.07	nr	**140.56**
Double inlet with back pressure valve, screwed or flanged to steel	143.45	157.79	1.28	14.55	nr	**172.34**
Quadrupie inlet with back pressure valve, flanged to steel	427.50	470.25	2.40	27.31	nr	**497.56**
Steel dry riser inlet box with hinged wire glazed door suitably lettered (fixing by others)						
610 x 460 x 325mm; double inlet	119.70	131.67	0.50	5.68	nr	**137.35**
610 x 610 x 356mm; quadruple inlet	119.70	131.67	0.50	5.68	nr	**137.35**
Bronze/gunmetal gate type outlet valve with 64mm dia. instantaneuous female coupling cap and chain; wheel head secured by padlock and leather strap						
Flanged BS Table D inlet (bolted connection to counter flanges measured separately)	119.70	131.67	0.80	9.09	nr	**140.76**
Bronze/gunmetal landing type outlet - valve, with 64mm dia. instantaneous female coupling; cap and chain; wheelhead secured by padlock and leather strap, (bolted connections to counter flanges measured separately						
Horizontal, flanged BS Table D inlet	104.50	114.95	0.80	9.09	nr	**124.04**
Oblique, flanged BS Table D inlet	99.75	109.72	0.80	9.09	nr	**118.81**
Air Valve, screwed joint to steel						
25mm dia.	11.40	12.54	0.80	9.09	nr	**21.63**

S:PIPED SUPPLY SYSTEMS

Item	Net Price £	Material £	Labour hours	Labour £	Unit	Total Rate £
S63 : SPRINKLERS : SPRINKLER HEADS						
Sprinkler Heads and Valves						
Sprinkler Heads						
15mm conventional; 68 degree; brass	2.83	3.11	0.15	1.71	nr	**4.82**
15mm sidewall; 68 degree; brass	3.04	3.35	0.15	1.70	nr	**5.05**
15mm conventional; 68 degree; chrome	3.14	3.45	0.15	1.70	nr	**5.15**
15mm sidewall; 68 degree; chrome	3.36	3.70	0.15	1.70	nr	**5.40**
Valves; wet alarm; flanged connections						
100mm dia.	289.05	317.96	2.18	24.80	nr	**342.76**
150mm dia.	323.90	356.29	2.69	30.54	nr	**386.83**
Alternate wet/dry system alarm station; flanged connections						
100mm dia.	1991.58	2190.73	4.00	45.44	nr	**2236.17**
150mm dia.	2340.08	2574.08	5.00	56.80	nr	**2630.88**
Water motor alarm and gong; including all connections	141.45	155.59	1.50	17.03	nr	**172.62**

S:PIPED SUPPLY SYSTEMS

Item	Net Price £	Material £	Labour hours	Labour £	Unit	Total Rate £
S65 : FIRE HYDRANTS : EXTINGUISHERS						
Fire Extinguishers; (placed in position.)						
Hand Held; lever operated, BS 5423; ready charged						
Water type, 9 litres capacity; 55grms CO2 cartridge; Class A fires (fire rating 13A)	40.40	44.44	0.50	5.68	nr	**50.12**
Foam type, 9 litres capacity; 75 grms CO2 cartridge; Class A and B fires (fire rating 13A, 113B)	50.49	55.54	0.50	5.68	nr	**61.22**
Dry Power Type for class A, B and electrical equipment fires						
Multipurpose, Dry Powder type 1kg capacity; compressed air; class A, B and C fires (fire rating 5A, 34B)	24.24	26.66	0.50	5.68	nr	**32.34**
Multipurpose, Dry Powder type 2kg capacity; compressed air; class A, B and C fires (fire rating 5A, 34B)	31.00	34.10	0.50	5.68	nr	**39.78**
Multipurpose, Dry Powder type 4kg capacity; compressed air; class A, B and C fires (fire rating 5A, 34B)	56.89	62.58	0.50	5.68	nr	**68.26**
Multipurpose, Dry Powder type 9kg capacity; compressed air; class A, B and C fires (fire rating 5A, 34B)	80.80	88.88	0.50	5.68	nr	**94.56**
Carbon Dioxide Type for class B and C fires						
Electrical equipment CO2 type 2kg capacity (fire rating 13A 345B)	56.68	62.35	0.50	5.68	nr	**68.03**
Electrical equipment CO2 type 2kg capacity (fire rating 13A 345B)	80.80	88.88	0.50	5.68	nr	**94.56**
Glass Fibre Blanket, in GRP container						
1100 x 1100 mm	11.42	12.57	0.50	5.68	nr	**18.25**
1200 x 1200 mm	15.90	17.49	0.50	5.68	nr	**23.17**

S:PIPED SUPPLY SYSTEMS

Item	Net Price £	Material £	Labour hours	Labour £	Unit	Total Rate £
S65 : FIRE HYDRANTS : HYDRANTS AND HYDRANT BOXES						
Fire Hydrants; (bolted connections.)						
100mm Diameter Pillar Hydrant						
Cast iron with sluice valve	877.80	965.58	1.00	11.36	nr	**976.94**
Steel with sluice valve	1425.00	1567.50	1.00	11.36	nr	**1578.86**
Underground Hydrants, complete with frost plug to BS 750						
Sluice valve pattern type 1	275.50	303.05	4.50	51.17	nr	**354.22**
Screw down pattern type 2	133.00	146.30	4.50	51.17	nr	**197.47**
Stand pipe for underground hydrant screwed base (light alloy)						
Single outlet	92.15	101.37	0.25	2.84	nr	**104.21**
Double outlet	125.40	137.94	0.25	2.84	nr	**140.78**
64mm Diameter bronze/gunmetal outlet valves						
Oblique flanged landing valve	92.15	101.37	0.80	9.09	nr	**110.46**
Oblique screwed landing valve	89.30	98.23	0.80	9.09	nr	**107.32**
Cast iron surface box (fixing by others)						
400 x 200 x 100 mm	52.25	57.47	0.50	5.68	nr	**63.15**
500 x 200 x 150 mm	72.20	79.42	0.50	5.68	nr	**85.10**
Frost Plug	15.20	16.72	0.25	2.84	nr	**19.56**

T:MECHANICAL HEATING/COOLING/REFRIGERATION SYSTEMS

Item	Net Price £	Material £	Labour hours	Labour £	Unit	Total Rate £
T10 : GAS/OIL FIRED BOILERS : BOILER PLANT AND ANCILLARIES						
Domestic Water Boilers; stove enamelled casing; electric controls; placing in position; assembling and connecting (Electrical work elsewhere.)						
Gas fired; floor standing; connected to conventional flue						
30,000 to 40,000 Btu/Hr	273.70	301.07	8.00	90.88	nr	**391.95**
40,000 to 55,000 Btu/Hr	299.60	329.56	8.00	90.88	nr	**420.44**
50,000 to 60,000 Btu/Hr	304.50	334.95	8.00	90.88	nr	**425.83**
60,000 to 70,000 Btu/Hr	304.50	334.95	9.01	102.34	nr	**437.29**
70,000 to 80,000 Btu/Hr	396.20	435.82	10.00	113.60	nr	**549.42**
80,000 to 100,000 Btu/Hr	511.00	562.10	10.00	113.60	nr	**675.70**
100,000 to 125,000 Btu/Hr	604.80	665.28	10.00	113.60	nr	**778.88**
125,000 to 140,000 Btu/Hr	633.50	696.85	10.00	113.60	nr	**810.45**
Gas fired; wall hung; connected to conventional flue						
20,000 to 30,000 Btu/Hr	251.30	276.43	8.00	90.88	nr	**367.31**
30,000 to 40,000 Btu/Hr	286.30	314.93	8.00	90.88	nr	**405.81**
40,000 to 50,000 Btu/Hr	299.60	329.56	8.00	90.88	nr	**420.44**
45,000 to 60,000 Btu/Hr	379.40	417.34	8.00	90.88	nr	**508.22**
Gas fired; floor standing; connected to balanced flue						
30,000 to 40,000 Btu/Hr	343.70	378.07	8.00	90.88	nr	**468.95**
40,000 to 55,000 Btu/Hr	375.20	412.72	10.00	113.60	nr	**526.32**
50,000 to 60,000 Btu/Hr	382.20	420.42	10.99	124.83	nr	**545.25**
60,000 to 70,000 Btu/Hr	464.10	510.51	12.05	136.87	nr	**647.38**
70,000 to 80,000 Btu/Hr	526.40	579.04	12.05	136.87	nr	**715.91**
80,000 to 100,000 Btu/Hr	669.20	736.12	14.08	160.00	nr	**896.12**
100,000 to 125,000 Btu/Hr	1274.00	1401.40	15.87	180.32	nr	**1581.72**
Gas fired; wall hung; connected to balanced flue						
20,000 to 30,000 Btu/Hr	251.30	276.43	8.00	90.88	nr	**367.31**
30,000 to 40,000 Btu/Hr	289.10	318.01	8.00	90.88	nr	**408.89**
40,000 to 50,000 Btu/Hr	331.80	364.98	8.00	90.88	nr	**455.86**
50,000 to 60,000 Btu/Hr	393.40	432.74	8.00	90.88	nr	**523.62**
60,000 to 75,000 Btu/Hr	550.90	605.99	8.00	90.88	nr	**696.87**
Oil fired; floor standing; connected to conventional flue						
40,000 to 50,000 Btu/Hr	609.00	669.90	10.00	113.60	nr	**783.50**
50,000 to 65,000 Btu/Hr	633.50	696.85	12.05	136.87	nr	**833.72**
70,000 to 85,000 Btu/Hr	735.00	808.50	14.08	160.00	nr	**968.50**
88,000 to 110,000 Btu/Hr	806.40	887.04	14.93	169.55	nr	**1056.59**
120,000 to 170,000 Btu/Hr	916.30	1007.93	20.00	227.20	nr	**1235.13**

T:MECHANICAL HEATING/COOLING/REFRIGERATION SYSTEMS

Item	Net Price £	Material £	Labour hours	Labour £	Unit	Total Rate £
T10 : GAS/OIL FIRED BOILERS : BOILER PLANT AND ANCILLARIES (contd)						
Cast Iron Sectional Water Boilers; controls enamelled jacket; insulation including thermometer, altitude gauge and safety valve - assembled on prepared base and commissioned by supplier; (Electrical work elsewhere.)						
Gas Fired (200mm Flue)						
4 sections 50-70kW Rating	1718.64	1890.50	21.28	241.70	nr	**2132.20**
5 sections 70-93kW Rating	1855.04	2040.54	25.00	284.00	nr	**2324.54**
6 sections 93-116kW Rating	1995.84	2195.42	27.78	315.56	nr	**2510.98**
7 sections 116-140kW Rating	2610.08	2871.09	43.48	493.91	nr	**3365.00**
8 sections 140-163kW Rating	2746.48	3021.13	55.56	631.11	nr	**3652.24**
9 sections 163-186W Rating	2985.84	3284.42	62.50	710.00	nr	**3994.42**
10 sections 186-210kW Rating	3549.04	3903.94	71.43	811.43	nr	**4715.37**
11 sections 210-233kW Rating	3791.04	4170.14	90.91	1032.73	nr	**5202.87**
Oil Fired						
4 sections 50-70kW Rating	1358.72	1494.59	21.28	241.70	nr	**1736.29**
5 sections 70-93kW Rating	1495.12	1644.63	25.00	284.00	nr	**1928.63**
6 sections 93-116kW Rating	1635.92	1799.51	27.78	315.56	nr	**2115.07**
7 sections 116-140kW Rating	1948.32	2143.15	43.48	493.91	nr	**2637.06**
8 sections 140-163kW Rating	2094.40	2303.84	55.56	631.11	nr	**2934.95**
9 sections 163-186W Rating	2298.56	2528.42	62.50	710.00	nr	**3238.42**
10 sections 186-210kW Rating	2634.72	2898.19	71.43	811.43	nr	**3709.62**
11 sections 210-233kW Rating	3007.84	3308.62	90.91	1032.73	nr	**4341.35**

T:MECHANICAL HEATING/COOLING/REFRIGERATION SYSTEMS

Item	Net Price £	Material £	Labour hours	Labour £	Unit	Total Rate £
T11 : COAL FIRED BOILERS : BOILER PLANT AND ANCILLARIES						
Domestic Water Boilers; stove enamelled casing; electric controls; placing in position; assembling and connecting (Electrical work elsewhere.)						
Solid fuel fired cast iron water boiler; floor standing stove enamelled casing; thermostat; draught stablizer; electric controls; conventional flue						
13.19kW Rating (45,000 Btu/Hr)	904.40	994.84	8.00	90.88	nr	**1085.72**
17.59kW Rating (60,000 Btu/Hr)	1052.10	1157.31	10.00	113.60	nr	**1270.91**
23.45kW Rating (80,000 Btu/Hr)	1311.10	1442.21	12.05	136.87	nr	**1579.08**
29.32kW Rating (100,000 Btu/Hr)	1543.50	1697.85	14.08	160.00	nr	**1857.85**
36.65kW Rating (125,000 Btu/Hr)	1748.60	1923.46	16.13	183.23	nr	**2106.69**
Fire place mounted natural gas fire and back boiler; cast iron water boiler; electric controls; fire output 3.5kW with wood surround						
11.72kW Rating (40,000 Btu/Hr)	585.90	644.49	8.00	90.88	nr	**735.37**
14.66kW Rating (50,000 Btu/Hr)	611.10	672.21	8.00	90.88	nr	**763.09**

T:MECHANICAL HEATING/COOLING/REFRIGERATION SYSTEMS

Item	Net Price £	Material £	Labour hours	Labour £	Unit	Total Rate £
T10 - T11 : GAS/OIL & COAL FIRED BOILERS : CHIMNEYS AND FLUES						
Chimney; domestic; medium sized industrial and multi-storey dwellings; (stainless steel twin wall insulated for gas, oil and solid fuel sited internally or externally.)						
Section 120 mm Long						
125 mm dia.	23.07	25.38	0.46	5.23	nr	**30.61**
150 mm dia.	26.15	28.76	0.49	5.57	nr	**34.33**
175 mm dia.	30.89	33.97	0.53	6.04	nr	**40.01**
200 mm dia.	35.65	39.22	0.58	6.60	nr	**45.82**
250 mm dia.	43.00	47.31	0.64	7.28	nr	**54.59**
300 mm dia.	53.49	58.83	0.70	8.00	nr	**66.83**
350 mm dia.	75.54	83.09	0.78	8.81	nr	**91.90**
Section 300 mm Long						
125 mm dia.	38.03	41.83	0.50	5.68	nr	**47.51**
150 mm dia.	43.34	47.67	0.50	5.68	nr	**53.35**
175 mm dia.	49.50	54.45	0.54	6.14	nr	**60.59**
200 mm dia.	57.36	63.10	0.66	7.52	nr	**70.62**
250 mm dia.	64.18	70.59	0.76	8.67	nr	**79.26**
300 mm dia.	76.33	83.96	0.86	9.79	nr	**93.75**
350 mm dia.	80.26	88.29	0.97	11.03	nr	**99.32**
400 mm dia.	87.37	96.11	1.08	12.28	nr	**108.39**
450 mm dia.	100.59	110.65	1.08	12.28	nr	**122.93**
500 mm dia.	107.49	118.24	1.08	12.28	nr	**130.52**
550 mm dia.	160.71	176.78	1.08	12.28	nr	**189.06**
600 mm dia.	131.93	145.13	1.08	12.28	nr	**157.41**
Section 500 mm Long						
125 mm dia.	45.48	50.03	0.54	6.14	nr	**56.17**
150 mm dia.	50.98	56.07	0.54	6.14	nr	**62.21**
175 mm dia.	57.77	63.55	0.65	7.39	nr	**70.94**
200 mm dia.	68.20	75.02	0.65	7.39	nr	**82.41**
250 mm dia.	80.11	88.12	0.86	9.79	nr	**97.91**
300 mm dia.	95.31	104.84	0.98	11.14	nr	**115.98**
350 mm dia.	103.74	114.11	1.09	12.39	nr	**126.50**
400 mm dia.	116.38	128.02	1.20	13.64	nr	**141.66**
450 mm dia.	135.24	148.77	1.20	13.64	nr	**162.41**
500 mm dia.	145.40	159.94	1.20	13.64	nr	**173.58**
550 mm dia.	160.71	176.78	1.20	13.64	nr	**190.42**
600 mm dia.	165.62	182.18	1.20	13.64	nr	**195.82**
Section 1000 mm Long						
125 mm dia.	84.83	93.32	0.64	7.27	nr	**100.59**
150 mm dia.	94.81	104.29	0.71	8.07	nr	**112.36**
175 mm dia.	107.04	117.75	0.80	9.09	nr	**126.84**
200 mm dia.	126.79	139.47	0.87	9.90	nr	**149.37**
250 mm dia.	144.78	159.26	0.87	9.89	nr	**169.15**
300 mm dia.	177.67	195.44	1.13	12.84	nr	**208.28**
350 mm dia.	187.59	206.35	1.26	14.32	nr	**220.67**

T:MECHANICAL HEATING/COOLING/REFRIGERATION SYSTEMS

Item	Net Price £	Material £	Labour hours	Labour £	Unit	Total Rate £
400 mm dia.	204.94	225.43	1.39	15.80	nr	**241.23**
450 mm dia.	218.88	240.77	1.39	15.80	nr	**256.57**
500 mm dia.	237.35	261.08	1.39	15.80	nr	**276.88**
550 mm dia.	261.44	287.58	1.39	15.80	nr	**303.38**
600 mm dia.	274.10	301.50	1.39	15.80	nr	**317.30**
Adjustable length; boiler removal; internal use only						
125 mm dia.	41.89	46.08	0.50	5.68	nr	**51.76**
150 mm dia.	47.95	52.75	0.54	6.14	nr	**58.89**
175 mm dia.	57.38	63.12	0.59	6.71	nr	**69.83**
200 mm dia.	63.50	69.85	0.66	7.50	nr	**77.35**
250 mm dia.	67.13	73.84	0.76	8.64	nr	**82.48**
300 mm dia.	77.31	85.04	0.86	9.78	nr	**94.82**
350 mm dia.	90.77	99.84	1.03	11.71	nr	**111.55**
400 mm dia.	205.46	226.00	0.93	10.52	nr	**236.52**
450 mm dia.	219.44	241.39	0.93	10.52	nr	**251.91**
500 mm dia.	239.48	263.43	0.93	10.52	nr	**273.95**
550 mm dia.	261.21	287.33	0.93	10.52	nr	**297.85**
600 mm dia.	272.81	300.09	0.93	10.52	nr	**310.61**

Chimney Fittings

90 Degree insulated tee

Item	Net Price £	Material £	Labour hours	Labour £	Unit	Total Rate £
125 mm dia.	93.34	102.67	1.94	22.06	nr	**124.73**
150 mm dia.	108.63	119.49	2.14	24.33	nr	**143.82**
175 mm dia.	119.13	131.04	2.42	27.51	nr	**158.55**
200 mm dia.	140.86	154.95	2.65	30.13	nr	**185.08**
250 mm dia.	142.68	156.94	2.98	33.81	nr	**190.75**
300 mm dia.	176.85	194.53	3.39	38.51	nr	**233.04**
350 mm dia.	221.06	243.17	3.88	44.03	nr	**287.20**

135 Degree insulated tee

Item	Net Price £	Material £	Labour hours	Labour £	Unit	Total Rate £
125 mm dia.	123.72	136.09	1.94	22.06	nr	**158.15**
150 mm dia.	134.02	147.42	2.14	24.33	nr	**171.75**
175 mm dia.	146.94	161.63	2.42	27.51	nr	**189.14**
200 mm dia.	188.83	207.71	2.65	30.13	nr	**237.84**
250 mm dia.	216.38	238.02	2.98	33.81	nr	**271.83**
300 mm dia.	258.32	284.15	3.39	38.51	nr	**322.66**
350 mm dia.	314.18	345.60	3.88	44.03	nr	**389.63**
400 mm dia.	451.08	496.19	4.33	49.18	nr	**545.37**
450 mm dia.	487.70	536.47	4.81	54.62	nr	**591.09**
500 mm dia.	570.62	627.68	5.29	60.11	nr	**687.79**
550 mm dia.	583.38	641.72	5.75	65.29	nr	**707.01**
600 mm dia.	614.74	676.21	6.25	71.00	nr	**747.20**

15 Degree insulated elbow complete with 2 locking bands

Item	Net Price £	Material £	Labour hours	Labour £	Unit	Total Rate £
125 mm dia.	62.03	68.24	1.61	18.29	nr	**86.53**
150 mm dia.	69.46	76.40	1.81	20.58	nr	**96.98**
175 mm dia.	74.47	81.91	2.06	23.42	nr	**105.33**
200 mm dia.	79.20	87.12	2.34	26.60	nr	**113.72**
250 mm dia.	81.65	89.82	2.79	31.64	nr	**121.46**
300 mm dia.	102.61	112.87	3.61	41.01	nr	**153.88**
350 mm dia.	133.78	147.16	5.13	58.26	nr	**205.42**

T:MECHANICAL HEATING/COOLING/REFRIGERATION SYSTEMS

Item	Net Price £	Material £	Labour hours	Labour £	Unit	Total Rate £
T10 - T11 : GAS/OIL & COAL FIRED BOILERS : CHIMNEYS AND FLUES (contd)						
Chimney; domestic; medium sized industrial and multi-storey dwellings; (stainless steel twin wall insulated for gas, oil and solid fuel sited internally or externally.) (contd)						
Chimney Fittings (contd)						
30 Degree insulated elbow complete with 2 locking bands						
125 mm dia.	62.03	68.24	1.44	16.37	nr	84.61
150 mm dia.	69.46	76.40	1.59	18.09	nr	94.49
175 mm dia.	74.47	81.91	1.85	21.04	nr	102.95
200 mm dia.	79.20	87.12	2.12	24.12	nr	111.24
250 mm dia.	81.65	89.82	2.40	27.31	nr	117.13
300 mm dia.	102.61	112.87	2.70	30.62	nr	143.49
350 mm dia.	133.78	147.16	3.08	34.95	nr	182.11
400 mm dia.	173.65	191.01	3.46	39.31	nr	230.32
450 mm dia.	180.24	198.27	3.83	43.52	nr	241.79
500 mm dia.	192.55	211.80	4.22	47.93	nr	259.73
550 mm dia.	210.44	231.49	4.61	52.35	nr	283.84
600 mm dia.	213.67	235.03	4.98	56.52	nr	291.55
45 Degree insulated elbow complete with 2 locking bands						
125 mm dia.	62.03	68.24	1.44	16.37	nr	84.61
150 mm dia.	69.46	76.40	1.44	16.37	nr	92.77
175 mm dia.	74.47	81.91	1.44	16.37	nr	98.28
200 mm dia.	79.20	87.12	1.44	16.37	nr	103.49
250 mm dia.	81.65	89.82	1.44	16.37	nr	106.19
300 mm dia.	102.61	112.87	1.44	16.37	nr	129.24
350 mm dia.	133.78	147.16	1.44	16.37	nr	163.53
400 mm dia.	173.65	191.01	1.44	16.37	nr	207.38
450 mm dia.	180.24	198.27	1.44	16.37	nr	214.64
500 mm dia.	192.55	211.80	1.44	16.37	nr	228.17
550 mm dia.	210.44	231.49	1.44	16.37	nr	247.86
600 mm dia.	213.67	235.03	1.44	16.37	nr	251.40
Heavy Duty Wall Support						
125 mm dia.	29.43	32.37	1.90	21.60	nr	53.97
150 mm dia.	26.15	28.76	2.18	24.80	nr	53.56
175 mm dia.	36.14	39.75	2.46	27.98	nr	67.73
200 mm dia.	38.20	42.02	2.89	32.83	nr	74.85
250 mm dia.	41.24	45.36	3.32	37.74	nr	83.10
300 mm dia.	48.62	53.49	3.76	42.71	nr	96.20
350 mm dia.	63.40	69.74	4.37	49.61	nr	119.35
400 mm dia.	143.62	157.98	4.74	53.84	nr	211.82
450 mm dia.	154.19	169.61	5.15	58.56	nr	228.17
500 mm dia.	165.03	181.54	5.56	63.11	nr	244.65
550 mm dia.	170.48	187.53	5.95	67.62	nr	255.15
600 mm dia.	183.75	202.12	6.37	72.36	nr	274.48

T:MECHANICAL HEATING/COOLING/REFRIGERATION SYSTEMS

Item	Net Price £	Material £	Labour hours	Labour £	Unit	Total Rate £
Insulated top stub						
125 mm dia.	35.39	38.93	1.44	16.37	nr	**55.30**
150 mm dia.	40.35	44.38	4.59	52.11	nr	**96.49**
175 mm dia.	43.58	47.94	1.85	21.04	nr	**68.98**
200 mm dia.	46.52	51.18	2.12	24.12	nr	**75.30**
250 mm dia.	49.64	54.61	2.40	27.24	nr	**81.85**
300 mm dia.	68.11	74.92	2.69	30.54	nr	**105.46**
350 mm dia.	88.43	97.28	3.07	34.85	nr	**132.13**
400 mm dia.	96.20	105.82	3.46	39.31	nr	**145.13**
450 mm dia.	100.46	110.51	3.83	43.52	nr	**154.03**
500 mm dia.	117.34	129.07	4.22	47.93	nr	**177.00**
550 mm dia.	122.11	134.33	4.61	52.35	nr	**186.68**
600 mm dia.	125.43	137.97	4.98	56.52	nr	**194.49**
Rain Cap						
125 mm dia.	18.43	20.27	1.44	16.37	nr	**36.64**
150 mm dia.	18.43	20.27	1.44	16.37	nr	**36.64**
175 mm dia.	21.42	23.57	1.59	18.09	nr	**41.66**
200 mm dia.	26.51	29.16	1.85	21.04	nr	**50.20**
250 mm dia.	36.29	39.92	2.40	27.31	nr	**67.23**
300 mm dia.	49.07	53.98	2.70	30.62	nr	**84.60**
350 mm dia.	63.86	70.25	3.08	34.95	nr	**105.20**
400 mm dia.	62.89	69.18	3.28	37.25	nr	**106.43**
450 mm dia.	68.16	74.97	3.83	43.52	nr	**118.49**
500 mm dia.	73.41	80.75	4.22	47.93	nr	**128.68**
550 mm dia.	78.67	86.53	4.61	52.35	nr	**138.88**
600 mm dia.	83.89	92.28	4.98	56.52	nr	**148.80**
Round top						
125 mm dia	39.71	43.68	1.44	16.37	nr	**60.05**
150 mm dia	43.48	47.83	1.59	18.06	nr	**65.89**
175 mm dia	50.05	55.06	1.85	21.04	nr	**76.10**
200 mm dia	59.68	65.65	2.12	24.12	nr	**89.77**
250 mm dia	71.24	78.37	2.40	27.31	nr	**105.68**
300 mm dia	93.52	102.87	2.70	30.62	nr	**133.49**
350 mm dia	122.84	135.12	3.08	34.95	nr	**170.07**
Coping cap						
125 mm dia.	20.08	22.09	1.44	16.37	nr	**38.46**
150 mm dia.	21.00	23.10	1.59	18.09	nr	**41.19**
175 mm dia.	23.38	25.72	1.85	21.04	nr	**46.76**
200 mm dia.	28.94	31.84	2.12	24.12	nr	**55.96**
250 mm dia.	36.29	39.92	2.40	27.24	nr	**67.16**
300 mm dia.	49.07	53.98	2.69	30.54	nr	**84.52**
350 mm dia.	63.86	70.25	3.08	34.95	nr	**105.20**
Storm collar						
125 mm dia.	4.76	5.23	0.50	5.68	nr	**10.91**
150 mm dia.	5.08	5.59	0.53	6.02	nr	**11.61**
175 mm dia.	5.63	6.19	0.55	6.25	nr	**12.44**
200 mm dia.	5.89	6.48	0.64	7.28	nr	**13.76**
250 mm dia.	7.37	8.11	0.64	7.27	nr	**15.38**
300 mm dia.	7.69	8.46	0.69	7.84	nr	**16.30**
350 mm dia.	7.69	8.46	0.74	8.41	nr	**16.87**

T:MECHANICAL HEATING/COOLING/REFRIGERATION SYSTEMS

Item	Net Price £	Material £	Labour hours	Labour £	Unit	Total Rate £
T10 - T11 : GAS/OIL & COAL FIRED BOILERS : CHIMNEYS AND FLUES (contd)						
Chimney; domestic; medium sized industrial and multi-storey dwellings; (stainless steel twin wall insulated for gas, oil and solid fuel sited internally or externally.) (contd)						
Chimney Fittings (contd)						
Storm collar (contd)						
400 mm dia.	21.78	23.96	0.79	8.98	nr	**32.94**
450 mm dia.	23.96	26.36	1.19	13.52	nr	**39.88**
500 mm dia.	26.14	28.75	0.89	10.12	nr	**38.87**
550 mm dia.	28.32	31.15	0.94	10.69	nr	**41.84**
600 mm dia.	30.49	33.54	0.99	11.25	nr	**44.79**
Flat flashing						
125 mm dia.	21.87	24.06	1.44	16.37	nr	**40.43**
150 mm dia.	22.42	24.66	1.59	18.09	nr	**42.75**
175 mm dia.	23.24	25.56	1.85	21.04	nr	**46.60**
200 mm dia.	25.71	28.28	2.12	24.12	nr	**52.40**
250 mm dia.	35.81	39.39	2.40	27.31	nr	**66.70**
300 mm dia.	43.91	48.30	2.70	30.62	nr	**78.92**
350 mm dia.	73.10	80.41	3.08	34.95	nr	**115.36**
400 mm dia.	84.08	92.48	3.46	39.31	nr	**131.79**
450 mm dia.	97.56	107.32	3.83	43.52	nr	**150.84**
500 mm dia.	105.56	116.12	4.22	47.93	nr	**164.05**
550 mm dia.	111.54	122.69	4.61	52.35	nr	**175.04**
600 mm dia.	114.50	125.95	4.98	56.52	nr	**182.47**
5 - 30 Degree adjustable flashing						
125 mm dia.	21.87	24.06	1.44	16.37	nr	**40.43**
150 mm dia.	22.42	24.66	1.59	18.09	nr	**42.75**
175 mm dia.	23.24	25.56	1.85	21.00	nr	**46.56**
200 mm dia.	25.71	28.28	2.12	24.12	nr	**52.40**
250 mm dia.	35.81	39.39	2.40	27.31	nr	**66.70**
300 mm dia.	43.91	48.30	2.70	30.62	nr	**78.92**
350 mm dia.	67.30	74.02	3.07	34.85	nr	**108.87**
400 mm dia.	108.31	119.15	3.46	39.31	nr	**158.46**
450 mm dia.	127.53	140.29	3.83	43.52	nr	**183.81**
500 mm dia.	137.50	151.25	4.22	47.93	nr	**199.18**
550 mm dia.	161.88	178.07	4.59	52.11	nr	**230.18**
600 mm dia.	185.11	203.62	4.98	56.52	nr	**260.14**
Ceiling Support						
125 mm dia.	15.04	16.54	1.90	21.60	nr	**38.14**
150 mm dia.	16.75	18.42	2.18	24.80	nr	**43.22**
175 mm dia.	15.04	16.54	1.90	21.60	nr	**38.14**
200 mm dia.	30.18	33.20	2.89	32.83	nr	**66.03**
250 mm dia.	34.41	37.86	3.32	37.74	nr	**75.60**
300 mm dia.	39.33	43.26	3.75	42.55	nr	**85.81**
350 mm dia.	46.88	51.57	4.37	49.61	nr	**101.18**

T:MECHANICAL HEATING/COOLING/REFRIGERATION SYSTEMS

Item	Net Price £	Material £	Labour hours	Labour £	Unit	Total Rate £
400 mm dia.	236.77	260.45	4.95	56.24	nr	**316.69**
450 mm dia.	250.43	275.48	5.56	63.11	nr	**338.59**
500 mm dia.	264.35	290.79	6.14	69.69	nr	**360.48**
550 mm dia.	289.39	318.33	6.76	76.76	nr	**395.09**
600 mm dia.	294.33	323.77	7.35	83.53	nr	**407.30**
Firestop spacer						
125 mm dia.	2.94	3.23	0.50	5.68	nr	**8.91**
150 mm dia.	3.27	3.59	0.53	6.04	nr	**9.63**
175 mm dia.	3.69	4.06	0.55	6.28	nr	**10.34**
200 mm dia.	4.37	4.81	0.59	6.71	nr	**11.52**
250 mm dia.	4.50	4.95	0.64	7.28	nr	**12.23**
300 mm dia.	5.43	5.97	0.69	7.83	nr	**13.80**
350 mm dia.	9.87	10.86	0.74	8.41	nr	**19.27**
Wall band; internal or external use						
125 mm dia.	17.72	19.49	0.98	11.14	nr	**30.63**
150 mm dia.	18.51	20.37	1.02	11.59	nr	**31.96**
175 mm dia.	19.25	21.17	1.06	12.05	nr	**33.22**
200 mm dia.	20.32	22.35	1.12	12.74	nr	**35.09**
250 mm dia.	21.15	23.27	1.24	14.09	nr	**37.36**
300 mm dia.	22.84	25.13	1.38	15.67	nr	**40.80**
350 mm dia.	24.31	26.74	1.57	17.86	nr	**44.60**
400 mm dia.	27.65	30.42	1.76	20.00	nr	**50.42**
450 mm dia.	32.12	35.34	2.43	27.57	nr	**62.91**
500 mm dia.	38.60	42.46	2.14	24.33	nr	**66.79**
550 mm dia.	40.44	44.49	2.33	26.48	nr	**70.97**
600 mm dia.	42.76	47.04	2.53	28.69	nr	**75.73**
Wall Support						
125 mm dia.	29.43	32.37	2.21	25.13	nr	**57.50**
150 mm dia.	32.92	36.21	2.37	26.92	nr	**63.13**
175 mm dia.	36.14	39.75	2.48	28.19	nr	**67.94**
200 mm dia.	38.20	42.02	2.72	30.95	nr	**72.97**
250 mm dia.	41.24	45.36	3.03	34.42	nr	**79.78**
300 mm dia.	48.62	53.49	3.46	39.31	nr	**92.80**
350 mm dia.	63.40	69.74	4.10	46.56	nr	**116.30**
400 mm dia.	143.62	157.98	4.76	54.10	nr	**212.08**
450 mm dia.	154.19	169.61	5.41	61.41	nr	**231.02**
500 mm dia.	165.03	181.54	6.06	68.85	nr	**250.39**
550 mm dia.	170.48	187.53	6.71	76.24	nr	**263.77**
600 mm dia.	183.75	202.12	7.35	83.53	nr	**285.65**
Twin wall galvanised steel flue box, 125mm dia.; (fitted where no chimney exists) for gas fire						
Free Standing	52.48	57.72	2.07	23.52	nr	**81.24**
Recess	52.48	57.72	2.07	23.52	nr	**81.24**
Back Boiler	64.35	70.79	2.40	27.31	nr	**98.10**

T:MECHANICAL HEATING/COOLING/REFRIGERATION SYSTEMS

Item	Net Price £	Material £	Labour hours	Labour £	Unit	Total Rate £
T10 - T11 : GAS/OIL & COAL FIRED BOILERS : CHIMNEYS AND FLUES (contd)						
Domestic and small comercial; Twin walled gas vent system suitable for gas fired aplliances; domestric gas boilers; small commercial boilers with internal or external flues.						
150 mm long						
100 mm dia.	4.04	4.45	0.50	5.68	nr	**10.13**
125 mm dia.	4.97	5.47	0.50	5.68	nr	**11.15**
150 mm dia.	5.38	5.92	0.50	5.68	nr	**11.60**
305 mm long						
100 mm dia.	6.13	6.75	0.50	5.68	nr	**12.43**
125 mm dia.	7.20	7.92	0.50	5.68	nr	**13.60**
150 mm dia.	8.53	9.38	0.50	5.68	nr	**15.06**
457 mm long						
100 mm dia.	6.79	7.47	0.54	6.14	nr	**13.61**
125 mm dia.	7.63	8.40	0.54	6.14	nr	**14.54**
150 mm dia.	9.44	10.39	0.54	6.14	nr	**16.53**
914 mm long						
100 mm dia.	12.13	13.34	0.64	7.27	nr	**20.61**
125 mm dia.	14.12	15.54	0.64	7.27	nr	**22.81**
150 mm dia.	16.20	17.82	0.64	7.27	nr	**25.09**
1524 mm long						
100 mm dia.	17.51	19.26	0.83	9.47	nr	**28.73**
125 mm dia.	21.55	23.70	0.83	9.47	nr	**33.17**
150 mm dia.	23.13	25.44	0.83	9.47	nr	**34.91**
Adjustable length 305 mm long						
100 mm dia.	7.75	8.53	0.56	6.31	nr	**14.84**
125 mm dia.	8.71	9.58	0.56	6.31	nr	**15.89**
150 mm dia.	23.13	25.44	0.56	6.31	nr	**31.75**
Adjustable length 457 mm long						
100 mm dia.	10.46	11.50	0.56	6.31	nr	**17.81**
125 mm dia.	12.69	13.95	0.56	6.31	nr	**20.26**
150 mm dia.	14.11	15.53	0.56	6.31	nr	**21.84**
Adjustable elbow 0 - 90 deg						
100 mm dia.	8.85	9.74	0.45	5.11	nr	**14.85**
125 mm dia.	10.46	11.51	0.45	5.11	nr	**16.62**
150 mm dia.	13.11	14.42	0.45	5.11	nr	**19.53**

T:MECHANICAL HEATING/COOLING/REFRIGERATION SYSTEMS

Item	Net Price £	Material £	Labour hours	Labour £	Unit	Total Rate £
Draughthood connector						
100 mm dia.	2.73	3.00	0.45	5.11	nr	**8.11**
125 mm dia.	3.08	3.38	0.45	5.11	nr	**8.49**
150 mm dia.	3.35	3.68	0.45	5.11	nr	**8.79**
Adaptor						
100 mm dia.	5.69	6.26	0.45	5.11	nr	**11.37**
125 mm dia.	6.16	6.78	0.45	5.11	nr	**11.89**
150 mm dia.	6.78	7.46	0.45	5.11	nr	**12.57**
Support plate						
100 mm dia.	4.78	5.26	0.45	5.11	nr	**10.37**
125 mm dia.	5.09	5.60	0.45	5.11	nr	**10.71**
150 mm dia.	5.45	5.99	0.45	5.11	nr	**11.10**
Wall band						
100 mm dia.	4.33	4.76	0.45	5.11	nr	**9.87**
125 mm dia.	4.62	5.08	0.45	5.11	nr	**10.19**
150 mm dia.	5.85	6.44	0.45	5.11	nr	**11.55**
Firestop						
100 mm dia.	1.88	2.07	0.45	5.11	nr	**7.18**
125 mm dia.	1.88	2.07	0.45	5.11	nr	**7.18**
150 mm dia.	2.15	2.37	0.45	5.11	nr	**7.48**
Flat flashing						
100 mm dia.	13.57	14.93	0.45	5.11	nr	**20.04**
125 mm dia.	14.77	16.24	0.45	5.11	nr	**21.35**
150 mm dia.	16.27	17.89	0.45	5.11	nr	**23.00**
Adjustable flashing 5-30 deg.						
100 mm dia.	13.57	14.93	0.45	5.11	nr	**20.04**
125 mm dia.	14.77	16.24	0.45	5.11	nr	**21.35**
150 mm dia.	16.27	17.89	0.45	5.11	nr	**23.00**
Adjustable flashing 30-45 deg.						
100 mm dia.	13.57	14.93	0.45	5.11	nr	**20.04**
125 mm dia.	14.77	16.24	0.45	5.11	nr	**21.35**
150 mm dia.	16.27	17.89	0.45	5.11	nr	**23.00**
Storm collar						
100 mm dia.	1.98	2.18	0.45	5.11	nr	**7.29**
125 mm dia.	2.36	2.60	0.45	5.11	nr	**7.71**
150 mm dia.	2.61	2.87	0.45	5.11	nr	**7.98**
Gas vent terminal						
100 mm dia.	10.11	11.12	0.45	5.11	nr	**16.23**
125 mm dia.	11.10	12.21	0.45	5.11	nr	**17.32**
150 mm dia.	14.24	15.67	0.45	5.11	nr	**20.78**

T:MECHANICAL HEATING/COOLING/REFRIGERATION SYSTEMS

Item	Net Price £	Material £	Labour hours	Labour £	Unit	Total Rate £
T 13 : PACKAGE STEAM GENERATORS						
Packaged Water Boilers; boiler mountings controls;enamelled casing; burner; insulation; all connections and commissioning						
Gas Fired						
88 kw rating	2778.75	3056.63	20.00	227.20	nr	**3283.83**
366 kw rating	5471.70	6018.87	32.26	366.45	nr	**6385.32**
586 kw rating	7026.83	7729.51	40.00	454.40	nr	**8183.91**
1465 kw rating	10769.85	11846.83	50.00	568.00	nr	**12414.83**
2930 kw rating	22752.60	25027.86	50.00	568.00	nr	**25595.86**
Oil Fired						
88 kw rating	2355.60	2591.16	20.00	227.20	nr	**2818.36**
366 kw rating	4095.00	4504.50	32.26	366.45	nr	**4870.95**
586 kw rating	5681.33	6249.46	40.00	454.40	nr	**6703.86**
1465 kw rating	8849.10	9734.01	50.00	568.00	nr	**10302.01**
2930 kw rating	21824.40	24006.84	50.00	568.00	nr	**24574.84**
Packaged Steam Boilers; boiler mountings centrifugal water feed pump; insulation; and sheet steel wrap around casing; plastic coated						
Gas Fired						
293 kw rating	14348.10	15782.91	83.33	946.67	nr	**16729.58**
1465 kw rating	23044.13	25348.54	142.86	1622.86	nr	**26971.40**
2930 kw rating	34650.52	38115.58	200.00	2272.00	nr	**40387.58**
Oil Fired						
293 kw rating	13097.18	14406.89	83.33	946.67	nr	**15353.56**
1465 kw rating	21660.60	23826.66	142.86	1622.86	nr	**25449.52**
2930 kw rating	33788.63	37167.49	200.00	2272.00	nr	**39439.49**

T:MECHANICAL HEATING/COOLING/REFRIGERATION SYSTEMS

Item	Net Price £	Material £	Labour hours	Labour £	Unit	Total Rate £
T30 : MEDIUM TEMPERATURE HOT WATER HEATING : NATURAL CONVECTORS						
Quality Sheet Steel Cased Units- extruded aluminium grilles for LTHW centrifugal fans;filter choice of 3 speeds; single phase thermostatic controls; 3/42" BSP connections; all 230mm deep complete with access locks						
Free standing flat top 695mm high medium speed rating						
E.A.T. at 18 degree						
695mm length 1 row 1.94 kw 75 l/sec 39C	410.00	451.00	2.38	27.05	nr	**478.05**
695mm length 2 row 2.64 kw 75 l/sec 47C	410.00	451.00	2.38	27.05	nr	**478.05**
895mm length 1 row 4.02 kw 150 l/sec 40C	462.00	508.20	2.38	27.05	nr	**535.25**
895mm length 2 row 5.62 kw 150 l/sec 49C	462.00	508.20	2.38	27.05	nr	**535.25**
1195mm length 1row 6.58 kw 250 l/sec 40C	526.00	578.60	2.38	27.05	nr	**605.65**
1195mm length 2row 9.27 kw 250 l/sec 48C	526.00	578.60	2.38	27.05	nr	**605.65**
1495mm length 1row 9.04kw 340 l/sec 40C	587.00	645.70	2.38	27.05	nr	**672.75**
1495mm length 2row 12.73kw 340 l/sec 49C	587.00	645.70	2.38	27.05	nr	**672.75**
Free standing flat top 695mm high medium speed rating c/w plinth						
E.A.T. at 18 degree						
695mm length 1 row 1.94 kw 75 l/sec 39C	425.00	467.50	2.38	27.05	nr	**494.55**
695mm length 2 row 2.64 kw 75 l/sec 47C	425.00	467.50	2.38	27.05	nr	**494.55**
895mm length 1 row 4.02 kw 150 l/sec 40C	477.00	524.70	2.38	27.05	nr	**551.75**
895mm length 2 row 5.62 kw 150 l/sec 49C	477.00	524.70	2.38	27.05	nr	**551.75**
1195mm length 1row 6.58 kw 250 l/sec 40C	541.00	595.10	2.38	27.05	nr	**622.15**
1195mm length 2row 9.27 kw 250 l/sec 48C	541.00	595.10	2.38	27.05	nr	**622.15**
1495mm length 1row 9.04kw 340 l/sec 40C	602.00	662.20	2.38	27.05	nr	**689.25**
1495mm length 2row 12.73kw 340 l/sec 49C	602.00	662.20	2.38	27.05	nr	**689.25**
Free standing sloping top 695mm high medium speed rating						
695mm length 1 row 1.94 kw 75 l/sec 39C	430.50	473.55	2.38	27.05	nr	**500.60**
695mm length 2 row 2.64 kw 75 l/sec 47C	430.50	473.55	2.38	27.05	nr	**500.60**
895mm length 1 row 4.02 kw 150 l/sec 40C	485.10	533.61	2.38	27.05	nr	**560.66**
895mm length 2 row 5.62 kw 150 l/sec 49C	485.10	533.61	2.38	27.05	nr	**560.66**
1195mm length 1row 6.58 kw 250 l/sec 40C	552.30	607.53	2.38	27.05	nr	**634.58**
1195mm length 2row 9.27 kw 250 l/sec 48C	552.30	607.53	2.38	27.05	nr	**634.58**
1495mm length 1row 9.04kw 340 l/sec 40C	616.35	677.99	2.38	27.05	nr	**705.04**
1495mm length 2row 12.73kw 340 l/sec 49C	616.35	677.99	2.38	27.05	nr	705.04
Free standing sloping top 695mm high medium speed rating c/w plinth						
695mm length 1 row 1.94 kw 75 l/sec 39C	445.50	490.05	2.38	27.05	nr	**517.10**
695mm length 2 row 2.64 kw 75 l/sec 47C	445.50	490.05	2.38	27.05	nr	**517.10**
895mm length 1 row 4.02 kw 150 l/sec 40C	500.10	550.11	2.38	27.05	nr	**577.16**
895mm length 2 row 5.62 kw 150 l/sec 49C	500.10	550.11	2.38	27.05	nr	**577.16**
1195mm length 1row 6.58 kw 250 l/sec 40C	567.30	624.03	2.38	27.05	nr	**651.08**
1195mm length 2row 9.27 kw 250 l/sec 48C	567.30	624.03	2.38	27.05	nr	**651.08**
1495mm length 1row 9.04kw 340 l/sec 40C	631.35	694.48	2.38	27.05	nr	**721.53**
1495mm length 2row 12.73kw 340 l/sec 49C	631.35	694.48	2.38	27.05	nr	**721.52**

T:MECHANICAL HEATING/COOLING/REFRIGERATION SYSTEMS

Item	Net Price £	Material £	Labour hours	Labour £	Unit	Total Rate £
T30 : MEDIUM TEMPERATURE HOT WATER HEATING : NATURAL CONVECTORS (contd)						
Quality Sheet Steel Cased Units- extruded aluminium grilles for LTHW centrifugal fans;filter choice of 3 speeds; single phase thermostatic controls; 3/42" BSP connections; all 230mm deep complete with access locks (contd)						
Wall mounted reversed air flow high level sloping discharge						
695mm length 1 row 1.94 kw 75 l/sec 39C	450.00	495.00	2.38	27.05	nr	**522.05**
695mm length 2 row 2.64 kw 75 l/sec 47C	450.00	495.00	2.38	27.05	nr	**522.05**
895mm length 1 row 4.02 kw 150 l/sec 40C	463.00	509.30	2.38	27.05	nr	**536.35**
895mm length 2 row 5.62 kw 150 l/sec 49C	463.00	509.30	2.38	27.05	nr	**536.35**
1195mm length 1row 6.58 kw 250 l/sec 40C	563.60	619.96	2.38	27.05	nr	**647.01**
1195mm length 2row 9.27 kw 250 l/sec 48C	563.60	619.96	2.38	27.05	nr	**647.01**
1495mm length 1row 9.04kw 340 l/sec 40C	611.00	672.10	2.38	27.05	nr	**699.15**
1495mm length 2row 12.73kw 340 l/sec 49C	611.00	672.10	2.38	27.05	nr	**699.15**
Ceiling mounted sloping inlet/outlet 665mm width						
895mm length 1 row 4.02 kw 150 l/sec 40C	509.00	559.90	4.00	45.44	nr	**605.34**
895mm length 2 row 5.62 kw 150 l/sec 49C	509.00	559.90	4.00	45.44	nr	**605.34**
1195mm length 1row 6.58 kw 250 l/sec 40C	572.00	629.20	4.00	45.44	nr	**674.64**
1195mm length 2row 9.27 kw 250 l/sec 48C	572.00	629.20	4.00	45.44	nr	**674.64**
1495mm length 1row 9.04kw 340 l/sec 40C	628.00	690.80	4.00	45.44	nr	**736.24**
1495mm length 2row 12.73kw 340 l/sec 49C	628.00	690.80	4.00	45.44	nr	**736.24**
Free standing unit extended height 1700/1900/2100mm						
895mm length 1 row 4.02 kw 150 l/sec 40C	589.00	647.90	3.00	34.11	nr	**682.01**
895mm length 2 row 5.62 kw 150 l/sec 49C	589.00	647.90	3.00	34.11	nr	**682.01**
1195mm length 1row 6.58 kw 250 l/sec 40C	688.00	756.80	3.00	34.11	nr	**790.91**
1195mm length 2row 9.27 kw 250 l/sec 48C	688.00	756.80	3.00	34.11	nr	**790.91**
1495mm length 1row 9.04kw 340 l/sec 40C	757.00	832.70	3.00	34.11	nr	**866.81**
1495mm length 2row 12.73kw 340 l/sec 49C	757.00	832.70	3.00	34.11	nr	**866.81**

T:MECHANICAL HEATING/COOLING/REFRIGERATION SYSTEMS

Item	Net Price £	Material £	Labour hours	Labour £	Unit	Total Rate £
T30-T32 : MEDIUM AND LOW TEMPERATURES HOT WATER HEATING : RADIANT PANELS						
Pressed Steel Panel Type Radiators; (fixed with and including brackets; taking down once for decoration; refixing.)						
300mm High; Single Panel						
500mm Length	16.81	18.49	1.40	15.91	nr	**34.40**
1000mm Length	31.96	35.16	1.40	15.91	nr	**51.07**
1500mm Length	46.64	51.31	1.40	15.91	nr	**67.22**
2000mm Length	60.96	67.06	2.00	22.72	nr	**89.78**
2500mm Length	74.99	82.49	3.00	34.11	nr	**116.60**
3000mm Length	88.90	97.79	3.21	36.41	nr	**134.20**
300mm High; Double Panel; Convector						
500mm Length	32.49	35.74	1.60	18.18	nr	**53.92**
1000mm Length	62.80	69.08	1.60	18.18	nr	**87.26**
1500mm Length	93.10	102.41	1.60	18.18	nr	**120.59**
2000mm Length	122.28	134.50	2.20	25.02	nr	**159.52**
2500mm Length	151.28	166.40	3.21	36.41	nr	**202.81**
3000mm Length	179.81	197.79	3.40	38.64	nr	**236.43**
450mm High; Single Panel						
500mm Length	15.04	16.54	1.50	17.06	nr	**33.60**
1000mm Length	29.53	32.49	1.50	17.06	nr	**49.55**
1600mm Length	45.46	50.00	2.40	27.31	nr	**77.31**
2000mm Length	56.23	61.85	3.00	34.11	nr	**95.96**
2400mm Length	74.04	81.44	4.00	45.44	nr	**126.88**
3000mm Length	91.15	100.26	4.41	50.04	nr	**150.30**
450mm High; Double Panel; Convector						
500mm Length	30.25	33.27	1.70	19.32	nr	**52.59**
1000mm Length	60.37	66.40	1.70	19.32	nr	**85.72**
1600mm Length	92.50	101.76	2.60	29.51	nr	**131.27**
2000mm Length	131.69	144.86	3.19	36.29	nr	**181.15**
2400mm Length	156.49	172.13	3.80	43.19	nr	**215.32**
3000mm Length	193.48	212.83	4.61	52.35	nr	**265.18**
600mm High; Single Panel						
500mm Length	19.57	21.52	1.70	19.32	nr	**40.84**
1000mm Length	37.46	41.21	2.20	25.02	nr	**66.23**
1600mm Length	58.12	63.93	3.60	40.86	nr	**104.79**
2000mm Length	80.32	88.35	4.61	52.35	nr	**140.70**
2400mm Length	95.17	104.69	5.21	59.17	nr	**163.86**
3000mm Length	117.07	128.77	6.99	79.44	nr	**208.21**
600mm High; Double Panel; Convector						
500mm Length	38.23	42.06	1.90	21.60	nr	**63.66**
1000mm Length	76.41	84.05	1.90	21.60	nr	**105.65**
1600mm Length	136.96	150.65	3.80	43.19	nr	**193.84**
2000mm Length	169.04	185.94	4.81	54.62	nr	**240.56**

T:MECHANICAL HEATING/COOLING/REFRIGERATION SYSTEMS

Item	Net Price £	Material £	Labour hours	Labour £	Unit	Total Rate £
T30-T32 : MEDIUM AND LOW TEMPERATURES HOT WATER HEATING : RADIANT PANELS (contd)						
Pressed Steel Panel Type Radiators; (fixed with and including brackets; taking down once for decoration; refixing.) (contd)						
600mm High; Double Panel; Convector (contd)						
2400mm Length	200.99	221.09	5.41	61.41	nr	**282.50**
3000mm Length	248.40	273.24	7.25	82.32	nr	**355.56**
700mm High; Single Panel						
500mm Length	23.14	25.46	1.80	20.47	nr	**45.93**
1000mm Length	44.16	48.57	3.00	34.11	nr	**82.68**
1600mm Length	74.51	81.97	4.81	54.62	nr	**136.59**
2000mm Length	92.15	101.37	5.99	68.02	nr	**169.39**
2400mm Length	109.14	120.05	6.02	68.43	nr	**188.48**
3000mm Length	134.30	147.72	7.25	82.32	nr	**230.04**
700mm High; Double Panel; Convector						
500mm Length	44.69	49.16	2.00	22.72	nr	**71.88**
1000mm Length	86.77	95.44	3.51	39.86	nr	**135.30**
1600mm Length	158.68	174.54	5.00	56.80	nr	**231.34**
2000mm Length	193.60	212.96	5.41	61.41	nr	**274.37**
2400mm Length	230.17	253.19	5.81	66.05	nr	**319.24**
3000mm Length	284.44	312.89	6.41	72.82	nr	**385.71**

T:MECHANICAL HEATING/COOLING/REFRIGERATION SYSTEMS

Item	Net Price £	Material £	Labour hours	Labour £	Unit	Total Rate £
T30-T32 MEDIUM AND LOW TEMPERATURE HOT WATER HEATING : RADIANT STRIP HEATERS						
Black 1.25" Steel tube, aluminium radiant plates including insulation, sliding brackets, cover plates, end closures weld or screwed BSP ends.						
One Tube						
1500mm Long	61.00	67.10	3.00	34.11	nr	101.21
3000mm Long	85.00	93.50	3.00	34.11	nr	127.61
4500mm Long	102.00	112.20	3.00	34.11	nr	146.31
6000mm Long	130.00	143.00	3.00	34.11	nr	177.11
Two Tube						
1500mm Long	112.00	123.20	4.00	45.44	nr	168.64
3000mm Long	157.00	172.70	4.00	45.44	nr	218.14
4500mm Long	188.00	206.80	4.00	45.44	nr	252.24
6000mm Long	239.00	262.90	4.00	45.44	nr	308.34

Keep your figures up to date, free of charge

This section, and most of the other information in this Price Book, is brought up to date every three months, until the next annual edition, in the *Price Book Update*.

The *Update* is available free to all Price Book purchasers.

To ensure you receive your copy, simply complete the reply card from the centre of the book and return it to us.

T:MECHANICAL HEATING/COOLING/REFRIGERATION SYSTEMS

Item	Net Price £	Material £	Labour hours	Labour £	Unit	Total Rate £
T42 : LOCAL HEATING UNITS : UNIT HEATERS						
Unit Heater; horizontal or vertical discharge; recirculating type for industrial and commercial user for heights up to 3m (normal speed); EAT 15C; fixed to existing suspension rods; complete with enclosures (includes connections or hot water services electrical work included elsewhere.)						
Low pressure hot water						
7.5Kw 265 l/sec	228.00	250.80	6.02	68.43	nr	**319.23**
15.4Kw 575 l/sec	276.00	303.60	6.99	79.44	nr	**383.04**
26.9Kw 1040 l/sec	374.00	411.40	8.47	96.27	nr	**507.67**
48.0Kw 1620 l/sec	493.00	542.30	9.01	102.34	nr	**644.64**
Steam, 2 bar						
9.2 Kw 2651 l/sec	326.00	350.00	6.02	68.43	nr	**418.43**
18.8 Kw 5751 l/sec	353.00	388.30	6.02	68.43	nr	**456.73**
34.4 Kw 10401 l/sec	406.00	446.00	6.02	68.43	nr	**514.43**
51.6 Kw 16251 l/sec	550.00	605.00	6.02	68.43	nr	**673.43**

U:VENTILATION/AIR CONDITIONING SYSTEMS

Item	Net Price £	Material £	Labour hours	Labour £	Unit	Total Rate £
U70 : AIR CURTAINS : AIR HEATERS						
The selection of an air curtain requires consideration of the particular conditions involved, such as, climatic conditions, wind influence, construction and position of. Consultation with a specialist manufacturer is therefore advisable.						
Vertical Discharge Type Curtains; recessed or exposed rigid sheet steel casing; high quality motor/centrifugal fan assembly three speed control.						
Water heated 240V Single phase supply; mounting height 2.20m						
1000 x 312mm; 3.80-7.10kW Output	1346.10	1480.71	12.05	136.87	nr	**1617.58**
1500 x 312mm; 5.80-10.70kW Output	1793.40	1972.74	12.05	136.87	nr	**2109.61**
2000 x 312mm; 7.80-14.60kW Output	2220.75	2442.82	12.05	136.87	nr	**2579.69**
Water heated 240V Single phase supply; mounting height 2.60m						
1000 x 312mm; 5.80-10.10kW Output	1597.05	1756.76	16.13	183.23	nr	**1939.99**
1500 x 312mm; 7.10-14.60kW Output	2103.15	2313.46	16.13	183.23	nr	**2496.69**
2000 x 312mm; 11.50-20.20kW Output	2679.60	2947.56	16.13	183.23	nr	**3130.79**
Water heated 240V Single phase supply; mounting height 3.00m						
1000 x 312mm; 9.90-18.90kW Output	2158.80	2374.68	17.24	195.86	nr	**2570.54**
1500 x 312mm; 15.10-28.40kW Output	3005.10	3305.61	17.24	195.86	nr	**3501.47**
2000 x 312mm; 20.20-37.70kW Output	3769.50	4146.45	17.24	195.86	nr	**4342.31**
Electrically heated 415V Three phase supply; mounting height 2.20m						
1000 x 312mm; 4.00-8.00kW Output	1781.85	1960.04	12.05	136.87	nr	**2096.91**
1500 x 312mm; 5.40-10.70kW Output	2249.10	2474.01	12.05	136.87	nr	**2610.88**
2000 x 312mm; 8.00-16.10kW Output	2714.25	2985.67	12.05	148.55	nr	**3134.22**
Electrically heated 415V Three phase supply; mounting height 2.60m						
1000 x 312mm; 5.80-10.10kW Output	2116.80	2328.48	16.13	198.87	nr	**2527.35**
1500 x 312mm; 7.10-14.60kW Output	2667.00	2933.70	16.13	183.23	nr	**3116.93**
2000 x 312mm; 11.50-20.20kW Output	3298.05	3627.85	16.13	183.23	nr	**3811.08**
Industrial Modular Curtain; recessed or exposed rigid sheet casing; high quality motor/centrifugal fan assembly; five speed control						
Electrically heated 415V Three phase supply; mounting height 3.50m						
1000 x 680mm; 34.00kW Output	2034.90	2238.39	17.24	195.86	nr	**2434.25**
1650 x 690mm; 51.00kW Output	2928.45	3221.30	17.24	195.86	nr	**3417.16**

Y:MECHANICAL AND ELECTRICAL SERVICES

Item	Net Price £	Material £	Labour hours	Labour £	Unit	Total Rate £
Y10 : PIPELINES : PLASTICS - VARIOUS						
UPVC Soil, Waste and Ventilating Pipe; solvent welded joints; BS 4514; fixed with PVC clips to backgrounds; BS 4514						
Pipe						
82mm dia.	3.15	3.47	0.31	3.52	m	**6.99**
110mm dia.	3.43	3.77	0.35	3.99	m	**7.76**
160mm dia.	7.88	8.67	0.45	5.11	m	**13.78**
Extra Over UPVC Pipe; solvent welded joints fittings; BS 4514						
92.5 degree d.s bend						
82mm	3.61	3.97	0.35	3.98	nr	**7.95**
110mm	4.27	4.69	0.38	4.32	nr	**9.01**
160mm	9.57	10.53	0.58	6.59	nr	**17.12**
135 degree d.s bend						
82mm	3.61	3.97	0.35	3.98	nr	**7.95**
110mm	4.27	4.69	0.38	4.32	nr	**9.01**
160mm	9.57	10.53	0.58	6.60	nr	**17.13**
92.5 degree t.s single branch						
82mm	5.04	5.55	0.41	4.66	nr	**10.21**
110mm	5.48	6.03	0.50	5.68	nr	**11.71**
160mm	18.16	19.97	0.58	6.59	nr	**26.56**
92.5 degree t.s single branch (access)						
110mm	11.80	12.99	0.50	5.68	nr	**18.67**
92.5 degree t.s single branch (4 boss)						
110mm	5.65	6.21	0.50	5.68	nr	**11.89**
135 degree t.s single branch						
110mm	5.60	6.16	0.50	5.68	nr	**11.84**
160mm	18.93	20.82	0.58	6.59	nr	**27.41**
92.5 degree unequal t.s single branch						
160mm	13.23	14.55	0.58	6.59	nr	**21.14**
135 degree unequal t.s single branch						
160mm	13.77	15.15	0.58	6.59	nr	**21.74**
S.s short access pipe						
110mm	5.62	6.19	0.25	2.84	nr	**9.03**

Y:MECHANICAL AND ELECTRICAL SERVICES

Item	Net Price £	Material £	Labour hours	Labour £	Unit	Total Rate £
D.s short access pipe						
110mm	5.62	6.19	0.25	2.84	nr	**9.03**
160mm	15.47	17.01	0.25	2.84	nr	**19.85**
110mm s.s boss pipe						
32mm	1.74	1.91	0.25	2.84	nr	**4.75**
40mm	1.74	1.91	0.25	2.84	nr	**4.75**
50mm	1.74	1.91	0.25	2.84	nr	**4.75**
110 d.s boss pipe						
32mm	1.74	1.91	0.25	2.84	nr	**4.75**
40mm	1.74	1.91	0.25	2.84	nr	**4.75**
50mm	1.74	1.91	0.25	2.84	nr	**4.75**
D.s pipe coupler						
82mm	3.10	3.41	0.25	2.84	nr	**6.25**
110mm	3.73	4.11	0.25	2.84	nr	**6.95**
160mm	6.91	7.60	0.25	2.84	nr	**10.44**
Reducer						
110mm	2.64	2.91	0.25	2.84	nr	**5.75**
Vent cowl						
82mm	0.95	1.04	0.10	1.14	nr	**2.18**
110mm	1.00	1.10	0.10	1.14	nr	**2.24**
160mm	2.42	2.67	0.10	1.14	nr	**3.81**
W.C. Connector						
110mm	1.75	1.92	0.25	2.84	nr	**4.76**
W.C. Connector 92.5 degree bend						
110mm	1.99	2.19	0.30	3.41	nr	**5.60**
Air Admittance valve						
82mm	13.99	15.39	0.17	1.89	nr	**17.28**
110mm	13.99	15.39	0.13	1.42	nr	**16.81**
Weathering slate						
82mm	13.77	15.15	1.00	11.36	nr	**26.51**
110mm	13.77	15.15	1.00	11.36	nr	**26.51**
UPVC Soil Waste and Ventilating Pipe; ring seal joints; fixed with PVC clips to backgrounds; BS 4514						
Pipe						
82mm dia.	3.15	3.47	0.31	3.52	m	**6.99**
110mm dia.	3.43	3.77	0.35	3.99	m	**7.76**
160mm dia.	7.88	8.67	0.45	5.11	m	**13.78**

Y:MECHANICAL AND ELECTRICAL SERVICES

Item	Net Price £	Material £	Labour hours	Labour £	Unit	Total Rate £
Y10 : PIPELINES : PLASTICS - VARIOUS (contd)						
Extra Over UPVC Soil and Ventilating Pipe; ring seal joint fittings; BS 4514						
S.s Coupling						
82mm	1.96	2.16	0.35	3.98	nr	**6.14**
110mm	2.11	2.32	0.35	3.98	nr	**6.30**
160mm	4.41	4.85	0.35	3.98	nr	**8.83**
D.s coupling						
82mm	3.10	3.41	0.35	3.98	nr	**7.39**
110mm	3.73	4.10	0.35	3.98	nr	**8.08**
160mm	6.91	7.59	0.35	3.98	nr	**11.57**
S.s 92.5 degree bend						
82mm	4.22	4.64	0.35	3.98	nr	**8.62**
110mm	5.04	5.54	0.35	3.98	nr	**9.53**
160mm	10.21.	11.22	0.35	3.98	nr	**15.20**
S.s 135 degree bend						
82mm	4.22	4.64	0.35	3.98	nr	**8.62**
110mm	5.04	5.58	0.35	3.98	nr	**9.53**
160mm	10.21	11.22	0.35	3.98	nr	**15.20**
D.s 92.5 degree bend						
110mm	5.60	6.16	0.35	3.98	nr	**10.14**
D.s 135 degree bend						
110mm	12.39	13.62	0.35	3.98	nr	**17.60**
High Temperature PVC Waste Pipe; solvent weldable joints; fixed with PVC clips to backgrounds						
Pipe						
32mm dia.	0.81	0.89	0.20	2.27	m	**3.16**
40mm dia.	0.99	1.09	0.20	2.27	m	**3.36**
50mm dia.	1.21	1.33	0.23	2.61	m	**3.94**
Extra Over High Temperature PVC Waste Pipe; solvent weldable fittings						
One Ended Fittings						
Access Plug						
32mm dia.	0.45	0.50	0.20	2.27	nr	**2.77**
40mm dia.	0.45	0.50	0.20	2.27	nr	**2.77**
50mm dia.	0.68	0.75	0.30	3.41	nr	**4.16**

Y:MECHANICAL AND ELECTRICAL SERVICES

Item	Net Price £	Material £	Labour hours	Labour £	Unit	Total Rate £
Socket plug						
32mm dia.	0.45	0.50	0.20	2.27	nr	**2.77**
40mm dia.	0.45	0.50	0.20	2.27	nr	**2.77**
40mm dia.	0.68	0.75	0.30	3.41	nr	**4.16**
Two Ended Fittings						
Straight coupling						
32mm dia.	0.40	0.44	0.25	2.84	nr	**3.28**
40mm dia.	0.40	0.44	0.25	2.84	nr	**3.28**
50mm dia.	0.68	0.75	0.25	2.84	nr	**3.59**
Expansion coupling						
32mm dia.	0.45	0.50	0.25	2.84	nr	**3.34**
40mm dia.	0.45	0.50	0.25	2.84	nr	**3.34**
50mm dia.	0.68	0.75	0.25	2.84	nr	**3.59**
Threaded coupling						
32mm dia.	0.45	0.50	0.25	2.84	nr	**3.34**
40mm dia.	0.45	0.50	0.25	2.84	nr	**3.34**
Swept bend; 90 Degree						
32mm dia.	0.45	0.50	0.25	2.84	nr	**3.34**
40mm dia.	0.45	0.50	0.25	2.84	nr	**3.34**
50mm dia.	0.45	0.50	0.30	3.41	nr	**3.91**
Knuckle Bend; 90 Degree						
32mm dia.	0.45	0.50	0.25	2.84	nr	**3.34**
40mm dia.	0.45	0.50	0.25	2.84	nr	**3.34**
50mm dia.	0.68	0.75	0.30	3.41	nr	**4.16**
Obtuse bend; 45 Degree						
40mm dia.	0.45	0.50	0.25	2.84	nr	**3.34**
50mm dia.	0.68	0.75	0.30	3.41	nr	**4.16**
Spigot; 45 degree						
32mm dia.	0.45	0.50	0.25	2.84	nr	**3.34**
40mm dia.	0.45	0.50	0.25	2.84	nr	**3.34**
50mm dia.	0.68	0.75	0.30	3.41	nr	**4.16**
Reducer						
40mm dia.	0.42	0.46	0.25	2.84	nr	**3.30**
50mm dia.	0.67	0.74	0.25	2.84	nr	**3.58**
Branch; 45 degree						
32mm dia.	0.45	0.50	0.25	2.84	nr	**3.34**
40mm dia.	0.45	0.50	0.25	2.84	nr	**3.34**
50mm dia.	0.68	0.75	0.30	3.41	nr	**4.16**

Y:MECHANICAL AND ELECTRICAL SERVICES

Item	Net Price £	Material £	Labour hours	Labour £	Unit	Total Rate £
Y10 : PIPELINES : PLASTICS - VARIOUS (contd)						
Extra Over High Temperature PVC Waste Pipe; solvent weldable fittings (contd)						
Two Ended Fittings (contd)						
Adaptor						
32mm dia.	0.45	0.50	0.25	2.84	nr	**3.34**
40mm dia.	0.45	0.50	0.25	2.84	nr	**3.34**
50mm dia.	0.68	0.75	0.30	3.41	nr	**4.16**
Tank connector						
32mm dia.	0.45	0.50	0.33	3.75	nr	**4.25**
40mm dia.	0.45	0.50	0.33	3.75	nr	**4.25**
Three Ended Fittings						
Swept tee; 92.5 Degree						
32mm dia.	0.45	0.50	0.30	3.41	nr	**3.91**
40mm dia.	0.45	0.50	0.30	3.41	nr	**3.91**
50mm dia.	0.68	0.75	0.30	3.41	nr	**4.16**
Four Ended Fittings						
Cross tee						
40mm dia.	0.45	0.50	0.50	5.68	nr	**6.18**
50mm dia.	0.68	0.75	0.50	5.68	nr	**6.43**
High Temperature PVC Waste Pipe; push fit joints; fixed with PVC clips to backgrounds						
Pipe						
32mm dia.	0.42	0.47	0.18	1.99	m	**2.46**
40mm dia.	0.52	0.58	0.18	1.99	m	**2.57**
50mm dia.	0.87	0.96	0.25	2.84	m	**3.80**
Extra Over High Temperature PVC waste Pipe; push fit fittings						
One Ended Fittings						
Access Plug						
32mm dia.	0.40	0.44	0.18	1.99	nr	**2.43**
40mm dia.	0.40	0.44	0.18	1.99	nr	**2.43**
Two Ended Fittings						
Straight coupler						
32mm dia.	0.40	0.44	0.13	1.48	nr	**1.92**
40mm dia.	0.40	0.44	0.13	1.48	nr	**1.92**
50mm dia.	0.73	0.80	0.15	1.70	nr	**2.50**

Y:MECHANICAL AND ELECTRICAL SERVICES

Item	Net Price £	Material £	Labour hours	Labour £	Unit	Total Rate £
Universial coupler						
32mm dia.	0.46	0.51	0.13	1.48	nr	**1.99**
40mm dia.	0.48	0.53	0.13	1.48	nr	**2.01**
50mm dia.	0.46	0.51	0.15	1.70	nr	**2.21**
Bend; 45 degree						
32mm dia.	0.40	0.44	0.13	1.48	nr	**1.92**
40mm dia.	0.40	0.44	0.13	1.48	nr	**1.92**
50mm dia.	0.73	0.80	0.15	1.70	nr	**2.50**
Bend; 90 degree						
32mm dia.	0.40	0.44	0.13	1.48	nr	**1.92**
40mm dia.	0.40	0.44	0.13	1.48	nr	**1.92**
50mm dia.	0.73	0.80	0.15	1.70	nr	**2.50**
Swept bend; 92.5 Degree						
32mm dia.	0.40	0.44	0.13	1.48	nr	**1.92**
40mm dia.	0.40	0.44	0.13	1.48	nr	**1.92**
50mm dia.	0.40	0.44	0.15	1.70	nr	**2.14**
Tank connector						
32mm dia.	0.40	0.44	0.13	1.48	nr	**1.92**
40mm dia.	0.40	0.44	0.13	1.48	nr	**1.92**
Three Ended Fitting						
Equal tee						
32mm dia.	0.40	0.44	0.16	1.82	nr	**2.26**
40mm dia.	0.40	0.44	0.16	1.82	nr	**2.26**
50mm dia.	0.73	0.80	0.18	2.04	nr	**2.84**
Ancillaries						
Vent cowl						
32mm dia.	0.73	0.80	0.10	1.14	nr	**1.94**
Polypropylene Waste Pipe and Fittings; push fit joints; fixed with PVC clips to backgrounds; BS 5255						
Pipe (3m length)						
22mm dia.	0.24	0.27	0.20	2.27	m	**2.54**

Y:MECHANICAL AND ELECTRICAL SERVICES

Item	Net Price £	Material £	Labour hours	Labour £	Unit	Total Rate £
Y10 : PIPELINES : PLASTICS - VARIOUS (contd)						
Extra Over Polypropylene Waste Pipe and Fittings; push fit joints; fittings; BS 5255						
Straight connector	0.29	0.32	0.20	2.27	nr	**2.59**
Bend; 45 degree	0.29	0.32	0.20	2.27	nr	**2.59**
Knuckle bend; 90 degree	0.29	0.32	0.20	2.27	nr	**2.59**
Tee; 90 degree	0.29	0.32	0.20	2.27	nr	**2.59**
Straight Adaptor	0.30	0.33	0.20	2.27	nr	**2.60**
Bent adaptor	0.30	0.33	0.20	2.27	nr	**2.60**
Straight tank connector	0.29	0.32	0.20	2.27	nr	**2.59**
Bent tank connector	0.29	0.32	0.20	2.27	nr	**2.59**
UPVC Ventilating and Overflow Pipe and Fittings; solvent welded joints; fixed with PVC clips to backgrounds; BS 4514						
Pipe (4m lengths)						
22mm dia.	0.24	0.27	0.20	2.27	m	**2.54**
Extra Over for;						
Straight connector	0.28	0.31	0.20	2.27	nr	**2.58**
Bend; 45 degree	0.28	0.31	0.20	2.27	nr	**2.58**
Knuckle bend; 90 degree	0.28	0.31	0.20	2.27	nr	**2.58**
Tee; 90 degree	0.28	0.31	0.20	2.27	nr	**2.58**
Straight adaptor	0.28	0.31	0.20	2.27	nr	**2.58**
Bent adaptor	0.28	0.31	0.20	2.27	nr	**2.58**
Straight tank connector	0.29	0.32	0.20	2.27	nr	**2.59**
Bent tank coonector	0.29	0.32	0.20	2.27	nr	**2.59**
Polypropylene Traps; fixed and connected to pipework; BS 3943						
Tubular trap 38mm seal with cleaning eye						
32mm dia.	1.13	1.24	0.20	2.27	nr	**3.51**
40mm dia.	1.25	1.37	0.20	2.27	nr	**3.64**
Tubular trap 75mm seal with cleaning eye						
32mm dia.	1.18	1.29	0.20	2.27	nr	**3.56**
40mm dia.	1.30	1.43	0.20	2.27	nr	**3.70**
Bottle trap 75mm seal						
32mm dia.	0.98	1.08	0.20	2.27	nr	**3.35**
40mm dia.	1.13	1.24	0.20	2.27	nr	**3.51**
Bottle trap 75mm seal						
32mm dia.	1.08	1.19	0.20	2.27	nr	**3.46**
40mm dia.	1.25	1.37	0.25	2.84	nr	**4.21**

Y:MECHANICAL AND ELECTRICAL SERVICES

Item	Net Price £	Material £	Labour hours	Labour £	Unit	Total Rate £
Resealing bottle trap; 75mm seal						
32mm dia.	1.46	1.61	0.20	2.27	nr	**3.88**
40mm dia.	1.70	1.87	0.25	2.84	nr	**4.71**
Bath trap 20mm seal with cleaning eye						
32mm dia.	0.98	1.08	0.20	2.27	nr	**3.35**
40mm dia.	1.01	1.11	0.25	2.84	nr	**3.95**
Low level bath trap 38mm seal						
40mm dia.	1.39	1.53	0.25	2.84	nr	**4.37**
Low level bath trap 38mm seal c/w overflow						
40mm dia.	2.93	3.22	0.25	2.84	nr	**6.06**
Low level bath trap 75mm seal						
40mm dia.	1.66	1.82	0.25	2.84	nr	**4.66**
Low level bath trap 75mm seal c/w overflow						
40mm dia.	3.22	3.54	0.20	2.27	nr	**5.81**
Tubular swivel trap 'S' 38mm seal						
32mm dia.	1.32	1.45	0.20	2.27	nr	**3.72**
40mm dia.	1.39	1.53	0.25	2.84	nr	**4.37**
Tubular swivel trap 'S' 75mm seal						
32mm dia.	1.32	1.45	0.20	2.27	nr	**3.72**
40mm dia.	1.46	1.61	0.25	2.84	nr	**4.45**
Washing Machine trap 75mm seal						
40mm dia.	2.09	2.30	0.25	2.84	nr	**5.14**
Washing Machine half trap 75mm seal						
40mm dia.	2.16	2.38	0.25	2.84	nr	**5.22**
MDPE (Medium Density Polyethylene) Pipes for water distribution; laid underground; electrofusion joints in the running length; BS 6572						
MDPE Pipe						
20mm dia.	0.38	0.44	0.37	4.20	m	**4.64**
25mm dia.	0.49	0.55	0.41	4.66	m	**5.21**
32mm dia.	0.81	0.92	0.47	5.34	m	**6.26**
50mm dia.	1.96	2.22	0.53	6.02	m	**8.24**
63mm dia.	3.02	3.43	0.60	6.82	m	**10.25**
90mm dia.	5.56	6.30	0.90	10.23	m	**16.53**
125mm dia.	10.70	12.13	1.20	13.64	m	**25.77**
180mm dia.	22.12	25.06	1.50	17.06	m	**42.12**
250mm dia.	42.60	48.27	1.75	19.89	m	**68.16**

Y:MECHANICAL AND ELECTRICAL SERVICES

Item	Net Price £	Material £	Labour hours	Labour £	Unit	Total Rate £
Y10 : PIPELINES : PLASTICS - VARIOUS (contd)						
MDPE (Medium Density Polyethylene) Pipes gas distribution; laid underground; electrofusion joints in the running length; BS 6572						
MDPE Pipe						
20mm dia.	0.35	0.40	0.37	4.20	m	**4.60**
25mm dia.	0.45	0.51	0.41	4.66	m	**5.17**
32mm dia.	0.75	0.84	0.47	5.34	m	**6.18**
63mm dia.	2.74	3.10	0.53	6.02	m	**9.12**
90mm dia.	7.72	8.75	0.90	10.23	m	**18.98**
125mm dia.	12.83	14.54	1.20	13.64	m	**28.18**
Extra Over Medium Density Polyethylene Pipes DZR compression fittings; BS 864						
Two Ended Fittings						
Straight Connector						
20mm dia.	1.84	2.02	0.38	4.32	nr	**6.34**
25mm dia.	2.05	2.26	0.45	5.11	nr	**7.37**
32mm dia.	4.71	5.18	0.50	5.68	nr	**10.86**
50mm dia.	11.27	12.39	0.68	7.73	nr	**20.12**
63mm dia.	16.35	17.99	0.85	9.66	nr	**27.65**
Straight Connector; Polythene to MI (BSP)						
20mm dia.	1.74	1.91	0.31	3.52	nr	**5.43**
25mm dia.	2.22	2.45	0.35	3.98	nr	**6.43**
32mm dia.	3.36	3.69	0.40	4.54	nr	**8.23**
50mm dia.	8.13	8.94	0.55	6.25	nr	**15.19**
63mm dia.	11.13	12.25	0.65	7.39	nr	**19.64**
Straight Connector; Polythene to FI (BSP)						
20mm dia.	2.38	2.61	0.31	3.52	nr	**6.13**
25mm dia.	2.56	2.82	0.35	3.98	nr	**6.80**
32mm dia.	3.16	3.47	0.40	4.54	nr	**8.01**
50mm dia.	9.68	10.65	0.55	6.25	nr	**16.90**
63mm dia.	13.27	14.59	0.75	8.52	nr	**23.11**
Elbow Polythene to Polythene						
20mm dia.	2.51	2.76	0.38	4.32	nr	**7.08**
25mm dia.	3.78	4.16	0.45	5.11	nr	**9.27**
32mm dia.	5.65	6.22	0.50	5.68	nr	**11.90**
50mm dia.	12.63	13.89	0.68	7.73	nr	**21.62**
63mm dia.	16.55	18.21	0.80	9.09	nr	**27.30**
Male Elbow Polythene to MI (BSP)						
25mm dia.	3.23	3.55	0.35	3.98	nr	**7.53**

Y:MECHANICAL AND ELECTRICAL SERVICES

Item	Net Price £	Material £	Labour hours	Labour £	Unit	Total Rate £
Female Elbow Polythene to FI (BSP)						
20mm dia.	2.35	2.58	0.31	3.52	nr	**6.10**
25mm dia.	3.23	3.55	0.35	3.98	nr	**7.53**
32mm dia.	4.90	5.40	0.42	4.77	nr	**10.17**
50mm dia.	11.22	12.35	0.50	5.68	nr	**18.03**
63mm dia.	14.76	16.24	0.55	6.25	nr	**22.49**
Reducer						
20mm dia.	3.96	4.36	0.35	3.98	nr	**8.34**
25mm dia.	3.85	4.24	0.38	4.32	nr	**8.56**
32mm dia.	11.38	12.51	0.45	5.11	nr	**17.62**
50mm dia.	17.53	19.29	0.50	5.68	nr	**24.97**
63mm dia.	23.30	25.63	0.62	7.05	nr	**32.68**
Tank Coupling						
25mm dia.	4.18	4.60	0.42	4.77	nr	**9.37**
Three Ended Fittngs						
Equal Tee						
20mm dia.	3.36	3.69	0.53	6.02	nr	**9.71**
25mm dia.	5.38	5.92	0.55	6.25	nr	**12.17**
32mm dia.	6.96	7.65	0.64	7.27	nr	**14.92**
50mm dia.	15.44	16.99	0.75	8.52	nr	**25.51**
63mm dia.	23.05	25.35	0.87	9.89	nr	**35.24**
Tee with FI Branch						
20mm dia.	3.23	3.55	0.45	5.11	nr	**8.66**
25mm dia.	5.25	5.77	0.50	5.68	nr	**11.45**
32mm dia.	6.74	7.42	0.60	6.82	nr	**14.24**
50mm dia.	14.76	16.24	0.68	7.73	nr	**23.97**
63mm dia.	20.09	22.10	0.81	9.21	nr	**31.31**
Extra Over Medium Density Polyethylene Pipes plastic compression fittings; BS 6572						
One Ended Fittings						
End connector polythene to MI BSP						
20mm dia.	1.74	1.91	0.07	0.80	nr	**2.71**
25mm dia.	2.22	2.45	0.10	1.14	nr	**3.59**
32mm dia.	3.36	3.69	0.15	1.70	nr	**5.39**
50mm dia.	8.13	8.94	0.18	2.04	nr	**10.98**
63mm dia.	11.13	12.25	0.25	2.84	nr	**15.09**
End connector polythene to FI BSP						
20mm dia.	2.38	2.61	0.10	1.14	nr	**3.75**
25mm dia.	2.56	2.82	0.12	1.36	nr	**4.18**
32mm dia.	3.16	3.47	0.15	1.70	nr	**5.17**
50mm dia.	9.68	10.65	0.18	2.04	nr	**12.69**
63mm dia.	13.27	14.59	0.25	2.84	nr	**17.43**

Y:MECHANICAL AND ELECTRICAL SERVICES

Item	Net Price £	Material £	Labour hours	Labour £	Unit	Total Rate £
Y10 : PIPELINES : PLASTICS - VARIOUS (contd)						
Extra Over Medium Density Polyethylene Pipes plastic compression fittings; BS 6572 (contd)						
Two Ended Fittings						
Straight connector						
20mm dia.	1.84	2.02	0.10	1.14	nr	3.16
25mm dia.	2.05	2.26	0.12	1.36	nr	3.62
32mm dia.	4.71	5.18	0.15	1.70	nr	6.88
50mm dia.	11.27	12.39	0.18	2.04	nr	14.43
63mm dia.	16.35	17.99	0.25	2.84	nr	20.83
Reducer						
25mm dia.	3.85	4.24	0.19	2.16	nr	6.40
32mm dia.	11.38	12.51	0.19	2.16	nr	14.67
50mm dia.	17.53	19.29	0.21	2.39	nr	21.68
63mm dia.	23.30	25.63	0.25	2.84	nr	28.47
Elbow						
20mm dia.	2.51	2.76	0.07	0.80	nr	3.56
25mm dia.	3.78	4.16	0.08	0.91	nr	5.07
32mm dia.	5.65	6.22	0.20	2.27	nr	8.49
50mm dia.	12.63	13.89	0.25	2.84	nr	16.73
63mm dia.	16.55	18.21	0.37	4.20	nr	22.41
Elbow x FI BSF						
20mm dia.	2.35	2.58	0.07	0.80	nr	3.38
25mm dia.	3.23	3.55	0.08	0.91	nr	4.46
32mm dia.	4.90	5.40	0.20	2.27	nr	7.67
50mm dia.	11.22	12.35	0.25	2.84	nr	15.19
63mm dia.	14.76	16.24	0.37	4.20	nr	20.44
Elbow x MI BSF						
25mm dia.	3.23	3.55	0.08	0.91	nr	4.46
Tank connector assembly c/w seal washers						
25mm dia.	4.18	4.60	0.20	2.27	nr	6.87
Three Ended Fittings						
Equal tee						
20mm dia.	3.36	3.69	0.14	1.59	nr	5.28
25mm dia.	5.38	5.92	0.17	1.93	nr	7.85
32mm dia.	6.96	7.65	0.31	3.52	nr	11.17
50mm dia.	15.44	16.99	0.39	4.43	nr	21.42
63mm dia.	23.05	25.35	0.56	6.36	nr	31.71

Y:MECHANICAL AND ELECTRICAL SERVICES

Item	Net Price £	Material £	Labour hours	Labour £	Unit	Total Rate £
Tee with FI BSP						
20mm dia.	3.23	3.55	0.14	1.59	nr	**5.14**
25mm dia.	5.25	5.77	0.17	1.93	nr	**7.70**
32mm dia.	6.74	7.42	0.31	3.52	nr	**10.94**
50mm dia.	14.76	16.24	0.39	4.43	nr	**20.67**
63mm dia.	20.09	22.10	0.56	6.36	nr	**28.46**
Tee with MI BSP						
25mm dia.	5.14	5.65	0.14	1.59	nr	**7.24**
Unplasticised PVC Pipes; compression joints in the running length; BS 3505						
Class C						
50mm dia.	2.66	3.01	0.39	4.43	m	**7.44**
80mm dia.	4.83	5.47	0.45	5.11	m	**10.58**
100mm dia.	7.15	8.10	0.52	5.91	m	**14.01**
150mm dia.	15.43	17.48	0.75	8.52	m	**26.00**
200mm dia.	24.09	27.29	0.88	10.00	m	**37.29**
Class D						
32mm dia.	1.87	2.12	0.41	4.66	m	**6.78**
40mm dia.	2.27	2.57	0.42	4.77	m	**7.34**
50mm dia.	3.27	3.70	0.45	5.11	m	**8.81**
80mm dia.	6.19	7.01	0.47	5.34	m	**12.35**
100mm dia.	9.26	10.49	0.55	6.25	m	**16.74**
150mm dia.	20.26	22.95	0.79	8.98	m	**31.93**
Class E						
25mm dia.	1.89	2.15	0.34	3.86	m	**6.01**
50mm dia.	4.69	5.32	0.43	4.89	m	**10.21**
80mm dia.	9.47	10.73	0.49	5.57	m	**16.30**
100mm dia.	15.22	17.25	0.57	6.48	m	**23.73**
150mm dia.	25.49	28.88	0.70	7.96	m	**36.84**
ABS (Acrylonitrile Butadiene Styrene) Pipes; solvent welded joints in the running length						
Class C (9 Bar)						
25mm dia.	1.24	1.40	0.30	3.41	m	**4.81**
32mm dia.	2.04	2.31	0.33	3.75	m	**6.06**
40mm dia.	2.62	2.97	0.36	4.09	m	**7.06**
50mm dia.	3.57	4.05	0.39	4.43	m	**8.48**
80mm dia.	7.57	8.58	0.45	5.11	m	**13.69**
100mm dia.	12.02	13.62	0.52	5.91	m	**19.53**
150mm dia.	23.12	26.19	0.75	8.52	m	**34.71**
200mm dia.	41.50	47.02	0.95	10.80	m	**57.82**

Y:MECHANICAL AND ELECTRICAL SERVICES

Item	Net Price £	Material £	Labour hours	Labour £	Unit	Total Rate £
Y10 : PIPELINES : PLASTICS - VARIOUS (contd)						
ABS (Acrylonitrile Butadiene Styrene) Pipes; solvent welded joints in the running length (contd)						
Class D (12 bar)						
32mm dia.	2.06	2.34	0.33	3.75	m	**6.09**
40mm dia.	2.45	2.77	0.36	4.09	m	**6.86**
50mm dia.	3.51	3.98	0.39	4.43	m	**8.41**
80mm dia.	6.43	7.29	0.45	5.11	m	**12.40**
100mm dia.	9.82	11.13	0.52	5.91	m	**17.04**
150mm dia.	21.04	23.84	0.75	8.52	m	**32.36**
Class E (15 bar)						
25mm dia.	1.56	1.76	0.30	3.41	m	**5.17**
32mm dia.	2.28	2.59	0.33	3.75	m	**6.34**
40mm dia.	2.88	3.26	0.36	4.09	m	**7.35**
50mm dia.	4.28	4.85	0.39	4.43	m	**9.28**
80mm dia.	8.63	9.77	0.49	5.57	m	**15.34**
100mm dia.	13.86	15.70	0.57	6.48	m	**22.18**
Extra Over ABS Pipes; solvent welded fittings						
One Ended Fittings						
Cap						
32mm dia.	1.04	1.14	0.26	2.95	nr	**4.09**
40mm dia.	1.68	1.85	0.29	3.29	nr	**5.14**
50mm dia.	1.04	1.14	0.32	3.64	nr	**4.78**
80mm dia.	1.04	1.14	0.37	4.20	nr	**5.34**
100mm dia.	1.04	1.14	0.46	5.23	nr	**6.37**
Two Ended Fittings						
Elbow 90 Degree						
32mm dia.	2.04	2.31	0.47	5.34	nr	**7.65**
40mm dia.	2.62	2.97	0.53	6.02	nr	**8.99**
50mm dia.	3.57	4.05	0.58	6.59	nr	**10.64**
80mm dia.	7.57	8.58	0.67	7.61	nr	**16.19**
100mm dia.	12.02	13.62	0.83	9.44	nr	**23.06**
150mm dia.	23.12	26.19	1.25	14.20	nr	**40.39**
200mm dia.	41.50	47.02	1.50	17.06	nr	**64.08**
Elbow 45 Degree						
32mm dia.	2.40	2.64	0.47	5.34	nr	**7.98**
40mm dia.	3.00	3.30	0.53	6.02	nr	**9.32**
50mm dia.	4.16	4.58	0.58	6.59	nr	**11.17**
80mm dia.	9.76	10.74	0.67	7.61	nr	**18.35**
100mm dia.	20.24	22.26	0.83	9.44	nr	**31.70**
150mm dia.	42.00	46.20	1.25	14.20	nr	**60.40**
200mm dia.	94.72	104.19	1.50	17.06	nr	**121.25**

Y:MECHANICAL AND ELECTRICAL SERVICES

Item	Net Price £	Material £	Labour hours	Labour £	Unit	Total Rate £
Reducing Bush						
40 x 32mm dia.	1.12	1.23	0.53	6.02	nr	**7.25**
50 x 32mm dia.	1.48	1.63	0.58	6.59	nr	**8.22**
50 x 40mm dia.	1.48	1.63	0.58	6.59	nr	**8.22**
80 x 40mm dia.	4.16	4.58	0.67	7.61	nr	**12.19**
80 x 50mm dia.	4.16	4.58	0.67	7.61	nr	**12.19**
100 x 80mm dia.	5.76	6.34	0.83	9.44	nr	**15.78**
150 x 100mm dia.	17.60	19.36	1.25	14.20	nr	**33.56**
Union						
32mm dia.	4.76	5.24	0.52	5.91	nr	**11.15**
40mm dia.	6.56	7.22	0.59	6.71	nr	**13.93**
50mm dia.	8.64	9.50	0.64	7.27	nr	**16.77**
Sockets						
32mm dia.	1.12	1.23	0.52	5.91	nr	**7.14**
40mm dia.	1.36	1.50	0.59	6.71	nr	**8.21**
50mm dia.	1.92	2.11	0.64	7.27	nr	**9.38**
80mm dia.	7.60	8.36	0.73	8.30	nr	**16.66**
100mm dia.	10.88	11.97	0.73	8.30	nr	**20.27**
150mm dia.	27.20	29.92	1.30	14.77	nr	**44.69**
200mm dia.	56.88	62.57	1.60	18.18	nr	**80.75**
Barrel Nipple						
32mm dia.	2.08	2.29	0.52	5.91	nr	**8.20**
40mm dia.	2.24	2.46	0.59	6.71	nr	**9.17**
50mm dia.	2.88	3.17	0.64	7.27	nr	**10.44**
80mm dia.	7.64	8.40	0.73	8.30	nr	**16.70**
Extra Over ABS Pipes; solvent welded flanges						
Full Face Flange						
32mm dia.	3.96	4.36	0.19	2.16	nr	**6.52**
40mm dia.	4.08	4.49	0.22	2.50	nr	**6.99**
50mm dia.	5.32	5.85	0.30	3.41	nr	**9.26**
80mm dia.	10.16	11.18	0.38	4.32	nr	**15.50**
100mm dia.	13.36	14.70	0.42	4.77	nr	**19.47**
High Impact PVC Pipes; solvent welded joints in the running length						
Class C (130 psi)						
2" dia.	5.31	6.14	0.42	4.77	m	**10.91**
3" dia.	10.98	12.68	0.49	5.57	m	**18.25**
4" dia.	18.07	20.87	0.52	5.91	m	**26.78**
6" dia.	39.10	45.16	1.82	20.65	m	**65.81**

Y:MECHANICAL AND ELECTRICAL SERVICES

Item	Net Price £	Material £	Labour hours	Labour £	Unit	Total Rate £
Y10 : PIPELINES : PLASTICS - VARIOUS (contd)						
High Impact PVC Pipes; solvent welded joints in the running length (contd)						
Class D (173 psi)						
1 1/4" dia.	3.10	3.58	0.42	4.77	m	8.35
1 1/2" dia.	4.25	4.91	0.44	5.00	m	9.91
2" dia.	6.60	7.62	0.47	5.34	m	12.96
3" dia.	14.13	16.31	0.50	5.68	m	21.99
4" dia.	23.65	27.31	0.55	6.25	m	33.56
6" dia.	51.76	59.79	0.60	6.82	m	66.61
Class E (217 psi)						
1/5" dia.	1.51	1.74	0.39	4.43	m	6.17
3/4" dia.	2.17	2.51	0.41	4.66	m	7.17
1" dia.	2.52	2.92	0.42	4.77	m	7.69
1 1/5" dia.	4.83	5.57	0.44	5.00	m	10.57
2" dia.	7.53	8.69	0.47	5.34	m	14.03
3" dia.	16.30	18.82	0.49	5.57	m	24.39
4" dia.	26.79	30.94	0.52	5.91	m	36.85
6" dia.	57.87	66.84	0.55	6.25	m	73.09
Class 7						
1/5" dia.	2.66	3.07	0.33	3.75	m	6.82
3/4" dia.	3.81	4.40	0.34	3.86	m	8.26
1" dia.	5.71	6.60	0.41	4.66	m	11.26
1 1/4" dia.	7.88	9.10	0.41	4.66	m	13.76
1 1/5" dia.	9.74	11.25	0.42	4.77	m	16.02
2" dia.	16.21	18.72	0.45	5.11	m	23.83
Extra Over PVC Pipes; solvent welded pressure fittings						
One Ended Fittings						
End Cap						
1/5" dia.	0.50	0.55	0.18	2.04	nr	2.59
3/4" dia.	0.59	0.65	0.20	2.27	nr	2.92
1" dia.	0.66	0.73	0.23	2.61	nr	3.34
1 1/4" dia.	1.07	1.17	0.26	2.95	nr	4.12
1 1/5" dia.	1.77	1.95	0.29	3.29	nr	5.24
2" dia.	2.17	2.39	0.32	3.64	nr	6.03
3" dia.	6.54	7.20	0.37	4.20	nr	11.40
4" dia.	10.07	11.08	0.46	5.23	nr	16.31
6" dia.	24.54	26.99	0.69	7.84	nr	34.83

Y:MECHANICAL AND ELECTRICAL SERVICES

Item	Net Price £	Material £	Labour hours	Labour £	Unit	Total Rate £
Two Ended Fittings						
Socket						
1/5" dia.	0.53	0.59	0.32	3.64	nr	**4.23**
3/4" dia.	0.59	0.65	0.36	4.09	nr	**4.74**
1" dia.	0.69	0.76	0.43	4.89	nr	**5.65**
1 1/4" dia.	1.25	1.37	0.47	5.34	nr	**6.71**
1 1/5" dia.	1.46	1.61	0.53	6.02	nr	**7.63**
2" dia.	2.07	2.27	0.58	6.59	nr	**8.86**
3" dia.	7.92	8.71	0.67	7.61	nr	**16.32**
4" dia.	11.45	12.60	0.83	9.44	nr	**22.04**
6" dia.	28.84	31.73	1.25	14.20	nr	**45.93**
Reducing Socket						
3/4" dia.	0.77	0.85	0.36	4.09	nr	**4.94**
1" dia.	1.51	1.66	0.43	4.89	nr	**6.55**
1 1/4" dia.	1.68	1.85	0.47	5.34	nr	**7.19**
1 1/2" dia.	2.54	2.79	0.53	6.02	nr	**8.81**
2" dia.	7.75	8.52	0.58	6.59	nr	**15.11**
3" dia.	11.37	12.50	0.67	7.61	nr	**20.11**
4" dia.	41.59	45.75	0.83	9.44	nr	**55.19**
6" dia.	67.51	74.26	1.25	14.20	nr	**88.46**
Elbow 90 Degree						
1/2" dia.	0.73	0.81	0.32	3.64	nr	**4.45**
3/4" dia.	0.86	0.95	0.36	4.09	nr	**5.04**
1" dia.	1.16	1.28	0.43	4.89	nr	**6.17**
1 1/4" dia.	2.07	2.27	0.47	5.34	nr	**7.61**
1 1/2" dia.	2.67	2.94	0.47	5.34	nr	**8.28**
2" dia.	3.96	4.36	0.58	6.59	nr	**10.95**
3" dia.	11.37	12.50	0.67	7.61	nr	**20.11**
4" dia.	17.22	18.94	0.83	9.44	nr	**28.38**
6" dia.	68.11	74.92	1.25	14.20	nr	**89.12**
Elbow 45 Degree						
1/2" dia.	1.33	1.47	0.32	3.64	nr	**5.11**
3/4" dia.	1.42	1.56	0.36	4.09	nr	**5.65**
1" dia.	1.77	1.94	0.47	5.34	nr	**7.28**
1 1/4" dia.	2.50	2.75	0.47	5.34	nr	**8.09**
1 1/2" dia.	3.14	3.46	0.53	6.02	nr	**9.48**
2" dia.	6.97	7.67	0.58	6.59	nr	**14.26**
3" dia.	21.44	23.58	0.67	7.61	nr	**31.19**
4" dia.	36.34	39.97	0.83	9.44	nr	**49.41**
6" dia.	44.26	48.68	1.25	14.20	nr	**62.88**
Bend 90 Degree (Long Radius)						
1" dia.	4.69	5.16	0.36	4.09	nr	**9.25**
1 1/2" dia.	7.79	8.57	0.53	6.02	nr	**14.59**
2" dia.	13.78	15.15	0.53	6.02	nr	**21.17**
3" dia.	31.86	35.04	0.67	7.61	nr	**42.65**
4" dia.	67.51	74.26	0.83	9.44	nr	**83.70**
6" dia.	148.96	163.86	1.25	14.20	nr	**178.06**

Y:MECHANICAL AND ELECTRICAL SERVICES

Item	Net Price £	Material £	Labour hours	Labour £	Unit	Total Rate £
Y10 : PIPELINES : PLASTICS - VARIOUS (contd)						
Extra Over PVC Pipes; solvent welded pressure fittings (contd)						
Two Ended Fittings (contd)						
Bend 45 Degree (Long Radius)						
1" dia.	4.69	5.16	0.47	5.34	nr	**10.50**
1 1/2" dia.	7.92	8.71	0.53	6.02	nr	**14.73**
2" dia.	13.09	14.40	0.58	6.59	nr	**20.99**
3" dia.	27.38	30.12	0.67	7.61	nr	**37.73**
4" dia.	53.90	59.29	0.83	9.44	nr	**68.73**
6" dia.	124.08	136.48	1.25	14.20	nr	**150.68**
Socket Union						
1/2" dia.	2.76	3.03	0.35	3.98	nr	**7.01**
3/4" dia.	3.14	3.46	0.40	4.54	nr	**8.00**
1" dia.	4.09	4.50	0.47	5.34	nr	**9.84**
1 1/4" dia.	5.04	5.54	0.52	5.91	nr	**11.45**
1 1/2" dia.	6.97	7.67	0.59	6.71	nr	**14.38**
2" dia.	9.04	9.95	0.64	7.27	nr	**17.22**
3" dia.	33.49	36.84	0.73	8.30	nr	**45.14**
4" dia.	45.29	49.82	0.92	10.45	nr	**60.27**
Saddle Plain						
2" x 1 1/4" dia.	7.31	8.44	0.44	5.00	nr	**13.44**
3" x 1 1/2" dia.	10.18	11.76	0.50	5.68	nr	**17.44**
6" x 2" dia.	13.46	15.55	0.94	10.69	nr	**26.24**
Straight Tank Connector						
1/2" dia.	1.90	2.20	0.13	1.48	nr	**3.68**
1" dia.	4.52	5.22	0.15	1.70	nr	**6.92**
90 Degree Bent Tank Connector						
1/2" dia.	10.18	11.76	0.13	1.48	nr	**13.24**
3/4" dia.	10.18	11.76	0.14	1.59	nr	**13.35**
Three Ended Fittings						
Equal Tee						
1/2" dia.	0.83	0.91	0.46	5.23	nr	**6.14**
3/4" dia.	1.08	1.18	0.50	5.68	nr	**6.86**
1" dia.	1.55	1.70	0.56	6.36	nr	**8.06**
1 1/4" dia.	2.20	2.42	0.73	8.30	nr	**10.72**
1 1/2" dia.	3.23	3.55	0.77	8.75	nr	**12.30**
2" dia.	5.04	5.54	0.83	9.44	nr	**14.98**
3" dia.	14.72	16.20	1.08	12.28	nr	**28.48**
4" dia.	21.61	23.77	1.33	15.15	nr	**38.92**
6" dia.	75.08	82.59	2.00	22.72	nr	**105.31**

Y:MECHANICAL AND ELECTRICAL SERVICES

Item	Net Price £	Material £	Labour hours	Labour £	Unit	Total Rate £
Y10 : PIPELINES : CAST AND DUCTILE IRON						
Cast Iron Soil Pipe; mechanical joints; nitrile rubber gasket; fixed with holderbats; BS 416						
Pipe						
50mm dia.	25.86	14.22	0.25	2.84	m	**17.06**
75mm dia.	25.86	14.22	0.45	5.11	m	**19.33**
100mm dia.	38.51	14.11	0.60	6.82	m	**20.93**
150mm dia.	72.54	26.57	0.70	7.96	m	**34.53**
Extra Over Cast Iron Pipe with Mechanical Joints; fittings; BS 416						
Two Ended Fittings						
Standard coupling						
50mm dia.	6.94	7.64	0.50	5.68	nr	**13.32**
75mm dia.	6.94	7.64	0.60	6.82	nr	**14.46**
100mm dia.	8.21	9.03	0.67	7.61	nr	**16.64**
150mm dia.	9.94	10.93	0.83	9.44	nr	**20.37**
Stepped; Coupling						
50mm dia.	7.26	7.98	0.50	5.68	nr	**13.66**
75mm dia.	7.26	7.98	0.60	6.82	nr	**14.80**
100mm dia.	8.52	9.37	0.67	7.61	nr	**16.98**
150mm dia.	10.37	11.41	0.83	9.44	nr	**20.85**
45 degree bend						
50mm dia.	11.75	12.92	0.50	5.68	nr	**18.60**
75mm dia.	11.75	12.92	0.60	6.82	nr	**19.74**
100mm dia.	13.05	14.35	0.67	7.61	nr	**21.96**
150mm dia.	26.48	29.13	0.83	9.44	nr	**38.57**
87.5 degree bend						
50mm dia.	12.46	13.71	0.50	5.68	nr	**19.39**
75mm dia.	12.46	13.71	0.60	6.82	nr	**20.53**
100mm dia.	15.35	16.89	0.67	7.61	nr	**24.50**
150mm dia.	30.98	34.08	0.83	9.44	nr	**43.52**
87.5 degree bend with heel rest						
100mm dia.	17.62	19.38	0.67	7.62	nr	**27.00**
150mm dia.	34.82	38.30	0.83	9.44	nr	**47.74**
87.5 degree bend with access						
100mm dia.	33.07	36.38	0.67	7.61	nr	**43.99**
150mm dia.	74.92	82.41	0.83	9.44	nr	**91.85**
87.5 degree long radius bend						
100mm dia.	18.83	20.71	0.67	7.61	nr	**28.32**
150mm dia.	40.33	44.37	0.83	9.44	nr	**53.81**

Y:MECHANICAL AND ELECTRICAL SERVICES

Item	Net Price £	Material £	Labour hours	Labour £	Unit	Total Rate £
Y10 : PIPELINES : CAST AND DUCTILE IRON (contd)						
Extra Over Cast Iron Pipe with Mechanical Joints; fittings; BS 416 (contd)						
Two Ended Fittings (contd)						
87.5 degree long tail bend						
100mm dia.	37.94	40.08	0.70	7.96	nr	**48.04**
Access pipe						
100mm dia.	37.94	41.73	0.67	7.62	nr	**49.35**
150mm dia.	69.28	76.20	0.83	9.44	nr	**85.64**
Reducer						
75mm dia.	7.46	8.21	0.60	6.82	nr	**15.03**
100mm dia.	13.05	14.35	0.67	7.61	nr	**21.96**
150mm dia.	20.02	22.02	0.83	9.44	nr	**31.46**
Three Ended Fittings						
Branch						
75mm dia.	17.32	19.06	0.78	8.86	nr	**27.92**
75mm dia.	17.32	19.06	0.85	9.66	nr	**28.72**
100mm dia.	18.16	19.97	1.00	11.36	nr	**31.33**
150mm dia.	44.09	48.50	1.20	13.64	nr	**62.14**
Access branch						
100mm dia.	46.99	51.69	1.02	11.59	nr	**63.28**
150mm dia.	86.89	95.58	1.20	13.64	nr	**109.22**
Four Ended Fittings						
Double branch						
100mm dia.	30.69	33.76	1.30	14.77	nr	**48.53**
Double access branch						
100mm dia.	56.69	62.36	1.43	16.25	nr	**78.61**
Ancillaries						
Gully trap						
75mm dia.	13.05	14.35	0.90	10.23	nr	**24.58**
100mm dia.	20.37	22.41	1.00	11.36	nr	**33.77**
150mm dia.	50.70	55.77	1.20	13.64	nr	**69.41**
Gully trap with Access; 100mm dia.	40.64	44.70	1.16	13.18	nr	**57.88**
Garage gully; 100mm dia.	214.18	235.60	1.50	17.06	nr	**252.66**

All four Spon Price Books – *Architects' and Builders', Civil Engineering and Highway Works, Landscape and External Works* and *Mechanical and Electrical Services* – are supported by an updating service. Three Updates are issued during the year, in November, February and May. Each gives details of changes in prices of materials, wage rates and other significant items, with regional price level adjustments for Northern Ireland, Scotland and Wales and regions of England.

As a purchaser of a Spon Price Book you are entitled to this updating service for the 1996 edition – *free of charge.* Simply complete this registration card and return it to us.

The updating service terminates with the publication of the next annual edition.

REGISTRATION CARD for Spon's Price Book Update   1996
Please print your details clearly

Name ....
(Please indicate membership of any professional body e.g. RICS, RIBA, CIoB etc.)

Position/Dept ....

Organisation ....

Address ....

....

.... Postcode ....

Tel .... Fax ....

ELECTRONIC DELIVERY OF PRICE DATA

Price data from Spon's Price Books will be available as cost libraries in reSPONse, a new Windows™ based estimating system.

If you would like to receive a free information pack about this software, please tick the box below and indicate the address to which it should be sent.

☐ Please send me your free information pack on Spon's new estimating software.

☐ Send the information to the address above.

☐ Send the information to the following address.

Name ....

Position/Dept ....

Organisation ....

Address ....

....

.... Postcode ....

Tel .... Fax .... Date ....

AFFIX
STAMP
HERE

E. & F.N. SPON
2-6 Boundary Row
London
SE1 8HN

Y:MECHANICAL AND ELECTRICAL SERVICES

Item	Net Price £	Material £	Labour hours	Labour £	Unit	Total Rate £
Y10 : PIPELINES : BLACK STEEL						
Black Steel Pipes; screwed joints in the running length; BS 1387						
Light weight pipes 6400 to 6550mm Long						
15mm dia.	0.94	1.09	0.53	6.02	m	7.11
20mm dia.	1.22	1.41	0.54	6.14	m	7.55
25mm dia.	1.67	1.93	0.59	6.71	m	8.64
32mm dia.	2.07	2.39	0.66	7.50	m	9.89
40mm dia.	2.56	2.95	0.72	8.18	m	11.13
50mm dia.	3.33	3.85	0.78	8.86	m	12.71
65mm dia.	4.90	5.66	0.88	10.00	m	15.66
80mm dia.	5.85	6.76	0.98	11.14	m	17.90
100mm dia.	8.49	9.80	1.32	15.01	m	24.81
Medium weight pipes 6400 to 6550mm Long						
15mm dia.	1.09	1.26	0.53	6.02	m	7.28
20mm dia.	1.29	1.49	0.54	6.14	m	7.63
25mm dia.	1.84	2.13	0.59	6.71	m	8.84
32mm dia.	2.29	2.64	0.66	7.50	m	10.14
40mm dia.	2.66	3.07	0.72	8.18	m	11.25
50mm dia.	3.74	4.32	0.78	8.86	m	13.18
65mm dia.	5.08	5.86	0.88	10.00	m	15.86
80mm dia.	6.59	7.61	0.98	11.14	m	18.75
100mm dia.	9.35	10.80	1.32	15.01	m	25.81
150mm dia.	16.01	18.49	1.70	19.32	m	37.81
Heavy weight pipes 6400 to 6550mm Long						
15mm dia.	1.28	1.48	0.53	6.02	m	7.50
20mm dia.	1.53	1.76	0.54	6.14	m	7.90
25mm dia.	2.22	2.57	0.59	6.71	m	9.28
32mm dia.	2.76	3.18	0.66	7.50	m	10.68
40mm dia.	3.21	3.71	0.72	8.18	m	11.89
50mm dia.	4.47	5.16	0.78	8.86	m	14.02
65mm dia.	6.08	7.02	0.88	10.00	m	17.02
80mm dia.	7.73	8.93	0.98	11.14	m	20.07
100mm dia.	10.79	12.46	1.32	15.01	m	27.47
150mm dia.	17.19	19.85	1.70	19.32	m	39.17
One Ended Fittings						
Blank flange; PN 6						
15mm dia.	4.51	4.96	0.25	2.84	nr	7.80
20mm dia.	5.06	5.56	0.25	2.84	nr	8.40
25mm dia.	5.82	6.40	0.25	2.84	nr	9.24
32mm dia.	7.00	7.69	0.30	3.41	nr	11.10
40mm dia.	7.76	8.54	0.30	3.41	nr	11.95
50mm dia.	8.18	8.99	0.30	3.41	nr	12.40
65mm dia.	9.88	10.86	0.33	3.75	nr	14.61
80mm dia.	11.65	12.82	0.33	3.75	nr	16.57
100mm dia.	13.63	14.99	0.38	4.32	nr	19.31
150mm dia.	23.36	25.69	0.57	6.48	nr	32.17

Y:MECHANICAL AND ELECTRICAL SERVICES

Item	Net Price £	Material £	Labour hours	Labour £	Unit	Total Rate £
Y10 : PIPELINES : BLACK STEEL (contd)						
Black Steel Pipes; screwed joints in the running length; BS 1387 (contd)						
One Ended Fittings (contd)						
Blank flange; PN 10						
15mm dia.	4.57	5.03	0.25	2.84	nr	**7.87**
20mm dia.	5.30	5.82	0.25	2.84	nr	**8.66**
25mm dia.	5.85	6.43	0.25	2.84	nr	**9.27**
32mm dia.	7.21	7.93	0.30	3.41	nr	**11.34**
40mm dia.	7.76	8.54	0.30	3.41	nr	**11.95**
50mm dia.	8.38	9.22	0.30	3.41	nr	**12.63**
65mm dia.	9.98	10.98	0.33	3.75	nr	**14.73**
80mm dia.	12.51	13.76	0.47	5.34	nr	**19.10**
100mm dia.	14.35	15.78	0.52	5.91	nr	**21.69**
150mm dia.	23.84	26.23	0.57	6.48	nr	**32.71**
Blank flange; PN 16						
15mm dia.	4.57	5.03	0.25	2.84	nr	**7.87**
20mm dia.	5.30	5.82	0.25	2.84	nr	**8.66**
25mm dia.	5.85	6.43	0.25	2.84	nr	**9.27**
32mm dia.	7.21	7.93	0.30	3.41	nr	**11.34**
40mm dia.	7.76	8.54	0.30	3.41	nr	**11.95**
50mm dia.	8.38	9.22	0.30	3.41	nr	**12.63**
65mm dia.	9.98	10.98	0.47	5.34	nr	**16.32**
80mm dia.	12.51	13.76	0.47	5.34	nr	**19.10**
100mm dia.	14.35	15.78	0.52	5.91	nr	**21.69**
150mm dia.	23.84	26.23	0.57	6.48	nr	**32.71**
Extra Over Black Steel Medium/Heavy Screwed Pipes; screwed flanges; Metric; BS 4504						
One Ended Fittings						
Steel bossed flanges; PN 6						
15mm dia.	7.45	8.20	0.37	4.20	nr	**12.40**
20mm dia.	7.45	8.20	0.49	5.57	nr	**13.77**
25mm dia.	7.88	8.67	0.56	6.36	nr	**15.03**
32mm dia.	10.20	11.22	0.63	7.16	nr	**18.38**
40mm dia.	10.20	11.22	0.73	8.30	nr	**19.52**
50mm dia.	12.30	13.53	0.88	10.00	nr	**23.53**
65mm dia.	15.83	17.41	1.08	12.28	nr	**29.69**
80mm dia.	18.87	20.76	1.29	14.66	nr	**35.42**
100mm dia.	22.46	24.70	1.42	16.14	nr	**40.84**
125mm dia.	54.97	60.47	1.70	19.32	nr	**79.79**
150mm dia.	54.97	60.47	2.08	23.67	nr	**84.14**

Y:MECHANICAL AND ELECTRICAL SERVICES

Item	Net Price £	Material £	Labour hours	Labour £	Unit	Total Rate £
Steel bossed flanges; PN 16						
15mm dia.	7.11	7.83	0.37	4.20	nr	**12.03**
20mm dia.	7.11	7.83	0.49	5.57	nr	**13.40**
25mm dia.	7.52	8.27	0.56	6.36	nr	**14.63**
32mm dia.	10.11	11.12	0.63	7.16	nr	**18.28**
40mm dia.	10.11	11.12	0.73	8.30	nr	**19.42**
50mm dia.	12.06	13.27	0.88	10.00	nr	**23.27**
65mm dia.	14.93	16.43	1.08	12.28	nr	**28.71**
80mm dia.	18.69	20.56	1.29	14.66	nr	**35.22**
100mm dia.	21.82	24.00	1.42	16.14	nr	**40.14**
150mm dia.	54.59	60.05	2.08	23.67	nr	**83.72**
Steel bossed flanges; PN 40						
15mm dia.	15.97	17.57	0.37	4.20	nr	**21.77**
20mm dia.	15.97	17.57	0.49	5.57	nr	**23.14**
25mm dia.	15.97	17.57	0.56	6.36	nr	**23.93**
32mm dia.	23.02	25.32	0.63	7.16	nr	**32.48**
40mm dia.	23.02	25.32	0.73	8.30	nr	**33.62**
50mm dia.	27.89	30.68	0.88	10.00	nr	**40.68**
65mm dia.	43.45	47.80	1.08	12.28	nr	**60.08**
80mm dia.	43.45	47.80	1.29	14.66	nr	**62.46**
100mm dia.	56.84	62.52	1.42	16.14	nr	**78.66**
150mm dia.	98.74	108.61	2.08	23.67	nr	**132.28**
Extra Over Black Steel Medium/Heavy Screwed Pipes; flange connections; Metric; BS 4504						
Bolted connection between pair of flanges; including corrugated brass joint ring, bolts and nuts; flanges given separately; PN 6						
15mm dia.	1.75	2.17	0.25	2.84	nr	**5.01**
20mm dia.	1.75	2.17	0.25	2.84	nr	**5.01**
25mm dia.	1.82	2.26	0.25	2.84	nr	**5.10**
32mm dia.	1.82	2.26	0.30	3.41	nr	**5.67**
40mm dia.	1.82	2.26	0.30	3.41	nr	**5.67**
50mm dia.	1.82	2.26	0.30	3.41	nr	**5.67**
65mm dia.	1.82	2.26	0.33	3.79	nr	**6.05**
80mm dia.	2.92	3.62	0.33	3.79	nr	**7.41**
100mm dia.	2.92	3.62	0.38	4.32	nr	**7.94**
150mm dia.	5.46	6.78	0.57	6.48	nr	**13.26**
Bolted connection between pair of flanges; including corrugated brass joint ring, bolts and nuts; Note: Flanges given separately; PN 16						
15mm dia.	1.82	2.26	0.25	2.84	nr	**5.10**
20mm dia.	1.82	2.26	0.25	2.84	nr	**5.10**
25mm dia.	1.82	2.26	0.25	2.84	nr	**5.10**
32mm dia.	2.92	3.62	0.30	3.41	nr	**7.03**
40mm dia.	2.92	3.62	0.30	3.41	nr	**7.03**
50mm dia.	3.03	3.76	0.30	3.41	nr	**7.17**
65mm dia.	3.03	3.76	0.33	3.79	nr	**7.55**
80mm dia.	5.46	6.78	0.47	5.34	nr	**12.12**
100mm dia.	5.57	6.91	0.52	5.91	nr	**12.82**
150mm dia.	9.99	12.41	0.57	6.48	nr	**18.89**

Y:MECHANICAL AND ELECTRICAL SERVICES

Item	Net Price £	Material £	Labour hours	Labour £	Unit	Total Rate £
Y10 : PIPELINES : BLACK STEEL (contd)						
Extra Over Black Steel Medium/Heavy Screwed Pipes; flange connections; Metric; BS 4504 (contd)						
Bolted connection between pair of flanges; including corrugated brass joint ring, bolts and nuts; Note: Flanges given separately; PN 40						
15mm dia.	1.75	2.17	0.25	2.84	nr	**5.01**
20mm dia.	1.82	2.26	0.25	2.84	nr	**5.10**
25mm dia.	1.82	2.26	0.25	2.84	nr	**5.10**
32mm dia.	1.82	2.26	0.30	3.41	nr	**5.67**
40mm dia.	2.92	3.62	0.30	3.41	nr	**7.03**
50mm dia.	2.92	3.62	0.30	3.41	nr	**7.03**
65mm dia.	5.57	6.91	0.47	5.34	nr	**12.25**
80mm dia.	9.99	12.41	0.47	5.34	nr	**17.75**
100mm dia.	9.99	12.41	0.52	5.91	nr	**18.32**
150mm dia.	9.99	12.41	0.57	6.48	nr	**18.89**
Extra Over Black Steel Medium/Heavy screwed pipes; black wrought iron fittings; BS 1740						
One Ended Fittings						
Cap						
15mm dia.	1.87	2.06	0.33	3.75	nr	**5.81**
20mm dia.	2.25	2.48	0.44	5.00	nr	**7.48**
25mm dia.	3.62	3.99	0.50	5.68	nr	**9.67**
32mm dia.	4.98	5.49	0.58	6.59	nr	**12.08**
40mm dia.	6.16	6.78	0.67	7.61	nr	**14.39**
50mm dia.	8.51	9.36	0.80	9.09	nr	**18.45**
65mm dia.	15.35	16.89	0.98	11.14	nr	**28.03**
80mm dia.	20.96	23.06	1.16	13.18	nr	**36.24**
100mm dia.	36.75	40.43	1.80	20.47	nr	**60.90**
Plug						
15mm dia.	1.20	1.33	0.33	3.75	nr	**5.08**
20mm dia.	1.47	1.62	0.44	5.00	nr	**6.62**
25mm dia.	1.84	2.03	0.50	5.68	nr	**7.71**
32mm dia.	2.45	2.71	0.58	6.59	nr	**9.30**
40mm dia.	2.89	3.19	0.67	7.61	nr	**10.80**
50mm dia.	4.21	4.63	0.80	9.09	nr	**13.72**
65mm dia.	10.23	11.26	0.98	11.14	nr	**22.40**
80mm dia.	18.04	19.84	1.16	13.18	nr	**33.02**
100mm dia.	38.59	42.45	1.80	20.47	nr	**62.92**
Elbow, Male and Female						
15mm dia.	4.62	5.09	0.66	7.50	nr	**12.59**
20mm dia.	7.01	7.71	0.88	10.00	nr	**17.71**
25mm dia.	10.05	11.06	1.00	11.36	nr	**22.42**
32mm dia.	13.30	14.64	1.16	13.18	nr	**27.82**
40mm dia.	17.18	18.90	1.34	15.23	nr	**34.13**
50mm dia.	32.47	35.72	1.60	18.18	nr	**53.90**
80mm dia.	62.62	68.89	2.32	26.36	nr	**95.25**

Y:MECHANICAL AND ELECTRICAL SERVICES

Item	Net Price £	Material £	Labour hours	Labour £	Unit	Total Rate £
Elbow Female, Equal						
15mm dia.	4.31	4.75	0.66	7.50	nr	**12.25**
20mm dia.	5.26	5.79	0.88	10.00	nr	**15.79**
25mm dia.	7.15	7.87	1.00	11.36	nr	**19.23**
32mm dia.	10.79	11.87	1.16	13.18	nr	**25.05**
40mm dia.	14.27	15.71	1.34	15.23	nr	**30.94**
50mm dia.	18.43	20.27	1.60	18.18	nr	**38.45**
65mm dia.	41.53	45.68	1.96	22.27	nr	**67.95**
80mm dia.	61.04	67.15	2.32	26.36	nr	**93.51**
100mm dia.	115.18	126.70	3.21	36.41	nr	**163.11**
Socket, Equal						
15mm dia.	0.68	0.75	0.66	7.50	nr	**8.25**
20mm dia.	0.78	0.85	0.88	10.00	nr	**11.86**
25mm dia.	1.01	1.11	1.00	11.36	nr	**12.47**
32mm dia.	1.48	1.63	1.16	13.18	nr	**14.81**
40mm dia.	1.90	2.08	1.34	15.23	nr	**17.31**
50mm dia.	2.84	3.12	1.60	18.18	nr	**21.30**
65mm dia.	4.87	5.36	1.96	22.27	nr	**27.63**
80mm dia.	6.81	7.49	2.32	26.36	nr	**33.85**
100mm dia.	13.00	14.30	3.21	36.41	nr	**50.71**
Socket, Reducing						
15mm dia.	2.27	2.50	0.66	7.50	nr	**10.00**
20mm dia.	2.56	2.83	0.88	10.00	nr	**12.83**
25mm dia.	3.64	4.00	1.00	11.36	nr	**15.36**
32mm dia.	5.04	5.55	1.16	13.18	nr	**18.73**
40mm dia.	6.49	7.14	1.34	15.23	nr	**22.37**
50mm dia.	9.01	9.92	1.60	18.18	nr	**28.10**
65mm dia.	20.72	22.80	1.96	22.27	nr	**45.07**
80mm dia.	28.56	31.42	2.32	26.36	nr	**57.78**
Three Ended Fittings						
Tee, Equal						
15mm dia.	4.86	5.35	0.95	10.80	nr	**16.15**
20mm dia.	5.98	6.58	1.27	14.43	nr	**21.01**
25mm dia.	9.64	10.61	1.45	16.49	nr	**27.10**
32mm dia.	13.92	15.32	1.68	19.09	nr	**34.41**
40mm dia.	17.57	19.33	1.94	22.06	nr	**41.39**
50mm dia.	25.70	28.28	2.31	26.30	nr	**54.58**
65mm dia.	47.68	52.45	2.83	32.18	nr	**84.63**
80mm dia.	77.19	84.91	3.36	38.12	nr	**123.03**
100mm dia.	133.41	146.75	4.63	52.59	nr	**199.34**
Four Ended Fittings						
Cross						
15mm dia.	14.64	16.11	1.05	11.93	nr	**28.04**
20mm dia.	17.40	19.14	1.40	15.91	nr	**35.05**
25mm dia.	22.36	24.60	1.60	18.18	nr	**42.78**
32mm dia.	30.74	33.82	1.86	21.15	nr	**54.97**
40mm dia.	37.25	40.98	2.14	24.33	nr	**65.31**
50mm dia.	76.97	84.68	2.56	29.13	nr	**113.81**

Y:MECHANICAL AND ELECTRICAL SERVICES

Item	Net Price £	Material £	Labour hours	Labour £	Unit	Total Rate £
Y10: PIPELINES : BLACK STEEL (contd)						
Black Steel Pipes; butt welded joints **in the running length; BS 1387**						
Lightweight Pipes 6400 to 6550mm long						
15mm dia.	0.92	1.01	0.55	6.25	m	**7.26**
20mm dia.	1.20	1.32	0.57	6.48	m	**7.80**
25mm dia.	1.84	2.02	0.63	7.16	m	**9.18**
32mm dia.	2.00	2.20	0.72	8.18	m	**10.38**
40mm dia.	2.50	2.75	0.80	9.09	m	**11.84**
50mm dia.	3.26	3.59	0.92	10.45	m	**14.04**
65mm dia.	4.83	5.31	1.07	12.16	m	**17.47**
80mm dia.	5.75	6.33	1.22	13.87	m	**20.20**
100mm dia.	8.34	9.17	1.56	17.72	m	**26.89**
Medium Weight Pipes 6400 to 6550mm long						
15mm dia.	1.15	1.26	0.55	6.25	m	**7.51**
20mm dia.	1.35	1.49	0.57	6.48	m	**7.97**
25mm dia.	1.95	2.14	0.63	7.16	m	**9.30**
32mm dia.	2.41	2.66	0.72	8.18	m	**10.84**
40mm dia.	2.80	3.08	0.80	9.09	m	**12.17**
50mm dia.	3.94	4.34	0.92	10.45	m	**14.79**
65mm dia.	5.36	5.89	1.07	12.16	m	**18.05**
80mm dia.	6.95	7.65	1.22	13.87	m	**21.52**
100mm dia.	9.85	10.84	1.58	17.97	m	**28.81**
150mm dia.	16.88	18.57	1.58	17.97	m	**36.54**
Heavy Weight Pipes 6400 to 6550mm long						
15mm dia.	1.35	1.49	0.55	6.25	m	**7.74**
20mm dia.	1.61	1.77	0.57	6.48	m	**8.25**
25mm dia.	2.35	2.58	0.63	7.16	m	**9.74**
32mm dia.	2.91	3.20	0.72	8.18	m	**11.38**
40mm dia.	3.39	3.73	0.80	9.09	m	**12.82**
50mm dia.	4.71	5.18	0.92	11.34	m	**16.52**
65mm dia.	6.41	7.05	1.07	13.20	m	**20.25**
80mm dia.	6.41	7.05	1.22	13.87	m	**20.92**
100mm dia.	8.16	8.98	1.58	17.97	m	**26.95**
150mm dia.	18.12	19.93	1.58	17.97	m	**37.90**
Extra Over Heavy Weight Black Steel **Welded pipes; welding flanges; Metric;** **BS 4504**						
One Ended Fittings						
Plate Flange; Slip on for Welding; PN 6						
15mm dia.	4.51	4.96	0.58	6.59	nr	**11.55**
20mm dia.	5.06	5.56	0.67	7.61	nr	**13.17**
25mm dia.	5.82	6.40	0.80	9.09	nr	**15.49**
32mm dia.	7.00	7.69	0.95	10.80	nr	**18.49**
40mm dia.	7.76	8.54	1.02	11.59	nr	**20.13**
50mm dia.	8.18	8.99	1.25	14.20	nr	**23.19**
65mm dia.	9.88	10.86	1.35	15.35	nr	**26.21**
80mm dia.	11.65	12.82	1.42	16.14	nr	**28.96**
100mm dia.	12.47	13.72	1.90	21.60	nr	**35.32**
150mm dia.	19.86	21.84	2.48	28.12	nr	**49.96**

Y:MECHANICAL AND ELECTRICAL SERVICES

Item	Net Price £	Material £	Labour hours	Labour £	Unit	Total Rate £
Plate Flange; Slip on for Welding; PN 10						
15mm dia.	4.57	5.03	0.58	6.59	nr	**11.62**
20mm dia.	5.30	5.82	0.67	7.61	nr	**13.43**
25mm dia.	5.85	6.43	0.80	9.09	nr	**15.52**
32mm dia.	7.21	7.93	0.95	10.80	nr	**18.73**
40mm dia.	7.76	8.54	1.02	11.59	nr	**20.13**
50mm dia.	8.38	9.22	1.25	14.20	nr	**23.42**
65mm dia.	9.98	10.98	1.35	15.35	nr	**26.33**
80mm dia.	12.51	13.76	1.42	16.14	nr	**29.90**
100mm dia.	12.99	14.29	1.90	21.60	nr	**35.89**
150mm dia.	20.79	22.87	2.48	28.12	nr	**50.99**
Plate Flange; Slip on for Welding; PN 16						
15mm dia.	4.57	5.03	0.67	7.61	nr	**12.64**
20mm dia.	5.30	5.82	0.77	8.75	nr	**14.57**
25mm dia.	5.85	6.43	0.93	10.57	nr	**17.00**
32mm dia.	7.21	7.93	1.09	12.39	nr	**20.32**
40mm dia.	7.76	8.54	1.17	13.30	nr	**21.84**
50mm dia.	8.38	9.22	1.44	16.37	nr	**25.59**
65mm dia.	9.98	10.98	1.55	17.61	nr	**28.59**
80mm dia.	12.51	13.76	1.63	18.53	nr	**32.29**
100mm dia.	12.99	14.29	2.19	24.91	nr	**39.20**
150mm dia.	20.79	22.87	2.84	32.27	nr	**55.14**
Weldneck Flange; PN 6						
15mm dia.	7.33	8.06	0.67	7.61	nr	**15.67**
20mm dia.	7.33	8.06	0.77	8.75	nr	**16.81**
25mm dia.	7.73	8.50	0.93	10.57	nr	**19.07**
32mm dia.	7.90	8.69	1.09	12.39	nr	**21.08**
40mm dia.	7.90	8.69	1.17	13.30	nr	**21.99**
50mm dia.	9.37	10.30	1.44	16.37	nr	**26.67**
65mm dia.	11.12	12.23	1.55	17.61	nr	**29.84**
80mm dia.	15.57	17.13	1.63	18.53	nr	**35.66**
100mm dia.	17.55	19.31	2.19	24.91	nr	**44.22**
150mm dia.	29.19	32.11	2.84	32.27	nr	**64.38**
Weldneck Flange; PN 16						
15mm dia.	7.22	7.95	0.67	7.61	nr	**15.56**
20mm dia.	7.22	7.95	0.77	8.75	nr	**16.70**
25mm dia.	7.60	8.36	0.93	10.57	nr	**18.93**
32mm dia.	10.96	12.06	1.09	12.39	nr	**24.45**
40mm dia.	10.96	12.06	1.17	13.30	nr	**25.36**
50mm dia.	13.07	14.38	1.44	16.37	nr	**30.75**
65mm dia.	14.81	16.29	1.55	17.61	nr	**33.90**
80mm dia.	19.63	21.59	1.63	18.53	nr	**40.12**
100mm dia.	22.99	25.29	2.19	24.91	nr	**50.20**
150mm dia.	35.92	39.51	2.84	32.27	nr	**71.78**

Y:MECHANICAL AND ELECTRICAL SERVICES

Item	Net Price £	Material £	Labour hours	Labour £	Unit	Total Rate £
Y10: PIPELINES : BLACK STEEL (contd)						
Extra Over Heavy Weight Black Steel Welded pipes; welding flanges; Metric; BS 4504 (contd)						
One Ended Fitting (contd)						
Weldneck Flange; PN 40						
15mm dia.	10.75	11.83	0.67	7.61	nr	**19.44**
20mm dia.	10.75	11.83	0.77	8.75	nr	**20.58**
25mm dia.	10.75	11.83	0.93	10.57	nr	**22.40**
32mm dia.	15.50	17.05	1.09	12.39	nr	**29.44**
40mm dia.	15.50	17.05	1.17	13.30	nr	**30.35**
50mm dia.	18.76	20.64	1.44	16.37	nr	**37.01**
65mm dia.	29.24	32.16	1.55	17.61	nr	**49.77**
80mm dia.	29.24	32.16	1.63	18.53	nr	**50.69**
100mm dia.	38.24	42.07	2.19	24.91	nr	**66.98**
150mm dia.	65.18	71.70	2.84	32.27	nr	**103.97**
Extra Over Heavy Weight Black Steel Welded pipes; welding flanges; Metric; BS 4504						
One Ended Fittings						
Bossed Flange; Slip on for Welding; PN 6						
15mm dia.	5.51	6.06	0.58	6.59	nr	**12.65**
20mm dia.	5.51	6.06	0.67	7.61	nr	**13.67**
25mm dia.	5.78	6.36	0.80	9.09	nr	**15.45**
32mm dia.	6.04	6.65	0.95	10.80	nr	**17.45**
40mm dia.	6.04	6.65	1.02	11.59	nr	**18.24**
50mm dia.	7.03	7.73	1.27	14.43	nr	**22.16**
65mm dia.	8.36	9.19	1.35	15.35	nr	**24.54**
80mm dia.	11.68	12.85	1.42	16.14	nr	**28.99**
100mm dia.	13.18	14.49	1.90	21.60	nr	**36.09**
150mm dia.	22.31	24.54	2.48	28.12	nr	**52.66**
Bossed Flange; Slip on for Welding; PN 16						
15mm dia.	5.29	5.82	0.58	6.59	nr	**12.41**
20mm dia.	5.29	5.82	0.67	7.61	nr	**13.43**
25mm dia.	5.58	6.13	0.80	9.09	nr	**15.22**
32mm dia.	8.03	8.84	0.95	10.80	nr	**19.64**
40mm dia.	8.03	8.84	1.02	11.59	nr	**20.43**
50mm dia.	9.79	10.77	1.25	14.20	nr	**24.97**
65mm dia.	11.23	12.35	1.35	15.35	nr	**27.70**
80mm dia.	14.39	15.83	1.42	16.14	nr	**31.97**
100mm dia.	16.87	18.56	1.90	21.60	nr	**40.16**
150mm dia.	29.64	32.60	2.48	28.12	nr	**60.72**

Y:MECHANICAL AND ELECTRICAL SERVICES

Item	Net Price £	Material £	Labour hours	Labour £	Unit	Total Rate £
Bossed Flange; Slip on for Welding; PN 40						
15mm dia.	5.29	5.82	0.58	6.59	nr	**12.41**
20mm dia.	8.07	8.88	0.67	7.61	nr	**16.49**
25mm dia.	8.07	8.88	0.80	9.09	nr	**17.97**
32mm dia.	11.40	12.54	0.95	10.80	nr	**23.34**
40mm dia.	11.62	12.78	1.02	11.59	nr	**24.37**
50mm dia.	14.08	15.48	1.25	14.20	nr	**29.68**
65mm dia.	21.94	24.13	1.35	15.35	nr	**39.48**
80mm dia.	21.94	24.13	1.42	16.14	nr	**40.27**
100mm dia.	28.70	31.57	1.90	21.60	nr	**53.17**
150mm dia.	48.89	53.78	2.48	28.12	nr	**81.90**
Extra Over Heavy Weight Black Steel Welded pipe flange connections; Metric BS 4504						
Bolted connection between pair of flanges; including corrugated brass joint ring bolts and nuts; PN 6 Note: Flanges given separately						
15mm dia.	1.75	2.17	0.50	5.68	nr	**7.85**
20mm dia.	1.75	2.17	0.50	5.68	nr	**7.85**
25mm dia.	1.82	2.26	0.50	5.68	nr	**7.94**
32mm dia.	1.82	2.26	0.50	5.68	nr	**7.94**
40mm dia.	1.82	2.26	0.50	5.68	nr	**7.94**
50mm dia.	1.82	2.26	0.50	5.68	nr	**7.94**
65mm dia.	1.82	2.26	0.50	5.68	nr	**7.94**
80mm dia.	2.92	3.62	0.50	5.68	nr	**9.30**
100mm dia.	2.92	3.62	0.50	5.68	nr	**9.30**
150mm dia.	5.46	6.78	0.88	10.00	nr	**16.78**
Bolted connection between pair of flanges; including corrugated brass joint ring bolts and nuts; PN 16 Note: Flanges given separately						
15mm dia.	1.82	2.26	0.50	5.68	nr	**7.94**
20mm dia.	1.82	2.26	0.50	5.68	nr	**7.94**
25mm dia.	1.82	2.26	0.50	5.68	nr	**7.94**
32mm dia.	2.92	3.62	0.50	5.68	nr	**9.30**
40mm dia.	2.92	3.62	0.50	5.68	nr	**9.30**
50mm dia.	3.03	3.76	0.50	5.68	nr	**9.44**
65mm dia.	3.03	3.76	0.50	5.68	nr	**9.44**
80mm dia.	5.46	6.78	0.50	5.68	nr	**12.46**
100mm dia.	5.57	6.91	0.50	5.68	nr	**12.59**
150mm dia.	9.99	12.41	0.88	10.00	nr	**22.41**
Bolted connection between pair of flanges; including corrugated brass joint ring bolts and nuts; PN 40 Note: Flanges given separately						
15mm dia.	1.75	2.17	0.50	5.68	nr	**7.85**
20mm dia.	1.82	2.26	0.50	5.68	nr	**7.94**
25mm dia.	1.82	2.26	0.50	5.68	nr	**7.94**
32mm dia.	1.82	2.26	0.50	5.68	nr	**7.94**
40mm dia.	2.92	3.62	0.50	5.68	nr	**9.30**

Y:MECHANICAL AND ELECTRICAL SERVICES

Item	Net Price £	Material £	Labour hours	Labour £	Unit	Total Rate £
Y10: PIPELINES : BLACK STEEL (contd)						
Extra Over Heavy Weight Black Steel Welded pipe flange connections; Metric BS 4504 (contd)						
Bolted connection between pair of flanges; including corrugated brass joint ring bolts and nuts; PN 40 Note: Flanges given separately (contd)						
50mm dia.	2.92	3.62	0.50	5.68	nr	**9.30**
65mm dia.	5.57	6.91	0.88	10.00	nr	**16.91**
80mm dia.	9.99	12.41	0.88	10.00	nr	**22.41**
100mm dia.	9.99	12.41	0.88	10.00	nr	**22.41**
150mm dia.	9.99	12.41	0.88	10.00	nr	**22.41**
Extra Over Medium Weight Black Steel Welded pipes; weldable fittings; BS 1965						
Two Ended Fittings						
45 degree elbow, long radius						
15mm dia.	6.81	7.49	0.52	5.91	nr	**13.40**
20mm dia.	6.81	7.49	0.69	7.84	nr	**15.33**
25mm dia.	7.40	8.14	0.85	9.66	nr	**17.80**
32mm dia.	7.40	8.14	1.07	12.16	nr	**20.30**
40mm dia.	7.40	8.14	1.25	14.20	nr	**22.34**
50mm dia.	8.49	9.34	1.86	21.15	nr	**30.49**
65mm dia.	10.31	11.34	2.60	29.58	nr	**40.92**
80mm dia.	12.05	13.26	3.24	36.76	nr	**50.02**
100mm dia.	16.67	18.33	3.92	44.55	nr	**62.88**
150mm dia.	43.07	47.38	5.41	61.41	nr	**108.79**
90 degree elbow, long radius						
15mm dia.	6.68	7.35	0.52	5.91	nr	**13.26**
20mm dia.	6.68	7.35	0.69	7.84	nr	**15.19**
25mm dia.	7.89	8.68	0.85	9.66	nr	**18.34**
32mm dia.	7.89	8.68	1.07	12.16	nr	**20.84**
40mm dia.	7.89	8.68	1.33	15.13	nr	**23.81**
50mm dia.	9.08	9.99	1.86	21.15	nr	**31.14**
65mm dia.	11.87	13.06	2.60	29.58	nr	**42.64**
80mm dia.	13.63	15.00	3.24	36.76	nr	**51.76**
100mm dia.	20.42	22.46	3.92	44.55	nr	**67.01**
150mm dia.	51.02	56.13	5.41	61.41	nr	**117.54**
Branch bend						
15mm dia.	17.73	19.51	0.71	8.07	nr	**27.58**
20mm dia.	17.73	19.51	0.94	10.69	nr	**30.20**
25mm dia.	22.30	24.53	1.19	13.52	nr	**38.05**
32mm dia.	22.82	25.10	1.52	17.29	nr	**42.39**
40mm dia.	22.30	24.53	1.73	19.65	nr	**44.18**
50mm dia.	22.20	24.42	2.48	28.19	nr	**52.61**
65mm dia.	33.64	37.00	3.30	37.49	nr	**74.49**
80mm dia.	34.75	38.22	4.13	46.94	nr	**85.16**
100mm dia.	61.85	68.03	4.98	56.52	nr	**124.55**
150mm dia.	166.58	183.24	6.71	76.24	nr	**259.48**

Y:MECHANICAL AND ELECTRICAL SERVICES

Item	Net Price £	Material £	Labour hours	Labour £	Unit	Total Rate £
Eccentric Reducer						
20 x 15mm dia.	11.32	12.45	0.61	6.93	nr	**19.38**
25 x 20mm dia.	20.01	22.01	0.77	8.74	nr	**30.75**
32 x 25mm dia.	22.39	24.62	0.96	10.91	nr	**35.53**
40 x 32mm dia.	24.62	27.08	1.23	13.97	nr	**41.05**
50 x 40mm dia.	24.62	27.08	1.66	18.87	nr	**45.95**
65 x 50mm dia.	27.04	29.74	2.36	26.86	nr	**56.60**
80 x 65mm dia.	29.99	32.99	3.13	35.61	nr	**68.60**
100 x 80mm dia.	21.01	23.11	3.91	44.38	nr	**67.49**
150 x 100mm dia.	83.12	91.43	4.81	54.62	nr	**146.05**
Three Ended Fittings						
Equal Tee						
25mm dia.	35.61	39.17	1.23	13.97	nr	**53.14**
32mm dia.	35.61	39.17	1.44	16.37	nr	**55.54**
40mm dia.	37.06	40.77	1.93	21.93	nr	**62.70**
50mm dia.	37.99	41.79	2.87	32.55	nr	**74.34**
65mm dia.	62.67	68.94	3.32	37.74	nr	**106.68**
80mm dia.	68.55	75.41	3.79	43.03	nr	**118.44**
100mm dia.	92.76	102.04	4.69	53.33	nr	**155.37**
150mm dia.	217.91	239.71	7.81	88.75	nr	**328.46**
Reducing Tee						
50 x 40mm dia.	48.20	53.02	7.81	88.75	nr	**141.77**
65 x 50mm dia.	80.54	88.60	7.81	88.75	nr	**177.35**
100 x 80mm dia.	119.92	131.92	7.81	88.75	nr	**220.67**
150 x 100mm dia.	281.53	309.68	7.81	88.75	nr	**398.43**
Extra Over Black Steel Welded Pipes; Labours						
Made Bend						
15mm dia.			0.40	4.54	nr	**4.54**
20mm dia.			0.40	4.54	nr	**4.54**
25mm dia.			0.48	5.45	nr	**5.45**
32mm dia.			0.64	7.27	nr	**7.27**
40mm dia.			0.80	9.09	nr	**9.09**
Screwed Joint To Fitting						
15mm dia.			0.09	1.02	nr	**1.02**
20mm dia.			0.09	1.02	nr	**1.02**
25mm dia.			0.09	1.02	nr	**1.02**
32mm dia.			0.09	1.02	nr	**1.02**
40mm dia.			0.12	1.36	nr	**1.36**
50mm dia.			0.15	1.70	nr	**1.70**
65mm dia.			0.18	2.04	nr	**2.04**
80mm dia.			0.21	2.39	nr	**2.39**
100mm dia.			0.31	3.52	nr	**3.52**

Y:MECHANICAL AND ELECTRICAL SERVICES

Item	Net Price £	Material £	Labour hours	Labour £	Unit	Total Rate £
Y10: PIPELINES : BLACK STEEL (contd)						
Extra Over Black Steel Welded Pipes; Labours (contd)						
Straight Butt Weld						
15mm dia.			0.30	3.41	nr	**3.41**
20mm dia.			0.40	4.54	nr	**4.54**
25mm dia.			0.50	5.68	nr	**5.68**
32mm dia.			0.66	7.50	nr	**7.50**
40mm dia.			0.80	9.09	nr	**9.09**
50mm dia.			1.20	13.64	nr	**13.64**
65mm dia.			1.60	18.18	nr	**18.18**
80mm dia.			2.00	22.72	nr	**22.72**
100mm dia.			2.40	27.31	nr	**27.31**
150mm dia.			3.21	36.41	nr	**36.41**
Branch Weld						
15mm dia.			0.45	5.11	nr	**5.11**
20mm dia.			0.60	6.82	nr	**6.82**
25mm dia.			0.76	8.64	nr	**8.64**
32mm dia.			0.99	11.25	nr	**11.25**
40mm dia.			1.07	12.16	nr	**12.16**
50mm dia.			1.50	17.06	nr	**17.06**
65mm dia.			2.00	22.72	nr	**22.72**
80mm dia.			2.50	28.40	nr	**28.40**
100mm dia.			3.00	34.11	nr	**34.11**
150mm dia.			4.00	45.44	nr	**45.44**
Welded Reducing Joint						
15mm dia.			0.60	6.82	nr	**6.82**
20mm dia.			0.80	9.09	nr	**9.09**
25mm dia.			1.00	11.36	nr	**11.36**
32mm dia.			1.32	15.01	nr	**15.01**
40mm dia.			1.60	18.18	nr	**18.18**
50mm dia.			2.40	27.31	nr	**27.31**
65mm dia.			3.21	36.41	nr	**36.41**
80mm dia.			4.00	45.44	nr	**45.44**
100mm dia.			4.41	50.04	nr	**50.04**
150mm dia.			5.21	59.17	nr	**59.17**
Extra Over Heavy Weight Black Steel Welded pipes; weldable fittings; BS 1965						
Two Ended Fittings						
Branch Bend						
25mm dia.	17.73	19.51	1.19	13.52	nr	**33.03**
32mm dia.	17.73	19.51	1.52	17.21	nr	**36.72**
40mm dia.	26.51	29.16	1.73	19.65	nr	**48.81**
50mm dia.	26.51	29.16	2.48	28.19	nr	**57.35**
65mm dia.	39.34	43.27	3.30	37.49	nr	**80.76**
80mm dia.	43.40	47.74	4.13	46.94	nr	**94.68**
100mm dia.	69.76	76.74	4.95	56.24	nr	**132.98**
150mm dia.	186.12	204.73	6.71	76.24	nr	**280.97**

Y:MECHANICAL AND ELECTRICAL SERVICES

Item	Net Price £	Material £	Labour hours	Labour £	Unit	Total Rate £
Concentric Reducer						
20 x 15mm dia.	11.32	12.45	0.61	6.93	nr	**19.38**
25 x 20mm dia.	11.75	12.92	0.77	8.74	nr	**21.66**
32 x 25mm dia.	15.23	16.75	0.96	10.90	nr	**27.65**
40 x 32mm dia.	14.84	16.33	1.23	13.97	nr	**30.30**
50 x 40mm dia.	16.20	17.81	1.66	18.87	nr	**36.68**
65 x 50mm dia.	18.57	20.43	2.36	26.86	nr	**47.29**
80 x 65mm dia.	18.47	20.32	3.13	35.50	nr	**55.82**
100 x 80mm dia.	21.01	23.11	3.91	44.38	nr	**67.49**
150 x 100mm dia.	42.54	46.79	4.81	54.62	nr	**101.41**

Keep your figures up to date, free of charge

This section, and most of the other information in this Price Book, is brought up to date every three months, until the next annual edition, in the *Price Book Update.*

The *Update* is available free to all Price Book purchasers.

To ensure you receive your copy, simply complete the reply card from the centre of the book and return it to us.

Y:MECHANICAL AND ELECTRICAL SERVICES

Item	Net Price £	Material £	Labour hours	Labour £	Unit	Total Rate £
Y10 : PIPELINES CARBON STEEL						
Malleable Grooved Jointing System; to comply with water Bylaws; gasket to BS 2494; earth continuity clips; for plain ended pipes to BS 1387 and BS 3601 (Note: The following are full value items not extra over).						
Grooved Joints; One Ended Fittings;						
Blank End						
2" dia.(nom)	4.47	4.92	0.65	7.39	nr	**12.31**
2.5" dia. (nom)	5.29	5.82	0.75	8.52	nr	**14.34**
3" dia. (nom)	6.67	7.33	1.00	11.36	nr	**18.69**
4" dia. (nom)	7.20	7.92	1.50	17.06	nr	**24.98**
6" dia. (nom)	16.78	18.46	1.75	19.89	nr	**38.35**
Grooved Joints; Two Ended Fittings;						
Malleable Iron Coupling; Standard Weight						
2" dia. (nom)	6.02	6.62	0.75	8.52	nr	**15.14**
2.5" dia. (nom)	6.71	7.38	0.80	9.09	nr	**16.47**
3" dia. (nom)	7.01	7.71	1.05	11.93	nr	**19.64**
4" dia. (nom)	10.67	11.74	1.55	17.61	nr	**29.35**
6" dia. (nom)	18.71	20.59	1.55	17.61	nr	**38.20**
Malleable Iron Coupling; Heavy Weight						
2" dia. (nom)	9.14	10.05	0.80	9.09	nr	**19.14**
2.5" dia. (nom)	10.48	11.53	0.85	9.66	nr	**21.19**
3" dia. (nom)	11.39	12.53	1.10	12.50	nr	**25.03**
4" dia. (nom)	15.59	17.15	1.75	19.89	nr	**37.04**
6" dia. (nom)	23.88	26.27	2.05	23.33	nr	**49.60**
Elbow 90 degree						
2" dia. (nom)	9.76	10.73	0.80	9.09	nr	**19.82**
2.5" dia. (nom)	12.30	13.53	0.90	10.23	nr	**23.76**
3" dia. (nom)	12.56	13.82	1.20	13.64	nr	**27.46**
4" dia. (nom)	16.82	18.50	1.65	18.75	nr	**37.25**
6" dia. (nom)	35.98	39.58	2.45	27.84	nr	**67.42**
Elbow 45 Degree						
2" dia. (nom)	8.28	9.10	0.80	9.09	nr	**18.19**
2.5" dia. (nom)	10.56	11.62	0.90	10.23	nr	**21.85**
3" dia. (nom)	11.86	13.04	1.20	13.64	nr	**26.68**
4" dia. (nom)	14.72	16.19	1.65	18.75	nr	**34.94**
6" dia. (nom)	27.39	30.13	2.30	26.17	nr	**56.30**
22.5 Degree Setting Piece						
2" dia. (nom)	8.46	9.31	0.80	9.09	nr	**18.40**
2.5 dia. (nom)	10.93	12.03	0.90	10.23	nr	**22.26**
3" dia. (nom)	13.56	14.91	1.20	13.64	nr	**28.55**
4" dia. (nom)	17.81	19.59	1.65	18.75	nr	**38.34**
6" dia. (nom)	27.99	30.79	2.30	26.17	nr	**56.96**

Y:MECHANICAL AND ELECTRICAL SERVICES

Item	Net Price £	Material £	Labour hours	Labour £	Unit	Total Rate £
Grooved or Threaded Nipple						
2" dia. (nom)	10.17	11.18	0.60	6.82	nr	**18.00**
2.5" dia. (nom)	17.23	18.96	0.75	8.52	nr	**27.48**
3" dia. (nom)	15.50	17.05	0.75	8.52	nr	**25.57**
4" dia. (nom)	21.46	23.60	0.80	9.09	nr	**32.69**
6" dia. (nom)	26.14	28.75	1.65	18.75	nr	**47.50**
Reducers Concentric						
3" x 2" dia. (nom)	11.74	12.92	1.10	12.50	nr	**25.42**
3" x 2.5" dia. (nom)	18.78	20.66	1.10	12.50	nr	**33.16**
4" x 3" dia. (nom)	23.58	25.93	1.50	17.06	nr	**42.99**
6" x 4" dia. (nom)	27.64	30.40	1.70	19.32	nr	**49.72**
Reducers Eccentric						
3" x 2" dia. (nom)	42.18	46.39	1.10	12.50	nr	**58.89**
4" x 3" dia. (nom)	51.77	56.95	1.40	15.91	nr	**72.86**
6" x 4" dia. (nom)	86.74	95.42	1.70	19.32	nr	**114.74**
Three Ended Fittings						
Equal Tee						
2" dia. (nom)	19.59	21.54	1.15	13.07	nr	**34.61**
2.5" dia. (nom)	22.16	24.37	1.25	14.20	nr	**38.57**
3" dia. (nom)	23.48	25.83	1.75	19.89	nr	**45.72**
4" dia. (nom)	26.24	28.87	2.30	26.17	nr	**55.04**
6" dia. (nom)	64.52	70.98	4.00	45.44	nr	**116.42**
Unequal Tee						
3" x 2" dia. (nom)	37.14	40.86	1.75	19.89	nr	**60.75**
4" x 3" dia. (nom)	46.72	51.39	2.30	26.17	nr	**77.56**
6" x 4" dia. (nom)	80.79	88.87	4.00	45.44	nr	**134.31**

Y:MECHANICAL AND ELECTRICAL SERVICES

Item	Net Price £	Material £	Labour hours	Labour £	Unit	Total Rate £
Y10 : PIPELINES : STAINLESS STEEL						
Stainless Steel Pipes; capillary or compression joints in the running length; BS 4127						
Grade 304; Satin Finish						
15mm dia.	1.13	1.31	0.41	4.66	m	**5.97**
22mm dia.	1.70	1.96	0.51	5.80	m	**7.76**
28mm dia.	2.40	2.77	0.58	6.59	m	**9.36**
35mm dia.	3.62	4.19	0.65	7.39	m	**11.58**
42mm dia.	4.58	5.29	0.71	8.07	m	**13.36**
54mm dia.	6.37	7.36	0.80	9.09	m	**16.45**
Grade 316 Satin Finish						
15mm dia.	1.66	1.91	0.61	6.93	m	**8.84**
22mm dia.	2.37	2.74	0.76	8.64	m	**11.38**
28mm dia.	3.63	4.20	0.87	9.89	m	**14.09**
35mm dia.	4.23	4.88	0.98	11.14	m	**16.02**
42mm dia.	5.38	6.22	1.06	12.05	m	**18.27**
54mm dia.	7.31	8.44	1.16	13.18	m	**21.62**
Extra Over Stainless Steel Pipes; capillary fittings						
Two Ended Fittings						
Coupling, Female						
15mm dia.	0.51	0.56	0.25	2.84	nr	**3.40**
22mm dia.	0.81	0.89	0.28	3.18	nr	**4.07**
28mm dia.	1.07	1.18	0.33	3.75	nr	**4.93**
35mm dia.	2.27	2.50	0.37	4.20	nr	**6.70**
42mm dia.	3.71	4.08	0.42	4.77	nr	**8.85**
54mm dia.	5.58	6.14	0.45	5.11	nr	**11.25**
45 Degree Bend						
15mm dia.	1.44	1.58	0.25	2.84	nr	**4.42**
22mm dia.	2.01	2.21	0.30	3.37	nr	**5.58**
28mm dia.	2.68	2.95	0.33	3.75	nr	**6.70**
35mm dia.	7.55	8.31	0.37	4.20	nr	**12.51**
42mm dia.	10.04	11.04	0.42	4.77	nr	**15.81**
54mm dia.	14.10	15.51	0.45	5.11	nr	**20.62**
90 Degree Bend						
15mm dia.	1.25	1.37	0.28	3.18	nr	**4.55**
22mm dia.	1.88	2.07	0.28	3.18	nr	**5.25**
28mm dia.	2.63	2.89	0.33	3.75	nr	**6.64**
35mm dia.	8.43	9.28	0.37	4.20	nr	**13.48**
42mm dia.	11.61	12.77	0.42	4.77	nr	**17.54**
54mm dia.	15.73	17.31	0.45	5.11	nr	**22.42**

Y:MECHANICAL AND ELECTRICAL SERVICES

Item	Net Price £	Material £	Labour hours	Labour £	Unit	Total Rate £
Reducer						
22mm dia.	3.84	4.23	0.28	3.18	nr	**7.41**
28mm dia.	3.84	4.23	0.33	3.75	nr	**7.98**
35mm dia.	3.84	4.23	0.37	4.20	nr	**8.43**
42mm dia.	4.39	4.83	0.42	4.77	nr	**9.60**
54mm dia.	15.96	17.56	0.48	5.46	nr	**23.02**
Tap Connector						
15mm dia.	9.49	10.44	0.13	1.48	nr	**11.92**
22mm dia.	12.54	13.79	0.14	1.59	nr	**15.38**
28mm dia.	17.40	19.14	0.17	1.93	nr	**21.07**
Tank Connectors						
15mm dia.	12.26	13.48	0.13	1.48	nr	**14.96**
22mm dia.	18.24	20.06	0.13	1.48	nr	**21.54**
35mm dia.	32.45	35.70	0.18	2.04	nr	**37.74**
42mm dia.	42.87	47.15	0.21	2.39	nr	**49.54**
54mm dia.	64.89	71.38	0.24	2.73	nr	**74.11**
Three Ended Fittings						
Tee Equal						
15mm dia.	1.81	1.99	0.37	4.20	nr	**6.19**
22mm dia.	2.87	3.15	0.40	4.54	nr	**7.69**
28mm dia.	3.76	4.14	0.45	5.11	nr	**9.25**
35mm dia.	11.76	12.94	0.59	6.71	nr	**19.65**
42mm dia.	14.51	15.96	0.62	7.05	nr	**23.01**
54mm dia.	29.30	32.23	0.67	7.61	nr	**39.84**
Unequal Tee						
22 x 15mm dia.	5.73	6.31	0.37	4.20	nr	**10.51**
22 x 18mm dia.	5.73	6.31	0.37	4.20	nr	**10.51**
28 x 15mm dia.	7.43	8.18	0.45	5.11	nr	**13.29**
28 x 22mm dia.	7.43	8.18	0.45	5.11	nr	**13.29**
35 x 22mm dia.	12.98	14.28	0.59	6.71	nr	**20.99**
35 x 28mm dia.	12.98	14.28	0.59	6.71	nr	**20.99**
42 x 28mm dia.	15.96	17.56	0.62	7.05	nr	**24.61**
42 x 35mm dia.	15.96	17.56	0.62	7.05	nr	**24.61**
54 x 35mm dia.	60.24	66.26	0.67	7.61	nr	**73.87**
54 x 42mm dia.	65.97	72.57	0.67	7.63	nr	**80.20**
Union (Conical Seat)						
15mm dia.	16.41	18.05	0.25	2.84	nr	**20.89**
22mm dia.	24.73	27.21	0.28	3.18	nr	**30.39**
28mm dia.	32.03	35.23	0.33	3.75	nr	**38.98**
35mm dia.	41.98	46.17	0.37	4.20	nr	**50.37**
42mm dia.	52.95	58.24	0.42	4.77	nr	**63.01**
54mm dia.	70.05	77.05	0.45	5.11	nr	**82.16**

Y:MECHANICAL AND ELECTRICAL SERVICES

Item	Net Price £	Material £	Labour hours	Labour £	Unit	Total Rate £
Y10 : PIPELINES : STAINLESS STEEL (contd)						
Extra Over Stainless Steel Pipes; capillary fittings (contd)						
Three Ended Fittings (contd)						
Union (Flat Seat)						
15mm dia.	16.41	18.05	0.25	2.84	nr	**20.89**
22mm dia.	25.55	28.10	0.28	3.18	nr	**31.28**
28mm dia.	33.02	36.32	0.33	3.75	nr	**40.07**
35mm dia.	43.14	47.45	0.37	4.20	nr	**51.65**
42mm dia.	54.34	59.77	0.42	4.77	nr	**64.54**
54mm dia.	72.29	79.52	0.45	5.11	nr	**84.63**
Extra Over Stainless Steel Pipes; compression fittings						
Two Ended Fittings						
Straight Coupling, Female Both Ends						
15mm dia.	10.93	12.02	0.18	2.04	nr	**14.06**
22mm dia.	20.80	22.88	0.22	2.50	nr	**25.38**
28mm dia.	28.00	30.80	0.25	2.84	nr	**33.64**
35mm dia.	43.24	47.56	0.30	3.41	nr	**50.97**
42mm dia.	55.29	60.82	0.40	4.54	nr	**65.36**
90 Degree Elbow Coupling						
15mm dia.	13.77	15.14	0.18	2.04	nr	**17.18**
22mm dia.	27.34	30.07	0.22	2.50	nr	**32.57**
28mm dia.	37.29	41.02	0.25	2.84	nr	**43.86**
35mm dia.	75.51	83.06	0.33	3.75	nr	**86.81**
42mm dia.	110.36	121.40	0.35	3.98	nr	**125.38**
Reducing Coupling						
15mm dia.	13.28	14.61	0.42	4.77	nr	**19.38**
22mm dia.	19.82	21.80	0.28	3.18	nr	**24.98**
28mm dia.	27.13	29.84	0.28	3.18	nr	**33.02**
35mm dia.	39.66	43.63	0.30	3.41	nr	**47.04**
42mm dia.	52.75	58.03	0.37	4.20	nr	**62.23**
Stud Coupling						
15mm dia.	9.50	10.45	0.42	4.77	nr	**15.22**
22mm dia.	15.39	16.92	0.25	2.84	nr	**19.76**
28mm dia.	8.20	9.02	0.25	2.84	nr	**11.86**
35mm dia.	34.78	38.26	0.37	4.20	nr	**42.46**
42mm dia.	45.02	49.52	0.42	4.77	nr	**54.29**

Y:MECHANICAL AND ELECTRICAL SERVICES

Item	Net Price £	Material £	Labour hours	Labour £	Unit	Total Rate £
Three Ended Fittings;						
Equal Tee						
15mm dia.	19.39	21.32	0.37	4.20	nr	**25.52**
22mm dia.	40.04	44.05	0.40	4.54	nr	**48.59**
28mm dia.	54.77	60.24	0.45	5.11	nr	**65.35**
35mm dia.	108.91	119.80	0.59	6.71	nr	**126.51**
42mm dia.	151.01	166.12	0.62	7.05	nr	**173.17**
54mm dia.	151.01	166.12	0.67	7.61	nr	**173.73**
Running Tee Coupling						
15mm dia.	23.85	26.24	0.37	4.20	nr	**30.44**
22mm dia.	42.98	47.28	0.40	4.54	nr	**51.82**
28mm dia.	72.78	80.05	0.59	6.71	nr	**86.76**

Y:MECHANICAL AND ELECTRICAL SERVICES

Item	Net Price £	Material £	Labour hours	Labour £	Unit	Total Rate £
Y10 : PIPELINES : COPPER						
Microbore Copper Pipe; capillary or compression joints in the running length; BS 2871						
Table W						
6mm dia.	0.49	0.57	0.41	4.66	m	**5.23**
8mm dia.	0.59	0.69	0.41	4.66	m	**5.35**
10mm dia.	0.88	1.02	0.41	4.66	m	**5.68**
Seamless Polythene Covered Microbore Copper Pipe; capillary or compression joints in the running length; BS 2871						
Table W						
6mm dia.	0.67	0.78	0.45	5.11	m	**5.89**
8mm dia.	0.83	0.96	0.45	5.11	m	**6.07**
10mm dia.	1.10	1.28	0.50	5.68	m	**6.96**
Seamless Polythene Covered Microbore Copper Pipe; with inner sheath of polythene contoured for cushioning and heat retention; capillary or compression joints in the running length; BS 2871						
Table W						
8mm dia.	8.37	0.97	0.45	5.11	m	**6.08**
10mm dia.	1.09	1.26	0.50	5.68	m	**6.94**
Manifold Connectors; Side Entry One Way Flow 22mm Body						
4 x 8mm connections	9.51	10.46	0.58	6.59	nr	**17.05**
6 x 8mm connections	11.35	12.48	0.87	9.89	nr	**22.37**
2 x 10mm connections	6.21	6.83	0.29	3.29	nr	**10.12**
4 x 10mm connections	9.95	10.94	0.58	6.59	nr	**17.53**
Manifold Connectors; Linear Flow 22mm Body						
4 x 8mm connections	8.18	9.00	0.58	6.59	nr	**15.59**
4 x 10mm connections	10.50	11.55	0.58	6.59	nr	**18.14**
Manifold Connectors; Linear Flow 28mm Body						
6 x 8mm connections	11.70	12.87	0.87	9.89	nr	**22.76**

Y:MECHANICAL AND ELECTRICAL SERVICES

Item	Net Price £	Material £	Labour hours	Labour £	Unit	Total Rate £
Copper Pipes; capillary or compression joints in the running length; BS 2871						
Table X						
15mm dia.	1.09	1.27	0.41	4.66	m	**5.93**
22mm dia.	2.17	2.52	0.47	5.35	m	**7.87**
28mm dia.	2.88	3.33	0.51	5.79	m	**9.12**
35mm dia.	6.02	6.95	0.58	6.59	m	**13.54**
42mm dia.	7.33	8.40	0.65	7.38	m	15.86
54mm dia.	9.47	10.94	0.71	8.07	m	**19.01**
67mm dia.	12.46	14.40	0.74	8.41	m	**22.81**
76mm dia.	17.56	20.29	0.75	8.52	m	**28.81**
108mm dia.	25.10	28.99	0.76	8.63	m	**37.62**
Table Y						
8mm dia.	0.89	1.04	0.43	4.89	m	**5.93**
10mm dia.	1.11	1.29	0.43	4.89	m	**6.18**
12mm dia.	1.45	1.68	0.43	4.89	m	**6.57**
15mm dia.	2.12	2.46	0.43	4.89	m	**7.35**
22mm dia.	3.72	4.30	0.50	5.68	m	**9.98**
28mm dia.	5.00	5.78	0.54	6.14	m	**11.92**

Y:MECHANICAL AND ELECTRICAL SERVICES

Item	Net Price £	Material £	Labour hours	Labour £	Unit	Total Rate £
Y10 : PIPELINES : COPPER (contd)						
Seamless Polythene Covered Copper Pipe; capillary joints in the running length; BS 2871						
Table X						
15mm dia.	1.51	1.74	0.60	6.82	m	**8.56**
22mm dia.	2.85	3.29	0.68	7.73	m	**11.02**
28mm dia.	3.61	4.17	0.74	8.41	m	**12.58**
35mm dia.	6.73	7.78	0.85	9.66	m	**17.44**
42mm dia.	8.07	9.32	0.96	10.91	m	**20.23**
54mm dia.	10.27	11.86	1.05	11.93	m	**23.79**
Table Y						
15mm dia.	2.51	2.91	0.60	6.82	m	**9.73**
22mm dia.	4.35	5.03	0.68	7.73	m	**12.76**
28mm dia.	5.68	6.56	0.74	8.41	m	**14.97**
35mm dia.	8.58	9.91	0.85	9.66	m	**19.57**
42mm dia.	10.21	11.79	0.96	10.91	m	**22.70**
54mm dia.	17.03	19.67	1.05	11.93	m	**31.60**
Seamless Polythene Covered Copper Pipe; with inner sheath of polythene contoured for cushioning and heat retention; capillary joints in the running length; BS 2871						
Table X						
15mm dia.	1.67	1.93	0.60	6.82	m	**8.75**
22mm dia.	3.16	3.65	0.68	7.73	m	**11.38**
28mm dia.	3.94	4.56	0.74	8.41	m	**12.97**
35mm dia.	7.28	8.41	0.85	9.66	m	**18.07**
42mm dia.	8.80	10.17	0.96	10.91	m	**21.08**
54mm dia.	11.26	13.01	1.05	11.93	m	**24.94**
Table Y						
15mm dia.	2.68	3.10	0.60	6.82	m	**9.92**
22mm dia.	4.66	5.39	0.68	7.73	m	**13.12**
Extra Over Copper Pipes; capillary fittings; BS 864						
One Ended Fittings						
Stop End						
15mm dia.	0.99	1.09	0.13	1.48	nr	**2.57**
22mm dia.	1.41	1.55	0.14	1.59	nr	**3.14**
28mm dia.	2.19	2.41	0.17	1.93	nr	**4.34**
35mm dia.	2.19	2.41	0.18	2.04	nr	**4.45**
42mm dia.	2.19	2.41	0.21	2.39	nr	**4.80**
54mm dia.	2.19	2.41	0.23	2.61	nr	**5.02**

Y:MECHANICAL AND ELECTRICAL SERVICES

Item	Net Price £	Material £	Labour hours	Labour £	Unit	Total Rate £
Two Ended Fittings						
Straight Coupling; Copper to Copper						
6mm dia.	0.80	0.89	0.25	2.84	nr	**3.73**
8mm dia.	0.59	0.64	0.25	2.84	nr	**3.48**
10mm dia.	0.52	0.57	0.25	2.84	nr	**3.41**
15mm dia.	0.18	0.20	0.25	2.84	nr	**3.04**
22mm dia.	0.42	0.46	0.28	3.18	nr	**3.64**
28mm dia.	0.99	1.09	0.33	3.75	nr	**4.84**
35mm dia.	0.99	1.09	0.37	4.20	nr	**5.29**
42mm dia.	0.99	1.09	0.42	4.77	nr	**5.86**
54mm dia.	0.99	1.09	0.45	5.11	nr	**6.20**
67mm dia.	0.99	1.09	0.56	6.36	nr	**7.45**
Adaptor Coupling; Imperial to Metric						
15mm dia.	0.61	0.67	0.25	2.84	nr	**3.51**
22mm dia.	0.61	0.67	0.28	3.18	nr	**3.85**
28mm dia.	0.61	0.67	0.33	3.75	nr	**4.42**
35mm dia.	0.61	0.67	0.37	4.20	nr	**4.87**
42mm dia.	0.61	0.67	0.42	4.77	nr	**5.44**
Reducing Coupling						
15 x 10mm dia.	1.06	1.16	0.25	2.84	nr	**4.00**
22 x 10mm dia.	2.16	2.38	0.28	3.18	nr	**5.56**
22 x 15mm dia.	1.17	1.29	0.30	3.41	nr	**4.70**
28 x 15mm dia.	2.28	2.51	0.31	3.52	nr	**6.03**
28 x 22mm dia.	1.83	2.01	0.33	3.75	nr	**5.76**
35 x 28mm dia.	1.83	2.01	0.37	4.20	nr	**6.21**
42 x 35mm dia.	1.83	2.01	0.42	4.77	nr	**6.78**
54 x 35mm dia.	1.83	2.01	0.45	5.11	nr	**7.12**
54 x 42mm dia.	1.83	2.01	0.45	5.11	nr	**7.12**
Straight Female Connector						
15mm dia.	1.38	1.52	0.25	2.84	nr	**4.36**
22mm dia.	1.38	1.52	0.28	3.18	nr	**4.70**
28mm dia.	1.38	1.52	0.33	3.75	nr	**5.27**
35mm dia.	1.38	1.52	0.37	4.20	nr	**5.72**
42mm dia.	1.38	1.52	0.42	4.77	nr	**6.29**
54mm dia.	1.38	1.52	0.45	5.11	nr	**6.63**
Straight Male Connector						
15mm dia.	1.38	1.52	0.25	2.84	nr	**4.36**
22mm dia.	2.41	2.66	0.28	3.18	nr	**5.84**
28mm dia.	3.77	4.15	0.33	3.75	nr	**7.90**
35mm dia.	3.77	4.15	0.37	4.20	nr	**8.35**
42mm dia.	3.77	4.15	0.42	4.77	nr	**8.92**
54mm dia.	3.77	4.15	0.45	5.11	nr	**9.26**
67mm dia.	3.77	4.13	0.56	6.36	nr	**10.49**
Female Reducing Connector						
15mm dia.	2.82	3.10	0.25	2.84	nr	**5.94**
22mm dia.	4.07	4.47	0.28	3.18	nr	**7.65**

Y:MECHANICAL AND ELECTRICAL SERVICES

Item	Net Price £	Material £	Labour hours	Labour £	Unit	Total Rate £
Y10 : PIPELINES : COPPER (contd)						
Extra Over Copper Pipes; capillary fittings; BS 864						
Male Reducing Connector (contd)						
Two Ended Fiittings (contd)						
15mm dia.	1.67	1.84	0.25	2.84	nr	4.68
22mm dia.	3.57	3.93	0.28	3.18	nr	7.11
28mm dia.	5.13	5.64	0.33	3.75	nr	9.39
Lead Connector						
15mm dia.	1.24	1.37	0.25	2.84	nr	4.21
22mm dia.	1.84	2.02	0.28	3.18	nr	5.20
28mm dia.	2.60	2.86	0.33	3.75	nr	6.61
Flanged Connector						
28mm dia.	36.91	40.60	0.33	3.75	nr	44.35
35mm dia.	33.50	36.85	0.37	4.20	nr	41.05
42mm dia.	40.03	44.04	0.42	4.77	nr	48.81
54mm dia.	60.53	66.58	0.45	5.11	nr	71.69
67mm dia.	74.74	82.21	0.56	6.36	nr	88.57
Tank Connector						
15mm dia.	3.46	3.81	0.25	2.84	nr	6.65
22mm dia.	5.29	5.82	0.28	3.18	nr	9.00
28mm dia.	7.06	7.77	0.33	3.75	nr	11.52
35mm dia.	9.11	10.02	0.37	4.20	nr	14.22
42mm dia.	11.94	13.14	0.42	4.77	nr	17.91
54mm dia.	18.24	20.06	0.45	5.11	nr	25.17
Tank Connector with Long Thread						
15mm dia.	4.70	5.17	0.32	3.64	nr	8.81
22mm dia.	6.70	7.37	0.35	3.98	nr	11.35
28mm dia.	8.27	9.09	0.42	4.77	nr	13.86
Reducer						
15 x 10mm dia.	0.92	1.01	0.25	2.84	nr	3.85
22 x 15mm dia.	0.76	0.84	0.28	3.18	nr	4.02
28 x 15mm dia.	0.76	0.84	0.30	3.41	nr	4.25
28 x 22mm dia.	1.17	1.29	0.33	3.75	nr	5.04
35 x 22mm dia.	1.17	1.29	0.37	4.20	nr	5.49
42 x 22mm dia.	1.17	1.29	0.39	4.43	nr	5.72
42 x 35mm dia.	1.17	1.29	0.42	4.77	nr	6.06
54 x 35mm dia.	1.17	1.29	0.43	4.89	nr	6.18
54 x 42mm dia.	1.17	1.29	0.45	5.11	nr	6.40
67 x 54mm dia.	1.17	1.29	0.56	6.36	nr	7.65

Y:MECHANICAL AND ELECTRICAL SERVICES

Item	Net Price £	Material £	Labour hours	Labour £	Unit	Total Rate £
Adaptor Copper to Female Iron						
15mm dia.	2.72	2.99	0.25	2.84	nr	**5.83**
22mm dia.	3.63	3.99	0.28	3.18	nr	**7.17**
28mm dia.	5.40	5.94	0.33	3.75	nr	**9.69**
35mm dia.	8.33	9.16	0.37	4.20	nr	**13.36**
42mm dia.	12.77	14.05	0.42	4.77	nr	**18.82**
54mm dia.	15.11	16.62	0.45	5.11	nr	**21.73**
Adaptor Copper to Male Iron						
15mm dia.	2.62	2.89	0.25	2.84	nr	**5.73**
22mm dia.	3.53	3.89	0.28	3.18	nr	**7.07**
28mm dia.	5.40	5.94	0.33	3.75	nr	**9.69**
35mm dia.	8.33	9.16	0.37	4.20	nr	**13.36**
42mm dia.	12.77	14.05	0.42	4.77	nr	**18.82**
54mm dia.	15.11	16.62	0.45	5.11	nr	**21.73**
Union Coupling						
15mm dia.	4.16	4.57	0.37	4.20	nr	**8.77**
22mm dia.	6.87	7.55	0.40	4.54	nr	**12.09**
28mm dia.	9.77	10.75	0.45	5.11	nr	**15.86**
35mm dia.	13.02	14.32	0.59	6.71	nr	**21.03**
42mm dia.	18.70	20.56	0.62	7.05	nr	**27.61**
54mm dia.	30.70	33.77	0.67	7.61	nr	**41.38**
67mm dia.	54.70	60.16	0.83	9.44	nr	**69.60**
Elbow						
15mm dia.	0.34	0.38	0.25	2.84	nr	**3.22**
22mm dia.	0.80	0.89	0.28	3.18	nr	**4.07**
28mm dia.	1.65	1.82	0.33	3.75	nr	**5.57**
35mm dia.	1.65	1.82	0.37	4.20	nr	**6.02**
42mm dia.	1.65	1.82	0.42	4.77	nr	**6.59**
54mm dia.	1.65	1.82	0.45	5.11	nr	**6.93**
67mm dia.	1.65	1.82	0.56	6.36	nr	**8.18**
Backplate Elbow						
15mm dia.	3.39	3.72	0.50	5.68	nr	**9.40**
22mm dia.	6.91	7.60	0.53	6.02	nr	**13.62**
Overflow Bend						
15mm dia.	7.20	7.92	0.25	2.84	nr	**10.76**
15mm dia.	7.74	8.52	0.28	3.18	nr	**11.70**
Return Bend						
15mm dia.	3.66	4.02	0.25	2.84	nr	**6.86**
22mm dia.	5.87	6.46	0.28	3.18	nr	**9.64**
28mm dia.	10.56	11.62	0.33	3.75	nr	**15.37**

Y:MECHANICAL AND ELECTRICAL SERVICES

Item	Net Price £	Material £	Labour hours	Labour £	Unit	Total Rate £
Y10 : PIPELINES : COPPER (contd)						
Extra Over Copper Pipes; capillary fittings; BS 864						
Two Ended Fittings (contd)						
Obtuse Elbow						
15mm dia.	1.01	1.12	0.25	2.84	nr	3.96
22mm dia.	1.89	2.08	0.28	3.18	nr	5.26
28mm dia.	3.62	3.98	0.33	3.75	nr	7.73
35mm dia.	6.84	7.53	0.37	4.20	nr	11.73
42mm dia.	12.03	13.23	0.42	4.77	nr	18.00
54mm dia.	20.47	22.52	0.45	5.11	nr	27.63
67mm dia.	40.84	44.92	0.56	6.36	nr	51.28
Straight Tap Connector						
15mm dia.	1.38	1.52	0.13	1.48	nr	3.00
22mm dia.	1.88	2.07	0.14	1.59	nr	3.66
Bent Tap Connector						
15mm dia.	1.38	1.52	0.13	1.48	nr	3.00
Bent Male Union Connector						
15mm dia.	6.29	6.92	0.37	4.20	nr	11.12
22mm dia.	7.87	8.66	0.40	4.54	nr	13.20
28mm dia.	11.21	12.33	0.45	5.11	nr	17.44
35mm dia.	16.28	17.91	0.59	6.71	nr	24.62
42mm dia.	28.13	30.94	0.62	7.05	nr	37.99
54mm dia.	44.81	49.29	0.67	7.61	nr	56.90
Bent Female Union Connector						
15mm dia.	6.05	6.66	0.37	4.20	nr	10.86
22mm dia.	7.63	8.39	0.40	4.54	nr	12.93
28mm dia.	11.21	12.33	0.45	5.11	nr	17.44
35mm dia.	17.91	19.70	0.59	6.71	nr	26.41
42mm dia.	30.93	34.03	0.62	7.05	nr	41.08
54mm dia.	49.28	54.21	0.67	7.61	nr	61.82
Straight Union Adaptor						
15mm dia.	2.74	3.01	0.37	4.20	nr	7.21
22mm dia.	3.63	3.99	0.40	4.54	nr	8.53
28mm dia.	5.71	6.28	0.45	5.11	nr	11.39
35mm dia.	8.53	9.38	0.59	6.71	nr	16.09
42mm dia.	11.38	12.52	0.62	7.05	nr	19.57
54mm dia.	16.62	18.28	0.67	7.61	nr	25.89
67mm dia.	27.07	29.77	0.83	9.44	nr	39.21

Y:MECHANICAL AND ELECTRICAL SERVICES

Item	Net Price £	Material £	Labour hours	Labour £	Unit	Total Rate £
Straight Union Connector Male						
15mm dia.	5.21	5.74	0.37	4.20	nr	**9.94**
22mm dia.	6.78	7.46	0.40	4.54	nr	**12.00**
28mm dia.	9.93	10.92	0.45	5.11	nr	**16.03**
35mm dia.	13.38	14.71	0.59	6.71	nr	**21.42**
42mm dia.	21.16	23.28	0.62	7.05	nr	**30.33**
54mm dia.	32.54	35.80	0.67	7.61	nr	**43.41**
Straight Union Connector Female						
15mm dia.	5.34	5.87	0.37	4.20	nr	**10.07**
22mm dia.	6.61	7.28	0.40	4.54	nr	**11.82**
28mm dia.	9.93	10.92	0.45	5.11	nr	**16.03**
35mm dia.	13.38	14.71	0.59	6.71	nr	**21.42**
42mm dia.	21.16	23.28	0.62	7.04	nr	**30.32**
54mm dia.	31.08	34.19	0.67	7.61	nr	**41.80**
Male Nipple						
15mm dia.	3.45	3.79	0.25	2.84	nr	**6.63**
22mm dia.	2.56	2.82	0.28	3.18	nr	**6.00**
28mm dia.	2.97	3.26	0.33	3.75	nr	**7.01**
35mm dia.	4.15	4.56	0.37	4.20	nr	**8.76**
42mm dia.	6.04	6.64	0.42	4.77	nr	**11.41**
54mm dia.	12.25	13.47	0.45	5.11	nr	**18.58**
67mm dia.	16.36	18.00	0.56	6.36	nr	**24.36**
Female Nipple						
15mm dia.	3.45	3.79	0.25	2.84	nr	**6.63**
22mm dia.	2.56	2.82	0.28	3.18	nr	**6.00**
28mm dia.	2.97	3.26	0.33	3.75	nr	**7.01**
35mm dia.	4.15	4.56	0.37	4.20	nr	**8.76**
42mm dia.	6.04	6.64	0.42	4.77	nr	**11.41**
54mm dia.	12.25	13.47	0.45	5.11	nr	**18.58**
67mm dia.	16.36	18.00	0.56	6.36	nr	**24.36**
Three Ended Fittings						
Equal Tee						
10mm dia.	1.65	1.82	0.23	2.61	nr	**4.43**
15mm dia.	0.61	0.67	0.37	4.20	nr	**4.87**
22mm dia.	1.40	1.54	0.40	4.54	nr	**6.08**
28mm dia.	3.24	3.56	0.45	5.11	nr	**8.67**
35mm dia.	3.24	3.56	0.59	6.71	nr	**10.27**
42mm dia.	3.24	3.56	0.62	7.05	nr	**10.61**
54mm dia.	3.24	3.56	0.67	7.61	nr	**11.17**
67mm dia.	3.24	3.56	0.83	9.43	nr	**12.99**
Female Tee, Reducing Branch F1						
15mm dia.	3.86	4.24	0.37	4.20	nr	**8.44**
22mm dia.	2.77	3.05	0.40	4.54	nr	**7.59**
28mm dia.	9.62	10.59	0.45	5.11	nr	**15.70**
35mm dia.	13.47	14.82	0.45	5.11	nr	**19.93**
42mm dia.	16.20	17.82	0.62	7.04	nr	**24.86**

Y:MECHANICAL AND ELECTRICAL SERVICES

Item	Net Price £	Material £	Labour hours	Labour £	Unit	Total Rate £
Y10 : PIPELINES : COPPER (contd)						
Extra Over Copper Pipes; capillary fittings; BS 864						
Three Ended Fittings (contd)						
Back Plate Tee						
15mm dia.	7.20	7.92	0.62	7.04	nr	**14.96**
Heater Tee						
15mm dia.	6.47	7.12	0.37	4.20	nr	**11.32**
Union Heater Tee						
15mm dia.	8.70	9.58	0.37	4.20	nr	**13.78**
Sweep Tee - Equal						
15mm dia.	5.09	5.60	0.37	4.20	nr	**9.80**
22mm dia.	6.20	6.82	0.40	4.54	nr	**11.36**
28mm dia.	11.76	12.93	0.45	5.11	nr	**18.04**
35mm dia.	16.99	18.69	0.59	6.71	nr	**25.40**
42mm dia.	25.19	27.71	0.62	7.04	nr	**34.75**
54mm dia.	27.91	30.70	0.67	7.61	nr	**38.31**
67mm dia.	48.58	53.44	0.83	9.44	nr	**62.88**
Sweep Tee - Reducing						
22 x 22 x 15mm dia.	5.64	6.21	0.40	4.54	nr	**10.75**
28 x 28 x 22mm dia.	9.55	10.51	0.45	5.11	nr	**15.62**
35 x 35 x 22mm dia.	16.70	18.37	0.59	6.71	nr	**25.08**
Sweep Tee - Double						
15mm dia.	5.90	6.49	0.37	4.20	nr	**10.69**
22mm dia.	8.04	8.84	0.40	4.54	nr	**13.38**
28mm dia.	12.22	13.44	0.45	5.11	nr	**18.55**
Four Ended Fittings						
Cross						
15mm dia.	7.81	8.59	0.49	5.57	nr	**14.16**
22mm dia.	8.74	9.61	0.53	6.02	nr	**15.63**
28mm dia.	12.52	13.77	0.59	6.71	nr	**20.48**
Extra Over Copper Pipes; high duty capillary fittings; BS 864						
One Ended Fittings						
Stop End						
15mm dia.	4.18	4.60	0.13	1.48	nr	**6.08**

Y:MECHANICAL AND ELECTRICAL SERVICES

Item	Net Price £	Material £	Labour hours	Labour £	Unit	Total Rate £
Two Ended Fittings						
Straight Coupling Copper to Copper						
15mm dia.	1.93	2.12	0.25	2.84	nr	**4.96**
22mm dia.	3.07	3.38	0.28	3.18	nr	**6.56**
28mm dia.	4.35	4.79	0.33	3.75	nr	**8.54**
35mm dia.	7.65	8.42	0.37	4.20	nr	**12.62**
42mm dia.	8.38	9.22	0.42	4.77	nr	**13.99**
54mm dia.	12.31	13.54	0.45	5.11	nr	**18.65**
Reducing Coupling						
15mm dia.	3.62	3.98	0.25	2.84	nr	**6.82**
22mm dia.	4.18	4.60	0.28	3.18	nr	**7.78**
28mm dia.	5.77	6.34	0.33	3.75	nr	**10.09**
Straight Female Connector						
15mm dia.	4.71	5.18	0.25	2.84	nr	**8.02**
22mm dia.	5.33	5.86	0.28	3.18	nr	**9.04**
28mm dia.	7.84	8.63	0.33	3.75	nr	**12.38**
Straight Male Connector						
15mm dia.	4.59	5.05	0.25	2.84	nr	**7.89**
22mm dia.	5.33	5.86	0.28	3.18	nr	**9.04**
28mm dia.	7.84	8.63	0.33	3.75	nr	**12.38**
42mm dia.	15.31	16.84	0.42	4.77	nr	**21.61**
54mm dia.	24.90	27.39	0.45	5.11	nr	**32.50**
Reducer						
15 x 12mm dia.	2.33	2.57	0.25	2.84	nr	**5.41**
22 x 15mm dia.	2.33	2.57	0.28	3.18	nr	**5.75**
28 x 22mm dia.	4.18	4.60	0.33	3.75	nr	**8.35**
35 x 28mm dia.	5.33	5.86	0.37	4.20	nr	**10.06**
42 x 35mm dia.	6.86	7.54	0.42	4.77	nr	**12.31**
54 x 42mm dia.	11.10	12.21	0.25	2.84	nr	**15.05**
Straight Union Adaptor						
15mm dia.	3.82	4.20	0.25	2.84	nr	**7.04**
22mm dia.	5.19	5.71	0.28	3.18	nr	**8.89**
28mm dia.	6.86	7.54	0.33	3.75	nr	**11.29**
35mm dia.	12.45	13.70	0.37	4.20	nr	**17.90**
42mm dia.	15.75	17.33	0.42	4.77	nr	**22.10**
Bent Union Adaptor						
15mm dia.	9.97	10.97	0.25	2.84	nr	**13.81**
22mm dia.	13.40	14.74	0.28	3.18	nr	**17.92**
28mm dia.	18.12	19.93	0.33	3.75	nr	**23.68**
Adaptor Male Copper to FI						
15mm dia.	7.09	7.80	0.25	2.84	nr	**10.64**
22mm dia.	7.23	7.96	0.28	3.18	nr	**11.14**

Y:MECHANICAL AND ELECTRICAL SERVICES

Item	Net Price £	Material £	Labour hours	Labour £	Unit	Total Rate £
Y10 : PIPELINES : COPPER (contd)						
Extra Over Copper Pipes; high duty capillary fittings; BS 864						
Two Ended Fittings						
Union Coupling						
15mm dia.	8.63	9.50	0.37	4.20	nr	**13.70**
22mm dia.	11.10	12.21	0.40	4.54	nr	**16.75**
28mm dia.	15.31	16.84	0.45	5.11	nr	**21.95**
35mm dia.	26.78	29.46	0.59	6.71	nr	**36.17**
42mm dia.	31.49	34.64	0.62	7.04	nr	**41.68**
Elbow						
15mm dia.	5.55	6.11	0.25	2.84	nr	**8.95**
22mm dia.	5.93	6.52	0.28	3.18	nr	**9.70**
28mm dia.	8.82	9.70	0.33	3.75	nr	**13.45**
35mm dia.	13.78	15.16	0.37	4.20	nr	**19.36**
42mm dia.	17.17	18.89	0.42	4.77	nr	**23.66**
54mm dia.	29.90	32.89	0.42	4.77	nr	**37.66**
Return Bend						
28mm dia.	18.10	19.91	0.33	3.75	nr	**23.66**
35mm dia.	20.98	23.08	0.37	4.20	nr	**27.28**
Bent Male Union Connector						
15mm dia.	12.90	14.19	0.37	4.20	nr	**18.39**
22mm dia.	17.31	19.04	0.40	4.54	nr	**23.58**
28mm dia.	31.49	34.64	0.45	5.11	nr	**39.75**
Composite Flange						
22mm dia.	21.29	23.42	0.35	3.98	nr	**27.40**
28mm dia.	23.66	26.02	0.37	4.20	nr	**30.22**
35mm dia.	30.77	33.85	0.38	4.32	nr	**38.17**
42mm dia.	35.47	39.02	0.41	4.66	nr	**43.68**
54mm dia.	49.68	54.65	0.43	4.89	nr	**59.54**
Three Ended Fittings						
Equal Tee						
15mm dia.	6.38	7.01	0.37	4.20	nr	**11.21**
22mm dia.	8.05	8.85	0.40	4.54	nr	**13.39**
28mm dia.	10.56	11.62	0.45	5.11	nr	**16.73**
35mm dia.	18.12	19.93	0.59	6.71	nr	**26.64**
42mm dia.	23.02	25.32	0.76	8.64	nr	**33.96**
54mm dia.	36.27	39.90	0.67	7.61	nr	**47.51**
Reducing Tee						
15mm dia.	8.80	9.67	0.37	4.20	nr	**13.87**
22mm dia.	10.37	11.41	0.40	4.54	nr	**15.95**
28mm dia.	14.73	16.21	0.45	5.11	nr	**21.32**
35mm dia.	23.43	25.78	0.63	7.16	nr	**32.94**
42mm dia.	29.94	32.93	0.76	8.64	nr	**41.57**
54mm dia.	47.38	52.12	0.96	10.91	nr	**63.03**

Y:MECHANICAL AND ELECTRICAL SERVICES

Item	Net Price £	Material £	Labour hours	Labour £	Unit	Total Rate £
Extra Over Copper Pipes; Compression Fittings; BS 864						
One Ended Fitting						
Stop End						
15mm dia.	0.89	0.97	0.10	1.14	nr	**2.11**
22mm dia.	1.13	1.25	0.33	3.75	nr	**5.00**
28mm dia.	2.77	3.05	0.14	1.59	nr	**4.64**
Straight Swivel Connector Copper to Imperial Copper						
22mm dia.	2.82	3.11	0.22	2.50	nr	**5.61**
Two Ended Fittings						
Straight Connector Copper to Copper						
15mm dia.	0.62	0.68	0.18	2.04	nr	**2.72**
22mm dia.	1.04	1.14	0.22	2.50	nr	**3.64**
28mm dia.	2.54	2.79	0.25	2.84	nr	**5.63**
Straight Connector;Copper to Imperial Copper						
22mm dia.	2.35	2.58	0.22	2.50	nr	**5.08**
Male Coupling Copper to M1 (BSP)						
15mm dia.	0.55	0.60	0.18	2.04	nr	**2.64**
22mm dia.	0.82	0.90	0.22	2.50	nr	**3.40**
28mm dia.	1.47	1.62	0.25	2.84	nr	**4.46**
Male Coupling with Long Thread and Backnut						
15mm dia.	2.37	2.61	0.18	2.04	nr	**4.65**
22mm dia.	3.01	3.32	0.22	2.50	nr	**5.82**
Female Coupling Copper to Fl. BSP						
15mm dia.	0.62	0.68	0.18	2.04	nr	**2.72**
22mm dia.	0.91	1.01	0.22	2.50	nr	**3.51**
28mm dia.	1.98	2.18	0.25	2.84	nr	**5.02**
Elbow						
15mm dia.	0.73	0.80	0.18	2.04	nr	**2.84**
22mm dia.	1.26	1.38	0.22	2.50	nr	**3.88**
28mm dia.	3.15	3.46	0.25	2.84	nr	**6.30**
Male Elbow - Copper to Fl (BSP)						
15mm x 1/2" dia.	1.06	1.17	0.18	2.04	nr	**3.21**
22mm x 3/4" dia.	1.36	1.50	0.22	2.50	nr	**4.00**
28mm x 1" dia.	2.91	3.20	0.25	2.84	nr	**6.04**

Y:MECHANICAL AND ELECTRICAL SERVICES

Item	Net Price £	Material £	Labour hours	Labour £	Unit	Total Rate £
Y10 : PIPELINES : COPPER (contd)						
Extra Over Copper Pipes; Compression Fittings (contd)						
Two Ended Fittings (contd)						
Female Elbow - Copper to FI (BSP)						
15mm x 1/2" dia.	1.46	1.61	0.18	2.04	nr	**3.65**
22mm x 3/4" dia.	2.12	2.33	0.22	2.50	nr	**4.83**
28mm x 1" dia.	3.53	3.89	0.25	2.84	nr	**6.73**
Backplate Elbow						
15mm x 1/2" dia.	2.17	2.38	0.48	5.45	nr	**7.83**
Reducing Set (Internal)						
15mm dia.	0.86	0.94	0.18	2.04	nr	**2.98**
Tank Coupling - Long Thread						
22mm dia.	3.96	4.35	0.53	6.02	nr	**10.37**
Tee Equal						
15mm dia.	1.02	1.12	0.29	3.29	nr	**4.41**
22mm dia.	1.75	1.92	0.30	3.41	nr	**5.33**
28mm dia.	5.41	5.95	0.34	3.86	nr	**9.81**
Three Ended Fittings						
Tee Reducing						
22mm dia.	3.28	3.61	0.30	3.41	nr	**7.02**
Backplate Tee						
15mm dia.	3.82	4.20	0.59	6.71	nr	**10.91**
Extra Over Copper Pipes; dezincification resistant compression fittings; BS 864						
One Ended Fittings						
Stop End						
15mm dia.	1.22	1.34	0.10	1.14	nr	**2.48**
22mm dia.	1.84	2.02	0.12	1.36	nr	**3.38**
28mm dia.	3.06	3.37	0.14	1.59	nr	**4.96**
35mm dia.	4.47	4.92	0.17	1.93	nr	**6.85**
42mm dia.	7.78	8.55	0.19	2.16	nr	**10.71**

Y:MECHANICAL AND ELECTRICAL SERVICES

Item	Net Price £	Material £	Labour hours	Labour £	Unit	Total Rate £
Two Ended Fittings						
Straight Coupling Copper to Copper						
15mm dia.	1.05	1.15	0.18	2.04	nr	**3.19**
22mm dia.	1.62	1.79	0.22	2.50	nr	**4.29**
28mm dia.	3.11	3.42	0.25	2.84	nr	**6.26**
35mm dia.	6.04	6.65	0.30	3.41	nr	**10.06**
42mm dia.	7.75	8.53	0.35	3.98	nr	**12.51**
54mm dia.	11.59	12.75	0.40	4.54	nr	**17.29**
Male Coupling Copper to MI (BSP)						
15mm x 1/2" dia.	0.89	0.97	0.18	2.04	nr	**3.01**
22mm x 3/4" dia.	1.31	1.44	0.22	2.50	nr	**3.94**
28mm x 1" dia.	2.36	2.59	0.25	2.84	nr	**5.43**
35mm x 1 1/4" dia.	4.73	5.20	0.30	3.41	nr	**8.61**
42mm x 1 1/2" dia.	6.88	7.57	0.35	3.98	nr	**11.55**
54mm x 2" dia.	9.91	10.90	0.60	6.82	nr	**17.72**
Male Coupling with Long Thread and Backnuts						
22mm dia.	3.63	3.99	0.22	2.50	nr	**6.49**
28mm dia.	4.03	4.43	0.22	2.50	nr	**6.93**
Female Coupling Copper to FI (BSP)						
15mm x 1/2" dia.	1.01	1.11	0.18	2.04	nr	**3.15**
22mm x 3/4" dia.	1.52	1.67	0.22	2.50	nr	**4.17**
28mm x 1" dia.	2.55	2.80	0.25	2.84	nr	**5.64**
35mm x 1 1/4" dia.	5.19	5.70	0.30	3.41	nr	**9.11**
42mm x 1 1/2" dia.	7.04	7.74	0.35	3.98	nr	**11.72**
54mm x 2" dia.	10.19	11.21	0.40	4.54	nr	**15.75**
Elbow						
15mm dia.	1.22	1.34	0.18	2.04	nr	**3.38**
22mm dia.	1.98	2.17	0.22	2.50	nr	**4.67**
28mm dia.	4.02	4.43	0.25	2.84	nr	**7.27**
35mm dia.	7.59	8.35	0.30	3.41	nr	**11.76**
42mm dia.	11.03	12.13	0.35	3.98	nr	**16.11**
54mm dia.	18.13	19.94	0.40	4.54	nr	**24.48**
Male Elbow Copper to MI (BSP)						
15mm x 1/2" dia.	1.06	1.17	0.18	2.04	nr	**3.21**
22mm x 3/4" dia.	1.36	1.50	0.22	2.50	nr	**4.00**
28mm x 1" dia.	2.91	3.20	0.25	2.84	nr	**6.04**
Female Elbow Copper to FI (BSP)						
15mm x 1/2" dia.	1.46	1.61	0.18	2.04	nr	**3.65**
22mm x 3/4" dia.	2.12	2.33	0.22	2.50	nr	**4.83**
28mm x 1" dia.	3.53	3.89	0.25	2.84	nr	**6.73**
Backplate Elbow						
15mm x 1/2" dia.	2.17	2.38	0.48	5.45	nr	**7.83**

Y:MECHANICAL AND ELECTRICAL SERVICES

Item	Net Price £	Material £	Labour hours	Labour £	Unit	Total Rate £
Y10 : PIPELINES : COPPER (contd)						
Extra Over Copper Pipes; dezincification resistant compression fittings; BS 864 (contd)						
Two Ended Fittings (contd)						
Straight Tap Connector						
15mm dia.	2.35	2.58	0.14	1.59	nr	**4.17**
22mm dia.	2.82	3.11	0.15	1.70	nr	**4.81**
Tank Coupling						
15mm dia.	2.37	2.61	0.18	2.04	nr	**4.65**
22mm dia.	2.55	2.80	0.22	2.50	nr	**5.30**
28mm dia.	4.66	5.13	0.25	2.84	nr	**7.97**
35mm dia.	5.23	5.75	0.30	3.41	nr	**9.16**
42mm dia.	7.25	7.97	0.35	3.98	nr	**11.95**
54mm dia.	11.16	12.27	0.25	2.84	nr	**15.11**
Reducing Set (Internal)						
15mm dia.	0.86	0.95	0.18	2.04	nr	**2.99**
22mm dia.	0.97	1.06	0.22	2.50	nr	**3.56**
Three Ended fittings						
Tee Equal						
15mm dia.	1.68	1.85	0.29	3.29	nr	**5.14**
22mm dia.	2.85	3.13	0.30	3.41	nr	**6.54**
28mm dia.	5.65	6.22	0.34	3.86	nr	**10.08**
35mm dia.	10.36	11.39	0.45	5.11	nr	**16.50**
42mm dia.	17.24	18.97	0.47	5.34	nr	**24.31**
54mm dia.	27.03	29.73	0.56	6.36	nr	**36.09**
Tee Reducing						
22mm dia.	3.50	3.85	0.30	3.41	nr	**7.26**
28mm dia.	5.78	6.36	0.34	3.86	nr	**10.22**
35mm dia.	10.36	11.39	0.45	5.11	nr	**16.50**
42mm dia.	16.16	17.77	0.47	5.34	nr	**23.11**
54mm dia.	27.03	29.73	0.56	6.36	nr	**36.09**

Y:MECHANICAL AND ELECTRICAL SERVICES

Item	Net Price £	Material £	Labour hours	Labour £	Unit	Total Rate £
Y10 : PIPELINES : COPPER (contd)						
Extra Over Copper Pipes; bronze welding flanges; Imperial; BS 4504						
Bronze Flange; PN6						
15mm dia.	11.19	12.31	0.25	2.84	nr	**15.15**
22mm dia.	13.29	14.62	0.28	3.18	nr	**17.80**
28mm dia.	15.24	16.76	0.30	3.41	nr	**20.17**
35mm dia.	21.46	23.61	0.37	4.20	nr	**27.81**
42mm dia.	26.26	28.89	0.42	4.77	nr	**33.66**
54mm dia.	37.40	41.14	0.45	5.11	nr	**46.25**
67mm dia.	43.14	47.45	0.56	6.36	nr	**53.81**
76mm dia.	49.10	54.01	0.67	7.61	nr	**61.62**
108mm dia.	65.98	72.57	0.75	8.52	nr	**81.09**
133mm dia.	80.10	88.11	0.92	10.45	nr	**98.56**
159mm dia.	113.51	124.86	1.17	13.30	nr	**138.16**
Bronze Flange; PN10						
15mm dia.	14.64	16.10	0.25	2.84	nr	**18.94**
22mm dia.	17.05	18.75	0.28	3.18	nr	**21.93**
28mm dia.	17.14	18.86	0.33	3.78	nr	**22.64**
35mm dia.	23.41	25.75	0.37	4.20	nr	**29.95**
42mm dia.	27.94	30.74	0.42	4.77	nr	**35.51**
54mm dia.	39.74	43.71	0.45	5.11	nr	**48.82**
67mm dia.	43.14	47.45	0.56	6.36	nr	**53.81**
76mm dia.	54.60	60.06	0.67	7.61	nr	**67.67**
108mm dia.	80.85	88.94	0.75	8.52	nr	**97.46**
133mm dia.	92.92	102.22	0.92	10.45	nr	**112.67**
159mm dia.	142.15	156.37	1.17	13.30	nr	**169.67**
Extra Over Copper Pipes; bronze welding flanges; BS4504						
Bronze Flange; PN22						
15mm dia.	14.64	16.10	0.25	2.84	nr	**18.94**
22mm dia.	17.05	18.75	0.28	3.18	nr	**21.93**
28mm dia.	17.68	19.44	0.33	3.78	nr	**23.22**
35mm dia.	23.41	25.75	0.37	4.20	nr	**29.95**
42mm dia.	27.94	30.74	0.42	4.77	nr	**35.51**
54mm dia.	42.00	46.20	0.45	5.11	nr	**51.31**
67mm dia.	51.13	56.25	0.56	6.36	nr	**62.61**
76mm dia.	64.98	71.48	0.67	7.61	nr	**79.09**
108mm dia.	83.50	91.85	0.75	8.52	nr	**100.37**
133mm dia.	138.17	151.98	0.92	10.45	nr	**162.43**
159mm dia.	172.06	189.27	1.17	13.30	nr	**202.57**
One Ended Fittings; Bronze Blank Flange; Tables D and E						
Bronze Blank Flange; PN6						
15mm dia.	9.94	10.94	0.25	2.84	nr	**13.78**
22mm dia.	12.70	13.97	0.25	2.84	nr	**16.81**
28mm dia.	13.04	14.34	0.25	2.84	nr	**17.18**
35mm dia.	21.34	23.48	0.30	3.41	nr	**26.89**

Y:MECHANICAL AND ELECTRICAL SERVICES

Item	Net Price £	Material £	Labour hours	Labour £	Unit	Total Rate £
Y10 : PIPELINES : COPPER (contd)						
One Ended Fittings; Bronze Blank Flange; Tables D and E						
Bronze Blank Flange; PN6 (contd)						
42mm dia.	28.79	31.67	0.30	3.41	nr	**35.08**
54mm dia.	31.26	34.38	0.30	3.41	nr	**37.79**
67mm dia.	39.07	42.98	0.33	3.75	nr	**46.73**
76mm dia.	50.33	55.36	0.33	3.75	nr	**59.11**
108mm dia.	79.95	87.95	0.38	4.32	nr	**92.27**
133mm dia.	94.33	103.77	0.52	5.91	nr	**109.68**
159mm dia.	118.85	130.74	0.57	6.48	nr	**137.22**
One Ended Fittings; Bronze Blank Flange; Table F						
Bronze Blank Flange; PN10						
15mm dia.	12.04	13.24	0.25	2.84	nr	**16.08**
22mm dia.	15.56	17.12	0.25	2.84	nr	**19.96**
28mm dia.	17.23	18.95	0.25	2.84	nr	**21.79**
35mm dia.	21.34	23.48	0.30	3.41	nr	**26.89**
42mm dia.	40.01	44.01	0.30	3.41	nr	**47.42**
54mm dia.	45.61	50.17	0.30	3.41	nr	**53.58**
67mm dia.	48.93	53.82	0.47	5.34	nr	**59.16**
76mm dia.	62.87	69.15	0.47	5.34	nr	**74.49**
108mm dia.	96.01	105.62	0.52	5.91	nr	**111.53**
133mm dia.	101.37	111.51	0.52	5.91	nr	**117.42**
159mm dia.	185.39	203.93	0.70	7.96	nr	**211.89**
One Ended Fittings; Bronze Blank Flange; Table H						
Bronze Blank Flange; PN16						
15mm dia.	12.04	13.24	0.25	2.84	nr	**16.08**
22mm dia.	15.80	17.39	0.25	2.84	nr	**20.23**
28mm dia.	17.23	18.95	0.25	2.84	nr	**21.79**
35mm dia.	21.34	23.48	0.30	3.41	nr	**26.89**
42mm dia.	40.01	44.01	0.30	3.41	nr	**47.42**
54mm dia.	45.61	50.17	0.30	3.41	nr	**53.58**
67mm dia.	70.93	78.02	0.47	5.34	nr	**83.36**
76mm dia.	84.01	92.41	0.47	5.34	nr	**97.75**
108mm dia.	106.80	117.48	0.52	5.91	nr	**123.39**
133mm dia.	175.27	192.80	0.52	5.91	nr	**198.71**
159mm dia.	224.33	246.77	0.70	7.96	nr	**254.73**
One Ended Fittings; Bronze Screwed Flange; Tables D and E						
Bronze Screwed Flange; 6 BSP						
15mm dia.	9.94	10.94	0.37	4.20	nr	**15.14**
22mm dia.	11.50	12.65	0.49	5.57	nr	**18.22**
28mm dia.	11.98	13.18	0.55	6.25	nr	**19.43**
35mm dia.	16.15	17.76	0.63	7.16	nr	**24.92**

Y:MECHANICAL AND ELECTRICAL SERVICES

Item	Net Price £	Material £	Labour hours	Labour £	Unit	Total Rate £
42mm dia.	19.40	21.34	0.73	8.30	nr	**29.64**
54mm dia.	26.35	28.98	0.88	10.00	nr	**38.98**
67mm dia.	33.09	36.40	1.08	12.28	nr	**48.68**
76mm dia.	39.92	43.91	1.28	14.55	nr	**58.46**
108mm dia.	63.22	69.54	1.42	16.14	nr	**85.68**
133mm dia.	74.95	82.44	1.67	19.00	nr	**101.44**
159mm dia.	95.41	104.95	2.08	23.67	nr	**128.62**
One Ended Fittings; Bronze Screwed Flange; Table F						
Bronze Screwed Flange; 10 BSP						
15mm dia.	12.05	13.26	0.37	4.20	nr	**17.46**
22mm dia.	14.04	15.44	0.49	5.57	nr	**21.01**
28mm dia.	15.50	17.05	0.55	6.25	nr	**23.30**
35mm dia.	22.07	24.28	0.63	7.16	nr	**31.44**
42mm dia.	26.95	29.64	0.73	8.30	nr	**37.94**
54mm dia.	38.33	42.17	0.88	10.00	nr	**52.17**
67mm dia.	44.87	49.36	1.08	12.28	nr	**61.64**
76mm dia.	50.65	55.72	1.28	14.55	nr	**70.27**
108mm dia.	67.25	73.97	1.42	16.14	nr	**90.11**
133mm dia.	81.49	89.64	1.67	19.00	nr	**108.64**
159mm dia.	144.61	159.08	2.08	23.67	nr	**182.75**
One Ended Fittings; Bronze Screwed Flange; Table H						
Bronze Screwed Flange; 16 BSP						
15mm dia.	12.05	13.26	0.37	4.20	nr	**17.46**
22mm dia.	14.04	15.44	0.49	5.57	nr	**21.01**
28mm dia.	15.50	17.05	0.55	6.25	nr	**23.30**
35mm dia.	22.07	24.28	0.64	7.23	nr	**31.51**
42mm dia.	26.95	29.64	0.73	8.30	nr	**37.94**
54mm dia.	38.33	42.17	0.88	10.00	nr	**52.17**
67mm dia.	52.83	58.12	1.08	12.28	nr	**70.40**
76mm dia.	66.69	73.36	1.28	14.55	nr	**87.91**
108mm dia.	85.17	93.69	1.42	16.14	nr	**109.83**
133mm dia.	140.56	154.62	1.67	19.00	nr	**173.62**
159mm dia.	174.25	191.67	2.08	23.67	nr	**215.34**
Extra Over Copper Pipes; labours						
Made Bend						
15mm dia.			0.30	3.41	nr	**3.41**
22mm dia.			0.33	3.75	nr	**3.75**
28mm dia.			0.36	4.09	nr	**4.09**
35mm dia.			0.48	5.45	nr	**5.45**
42mm dia.			0.60	6.82	nr	**6.82**
54mm dia.			0.66	7.50	nr	**7.50**
67mm dia.			0.76	8.64	nr	**8.64**
76mm dia.			0.84	9.55	nr	**9.55**

Y:MECHANICAL AND ELECTRICAL SERVICES

Item	Net Price £	Material £	Labour hours	Labour £	Unit	Total Rate £
Y10 : PIPELINES : COPPER (contd)						
Extra Over Copper Pipes; labours (contd)						
Bronze Butt Weld						
15mm dia.			0.22	2.50	nr	**2.50**
22mm dia.			0.25	2.84	nr	**2.84**
28mm dia.			0.28	3.18	nr	**3.18**
35mm dia.			0.35	3.98	nr	**3.98**
42mm dia.			0.42	4.77	nr	**4.77**
54mm dia.			0.50	5.68	nr	**5.68**
67mm dia.			0.63	7.16	nr	**7.16**
76mm dia.			0.75	8.52	nr	**8.52**
108mm dia.			0.91	10.35	nr	**10.35**
133mm dia.			1.15	13.07	nr	**13.07**
159mm dia.			1.32	15.01	nr	**15.01**

Keep your figures up to date, free of charge

This section, and most of the other information in this Price Book, is brought up to date every three months, until the next annual edition, in the *Price Book Update*.

The *Update* is available free to all Price Book purchasers.

To ensure you receive your copy, simply complete the reply card from the centre of the book and return it to us.

Y:MECHANICAL AND ELECTRICAL SERVICES

Item	Net Price £	Material £	Labour hours	Labour £	Unit	Total Rate £
Y10 : PIPELINES : PIPE FIXINGS						
For Steel Pipes						
Single pipe bracket, screw on, black malleable, screwed to wood						
15mm dia.	0.62	0.68	0.15	1.70	nr	**2.38**
20mm dia.	0.69	0.76	0.15	1.70	nr	**2.46**
25mm dia.	0.79	0.87	0.20	2.27	nr	**3.14**
32mm dia.	1.08	1.18	0.20	2.27	nr	**3.45**
40mm dia.	1.43	1.58	0.25	2.84	nr	**4.42**
50mm dia.	1.91	2.10	0.25	2.84	nr	**4.94**
65mm dia.	2.51	2.76	0.30	3.41	nr	**6.17**
80mm dia.	3.44	3.79	0.35	3.98	nr	**7.77**
100mm dia.	5.02	5.52	0.40	4.54	nr	**10.06**
Single pipe bracket, screw on, galvanised, screwed to wood						
15mm dia.	0.62	0.68	0.15	1.70	nr	**2.38**
20mm dia.	0.69	0.76	0.15	1.70	nr	**2.46**
25mm dia.	0.79	0.87	0.20	2.27	nr	**3.14**
32mm dia.	1.08	1.18	0.20	2.27	nr	**3.45**
40mm dia.	1.43	1.58	0.25	2.84	nr	**4.42**
50mm dia.	1.91	2.10	0.25	2.84	nr	**4.94**
65mm dia.	2.51	2.76	0.30	3.41	nr	**6.17**
80mm dia.	3.44	3.79	0.35	3.98	nr	**7.77**
100mm dia.	5.02	5.52	0.40	4.54	nr	**10.06**
Single pipe bracket for building in, black malleable						
15mm dia.	0.62	0.68	0.11	1.25	nr	**1.93**
20mm dia.	0.69	0.76	0.12	1.36	nr	**2.12**
25mm dia.	0.83	0.91	0.13	1.48	nr	**2.39**
32mm dia.	1.08	1.18	0.14	1.59	nr	**2.77**
40mm dia.	1.43	1.58	0.15	1.70	nr	**3.28**
50mm dia.	1.69	1.86	0.16	1.82	nr	**3.68**
Single pipe bracket for building in, galvanised						
15mm dia.	0.62	0.68	0.11	1.25	nr	**1.93**
20mm dia.	0.69	0.76	0.12	1.36	nr	**2.12**
25mm dia.	0.83	0.91	0.13	1.48	nr	**2.39**
32mm dia.	1.08	1.18	0.14	1.59	nr	**2.77**
40mm dia.	1.43	1.58	0.15	1.70	nr	**3.28**
50mm dia.	1.69	1.86	0.16	1.82	nr	**3.68**
Single pipe bracket, screw on, black malleable, plugged and screwed						
15mm dia.	0.62	0.68	0.30	3.41	nr	**4.09**
20mm dia.	0.69	0.76	0.30	3.41	nr	**4.17**
25mm dia.	0.79	0.87	0.38	4.32	nr	**5.19**
32mm dia.	1.08	1.18	0.38	4.32	nr	**5.50**
40mm dia.	1.43	1.58	0.38	4.32	nr	**5.90**

Y:MECHANICAL AND ELECTRICAL SERVICES

Item	Net Price £	Material £	Labour hours	Labour £	Unit	Total Rate £
Y10 : PIPELINES : PIPE FIXINGS (contd)						
For Steel Pipes (contd)						
Single pipe bracket, screw on, black malleable, plugged and screwed (contd)						
50mm dia.	1.91	2.10	0.38	4.32	nr	**6.42**
65mm dia.	2.51	2.76	0.40	4.54	nr	**7.30**
80mm dia.	3.44	3.79	0.50	5.68	nr	**9.47**
100mm dia.	5.02	5.52	0.50	5.68	nr	**11.20**
Single pipe bracket, screw on, galvanised, plugged and screwed						
15mm dia.	0.87	0.96	0.30	3.41	nr	**4.37**
20mm dia.	0.97	1.07	0.30	3.41	nr	**4.48**
25mm dia.	1.13	1.24	0.38	4.32	nr	**5.56**
32mm dia.	1.54	1.69	0.38	4.32	nr	**6.01**
40mm dia.	2.05	2.26	0.38	4.32	nr	**6.58**
50mm dia.	2.72	2.99	0.38	4.32	nr	**7.31**
65mm dia.	3.59	3.95	0.40	4.54	nr	**8.49**
80mm dia.	4.92	5.41	0.50	5.68	nr	**11.09**
100mm dia.	7.18	7.89	0.50	5.68	nr	**13.57**
Pipe ring single socket, black malleable						
15mm dia.	0.97	1.07	0.11	1.25	nr	**2.32**
20mm dia.	1.13	1.24	0.12	1.36	nr	**2.60**
25mm dia.	1.23	1.35	0.13	1.48	nr	**2.83**
32mm dia.	1.28	1.41	0.14	1.59	nr	**3.00**
40mm dia.	1.64	1.80	0.15	1.70	nr	**3.50**
50mm dia.	2.10	2.31	0.16	1.82	nr	**4.13**
65mm dia.	3.02	3.33	0.30	3.41	nr	**6.74**
80mm dia.	3.64	4.00	0.35	3.98	nr	**7.98**
100mm dia.	5.54	6.09	0.40	4.54	nr	**10.63**
150mm dia.	12.45	13.70	0.60	6.82	nr	**20.52**
Pipe ring, single socket, galvanised						
15mm dia.	0.97	1.07	0.11	1.25	nr	**2.32**
20mm dia.	1.13	1.24	0.12	1.36	nr	**2.60**
25mm dia.	1.23	1.35	0.13	1.48	nr	**2.83**
32mm dia.	1.28	1.41	0.15	1.70	nr	**3.11**
40mm dia.	1.64	1.80	0.15	1.70	nr	**3.50**
50mm dia.	2.10	2.31	0.16	1.82	nr	**4.13**
65mm dia.	3.02	3.33	0.30	3.41	nr	**6.74**
80mm dia.	3.64	4.00	0.35	3.98	nr	**7.98**
100mm dia.	5.54	6.09	0.40	4.54	nr	**10.63**
150mm dia.	12.45	13.70	0.60	6.82	nr	**20.52**
Pipe ring, double socket, black malleable						
15mm dia.	1.18	1.30	0.11	1.25	nr	**2.55**
20mm dia.	1.33	1.47	0.12	1.36	nr	**2.83**
25mm dia.	1.49	1.63	0.13	1.48	nr	**3.11**
32mm dia.	1.74	1.92	0.14	1.59	nr	**3.51**
40mm dia.	2.05	2.26	0.15	1.70	nr	**3.96**
50mm dia.	2.31	2.54	0.16	1.82	nr	**4.36**

Y:MECHANICAL AND ELECTRICAL SERVICES

Item	Net Price £	Material £	Labour hours	Labour £	Unit	Total Rate £
Pipe ring, double socket, galvanised						
15mm dia.	1.18	1.30	0.11	1.25	nr	**2.55**
20mm dia.	1.33	1.47	0.12	1.36	nr	**2.83**
25mm dia.	1.49	1.63	0.13	1.48	nr	**3.11**
32mm dia.	1.74	1.92	0.14	1.59	nr	**3.51**
40mm dia.	2.05	2.26	0.15	1.70	nr	**3.96**
50mm dia.	2.31	2.54	0.16	1.82	nr	**4.36**
Screw on backplate, black malleable, screwed to wood						
15mm dia.	1.18	1.30	0.15	1.70	nr	**3.00**
Screw on backplate, galvanised, screwed to wood						
15mm dia.	1.18	1.30	0.15	1.70	nr	**3.00**
Screw on backplate, black malleable, plugged and screwed						
15mm dia.	1.18	1.30	0.30	3.41	nr	**4.71**
Screw on backplate, galvanised, plugged and screwed						
15mm dia.	1.18	1.30	0.30	3.41	nr	**4.71**
For Copper Pipes						
Saddle Band						
15mm dia.	0.06	0.07	0.15	1.70	nr	**1.77**
22mm dia.	0.06	0.07	0.15	1.70	nr	**1.77**
28mm dia.	0.08	0.09	0.20	2.27	nr	**2.36**
35mm dia.	0.12	0.13	0.20	2.27	nr	**2.40**
42mm dia.	0.26	0.28	0.25	2.84	nr	**3.12**
54mm dia.	0.35	0.39	0.25	2.84	nr	**3.23**
Single spacing clip						
15mm dia.	0.08	0.09	0.15	1.70	nr	**1.79**
22mm dia.	0.08	0.09	0.16	1.82	nr	**1.91**
28mm dia.	0.35	0.39	0.20	2.27	nr	**2.66**
Two Piece Spacing Clip						
15mm dia.	0.13	0.14	0.15	1.70	nr	**1.84**
22mm dia.	0.14	0.15	0.15	1.70	nr	**1.85**
28mm dia.	0.18	0.20	0.20	2.27	nr	**2.47**
35mm dia.	0.27	0.30	0.25	2.84	nr	**3.14**
42mm dia.	0.51	0.56	0.25	2.84	nr	**3.40**
54mm dia.	0.65	0.72	0.25	2.84	nr	**3.56**
Single Pipe Bracket						
15mm dia.	0.86	0.94	0.15	1.70	nr	**2.64**
22mm dia.	0.98	1.08	0.15	1.70	nr	**2.78**
28mm dia.	1.17	1.29	0.20	2.27	nr	**3.56**

Y:MECHANICAL AND ELECTRICAL SERVICES

Item	Net Price £	Material £	Labour hours	Labour £	Unit	Total Rate £
Y10 : PIPELINES : PIPE FIXINGS (contd)						
For Copper Pipes (contd)						
Single Pipe Bracket for Building In						
22mm dia.	1.52	1.68	0.05	0.57	nr	**2.25**
28mm dia.	1.63	1.79	0.05	0.57	nr	**2.36**
Single Pipe Ring						
15mm dia.	1.52	1.68	0.30	3.41	nr	**5.09**
22mm dia.	1.63	1.79	0.30	3.41	nr	**5.20**
28mm dia.	1.93	2.13	0.38	4.32	nr	**6.45**
35mm dia.	2.08	2.29	0.38	4.32	nr	**6.61**
42mm dia.	2.26	2.49	0.38	4.32	nr	**6.81**
54mm dia.	2.73	3.00	0.40	4.54	nr	**7.54**
67mm dia.	6.36	7.00	0.40	4.54	nr	**11.54**
76mm dia.	8.02	8.83	0.50	5.68	nr	**14.51**
108mm dia.	12.45	13.69	0.50	5.68	nr	**19.37**
Double Pipe Ring						
15mm dia.	1.73	1.91	0.30	3.41	nr	**5.32**
22mm dia.	1.85	2.03	0.30	3.41	nr	**5.44**
28mm dia.	2.47	2.72	0.38	4.32	nr	**7.04**
35mm dia.	2.56	2.82	0.38	4.32	nr	**7.14**
42mm dia.	2.80	3.08	0.38	4.32	nr	**7.40**
54mm dia.	3.35	3.69	0.40	4.54	nr	**8.23**
67mm dia.	7.21	7.93	0.40	4.54	nr	**12.47**
76mm dia.	8.97	9.87	0.50	5.68	nr	**15.55**
108mm dia.	15.06	16.57	0.50	5.68	nr	**22.25**
Wall Bracket						
15mm dia.	2.17	2.39	0.05	0.57	nr	**2.96**
22mm dia.	2.56	2.82	0.05	0.57	nr	**3.39**
28mm dia.	3.12	3.43	0.05	0.57	nr	**4.00**
35mm dia.	4.02	4.42	0.05	0.57	nr	**4.99**
42mm dia.	5.31	5.84	0.05	0.57	nr	**6.41**
54mm dia.	6.73	7.41	0.05	0.57	nr	**7.98**
Hospital Bracket						
15mm dia.	1.85	2.03	0.30	3.41	nr	**5.44**
22mm dia.	1.94	2.14	0.30	3.41	nr	**5.55**
28mm dia.	2.38	2.62	0.38	4.32	nr	**6.94**
35mm dia.	2.56	2.82	0.38	4.32	nr	**7.14**
42mm dia.	3.59	3.95	0.38	4.32	nr	**8.27**
54mm dia.	4.88	5.37	0.40	4.54	nr	**9.91**
Screw on Backplate, Female						
15mm dia.	0.82	0.90	0.30	3.41	nr	**4.31**
22mm dia.	0.82	0.90	0.30	3.41	nr	**4.31**
28mm dia.	0.82	0.90	0.38	4.32	nr	**5.22**
35mm dia.	0.82	0.90	0.38	4.32	nr	**5.22**
42mm dia.	0.82	0.90	0.38	4.32	nr	**5.22**

Y:MECHANICAL AND ELECTRICAL SERVICES

Item	Net Price £	Material £	Labour hours	Labour £	Unit	Total Rate £
54mm dia.	0.82	0.90	0.40	4.54	nr	**5.44**
67mm dia.	0.82	0.90	0.40	4.54	nr	**5.44**
76mm dia.	0.82	0.90	0.50	5.68	nr	**6.58**
108mm dia.	0.82	0.90	0.50	5.68	nr	**6.58**
Screw on Backplate, Male						
15mm dia.	0.72	0.79	0.30	3.41	nr	**4.20**
22mm dia.	0.72	0.79	0.30	3.41	nr	**4.20**
28mm dia.	0.72	0.79	0.38	4.32	nr	**5.11**
35mm dia.	0.72	0.79	0.38	4.32	nr	**5.11**
42mm dia.	0.72	0.79	0.38	4.32	nr	**5.11**
54mm dia.	0.72	0.79	0.40	4.54	nr	**5.33**
67mm dia.	0.72	0.79	0.40	4.54	nr	**5.33**
76mm dia.	0.72	0.79	0.50	5.68	nr	**6.47**
108mm dia.	0.82	0.90	0.50	5.68	nr	**6.58**
White Plastic Clip, Snap On						
15mm dia.	0.04	0.04	0.08	0.91	nr	**0.95**
22mm dia.	0.05	0.05	0.08	0.91	nr	**0.96**
28mm dia.	0.07	0.08	0.10	1.14	nr	**1.22**
White Plastic Clip, Hinged						
15mm dia.	0.08	0.09	0.08	0.91	nr	**1.00**
22mm dia.	0.27	0.30	0.08	0.91	nr	**1.21**
28mm dia.	0.22	0.24	0.10	1.14	nr	**1.38**
Fabricated Hangers and Brackets.						
(Note: It has been assumed there would be sufficient quantities required to gain the benefit of bulk purchase.)						
Various sizes cut to length; including handling and fixing						
Mild steel flats	1.46	1.60	0.13	1.48	kg	**3.08**
Mild steel angle	1.40	1.54	0.17	1.93	kg	**3.47**
Mild steel channel	1.40	1.54	0.17	1.93	kg	**3.47**
Mild steel rods up to 10mm	1.40	1.54	0.17	1.93	kg	**3.47**
Labours; cut other than initial cut on metal of thickness shown						
6mm thick			0.04	0.45	nr	**0.45**
8mm thick			0.05	0.57	nr	**0.57**
10mm thick			0.06	0.68	nr	**0.68**
13mm thick			0.07	0.80	nr	**0.80**
16mm thick			0.08	0.91	nr	**0.91**
Ragged ends of flat, ditto						
6mm thick			0.12	1.36	nr	**1.36**
8mm thick			0.13	1.48	nr	**1.48**
10mm thick			0.14	1.59	nr	**1.59**

Y:MECHANICAL AND ELECTRICAL SERVICES

Item	Net Price £	Material £	Labour hours	Labour £	Unit	Total Rate £
Y10 : PIPELINES : PIPE FIXINGS (contd)						
Fabricated Hangers and Brackets (contd)						
(Note: It has been assumed there would be sufficient quantities required to gain the benefit of bulk purchase.)						
Ragged ends of angle, cut other than initial cut on metal of thickness shown						
5mm thick			0.18	2.04	nr	**2.04**
6mm thick			0.18	2.04	nr	**2.04**
8mm thick			0.19	2.16	nr	**2.16**
10mm thick			0.20	2.27	nr	**2.27**
Bend on flat, ditto						
6mm thick			0.08	0.91	nr	**0.91**
8mm thick			0.09	1.02	nr	**1.02**
10mm thick			0.09	1.02	nr	**1.02**
Twist on flat, ditto						
6mm thick			0.08	0.91	nr	**0.91**
8mm thick			0.09	1.02	nr	**1.02**
10mm thick			0.09	1.02	nr	**1.02**
Fabricated Hangers and Brackets Labours						
Drill hole not exceeding 6mm dia. in steel, not exceeding thickness shown						
6mm thick			0.04	0.45	nr	**0.45**
13mm thick			0.05	0.57	nr	**0.57**
Ditto 13mm dia. ditto						
6mm thick			0.05	0.57	nr	**0.57**
13mm thick			0.06	0.68	nr	**0.68**
Ditto 25mm dia. ditto						
6mm thick			0.08	0.91	nr	**0.91**
13mm thick			0.09	1.02	nr	**1.02**
Thread end of rod						
6mm thick			0.07	0.80	nr	**0.80**
13mm thick			0.08	0.91	nr	**0.91**
10mm thick			0.09	1.02	nr	**1.02**
Bend in rod						
3mm thick			0.06	0.68	nr	**0.68**
6mm thick			0.06	0.68	nr	**0.68**
10mm thick			0.06	0.68	nr	**0.68**

Y:MECHANICAL AND ELECTRICAL SERVICES

Item	Net Price £	Material £	Labour hours	Labour £	Unit	Total Rate £
Form eye in rod						
3mm thick			0.06	0.68	nr	**0.68**
6mm thick			0.06	0.68	nr	**0.68**
10mm thick			0.06	0.68	nr	**0.68**
Butt welding mild steel						
5mm thick			0.10	1.14	nr	**1.14**
6mm thick			0.13	1.48	nr	**1.48**
10mm thick			0.19	2.16	nr	**2.16**
13mm thick			0.26	2.96	nr	**2.96**
Welding ends of rods to steel						
3mm thick			0.08	0.91	nr	**0.91**
6mm thick			0.11	1.25	nr	**1.25**
10mm thick			0.16	1.82	nr	**1.82**
Floor or Ceiling Cover Plates						
Plastic (White)						
15mm dia.	0.26	0.29	0.15	1.70	nr	**1.99**
20mm dia.	0.29	0.32	0.21	2.39	nr	**2.71**
25mm dia.	0.29	0.32	0.21	2.39	nr	**2.71**
32mm dia.	0.31	0.34	0.23	2.61	nr	**2.95**
40mm dia.	0.48	0.52	0.25	2.84	nr	**3.36**
50mm dia.	0.78	0.85	0.25	2.84	nr	**3.69**
Chromium Plated						
15mm dia.	2.00	2.20	0.15	1.70	nr	**3.90**
20mm dia.	2.11	2.32	0.15	1.70	nr	**4.02**
25mm dia.	2.23	2.45	0.20	2.27	nr	**4.72**
32mm dia.	2.39	2.63	0.20	2.27	nr	**4.90**
40mm dia.	2.54	2.79	0.25	2.84	nr	**5.63**
50mm dia.	3.05	3.35	0.25	2.84	nr	**6.19**
Pipe Roller and Chair						
Black Malleable						
Up to 50mm dia.	2.11	2.32	0.20	2.27	nr	**4.59**
Up to 80mm dia.	3.90	4.29	0.20	2.27	nr	**6.56**
Up to 100mm dia.	4.20	4.62	0.20	2.27	nr	**6.89**
Up to 150mm dia.	5.09	5.60	0.30	3.41	nr	**9.01**
Galvanised						
Up to 50mm dia.	3.16	3.48	0.20	2.27	nr	**5.75**
Up to 80mm dia.	5.85	6.43	0.20	2.27	nr	**8.70**
Up to 100mm dia.	6.30	6.93	0.20	2.27	nr	**9.20**
Up to 150mm dia.	7.64	8.40	0.30	3.41	nr	**11.81**
Roller bracket; black malleable						
Up to 50mm dia.	1.97	2.17	0.30	3.41	nr	**5.58**

Y:MECHANICAL AND ELECTRICAL SERVICES

Item	Net Price £	Material £	Labour hours	Labour £	Unit	Total Rate £
Y10 : PIPELINES : PIPE FIXINGS (contd)						
Pipe Roller and Chair (contd)						
Roller bracket; galvanised						
Up to 50mm dia.	2.96	3.26	0.30	3.41	nr	**6.67**
Medium Quality Black Pipe Sleeves; Mild Steel						
150mm long; to pass pipe of the diameter shown						
15mm dia.	1.16	1.28	0.20	2.27	nr	**3.55**
20mm dia.	1.33	1.47	0.27	3.07	nr	**4.54**
25mm dia.	1.33	1.47	0.29	3.29	nr	**4.76**
32mm dia.	1.44	1.58	0.34	3.86	nr	**5.44**
40mm dia.	2.23	2.46	0.34	3.86	nr	**6.32**
50mm dia.	2.57	2.82	0.40	4.54	nr	**7.36**
65mm dia.	3.43	3.78	0.46	5.23	nr	**9.01**
80mm dia.	4.61	5.07	0.52	5.91	nr	**10.98**
100mm dia.	6.44	7.09	0.76	8.64	nr	**15.73**
115mm dia.	7.67	8.43	1.00	11.36	nr	**19.79**
150mm dia.	12.06	13.27	1.32	15.01	nr	**28.28**
300mm ditto						
15mm dia.	1.76	1.94	0.20	2.27	nr	**4.21**
20mm dia.	1.94	2.14	0.27	3.07	nr	**5.21**
25mm dia.	1.94	2.14	0.29	3.29	nr	**5.43**
32mm dia.	2.14	2.35	0.34	3.86	nr	**6.21**
40mm dia.	3.24	3.56	0.34	3.86	nr	**7.42**
50mm dia.	3.94	4.33	0.40	4.54	nr	**8.87**
65mm dia.	5.15	5.66	0.46	5.23	nr	**10.89**
80mm dia.	7.00	7.70	0.52	5.91	nr	**13.61**
100mm dia.	9.70	10.67	0.76	8.64	nr	**19.31**
125mm dia.	10.01	11.01	1.00	11.36	nr	**22.37**
150mm dia.	12.60	13.86	1.32	16.29	nr	**30.15**
600mm ditto						
15mm dia.	3.02	3.33	0.20	2.47	nr	**5.80**
20mm dia.	3.14	3.46	0.27	3.07	nr	**6.53**
25mm dia.	3.14	3.46	0.29	3.29	nr	**6.75**
32mm dia.	3.61	3.97	0.34	3.86	nr	**7.83**
40mm dia.	4.55	5.00	0.34	3.86	nr	**8.86**
50mm dia.	5.72	6.30	0.40	4.93	nr	**11.23**
65mm dia.	7.45	8.20	0.46	5.67	nr	**13.87**
80mm dia.	10.20	11.22	0.52	5.91	nr	**17.13**
100mm dia.	13.50	14.85	0.76	8.64	nr	**23.49**
115mm dia.	14.70	16.17	1.00	11.36	nr	**27.53**
150mm dia.	20.89	22.98	1.32	15.01	nr	**37.99**

Y:MECHANICAL AND ELECTRICAL SERVICES

Item	Net Price £	Material £	Labour hours	Labour £	Unit	Total Rate £
Y11 : PIPELINE ANCILLARIES **: REGULATING VALVES**						
Regulators						
Gunmetal: self-acting two port thermostatic regulator; Single Seat; water or water ; normally closed or normally open; screwed ends; complete with sensing element, 2m long capillary tube						
15mm dia.	256.00	281.60	1.50	17.06	nr	**298.66**
20mm dia.	263.20	289.52	1.80	20.47	nr	**309.99**
25mm dia.	271.20	298.32	1.94	22.06	nr	**320.38**
Gunmetal: self-acting two port thermostatic regulator; Double seat; operation; water or steam; flanged ends; (BS 4504 PN25); with sensing element, 2m long capillary tube; steel body						
65mm dia.	873.60	960.96	1.00	12.33	nr	**973.29**
80mm dia.	1027.20	1129.92	1.20	14.80	nr	**1144.72**
Gunmetal: self-acting two port thermostat; single seat; screwed; normally closed; with adjustable or fixed bleed device						
25mm dia.	198.40	218.24	1.70	19.32	nr	**237.56**
32mm dia.	204.00	224.40	1.50	17.06	nr	**241.46**
40mm dia.	218.40	240.24	1.55	17.61	nr	**257.85**
50mm dia.	262.40	288.64	1.60	18.18	nr	**306.82**
Self acting temperature regulator for storage calorifier; integral sensing element and pocket; screwed ends						
15mm dia.	242.88	267.17	1.40	15.91	nr	**283.08**
25mm dia.	250.80	275.88	1.70	19.32	nr	**295.20**
32mm dia.	333.52	366.87	2.06	23.42	nr	**390.29**
40mm dia.	407.44	448.18	2.30	26.17	nr	**474.35**
50mm dia.	476.08	523.69	2.65	30.05	nr	**553.74**
Self acting temperature regulator for storage calorifier; integral sensing element and pocket; flanged ends; bolted connection						
15mm dia.	357.28	393.01	0.50	5.68	nr	**398.69**
25mm dia.	385.44	423.98	0.61	6.93	nr	**430.91**
32mm dia.	498.08	547.89	0.74	8.41	nr	**556.30**
40mm dia.	588.72	647.59	0.83	9.44	nr	**657.03**
50mm dia.	684.64	753.10	0.95	10.80	nr	**763.90**
Chrome plated thermostatic mixing valves including non-return valves and inlet swivel connections with strainers; copper compression fittings						
15mm dia.	95.92	105.51	0.50	5.68	nr	**111.19**

Y:MECHANICAL AND ELECTRICAL SERVICES

Item	Net Price £	Material £	Labour hours	Labour £	Unit	Total Rate £
Y11 : PIPELINE ANCILLARIES **: REGULATING VALVES (contd)**						
Regulators (contd)						
Chrome plated thermostatic mixing valves including non-return valves and inlet swivel connections with angle pattern combined isolating valves and strainers; copper compression fittings						
15mm dia.	100.38	110.42	0.50	5.68	nr	**116.10**
Gunmetal thermostatic mixing valves including non-return valves and inlet swivel connections with strainers; copper compression fittings						
15mm dia.	87.88	96.66	0.50	5.68	nr	**102.34**
Gunmetal thermostatic mixing valves including non-return valves and inlet swivel connections with angle pattern combined isolating valves and strainers; copper compression fittings						
15mm dia.	90.99	100.09	0.50	5.68	nr	**105.77**
Pipeline Ancillaries: Steam Traps and Accessories						
Cast iron; inverted bucket type; steam trap pressure range up to 17 bar at 210 degree celsius; screwed ends						
1/2" dia.	64.27	70.70	0.84	9.55	nr	**80.25**
3/4" dia.	93.94	103.33	1.14	12.95	nr	**116.28**
1" dia.	147.50	162.25	1.30	14.77	nr	**177.02**
11/2" dia.	273.57	300.92	1.71	19.45	nr	**320.37**
2" dia.	421.89	464.08	2.08	23.67	nr	**487.75**
Cast iron; inverted bucket type; steam trap pressure range up to 17 bar at 210 degree celsius; flanged ends; (BS 10 Table H) bolted connections						
15mm dia.	152.44	167.68	1.24	14.09	nr	**181.77**
20mm dia.	180.46	198.50	1.34	15.23	nr	**213.73**
25mm dia.	276.04	303.64	1.40	15.91	nr	**319.55**
40mm dia.	426.83	469.52	1.54	17.50	nr	**487.02**
50mm dia.	519.94	571.94	1.64	18.65	nr	**590.59**

Y:MECHANICAL AND ELECTRICAL SERVICES

Item	Net Price £	Material £	Labour hours	Labour £	Unit	Total Rate £
Y11 : PIPELINE ANCILLARIES : RADIATOR VALVES						
Radiator Valves						
Bronze, straight pattern; wheelhead or lockshield; matt finish; screwed joints to steel						
15mm dia.	13.48	14.98	0.84	9.55	nr	**24.53**
20mm dia.	17.67	19.63	1.14	12.95	nr	**32.58**
25mm dia.	22.19	24.66	1.30	14.77	nr	**39.43**
Bronze, angle pattern; wheelhead or lockshield; matt finish; screwed joints to steel						
15mm dia.	8.71	9.68	0.84	9.55	nr	**19.23**
20mm dia.	11.07	12.30	1.14	12.95	nr	**25.25**
25mm dia.	15.89	17.65	1.30	14.77	nr	**32.42**
32mm dia.	34.53	38.36	1.50	17.06	nr	**55.42**
Chromium plated, straight pattern; compression joint to copper						
15mm dia.	20.96	23.29	0.46	5.23	nr	**28.52**
20mm dia.	27.13	30.14	0.54	6.14	nr	**36.28**
Chromium plated, angle pattern; compression joint to copper						
15mm dia.	16.17	17.96	0.46	5.23	nr	**23.19**
20mm dia.	20.96	23.29	0.54	6.14	nr	**29.43**
Thermostatic angle valve with built in sensor						
10mm dia.	11.39	12.66	0.50	5.68	nr	**18.34**
15mm dia.	11.39	12.66	0.50	5.68	nr	**18.34**
Thermostatic straight valve with built in sensor						
15mm dia.	11.39	12.66	0.50	5.68	nr	**18.34**
Microbore, single entry; brass; wheel head/lockshield head						
8mm dia.	4.04	4.49	0.50	5.68	nr	**10.17**
10mm dia.	4.04	4.49	0.50	5.68	nr	**10.17**
15mm dia.	4.04	4.49	0.50	5.68	nr	**10.17**
Microbore, single entry; chromium plated; wheel head/lockshield head						
8mm dia.	4.31	4.79	0.50	5.68	nr	**10.47**
10mm dia.	4.31	4.79	0.50	5.68	nr	**10.47**
12mm dia.	4.31	4.79	0.50	5.68	nr	**10.47**
Microbore, twin entry; brass; wheel head/lockshield head						
8mm dia.	9.64	10.71	0.50	5.68	nr	**16.39**
10mm dia.	10.53	11.70	0.50	5.68	nr	**17.38**

Y:MECHANICAL AND ELECTRICAL SERVICES

Item	Net Price £	Material £	Labour hours	Labour £	Unit	Total Rate £
Y11 : PIPELINE ANCILLARIES : RADIATOR VALVES (contd)						
Radiator Valves (contd)						
Microbore, twin entry; chromium plated; wheel head/lockshield head						
8mm dia.	11.71	13.01	0.50	5.68	nr	**18.69**
10mm dia.	12.85	14.27	0.50	5.68	nr	**19.95**

Y:MECHANICAL AND ELECTRICAL SERVICES

Item	Net Price £	Material £	Labour hours	Labour £	Unit	Total Rate £
Y11 : PIPELINE ANCILLARIES						
: BALL VALVES						
Ball Valves						
Bronze, equilibrium; hydraulic (10 bar);						
flanged ends; BS 4504 Table 16/21;						
bolted connections						
25mm dia.	206.10	226.71	1.00	11.36	nr	**238.07**
32mm dia.	299.27	329.20	1.16	13.18	nr	**342.38**
40mm dia.	335.97	369.57	1.34	15.23	nr	**384.80**
50mm dia.	539.25	593.18	1.64	18.65	nr	**611.83**
65mm dia.	567.49	624.24	1.80	20.47	nr	**644.71**
80mm dia.	688.89	757.78	1.92	21.85	nr	**779.63**
Heavy, equilibrium; with long tail and						
backnut; copper float; screwed for iron						
25mm dia.	107.18	112.54	1.64	18.65	nr	**131.19**
32mm dia.	149.84	157.34	1.80	20.47	nr	**177.81**
40mm dia.	189.03	198.48	1.92	21.85	nr	**220.33**
50mm dia.	393.75	413.44	2.80	31.82	nr	**445.26**
Brass, ball valve; BS 1212; copper						
float; screwed						
15mm dia.	3.49	3.66	0.24	2.73	nr	**6.39**
22mm dia.	11.55	12.13	0.34	3.86	nr	**15.99**
28mm dia.	37.34	39.21	0.40	4.54	nr	**43.75**
35mm dia.	121.98	128.07	0.46	5.23	nr	**133.30**
42mm dia.	129.59	136.07	0.54	6.14	nr	**142.21**
54mm dia.	255.76	268.55	0.64	7.27	nr	**275.82**

Y:MECHANICAL AND ELECTRICAL SERVICES

Item	Net Price £	Material £	Labour hours	Labour £	Unit	Total Rate £
Y11 : PIPELINE ANCILLARIES : CHECK VALVES						
Check Valves						
Bronze, swing pattern; BS 5154 Series B, PN 25; working pressure for saturated steam up to 9 bar; screwed ends						
15mm dia.	14.39	15.83	0.84	9.55	nr	**25.38**
20mm dia.	17.58	19.34	1.14	12.95	nr	**32.29**
25mm dia.	23.88	26.27	1.30	14.77	nr	**41.04**
32mm dia.	37.25	40.97	1.50	17.06	nr	**58.03**
40mm dia.	45.54	50.09	1.74	19.79	nr	**69.88**
50mm dia.	65.94	72.53	2.08	23.67	nr	**96.20**
65mm dia.	123.28	135.60	2.60	29.58	nr	**165.18**
80mm dia.	171.88	189.07	3.00	34.11	nr	**223.18**
Bronze, renewable disc pattern; BS 5154 Series B; PN 25; water pressure for steam up to 10 bar and temperature 100 degree celsius, cold services up to 25 bar; flanged ends; bolted connections						
15mm dia.	53.14	58.45	1.24	14.09	nr	**72.54**
20mm dia.	68.49	75.34	1.34	15.23	nr	**90.57**
25mm dia.	86.76	95.43	1.40	15.91	nr	**111.34**
32mm dia.	107.77	118.55	1.54	17.50	nr	**136.05**
40mm dia.	145.29	159.82	1.54	17.50	nr	**177.32**
50mm dia.	208.75	229.63	1.64	18.65	nr	**248.28**
65mm dia.	317.90	349.69	1.80	20.47	nr	**370.16**
80mm dia.	461.27	507.40	1.92	21.85	nr	**529.25**
Cast iron, swing pattern; BS 4090; class 200; working pressure for cold services up to 7 bar; flanged ends (BS 10 Table D); bolted connections						
50mm dia.	186.22	204.84	1.64	18.65	nr	**223.49**
65mm dia.	211.81	232.99	1.80	20.47	nr	**253.46**
80mm dia.	230.83	253.91	1.92	21.85	nr	**275.76**
100mm dia.	315.23	346.75	2.80	31.82	nr	**378.57**
125mm dia.	457.00	502.70	9.01	102.34	nr	**605.04**
150mm dia.	509.08	559.98	12.05	136.87	nr	**696.85**
Cast iron, swing pattern; BS 4090; Class 200; working pressure for cold services up to 7 bar; flanged and drilled ends; (BS 10 table F)						
50mm dia.	211.44	232.58	1.64	18.65	nr	**251.23**
65mm dia.	244.05	268.46	1.80	20.47	nr	**288.93**
80mm dia.	299.28	329.21	1.92	21.85	nr	**351.06**
100mm dia.	388.38	427.22	2.80	31.82	nr	**459.04**
125mm dia.	495.15	544.66	9.01	102.34	nr	**647.00**
150mm dia.	552.64	607.90	12.05	136.87	nr	**744.77**

Y:MECHANICAL AND ELECTRICAL SERVICES

Item	Net Price £	Material £	Labour hours	Labour £	Unit	Total Rate £
Y11 : PIPELINE ANCILLARIES : COMMISSIONING VALVES						
Commissioning Valves						
Bronze commissioning set; LT, MTHW and chilled water; metering station; double regulating valve; screwed ends to steel						
15mm dia.	37.41	41.15	1.26	14.32	nr	**55.47**
20mm dia.	61.40	67.54	1.71	19.45	nr	**86.99**
25mm dia.	74.59	82.05	1.95	22.19	nr	**104.24**
32mm dia.	98.27	108.09	2.25	25.59	nr	**133.68**
40mm dia.	130.56	143.62	2.61	29.66	nr	**173.28**
50mm dia.	188.74	207.62	3.13	35.50	nr	**243.12**
Cast iron, double regulating valve; flanged ends; (BS 4504 Part 1 Table 16); bolted connections						
65mm dia.	416.51	458.16	1.80	20.47	nr	**478.63**
80mm dia.	466.10	512.71	1.92	21.85	nr	**534.56**
100mm dia.	565.27	621.79	2.80	31.82	nr	**653.61**
125mm dia.	743.77	818.15	9.01	102.34	nr	**920.49**
150mm dia.	882.61	970.87	12.05	136.87	nr	**1107.74**
Cast iron commissioning set; LT, MTHW and chilled water; metering station; double regulating valve; flanged ends; (BS 4504 Part 1 Table 16); bolted connections						
65mm dia.	379.25	417.18	1.80	20.47	nr	**437.65**
80mm dia.	582.78	641.06	1.92	21.85	nr	**662.91**
100mm dia.	704.76	775.24	2.80	31.82	nr	**807.06**
125mm dia.	991.69	1090.86	9.01	102.34	nr	**1193.20**
150mm dia.	1171.84	1289.03	12.05	136.87	nr	**1425.90**

Y:MECHANICAL AND ELECTRICAL SERVICES

Item	Net Price £	Material £	Labour hours	Labour £	Unit	Total Rate £
Y11 : PIPELINE ANCILLARIES : CONTROL VALVES						
Control Valves						
Pressure reducing valve for steam; maximum range of 17 bar and 232 degree Celsius; screwed ends to steel						
15mm dia.	242.72	266.99	0.84	9.55	nr	**276.54**
20mm dia.	262.40	288.64	0.88	9.96	nr	**298.60**
25mm dia.	284.54	312.99	1.30	14.77	nr	**327.76**
Pressure reducing valve for steam; maximum range of 17 bar and 232 degree Celsius; flanged ends (BS 10 Table H or BS 4504 Tables 25 and 16)						
25mm dia.	341.12	375.23	1.64	18.65	nr	**393.88**
32mm dia.	388.68	427.55	1.80	20.47	nr	**448.02**
40mm dia.	464.12	510.53	2.04	23.18	nr	**533.71**
50mm dia.	535.46	589.01	2.48	28.19	nr	**617.20**
Cast iron; butterfly type; two position electrically controlled motor for low pressure hot water; max 6 bar 120 degree Celsius; flanged ends; counter flanges						
25mm dia.	192.86	212.15	1.40	15.91	nr	**228.06**
32mm dia.	196.03	215.63	1.46	16.61	nr	**232.24**
40mm dia.	201.74	221.92	1.54	17.50	nr	**239.42**
50mm dia.	211.43	232.57	1.64	18.65	nr	**251.22**
65mm dia.	221.86	244.05	2.68	30.46	nr	**274.51**
80mm dia.	230.05	253.05	2.68	30.46	nr	**283.51**
100mm dia.	247.46	272.21	2.80	31.82	nr	**304.03**
125mm dia.	461.02	507.12	2.80	31.82	nr	**538.94**
150mm dia.	516.66	568.33	2.80	31.82	nr	**600.15**
200mm dia.	678.33	746.16	2.80	31.82	nr	**777.98**
Cast iron; three way motorized; low pressure hot water; max 6 bar 120 degree Celsius; flanged ends and drilled (BS 10 Table F)						
25mm dia.	292.96	322.25	1.90	21.60	nr	**343.85**
32mm dia.	305.97	336.57	1.96	22.27	nr	**358.84**
40mm dia.	323.34	355.67	2.04	23.18	nr	**378.85**
50mm dia.	353.20	388.52	3.45	39.17	nr	**427.69**
65mm dia.	395.93	435.52	3.45	39.17	nr	**474.69**
80mm dia.	511.93	563.13	3.57	40.57	nr	**603.70**
Control Valves; fitted to pipeline. (Electrical work elsewhere.)						
Two port normally closed motorised valve electric actuator; spring return; domestic usage						
22mm dia.	22.67	24.94	1.14	12.95	nr	**37.89**
28mm dia.	29.22	32.14	1.30	14.77	nr	**46.91**

Y:MECHANICAL AND ELECTRICAL SERVICES

Item	Net Price £	Material £	Labour hours	Labour £	Unit	Total Rate £
Two port on/off motorised valve electric actuator; spring return; domestic usage						
22mm dia.	34.94	38.43	1.14	12.95	nr	**51.38**
Three port motorised valve; electric actuator; spring return; domestic usage						
22mm dia.	30.71	33.78	1.14	12.95	nr	**46.73**

Y:MECHANICAL AND ELECTRICAL SERVICES

Item	Net Price £	Material £	Labour hours	Labour £	Unit	Total Rate £
Y11 : PIPELINE ANCILLARIES **: LUBRICATED PLUG VALVES**						
Lubricated Plug Valves						
Cast iron; lubricated plug valve; wrench operated with lever; screwed connections						
15mm dia.	49.83	55.36	0.94	10.69	nr	**66.05**
20mm dia.	50.86	56.51	1.12	12.74	nr	**69.25**
25mm dia.	58.74	65.26	1.30	14.77	nr	**80.03**
32mm dia.	69.66	77.39	1.50	17.06	nr	**94.45**
40mm dia.	69.66	77.39	1.74	19.79	nr	**97.18**
50mm dia.	107.58	119.52	2.08	23.67	nr	**143.19**
Cast iron; lubricated plug valve; wrench operated with lever; flanged ends to BS 4504 Table 16/11; bolted connections						
65mm dia.	138.37	153.73	1.80	20.47	nr	**174.20**
80mm dia.	167.14	185.69	1.92	21.85	nr	**207.54**
100mm dia.	275.71	306.31	2.80	31.82	nr	**338.13**
Cast iron; lubricated plug valve; gear operated with lever; flanged ends to BS 4504 Table 16/11; bolted connections						
100mm dia.	968.86	1076.40	7.04	80.00	nr	**1156.40**
150mm dia.	1652.08	1835.46	12.05	136.87	nr	**1972.33**

Y:MECHANICAL AND ELECTRICAL SERVICES

Item	Net Price £	Material £	Labour hours	Labour £	Unit	Total Rate £
Y11 : PIPELINE ANCILLARIES						
: GATE VALVES						
Gate Valves						
Bronze wedge non-rising stem; BS 5154; PN 32, Series B; wheelhead working pressure; saturated steam up to 9 bar; cold services up to 14 bar; screwed ends to steel						
15mm dia.	28.35	31.19	0.94	10.69	nr	**41.88**
20mm dia.	34.47	37.92	1.12	12.74	nr	**50.66**
25mm dia.	45.50	50.05	1.30	14.77	nr	**64.82**
32mm dia.	73.85	81.24	1.50	17.06	nr	**98.30**
40mm dia.	108.50	119.35	1.74	19.79	nr	**139.14**
50mm dia.	154.35	169.78	2.08	23.67	nr	**193.45**
DZR copper alloy wedge non-rising stem; BS 5154; PN 20; working pressure saturated steam up to 10 bar; cold services up to 21 bar; screwed ends						
15mm dia.	6.19	6.81	0.94	10.69	nr	**17.50**
20mm dia.	7.32	8.05	1.12	12.74	nr	**20.79**
25mm dia.	10.12	11.13	1.30	14.77	nr	**25.90**
32mm dia.	15.16	16.68	1.50	17.06	nr	**33.74**
40mm dia.	21.35	23.49	1.74	19.79	nr	**43.28**
50mm dia.	29.58	32.54	2.08	23.67	nr	**56.21**
Bronze wedge non-rising stem; BS 5154 PN 16; working pressure saturated steam up to 7 bar; cold services; up to 16 bar; flanged ends; bolted connections						
15mm dia.	40.25	44.28	1.36	15.46	nr	**59.74**
20mm dia.	47.95	52.74	1.40	15.91	nr	**68.65**
25mm dia.	70.35	77.39	1.46	16.61	nr	**94.00**
32mm dia.	88.90	97.79	1.54	17.50	nr	**115.29**
40mm dia.	102.90	113.19	1.64	18.65	nr	**131.84**
50mm dia.	151.55	166.71	1.64	18.65	nr	**185.36**
65mm dia.	229.25	252.17	1.80	20.47	nr	**272.64**
80mm dia.	311.50	342.65	1.92	21.85	nr	**364.50**
100mm dia.	563.50	619.85	2.80	31.82	nr	**651.67**
Cast iron, trim material bronze; non rising; inside screwed BS 5150; PN6 working pressure upto 6 bar at 120 degrees C; flanged end connections						
40mm dia.	103.43	113.77	1.64	18.65	nr	**132.42**
50mm dia.	104.46	114.91	1.64	18.65	nr	**133.56**
65mm dia.	114.39	125.83	1.80	20.47	nr	**146.30**
80mm dia.	132.21	145.43	1.92	21.85	nr	**167.28**
100mm dia.	174.68	192.15	2.80	31.82	nr	**223.97**
125mm dia.	250.02	275.02	9.01	102.34	nr	**377.36**
150mm dia.	287.70	316.47	12.00	136.87	nr	**453.34**
200mm dia.	548.00	602.80	13.16	149.47	nr	**752.27**
250mm dia.	804.88	885.37	14.08	160.00	nr	**1045.37**
300mm dia.	1000.10	1100.11	15.15	172.12	nr	**1272.23**

Y:MECHANICAL AND ELECTRICAL SERVICES

Item	Net Price £	Material £	Labour hours	Labour £	Unit	Total Rate £
Y11 : PIPELINE ANCILLARIES						
: GATE VALVES (contd)						
Gate Valves (contd)						
Cast iron, trim material bronze; non rising; inside screwed BS 5154; PN 10 working pressure upto 10 bar at 120 degrees C; flanged end connections						
50mm dia.	38.01	41.81	1.64	18.65	nr	**60.46**
65mm dia.	42.70	46.97	1.80	20.47	nr	**67.44**
80mm dia.	49.50	54.46	1.92	21.85	nr	**76.31**
100mm dia.	65.69	72.26	2.80	31.82	nr	**104.08**
125mm dia.	92.67	101.94	9.01	102.34	nr	**204.28**
150mm dia.	106.75	117.42	12.05	136.87	nr	**254.29**
200mm dia.	193.56	212.91	13.16	149.47	nr	**362.38**
250mm dia.	288.57	317.43	14.08	160.00	nr	**477.43**
300mm dia.	367.17	403.89	15.15	172.12	nr	**576.01**
350mm dia.	617.03	678.73	15.15	172.12	nr	**850.85**
Cast iron, trim material bronze; non rising; inside screwed BS 5154; PN 16 working pressure upto 16 bar at 120 degrees C; flanged end connections						
80mm dia.	158.23	174.05	1.92	21.85	nr	**195.90**
100mm dia.	210.64	231.70	2.80	31.82	nr	**263.52**
125mm dia.	304.83	335.31	9.01	102.34	nr	**437.65**
150mm dia.	359.62	395.58	12.05	136.87	nr	**532.45**
200mm dia.	623.35	685.68	13.16	149.47	nr	**835.15**
250mm dia.	935.03	1028.53	14.08	160.00	nr	**1188.53**
300mm dia.	1061.75	1167.93	15.15	172.12	nr	**1340.05**
350mm dia.	2085.82	2294.40	15.15	172.12	nr	**2466.52**
400mm dia.	2616.70	2878.37	16.13	183.23	nr	**3061.60**
450mm dia.	3688.72	4057.59	16.13	183.23	nr	**4240.82**

Y:MECHANICAL AND ELECTRICAL SERVICES

Item	Net Price £	Material £	Labour hours	Labour £	Unit	Total Rate £
Y11 : PIPELINE ANCILLARIES : GLOBE VALVES						
Globe Valves						
Bronze; renewable disc; BS 5154, Series B, PN 20 working pressure saturated steam up to 9 bar; cold services up to 14 bar; screwed ends to steel						
15mm dia.	29.27	32.20	0.84	9.55	nr	**41.75**
20mm dia.	35.59	39.15	1.14	12.95	nr	**52.10**
25mm dia.	51.95	57.14	1.30	14.77	nr	**71.91**
32mm dia.	67.45	74.19	1.50	17.06	nr	**91.25**
40mm dia.	82.94	91.24	1.74	19.79	nr	**111.03**
50mm dia.	121.11	133.23	1.64	18.65	nr	**151.88**
Bronze; renewable disc; BS 5154, Series B, PN 20 working pressure; saturated steam up to 9 bar; cold services up to 14 bar; flanged ends; (BS 10 Class 100); bolted connections						
15mm dia.	12.74	14.02	1.24	14.09	nr	**28.11**
20mm dia.	21.39	23.52	1.34	15.23	nr	**38.75**
25mm dia.	30.13	33.15	1.40	15.91	nr	**49.06**
32mm dia.	62.93	69.23	1.46	16.61	nr	**85.84**
40mm dia.	99.02	108.92	1.54	17.50	nr	**126.42**
50mm dia.	125.42	137.96	1.64	18.65	nr	**156.61**
Bronze; renewable disc; BS 2060, Class 250; working pressure saturated steam up to 14 bar; cold services up to 28 bar; flanged ends; (BS 10 Table H); bolted connections						
15mm dia.	79.79	87.76	1.24	14.09	nr	**101.85**
20mm dia.	92.99	102.29	1.34	15.23	nr	**117.52**
25mm dia.	127.43	140.17	1.40	15.91	nr	**156.08**
32mm dia.	166.46	183.11	1.46	16.61	nr	**199.72**
40mm dia.	210.94	232.04	1.54	17.50	nr	**249.54**
50mm dia.	309.96	340.96	1.64	18.65	nr	**359.61**
65mm dia.	430.50	473.55	1.80	20.47	nr	**494.02**
80mm dia.	614.18	675.60	1.92	21.85	nr	**697.45**

Y:MECHANICAL AND ELECTRICAL SERVICES

Item	Net Price £	Material £	Labour hours	Labour £	Unit	Total Rate £
Y11 : PIPELINE ANCILLARIES : SAFEY AND RELIEF VALVES						
Safety and Relief Valves						
Bronze 'pop' type; side outlet; screwed ends to steel						
15mm dia.	53.43	60.54	0.24	2.73	nr	**63.27**
20mm dia.	61.65	69.85	0.34	3.86	nr	**73.71**
25mm dia.	76.18	86.31	0.34	3.86	nr	**90.17**
32mm dia.	98.91	112.07	0.46	5.23	nr	**117.30**
40mm dia.	129.60	146.84	0.60	6.82	nr	**153.66**
50mm dia.	176.73	200.23	0.72	8.18	nr	**208.41**
65mm dia.	315.10	357.01	0.90	10.23	nr	**367.24**
80mm dia.	353.46	400.47	1.06	12.05	nr	**412.52**
Bronze spring type; side outlet; screwed ends to steel						
15mm dia.	29.86	33.84	0.24	2.73	nr	**36.57**
20mm dia.	36.72	41.60	0.34	3.86	nr	**45.46**
25mm dia.	49.59	56.19	0.34	3.86	nr	**60.05**
32mm dia.	49.59	56.19	0.46	5.23	nr	**61.42**
40mm dia.	87.14	98.73	0.60	6.82	nr	**105.55**
50mm dia.	121.66	137.84	0.72	8.18	nr	**146.02**
65mm dia.	193.17	218.86	0.90	10.23	nr	**229.09**
80mm dia.	253.45	287.16	1.06	12.05	nr	**299.21**

Y:MECHANICAL AND ELECTRICAL SERVICES

Item	Net Price £	Material £	Labour hours	Labour £	Unit	Total Rate £
Y11 : PIPELINE ANCILLARIES : DRAIN COCKS AND VENT COCKS						
Cocks; (capillary joints to copper.)						
Stopcock; Brass head with gun metal body						
15mm dia.	2.12	2.33	0.50	5.68	nr	**8.01**
22mm dia.	3.96	4.35	0.50	5.68	nr	**10.03**
28mm dia.	11.25	12.38	0.53	6.02	nr	**18.40**
Lockshield stop cocks; brass head with gun metal body						
15mm dia.	5.20	5.71	0.50	5.68	nr	**11.39**
22mm dia.	7.45	8.20	0.50	5.68	nr	**13.88**
28mm dia.	13.22	14.54	0.53	6.02	nr	**20.56**
DZR stopcock; brass head with gun metal body						
15mm dia.	5.32	5.85	0.50	5.68	nr	**11.53**
22mm dia.	9.22	10.14	0.50	5.68	nr	**15.82**
28mm dia.	15.36	16.89	0.53	6.02	nr	**22.91**
Gunmetal stopcock						
35mm dia.	24.10	26.50	0.75	8.52	nr	**35.02**
42mm dia.	32.01	35.21	0.75	8.52	nr	**43.73**
54mm dia.	47.83	52.61	0.83	9.44	nr	**62.05**
Lockshield DZR stopcock; brass head gun metal body						
15mm dia.	7.00	7.70	0.50	5.68	nr	**13.38**
22mm dia.	10.43	11.47	0.50	5.68	nr	**17.15**
28mm dia.	17.33	19.07	0.53	6.02	nr	**25.09**
Double union stopcock						
15mm dia.	6.85	7.53	0.66	7.50	nr	**15.03**
22mm dia.	9.68	10.65	0.66	7.50	nr	**18.15**
28mm dia.	17.21	18.93	0.70	7.96	nr	**26.89**
Double union lockshield stopcock						
15mm dia.	6.76	7.44	0.67	7.57	nr	**15.01**
22mm dia.	10.78	11.86	0.67	7.57	nr	**19.43**
28mm dia.	19.23	21.16	0.70	7.96	nr	**29.12**
Double union DZR stopcock						
15mm dia.	11.19	12.31	0.67	7.57	nr	**19.88**
22mm dia.	13.78	15.16	0.67	7.57	nr	**22.73**
28mm dia.	25.46	28.01	0.70	7.96	nr	**35.97**
Double union lockshield DZR stopcock						
15mm dia.	11.88	13.07	0.67	7.57	nr	**20.64**
22mm dia.	14.88	16.36	0.67	7.57	nr	**23.93**
28mm dia.	27.48	30.22	0.70	7.96	nr	**38.18**

Y:MECHANICAL AND ELECTRICAL SERVICES

Item	Net Price £	Material £	Labour hours	Labour £	Unit	Total Rate £
Y11 : PIPELINE ANCILLARIES : DRAIN COCKS AND VENT COCKS (contd)						
Cocks; (capillary joints to copper) (contd)						
Double union gun metal stopcock						
35mm dia.	42.48	46.73	0.70	7.96	nr	**54.69**
42mm dia.	58.26	64.09	0.75	8.52	nr	**72.61**
54mm dia.	91.68	100.85	0.91	10.29	nr	**111.14**
Stopcock with easy clean cover						
15mm dia.	5.46	6.00	0.70	7.96	nr	**13.96**
22mm dia.	7.75	8.53	0.67	7.57	nr	**16.10**
28mm dia.	13.64	15.00	0.70	7.96	nr	**22.96**
Union stopcock with easy clean cover						
15mm dia.	6.65	7.32	0.67	7.57	nr	**14.89**
22mm dia.	10.14	11.15	0.67	7.57	nr	**18.72**
28mm dia.	16.10	17.71	0.70	7.96	nr	**25.67**
Double union stopcock with easy clean cover						
15mm dia.	12.00	13.20	0.67	7.57	nr	**20.77**
22mm dia.	14.94	16.44	0.67	7.57	nr	**24.01**
28mm dia.	27.78	30.56	0.70	7.96	nr	**38.52**
Combined stopcock and drain						
15mm dia.	11.79	12.97	0.72	8.18	nr	**21.15**
22mm dia.	14.51	15.97	0.72	8.18	nr	**24.15**
Combined DZR stopcock and drain						
15mm dia.	15.62	17.18	0.72	8.18	nr	**25.36**
Cocks; (capillary/compression joints to copper.)						
Stopcock; brass head gun metal body						
15mm dia.	2.71	2.98	0.50	5.68	nr	**8.66**
22mm dia.	4.77	5.24	0.50	5.68	nr	**10.92**
28mm dia.	12.43	13.67	0.53	6.02	nr	**19.69**
Lockshield stopcock; brass head gun metal body						
15mm dia.	5.92	6.51	0.50	5.68	nr	**12.19**
22mm dia.	8.30	9.12	0.50	5.68	nr	**14.80**
28mm dia.	16.01	17.61	0.53	6.02	nr	**23.63**

Y:MECHANICAL AND ELECTRICAL SERVICES

Item	Net Price £	Material £	Labour hours	Labour £	Unit	Total Rate £
DZR stopcock						
15mm dia.	6.32	6.95	0.45	5.11	nr	**12.06**
22mm dia.	10.38	11.42	0.45	5.11	nr	**16.53**
28mm dia.	17.17	18.89	0.47	5.32	nr	**24.21**
35mm dia.	31.59	34.75	0.59	6.71	nr	**41.46**
42mm dia.	45.20	49.72	0.59	6.71	nr	**56.43**
54mm dia.	61.56	67.72	0.67	7.61	nr	**75.33**
DZR Lockshield stopcock						
15mm dia.	7.25	7.98	0.45	5.11	nr	**13.09**
22mm dia.	11.39	12.53	0.45	5.11	nr	**17.64**
Stopcock with easy clean cover						
15mm dia.	6.51	7.16	0.50	5.68	nr	**12.84**
22mm dia.	9.18	10.10	0.50	5.68	nr	**15.78**
28mm dia.	16.44	18.09	0.53	6.02	nr	**24.11**
Combined stop/draincock						
15mm dia.	9.11	10.02	0.20	2.27	nr	**12.29**
22mm dia.	14.02	15.42	0.50	5.68	nr	**21.10**
DZR Combined stop/draincock						
15mm dia.	11.74	12.91	0.45	5.11	nr	**18.02**
22mm dia.	17.19	18.91	0.45	5.11	nr	**24.02**
Stopcock to polythene						
15mm dia.	7.22	7.95	0.45	5.11	nr	**13.06**
22mm dia.	7.41	8.15	0.45	5.11	nr	**13.26**
28mm dia.	11.17	12.29	0.47	5.34	nr	**17.63**
Draw off coupling						
15mm dia.	3.93	4.32	0.45	5.11	nr	**9.43**
DZR Draw off coupling						
15mm dia.	5.42	5.97	0.45	5.11	nr	**11.08**
22mm dia.	6.28	6.91	0.45	5.11	nr	**12.02**
Draw off elbow						
15mm dia.	8.40	9.24	0.45	5.11	nr	**14.35**
22mm dia.	9.73	10.71	0.45	5.11	nr	**15.82**
Lockshield drain cock						
15mm dia.	4.28	4.71	0.45	5.11	nr	**9.82**

Y:MECHANICAL AND ELECTRICAL SERVICES

Item	Net Price £	Material £	Labour hours	Labour £	Unit	Total Rate £
Y11 : PIPELINE ANCILLARIES : AUTOMATIC AIR VENTS						
Air Vents; (Including regulating, adjusting and testing.)						
Automatic air vent; pressures up to 7 bar; 93 degree Celsius screwed ends to steel						
15mm dia.	53.51	58.86	0.80	9.09	nr	**67.95**
Automatic air vent; pressures up to 7 bar; 93 degree Celsius lockhead isolating valve; screwed ends to steel						
15mm dia.	61.35	67.48	0.80	9.09	nr	**76.57**
Automatic air vent; pressures up to 17 bar; 204 degree Celsius; flanged end (BS 10 Table H); bolted connections to counter flange (measured separately)						
15mm dia.	334.87	368.35	0.80	9.09	nr	**377.44**

Keep your figures up to date, free of charge

This section, and most of the other information in this Price Book, is brought up to date every three months, until the next annual edition, in the *Price Book Update*.

The *Update* is available free to all Price Book purchasers.

To ensure you receive your copy, simply complete the reply card from the centre of the book and return it to us.

Y:MECHANICAL AND ELECTRICAL SERVICES

Item	Net Price £	Material £	Labour hours	Labour £	Unit	Total Rate £
Y11 : PIPELINE ANCILLARIES **: STEAM TRAPS AND STRAINERS**						
Steam Traps and Strainers						
Stainless steel: thermodynamic trap with pressure range up to 42 bar; temperature range to 400 degree Celsius; screwed ends to steel						
15mm dia.	44.91	49.40	0.84	9.55	nr	**58.95**
20mm dia.	66.33	72.97	1.14	12.95	nr	**85.92**
Stainless steel; thermodynamic trap with pressure range up to 24 bar; temperature range to 288 degree Celsius; flanged ends (BS 10 Table H); bolted connections						
15mm dia.	171.39	188.53	1.24	14.09	nr	**202.62**
20mm dia.	176.34	193.97	1.34	15.23	nr	**209.20**
25mm dia.	190.34	209.38	1.40	15.91	nr	**225.29**
Malleable iron pipeline strainer; max steam working pressure 14 bar and temperature range to 230 degree Celsius; screwed ends to steel						
1/2" dia.	6.94	7.63	0.84	9.55	nr	**17.18**
3/4" dia.	9.27	10.20	1.14	12.95	nr	**23.15**
1" dia.	13.72	15.09	1.30	14.77	nr	**29.86**
11/2" dia.	22.70	24.97	1.50	17.06	nr	**42.03**
2" dia.	40.42	44.46	1.74	19.79	nr	**64.25**
Bronze pipeline strainer; max steam working pressure 17 bar; flanged ends (BS 19 Table H); bolted connections						
15mm dia.	75.40	82.94	1.24	14.09	nr	**97.03**
20mm dia.	78.28	86.11	1.34	15.23	nr	**101.34**
25mm dia.	95.58	105.14	1.40	15.91	nr	**121.05**
32mm dia.	140.08	154.09	1.46	16.61	nr	**170.70**
40mm dia.	174.69	192.16	1.54	17.50	nr	**209.66**
50mm dia.	261.21	287.33	1.64	18.65	nr	**305.98**
65mm dia.	333.72	367.09	2.50	28.40	nr	**395.49**
80mm dia.	418.59	460.45	2.91	33.02	nr	**493.47**
100mm dia.	749.02	823.92	3.51	39.86	nr	**863.78**
Balanced pressure thermostatic steam trap and strainer; max working pressure up to 5.5 bar; screwed ends to steel						
1/2" dia.	31.27	34.40	1.26	14.32	nr	**48.72**
3/4" dia.	36.13	39.75	1.71	19.45	nr	**59.20**

Y:MECHANICAL AND ELECTRICAL SERVICES

Item	Net Price £	Material £	Labour hours	Labour £	Unit	Total Rate £
Y11 : PIPELINE ANCILLARIES : SIGHT GLASSES						
Sight Glasses						
Pressed brass; straight; single window; screwed ends to steel						
15mm dia.	19.42	21.36	0.84	9.55	nr	**30.91**
20mm dia.	21.66	23.83	1.14	12.95	nr	**36.78**
25mm dia.	26.93	29.62	1.30	14.77	nr	**44.39**
Gunmetal; straight; double window; screwed ends to steel						
15mm dia.	31.50	34.65	0.84	9.55	nr	**44.20**
20mm dia.	34.39	37.83	1.14	12.95	nr	**50.78**
25mm dia.	42.84	47.12	1.30	14.77	nr	**61.89**
32mm dia.	70.18	77.19	1.35	15.35	nr	**92.54**
40mm dia.	70.18	77.19	1.74	19.79	nr	**96.98**
50mm dia.	84.86	93.35	2.08	23.67	nr	**117.02**
SG Iron flanged; BS 4504; PN 25						
15mm dia.	64.46	70.91	1.00	11.36	nr	**82.27**
20mm dia.	75.89	83.48	1.25	14.20	nr	**97.68**
25mm dia.	96.29	105.92	1.50	17.06	nr	**122.98**
32mm dia.	106.08	116.69	1.70	19.32	nr	**136.01**
40mm dia.	139.54	153.49	2.00	22.72	nr	**176.21**
50mm dia.	168.10	184.91	2.30	26.17	nr	**211.08**
Check Valve and sight glass; gun metal; screwed						
15mm dia.	31.66	34.83	0.84	9.55	nr	**44.38**
20mm dia.	33.33	36.67	1.14	12.95	nr	**49.62**
25mm dia.	56.30	61.93	1.30	14.77	nr	**76.70**

Y:MECHANICAL AND ELECTRICAL SERVICES

Item	Net Price £	Material £	Labour hours	Labour £	Unit	Total Rate £
Y11: PIPELINE ANCILLARIES : EXPANSION JOINTS						
Stainless Steel Sleeved Bellows Type						
Screwed ends for mild steel pipes with external shroud (10 bar)						
20mm dia.	58.61	64.47	1.28	14.55	nr	**79.02**
25mm dia.	62.43	68.68	1.50	17.06	nr	**85.74**
32mm dia.	87.87	96.66	1.76	20.00	nr	**116.66**
40mm dia.	96.17	105.78	2.00	22.72	nr	**128.50**
50mm dia.	104.18	114.60	2.28	25.94	nr	**140.54**
Flanged ends to BS 10 Table E (6 bar)						
20mm dia.	65.23	71.75	0.48	5.45	nr	**77.20**
25mm dia.	68.77	75.64	0.60	6.82	nr	**82.46**
32mm dia.	96.26	105.88	0.72	8.18	nr	**114.06**
40mm dia.	103.71	114.08	0.78	8.86	nr	**122.94**
50mm dia.	119.56	131.51	0.84	9.55	nr	**141.06**
65mm dia.	85.86	94.45	1.08	12.28	nr	**106.73**
80mm dia.	97.21	106.93	1.32	15.01	nr	**121.94**
100mm dia.	119.61	131.57	1.92	21.85	nr	**153.42**
150mm dia.	239.71	263.68	3.61	41.01	nr	**304.69**
Flanged ends to BS 10 Table F (10 bar)						
20mm dia.	65.23	71.75	0.48	5.45	nr	**77.20**
25mm dia.	68.77	75.64	0.60	6.82	nr	**82.46**
32mm dia.	96.26	105.88	0.72	8.18	nr	**114.06**
40mm dia.	103.71	114.08	0.78	8.86	nr	**122.94**
50mm dia.	119.56	131.51	0.84	9.55	nr	**141.06**
65mm dia.	85.86	94.45	1.08	12.28	nr	**106.73**
80mm dia.	97.21	106.93	1.32	15.01	nr	**121.94**
100mm dia.	119.61	131.57	1.92	21.85	nr	**153.42**
150mm dia.	239.71	263.68	3.61	41.01	nr	**304.69**
Screwed Union Ends for copper pipe						
22mm dia.	66.51	73.16	0.48	5.45	nr	**78.61**
25mm dia.	71.20	78.32	0.60	6.82	nr	**85.14**
32mm dia.	86.00	94.60	0.72	8.18	nr	**102.78**
40mm dia.	98.05	107.86	0.78	8.86	nr	**116.72**
50mm dia.	122.73	135.01	0.84	9.55	nr	**144.56**
Flanged Ends to BS 10 Table e for Copper						
65mm dia.	123.00	135.30	0.78	8.86	nr	**144.16**
80mm dia.	139.20	153.11	0.84	9.55	nr	**162.66**
100mm dia.	167.34	184.08	1.08	12.28	nr	**196.36**
150mm dia.	308.94	339.83	1.32	15.01	nr	**354.84**

Y:MECHANICAL AND ELECTRICAL SERVICES

Item	Net Price £	Material £	Labour hours	Labour £	Unit	Total Rate £
Y11: PIPELINE ANCILLARIES : EXPANSION JOINTS (contd)						
Reinforced Neoprene Flexible Connector; working pressure 10 Bar at 100 Degree C						
Twin sphere with BSP, threaded union ends						
20mm dia.	33.82	37.21	1.28	14.55	nr	**51.76**
25mm dia.	38.08	41.89	1.50	17.06	nr	**58.95**
32mm dia.	44.28	48.71	1.76	20.00	nr	**68.71**
40mm dia.	54.64	60.11	2.00	22.72	nr	**82.83**
50mm dia.	67.52	74.27	2.28	25.94	nr	**100.21**
Twin sphere flanged to BS 10 Table E						
40mm dia.	141.65	155.82	0.78	8.86	nr	**164.68**
50mm dia.	153.31	168.64	0.84	9.55	nr	**178.19**
65mm dia.	171.52	188.68	1.08	12.28	nr	**200.96**
80mm dia.	196.44	216.09	1.32	15.01	nr	**231.10**
100mm dia.	226.58	249.23	1.92	21.85	nr	**271.08**
150mm dia.	318.53	350.38	3.61	41.01	nr	**391.39**
Stainless Steel Single Hinged; angular joints						
Flanged ends to BS 4504 PN 16						
50mm dia.	346.01	380.61	0.60	6.82	nr	**387.43**
65mm dia.	393.22	432.54	0.72	8.18	nr	**440.72**
80mm dia.	415.63	457.19	0.78	8.86	nr	**466.05**
100mm dia.	527.15	579.86	0.84	9.55	nr	**589.41**
125mm dia.	548.69	603.56	1.08	12.28	nr	**615.84**
150mm dia.	580.80	638.88	1.08	12.28	nr	**651.16**
Stainless Steel Articulated Tied; angular joints (in 1m lengths; 0-100mm expansion)						
Flanged ends to BS 4505 PN16						
50mm dia.	282.11	310.32	0.84	9.55	nr	**319.87**
65mm dia.	403.33	443.66	1.08	12.28	nr	**455.94**
80mm dia.	424.70	467.17	1.32	15.01	nr	**482.18**
100mm dia.	551.34	606.47	1.92	21.85	nr	**628.32**
125mm dia.	600.59	660.65	3.61	41.01	nr	**701.66**
150mm dia.	676.06	743.67	3.61	41.01	nr	**784.68**
Stainless Steel Axial Expansion Joints						
BSP Female Union Ends						
15mm dia.	43.72	48.09	0.48	5.45	nr	**53.54**
20mm dia.	52.73	58.00	0.60	6.82	nr	**64.82**
25mm dia.	57.11	62.82	0.72	8.18	nr	**71.00**
32mm dia.	69.13	76.04	0.78	8.86	nr	**84.90**
40mm dia.	77.04	84.74	0.84	9.55	nr	**94.29**
50mm dia.	90.45	99.49	0.84	9.55	nr	**109.04**

Y:MECHANICAL AND ELECTRICAL SERVICES

Item	Net Price £	Material £	Labour hours	Labour £	Unit	Total Rate £
Flanged Ends to BS 4505-16						
50mm dia.	119.56	131.51	0.60	6.82	nr	**138.33**
65mm dia.	85.86	94.45	0.72	8.18	nr	**102.63**
80mm dia.	97.21	106.93	0.84	9.55	nr	**116.48**
100mm dia.	119.61	131.57	1.08	12.28	nr	**143.85**
150mm dia.	239.71	263.68	1.08	12.28	nr	**275.96**

Y:MECHANICAL AND ELECTRICAL SERVICES

Item	Net Price £	Material £	Labour hours	Labour £	Unit	Total Rate £
Y11 : PIPELINE ANCILLARIES : GAUGES						
Thermometers and Pressure Gauges						
Dial Pattern; mercury-in-steel thermometer; 15mm screwed tail; brass pockets; 20/60 degrees C; screwed ends to steel						
Back entry						
100mm dia.	88.96	97.86	0.81	9.21	nr	**107.07**
150mm dia.	101.67	111.84	0.81	9.21	nr	**121.05**
Bottom entry						
100mm dia.	88.96	97.86	0.81	9.21	nr	**107.07**
150mm dia.	101.67	111.84	0.81	9.21	nr	**121.05**
Dial thermometer; bimetal; brass pocket; 20/60 degrees C; 100mm dial; screwed ends to steel						
Horizontal pocket length 50mm	28.73	31.61	0.81	9.21	nr	**40.82**
Horizontal pocket length 100mm	33.96	37.36	0.81	9.21	nr	**46.57**
Vertical pocket length 50mm	31.87	35.06	0.81	9.21	nr	**44.27**
Vertical pocket length 100mm	36.84	40.53	0.81	9.21	nr	**49.74**
Dial pressure gauge; screwed ends to steel						
Direct 0/4 bar						
100mm	18.95	20.85	1.00	11.36	nr	**32.21**
150mm	32.92	36.22	1.00	11.36	nr	**47.58**
Direct 0/20 bar						
100mm	18.95	20.85	1.00	11.36	nr	**32.21**
150mm	32.92	36.22	1.00	11.36	nr	**47.58**
Dial pressure gauge; flanged ends						
Direct 0/4 bar						
100mm	24.96	27.46	1.30	14.77	nr	**42.23**
150mm	39.72	43.69	1.30	14.77	nr	**58.46**
Direct 0/20 bar						
100mm	24.96	27.46	1.30	14.77	nr	**42.23**
150mm	39.72	43.69	1.30	14.77	nr	**58.46**
Combined altitude gauge and thermometer; 20/120 degrees C; range 0/20m; 50mm tail; screwed ends						
Back entry; 100mm	19.92	21.91	1.00	11.36	nr	**33.27**
Bottom entry; 100mm	28.73	31.61	1.00	11.36	nr	**42.97**

Y:MECHANICAL AND ELECTRICAL SERVICES

Item	Net Price £	Material £	Labour hours	Labour £	Unit	Total Rate £
Y20 : PUMPS : PUMPS, CIRCULATORS AND ACCELERATORS						
Centrifugal Heating Pump; belt drive; 3 phase, 1450 rpm motor; max. pressure 400kN/m2; max. temperature 125 degree C; bed plate; coupling guard bolted connections; supply only mating flanges; (Includes fixing on prepared base; electrical work elsewhere.)						
Heating Pumps						
40mm pump size; 4.0 litre/sec maximum delivery; 70 KPa maximum head; 0.55kW maximum motor rating	631.00	694.10	9.01	102.34	nr	**796.44**
40mm pump size; 4.0 litre/sec maximum delivery; 130 KPa maximum head; 1.1kW maximum motor rating	676.00	743.60	9.01	102.34	nr	**845.94**
50mm pump size; 8.5 litre/sec maximum delivery; 90KPa maximum head; 1.1kW maximum motor rating	722.00	794.20	9.01	102.34	nr	**896.54**
50mm pump size; 8.5 litre/sec maximum delivery; 190KPa maximum head; 2.2kW maximum motor rating	832.00	915.20	14.08	160.00	nr	**1075.20**
50mm pump size; 8.5 litre/sec maximum delivery; 215 KPa maximum head; 3.0 kW maximum motor rating	886.00	974.60	14.08	160.00	nr	**1134.60**
65mm pump size; 14.0 litre/sec maximum delivery; 90 KPa maximum head; 1.50kW maximum motor rating	815.00	896.50	14.08	160.00	nr	**1056.50**
65mm pump size; 14.0 litre/sec maximum delivery; 160 KPa maximum head; 3.0 kW maximum motor rating	878.00	965.80	14.08	160.00	nr	**1125.80**
65mm pump size; 14.5 litre/sec maximum delivery; 210 KPa maximum head; 4.0 kW maximum motor rating	932.00	1025.20	14.08	160.00	nr	**1185.20**
80mm pump size; 22.0 litre/sec maximum delivery; 130 KPa maximum head; 4.0 kW maximum motor rating	1119.00	1230.90	15.15	172.12	nr	**1403.02**
80mm pump size; 22.0 litre/sec maximum delivery; 200 KPa maximum head; 5.5 kW maximum motor rating	1419.00	1560.90	15.15	172.12	nr	**1733.02**
80mm pump size; 22.0 litre/sec maximum delivery; 250 KPa maximum head; 7.50kW maximum motor rating	1569.00	1725.90	15.15	172.12	nr	**1898.02**
100mm pump size; 30.0 litre/sec maximum delivery; 100 KPa maximum head; 4.0 kW maximum motor rating	1361.00	1497.10	22.22	252.44	nr	**1749.54**

Y:MECHANICAL AND ELECTRICAL SERVICES

Item	Net Price £	Material £	Labour hours	Labour £	Unit	Total Rate £
Y20 : PUMPS : PUMPS, CIRCULATORS AND ACCELERATORS (contd)						
Centrifugal Heating Pump; belt drive; 3 phase, 1450 rpm motor; max. pressure 400kN/m2; max. temperature 125 degree C; bed plate; coupling guard bolted connections; supply only mating flanges; (Includes fixing on prepared base; electrical work elsewhere.) (contd)						
Heating Pumps (contd)						
100mm pump size; 36.0 litre/sec maximum delivery; 25 KPa maximum head; 11.0kW maximum motor rating	2084.00	2292.40	22.22	252.44	nr	**2544.84**
100mm pump size; 36.0 litre/sec maximum delivery; 550 KPa maximum head; 30.0 kW maximum motor rating	3266.00	3592.60	22.22	252.44	nr	**3845.04**
Extra for single phase motor						
0.37 kW	142.56	156.82	6.02	68.43	nr	**225.25**
0.55 kW	142.56	156.82	6.02	68.43	nr	**225.25**
0.75 kW	142.56	156.82	6.02	68.43	nr	**225.25**
1.10 kW	142.56	156.82	6.02	68.43	nr	**225.25**
Centrifugal Heating Pump; close coupled; 3 phase, 1450rpm motor; max. pressure 400kN/m2; max. temperature 110 degree C; bed plate; coupling guard; bolted connections; supply only mating flanges; (Includes fixing on prepared base; electrical work elsewhere.)						
Heating Pumps						
40mm pump size; 4.0 litre/sec maximum delivery; 4 KPa maximum head; 0.55kW maximum motor rating	438.00	481.80	9.01	102.34	nr	**584.14**
40mm pump size; 4.0 litre/sec maximum delivery; 23 KPa maximum head; 2.20kW maximum motor rating	503.00	553.30	9.01	102.34	nr	**655.64**
40mm pump size; 4.0 litre/sec maximum delivery; 75 KPa maximum head; 11 kW maximum motor rating	1504.00	1654.40	9.01	102.34	nr	**1756.74**
50mm pump size; 7.0 litre/sec maximum delivery; 65 KPa maximum head; 0.75 kW maximum motor rating	470.00	517.00	9.01	102.34	nr	**619.34**
50mm pump size; 10.0 litre/sec maximum delivery; 33 KPa maximum head; 5.5kW maximum motor rating	894.00	983.40	9.01	102.34	nr	**1085.74**
50mm pump size; 4.0 litre/sec maximum delivery; 120 KPa maximum head; 1.1 kW maximum motor rating	631.00	694.10	9.01	102.34	nr	**796.44**

Y:MECHANICAL AND ELECTRICAL SERVICES

Item	Net Price £	Material £	Labour hours	Labour £	Unit	Total Rate £
80mm pump size; 16.0 litre/sec maximum delivery; 80 KPa maximum head; 2.2 kW maximum motor rating	817.00	898.70	15.15	172.12	nr	**1070.82**
80mm pump size; 16.0 litre/sec maximum delivery; 120 KPa maximum head; 3.0 kW maximum motor rating	840.00	924.00	15.15	172.12	nr	**1096.12**
100mm pump size; 28.0 litre/sec maximum delivery; 40 KPa maximum head; 2.2 kW maximum motor rating	807.00	887.70	22.22	252.44	nr	**1140.14**
100mm pump size; 28.0 litre/sec maximum delivery; 90 KPa maximum head; 4.0 kW maximum motor rating	890.00	979.00	22.22	252.44	nr	**1231.44**
125mm pump size; 40.0 litre/sec maximum delivery; 50 KPa maximum head; 3.0 kW maximum motor rating	842.00	926.20	30.30	344.24	nr	**1270.44**
125mm pump size; 40.0 litre/sec maximum delivery; 120 Kpa maximum head; 7.5 kW maximum motor rating	1040.00	1144.00	30.30	344.24	nr	**1488.24**
150mm pump size; 70.0 litre/sec maximum delivery; 75 KPa maximum head; 7.5 kW maximum motor rating	1024.00	1126.40	38.46	436.92	nr	**1563.32**
150mm pump size; 70.0 litre/sec maximum delivery; 120 KPa maximum head; 11.0 kW maximum motor rating	1346.00	1480.60	38.46	436.92	nr	**1917.52**
150mm pump size; 70.0 litre/sec maximum delivery; 150 KPa maximum head; 15.0 kW maximum motor rating	1439.00	1582.90	38.46	436.92	nr	**2019.82**
Glandless Domestic Heating Pump; for low pressure domestic hot water heating systems; 240 volt; 50Hz electric motor; max working pressure 1000N/m2 and max temperature of 110 degree C; (Includes fixing; electrical work elsewhere).						
Single speed						
20mm BSP unions	90.95	100.04	1.50	17.06	nr	**117.10**
Three speed						
Pump Only						
40mm	214.20	235.62	1.50	17.06	nr	**252.68**

Y:MECHANICAL AND ELECTRICAL SERVICES

Item	Net Price £	Material £	Labour hours	Labour £	Unit	Total Rate £
Y20 : PUMPS : PUMPS, CIRCULATORS AND ACCELERATORS (contd)						
Automatic Sump Pump; self contained, totally enclosed electric motor, float switch and gear; (Includes fixing; electrical work elsewhere.)						
Small Portable						
Single Phase	142.00	156.20	1.50	17.06	nr	**173.26**
Fixed Installation						
Single Phase	280.00	308.00	2.00	22.72	nr	**330.72**
Three Phase	293.00	322.30	2.00	22.72	nr	**345.02**
Pipeline Mounted Circulator; for low and medium pressure heating and hot water services; silent running; 3 phase; 1450 rpm motor; max pressure 600 kN/m2; max temperature 110 degree C; bolted connections; supply only mating flanges; (Includes fixing; electrical elsewhere.)						
Circulator						
32mm pump size; 2.0 litre/sec maximum delivery; 17 KPa maximum head; 0.18kW maximum motor rating	345.00	379.50	9.01	102.34	nr	**481.84**
40mm pump size; 3.0 litre/sec maximum delivery; 20 KPa maximum head; 0.37kW maximum motor rating	422.00	464.20	9.01	102.34	nr	**566.54**
50mm pump size; 5.0 litre/sec maximum delivery; 30 KPa maximum head; 1.3 kW maximum motor rating	467.00	513.70	10.00	113.60	nr	**627.30**
65mm pump size; 8.0 litre/sec maximum delivery; 37 KPa maximum head; 0.6 kW maximum motor rating	488.00	536.80	10.00	113.60	nr	**650.40**
80mm pump size; 12.0 litre/sec maximum delivery; 42 KPa maximum head; 0.8 kW maximum motor rating	629.00	691.90	10.00	113.60	nr	**805.50**
100mm pump size; 25.0 litre/sec maximum delivery; 37 KPa maximum head; 3.2 kW maximum motor rating	827.00	909.70	11.11	126.22	nr	**1035.92**

Y:MECHANICAL AND ELECTRICAL SERVICES

Item	Net Price £	Material £	Labour hours	Labour £	Unit	Total Rate £
Dual Pipeline Mounted Circulator; for low and medium pressure heating and hot water services; silent running; 3 phase; 1450 rpm motor; max pressure 600kN/m2; max temperature 110 degree C; bolted connections; supply one mating flanges; (Includes fixing; electrical work elsewhere.)						
Circulator						
32mm pump size; 2.0 litre/sec maximum delivery; 17 KPa maximum head; 0.25kW maximum motor rating	811.00	892.10	9.01	102.34	nr	**994.44**
40mm pump size; 3.0 litre/sec maximum delivery; 20 KPa maximum head; 0.6 kW maximum motor rating	1064.00	1170.40	9.01	102.34	nr	**1272.74**
50mm pump size; 5.0 litre/sec maximum delivery; 30 KPa maximum head; 0.6 kW maximum motor rating	1064.00	1170.40	10.00	113.60	nr	**1284.00**
65mm pump size; 8.0 litre/sec maximum delivery; 37 KPa maximum head; 0.8 kW maximum motor rating	1092.00	1201.20	10.00	113.60	nr	**1314.80**
80mm pump size; 12.0 litre/sec maximum delivery; 42 KPa maximum head; 1.1 kW maximum motor rating	1200.00	1320.00	10.00	113.60	nr	**1433.60**
Glandless Accelerator Pumps; for low and medium pressure heating and hot water services; silent running; 3 phase; 1450 rpm motor; max pressure 600 kN/m2; max temperature 120 degree C; bolted connections; supply only mating flanges; (Includes fixing; electrical work elsewhere.)						
Accelerator Pump						
40mm pump size; 4.0 litre/sec maximum delivery; 15 KPa maximum head; 0.35kW maximum motor rating	302.00	332.20	10.00	113.60	nr	**445.80**
50mm pump size; 6.0 litre/sec maximum delivery; 20 KPa maximum head; 0.45kW maximum motor rating	380.00	418.00	10.00	113.60	nr	**531.60**
80mm pump size; 13.0 litre/sec maximum delivery; 28 KPa maximum head; 0.58kW maximum motor rating	447.00	491.70	10.00	113.60	nr	**605.30**

Y:MECHANICAL AND ELECTRICAL SERVICES

Item	Net Price £	Material £	Labour hours	Labour £	Unit	Total Rate £
Y20 : PUMPS : PRESSURISED COLD WATER SUPPLY SET						
Pressurised Cold Water Supply Set; packaged; fully automatic; pumps with 3 phase motors; membrane tank; valves; control panel; inter connecting pipework; fixed on steel frame; (Includes fixing; electrical work elsewhere.)						
Cold Water Supply Set						
3 litre/sec maximum delivery; 600kN/m2 maximum head	3969.00	4365.90	10.00	113.60	nr	**4479.50**
7 litre/sec maximum delivery; 600kN/m2 maximum head	5400.00	5940.00	10.00	113.60	nr	**6053.60**
16 litre/sec maximum delivery; 600kN/m2 maximum head	7632.00	8395.20	10.00	113.60	nr	**8508.80**

Y:MECHANICAL AND ELECTRICAL SERVICES

Item	Net Price £	Material £	Labour hours	Labour £	Unit	Total Rate £
Y21 : WATER TANKS/CISTERNS : SECTIONAL GLASS FIBRE TANKS						
Combination Hot and Cold Water Tanks						
Super seven high performance insulated copper storage units; direct pattern; standard connections; BS 3198 (Includes placing in position)						
450mm dia.; 1200mm height	151.17	166.29	1.50	17.06	nr	**183.35**
450mm dia.; 1400mm height	198.00	327.03	1.00	11.36	nr	**338.39**
450mm dia.; 1850mm height	235.80	389.46	1.00	11.36	nr	**400.82**

Y:MECHANICAL AND ELECTRICAL SERVICES

Item	Net Price £	Material £	Labour hours	Labour £	Unit	Total Rate £
Y21 : WATER TANKS/CISTERNS **: MOULDED FIBREGLASS CISTERNS**						
Cisterns; fibreglass; complete with ball valve, fixing plate and fitted covers; compiles to bye law 30.						
Rectangular						
70 litres capacity	132.23	145.45	1.00	11.36	nr	**156.81**
110 litres capacity	160.93	177.02	1.00	11.36	nr	**188.38**
170 litres capacity	185.52	204.08	1.00	11.36	nr	**215.44**
280 litres capacity	239.85	263.84	1.00	11.36	nr	**275.20**
420 litres capacity	339.27	373.20	1.50	17.06	nr	**390.26**
710 litres capacity	448.95	493.85	2.50	28.40	nr	**522.25**
840 litres capacity	522.75	575.03	2.50	28.40	nr	**603.43**
1590 litres capacity	733.90	807.29	13.16	149.47	nr	**956.76**
2275 litres capacity	944.03	1038.43	20.00	227.20	nr	**1265.63**
3365 litres capacity	1411.43	1552.57	25.00	284.00	nr	**1836.57**
4545 litres capacity	1634.87	1798.36	30.30	344.24	nr	**2142.60**

Y:MECHANICAL AND ELECTRICAL SERVICES

Item	Net Price £	Material £	Labour hours	Labour £	Unit	Total Rate £
Y21 : WATER TANKS/CISTERNS **: MOULDED PLASTIC CISTERNS**						
Cisterns; polypropylene; complete with ball valve, fixing plate and cover; (Includes hoisting and placing in position.)						
Rectangular						
18 litres capacity	3.44	3.79	1.00	11.36	nr	**15.15**
68 litres capacity	12.20	13.42	1.00	11.36	nr	**24.78**
114 litres capacity	16.73	18.40	1.00	11.36	nr	**29.76**
182 litres capacity	30.49	33.54	1.00	11.36	nr	**44.90**
227 litres capacity	30.49	33.54	1.00	11.36	nr	**44.90**
Circular						
114 litres capacity	14.07	15.48	1.00	11.36	nr	**26.84**
227 litres capacity	21.06	23.16	1.00	11.36	nr	**34.52**
Extra over rectangular cisterns; bye law 30 kit						
18 litres capacity	7.42	8.17	0.50	5.68	nr	**13.85**
68 litres capacity	12.98	14.28	0.50	5.68	nr	**19.96**
114 litres capacity	15.39	16.92	0.50	5.68	nr	**22.60**
182 litres capacity	21.88	24.06	0.75	8.52	nr	**32.58**
127 litres capacity	22.96	25.25	0.75	8.52	nr	**33.77**
Extra over circular cisterns; bye law 30 kit						
114 litres capacity	15.26	16.79	0.50	5.68	nr	**22.47**
127 litres capacity	21.54	23.69	0.75	8.52	nr	**32.21**

Y:MECHANICAL AND ELECTRICAL SERVICES

Item	Net Price £	Material £	Labour hours	Labour £	Unit	Total Rate £
Y22 : HEAT EXCHANGERS **: ROTARY AIR TO AIR WHEEL**						
Thermal Wheels; fixed in operrational location. (Electrical work elsewhere.)						
Thermal wheels; rotary air to air energy exchanger comprising supporting frame; rotor wheel and drive motor; (wheel constructed of alternate flat and corrugated foils of inorganic material glued together to form axial flutes; coated with a dessicant or similar treatment) placing in ductwork system						
Fixed speed drive						
1m3/sec. flow rate	2118.00	2329.80	18.18	206.55	nr	**2536.35**
2m3/sec. flow rate	2457.00	2702.70	25.00	284.00	nr	**2986.70**
Variable speed drive						
4m3/sec. flow rate	3246.00	3570.60	35.71	405.71	nr	**3976.31**
7m/3sec. flow rate	4494.00	4943.40	50.00	568.00	nr	**5511.40**
12m3/sec. flow rate	8969.00	9865.90	76.92	873.85	nr	**10739.75**

Y:MECHANICAL AND ELECTRICAL SERVICES

Item	Net Price £	Material £	Labour hours	Labour £	Unit	Total Rate £
Y23 : STORAGE CYLINDERS/CALORIFERS : COPPER DIRECT CYLINDERS						
Insulated copper storage cylinders; BS699; (Includes placing in position)						
Grade 3 (maximum 10m working head)						
BS size 6; 115 litres capacity; 400mm dia.; 1050mm height	54.51	59.96	1.50	17.06	nr	**77.02**
BS size 7; 120 litres capacity; 450mm dia.; 900mm height	64.45	70.90	2.00	22.72	nr	**93.62**
BS size 8; 144 litres capacity; 450mm dia.; 1050mm height	60.36	66.40	2.80	31.82	nr	**98.22**
Grade 4 (maximum 6m working head)						
BS size 2; 96 litres capacity; 400mm dia.; 900mm height	51.00	56.10	1.50	17.06	nr	**73.16**
BS size 7; 120 litres capacity; 450mm dia.; 900mm height	42.20	46.42	1.50	17.06	nr	**63.48**
BS size 8; 144 litres capacity; 450mm dia.; 1050mm height	52.95	58.25	1.50	17.06	nr	**75.31**
BS size 9; 166 litres capacity; 450mm dia.; 1200mm height	77.42	85.16	1.50	17.06	nr	**102.22**
Storage Cylinders; brazed copper construction; to BS 699; screwed bosses (Includes placing in position)						
Tested 2.2 bar, 15m maximum head						
144 litres	230.00	253.00	3.00	34.11	nr	**287.11**
160 litres	260.00	286.00	3.00	34.11	nr	**320.11**
200 litres	268.00	294.80	3.76	42.71	nr	**337.51**
255 litres	305.00	335.50	3.76	42.71	nr	**378.21**
290 litres	412.00	453.20	3.76	42.71	nr	**495.91**
370 litres	472.00	519.20	4.50	51.17	nr	**570.37**
450 litres	628.00	690.80	5.00	56.80	nr	**747.60**
Tested to 2.55 bar, 17m maximum head						
550 litres	680.00	748.00	5.00	56.80	nr	**804.80**
700 litres	795.00	874.50	6.02	68.43	nr	**942.93**
800 litres	950.00	1045.00	6.54	74.25	nr	**1119.25**
900 litres	995.00	1094.50	8.00	90.88	nr	**1185.38**
1000 litres	1050.00	1155.00	8.00	90.88	nr	**1245.88**
1250 litres	1150.00	1265.00	13.16	149.47	nr	**1414.47**
1500 litres	1750.00	1925.00	15.15	172.12	nr	**2097.12**
2000 litres	2100.00	2310.00	17.24	195.86	nr	**2505.86**
3000 litres	2950.00	3245.00	24.39	277.07	nr	**3522.07**

Y:MECHANICAL AND ELECTRICAL SERVICES

Item	Net Price £	Material £	Labour hours	Labour £	Unit	Total Rate £
Y23 : STORAGE CYLINDERS/CALORIFIERS : STEEL INDIRECT CYLINDERS						
Indirect Cylinders; mild steel; welded throughout; galvanised after made; with bolted connections. (Includes placing in position.)						
3.2mm plate						
136 litres capacity	325.00	357.50	2.50	28.40	nr	**385.90**
159 litres capacity	350.00	385.00	2.80	31.82	nr	**416.82**
182 litres capacity	430.00	473.00	3.00	34.11	nr	**507.11**
227 litres capacity	530.00	583.00	3.00	34.11	nr	**617.11**
273 litres capacity	625.00	687.50	4.00	45.44	nr	**732.94**
364 litres capacity	720.00	792.00	4.50	51.17	nr	**843.17**
455 litres capacity	830.00	913.00	5.00	56.80	nr	**969.80**
683 litres capacity	1270.00	1397.00	6.02	68.43	nr	**1465.43**
910 litres capacity	1560.00	1716.00	7.04	80.00	nr	**1796.00**
Welded Cylinders; mild steel; galvanised after made; bolted top; BS 417; (Includes hoisting and placing in position.)						
3.2mm Gauge; 30m Working Head						
YM141-123 litres capacity	160.00	176.00	2.00	22.72	nr	**198.72**
YM150-136 litres capacity	170.00	187.00	2.50	28.40	nr	**215.40**
YM177-159 litres capacity	200.00	220.00	2.50	28.40	nr	**248.40**

Y:MECHANICAL AND ELECTRICAL SERVICES

Item	Net Price £	Material £	Labour hours	Labour £	Unit	Total Rate £
Y23 : STORAGE CYLINDERS/CALORIFIERS : COPPER INDIRECT CYLINDERS						
Indirect Cylinders; copper; bolted top; up to 5 tappings for connections; BS 1586; (includes placing in position)						
Grade 3, Tested 1.45 bar, 10m maximum head						
74 litres capacity	110.00	121.00	1.50	17.06	nr	**138.06**
96 litres capacity	112.00	123.20	1.50	17.06	nr	**140.26**
114 litres capacity	115.00	126.50	1.50	17.06	nr	**143.56**
117 litres capacity	120.00	132.00	2.00	22.72	nr	**154.72**
140 litres capacity	123.00	135.30	2.50	28.40	nr	**163.70**
162 litres capacity	172.00	189.20	3.00	34.11	nr	**223.31**
190 litres capacity	188.00	206.80	3.51	39.86	nr	**246.66**
245 litres capacity	220.00	242.00	3.80	43.19	nr	**285.19**
280 litres capacity	390.00	429.00	4.00	45.44	nr	**474.44**
360 litres capacity	422.00	464.20	4.50	51.17	nr	**515.37**
440 litres capacity	490.00	539.00	4.50	51.17	nr	**590.17**
Grade 2, Tested to 2.2 bar, 15m maximum head						
117 litres capacity	159.00	174.90	2.00	22.72	nr	**197.62**
140 litres capacity	173.00	190.30	2.50	28.40	nr	**218.70**
162 litres capacity	198.00	217.80	2.80	31.82	nr	**249.62**
190 litres capacity	230.00	253.00	3.00	34.11	nr	**287.11**
245 litres capacity	278.00	305.80	4.00	45.44	nr	**351.24**
280 litres capacity	444.00	488.40	4.00	45.44	nr	**533.84**
360 litres capacity	490.00	539.00	4.50	51.17	nr	**590.17**
440 litres capacity	580.00	638.00	4.50	51.17	nr	**689.17**
Grade 1, Tested 3.65 bar, 25m maximum head						
190 litres capacity	342.00	376.20	3.00	34.11	nr	**410.31**
245 litres capacity	389.00	427.90	3.00	34.11	nr	**462.01**
280 litres capacity	550.00	605.00	4.00	45.44	nr	**650.44**
360 litres capacity	696.00	765.60	4.50	51.17	nr	**816.77**
440 litres capacity	845.00	929.50	4.50	51.17	nr	**980.67**
Indirect cylinders, including manhole, to BS 853.						
Grade 3, tested to 1.5 bar, 10m maximum head						
550 litres capacity	700.00	770.00	5.21	59.17	nr	**829.17**
700 litres capacity	775.00	852.50	6.02	68.43	nr	**920.93**
800 litres capacity	900.00	990.00	6.54	74.25	nr	**1064.25**
1000 litres capacity	1125.00	1237.50	7.04	80.00	nr	**1317.50**
1500 litres capacity	1300.00	1430.00	10.00	113.60	nr	**1543.60**
2000 litres capacity	1800.00	1980.00	16.13	183.23	nr	**2163.23**

Y:MECHANICAL AND ELECTRICAL SERVICES

Item	Net Price £	Material £	Labour hours	Labour £	Unit	Total Rate £
Y23 : STORAGE CYLINDERS/CALORIFIERS **: COPPER INDIRECT CYLINDERS (contd)**						
Indirect cylinders, including manhole, **to BS 853. (contd)**						
Grade 2, tested to 2.55 bar, 15m maximum head						
550 litres capacity	795.00	874.50	5.21	59.17	nr	**933.67**
700 litres capacity	995.00	1094.50	6.02	68.43	nr	**1162.93**
800 litres capacity	1050.00	1155.00	6.54	74.25	nr	**1229.25**
1000 litres capacity	1300.00	1430.00	7.04	80.00	nr	**1510.00**
1500 litres capacity	1600.00	1760.00	10.00	113.60	nr	**1873.60**
2000 litres capacity	2000.00	2200.00	16.13	183.23	nr	**2383.23**
Grade 1, tested to 4 bar, 25m maximum head						
550 litres capacity	925.00	1017.50	5.21	59.17	nr	**1076.67**
700 litres capacity	1050.00	1155.00	6.02	68.43	nr	**1223.43**
800 litres capacity	1125.00	1237.50	6.54	74.25	nr	**1311.75**
1000 litres capacity	1500.00	1650.00	7.04	80.00	nr	**1730.00**
1500 litres capacity	1800.00	1980.00	10.00	113.60	nr	**2093.60**
2000 litres capacity	2200.00	2420.00	16.13	183.23	nr	**2603.23**

Y:MECHANICAL AND ELECTRICAL SERVICES

Item	Net Price £	Material £	Labour hours	Labour £	Unit	Total Rate £
Y23 : STORAGE CYLINDERS/CALORIFIERS : STEEL NON STORAGE CALORIFIER						
Non-Storage Calorifiers; mild steel; heater battery duty 82.71 degrees C to BS 853, maximum test on shell 11.55 bar, tubes 26.25 bar.						
Horizontal; Steam at 3.2 bar						
88 kW capacity	350.00	385.00	8.00	90.88	nr	**475.88**
176 kW capacity	525.00	577.50	12.05	136.87	nr	**714.37**
293 kW capacity	550.00	605.00	14.08	160.00	nr	**765.00**
586 kW capacity	720.00	792.00	37.04	420.74	nr	**1212.74**
879 kW capacity	895.00	984.50	40.00	454.40	nr	**1438.90**
1465 kW capacity	1425.00	1567.50	45.45	516.36	nr	**2083.86**
Horizontal; Steam at 4.8 bar						
88 kW capacity	350.00	385.00	5.00	56.80	nr	**441.80**
176 kW capacity	525.00	577.50	8.00	90.88	nr	**668.38**
293 kW capacity	545.00	599.50	12.05	136.87	nr	**736.37**
586 kW capacity	710.00	781.00	22.22	252.44	nr	**1033.44**
879 kW capacity	885.00	973.50	28.57	324.57	nr	**1298.07**
1465 kW capacity	1425.00	1567.50	40.00	454.40	nr	**2021.90**
Vertical; Steam at 3.2 bar						
88 kW capacity	360.00	396.00	8.00	90.88	nr	**486.88**
176 kW capacity	540.00	594.00	12.05	136.87	nr	**730.87**
293 kW capacity	570.00	627.00	14.08	160.00	nr	**787.00**
586 kW capacity	725.00	797.50	37.04	420.74	nr	**1218.24**
879 kW capacity	900.00	990.00	40.00	454.40	nr	**1444.40**
1465 kW capacity	1450.00	1595.00	45.45	516.36	nr	**2111.36**
Vertical; Steam at 4.8 bar						
88 kW capacity	360.00	396.00	5.00	56.80	nr	**452.80**
176 kW capacity	540.00	594.00	12.05	136.87	nr	**730.87**
293 kW capacity	570.00	627.00	12.05	136.87	nr	**763.87**
586 kW capacity	725.00	797.50	22.22	252.44	nr	**1049.94**
879 kW capacity	900.00	990.00	28.57	324.57	nr	**1314.57**
1465 kW capacity	1450.00	1595.00	40.00	454.40	nr	**2049.40**
Horizontal; Primary water at 116 degree C (on) 90 degree C (off)						
88 kW capacity	380.00	418.00	5.00	56.80	nr	**474.80**
176 kW capacity	650.00	715.00	7.04	80.00	nr	**795.00**
293 kW capacity	800.00	880.00	9.01	102.34	nr	**982.34**
586 kW capacity	1350.00	1485.00	22.22	252.44	nr	**1737.44**
879 kW capacity	1580.00	1738.00	28.57	324.57	nr	**2062.57**
1465 kW capacity	2200.00	2420.00	50.00	568.00	nr	**2988.00**

Y:MECHANICAL AND ELECTRICAL SERVICES

Item	Net Price £	Material £	Labour hours	Labour £	Unit	Total Rate £
Y23 : STORAGE CYLINDERS/CALORIFIERS : STEEL NON STORAGE CALORIFIER (contd)						
Non-Storage Calorifiers; mild steel; heater battery duty 82.71 degrees C to BS 853, maximum test on shell 11.55 bar, tubes 26.25 bar. (contd)						
Vertical; Primary water at 116 degree C (on)/93 degree C (off)						
88 kW capacity	390.00	429.00	5.00	56.80	nr	**485.80**
176 kW capacity	665.00	731.50	7.04	80.00	nr	**811.50**
293 kW capacity	840.00	924.00	9.01	102.34	nr	**1026.34**
586 kW capacity	1350.00	1485.00	22.22	252.44	nr	**1737.44**
879 kW capacity	1600.00	1760.00	28.57	324.57	nr	**2084.57**
1465 kW capacity	2250.00	2475.00	50.00	568.00	nr	**3043.00**

Keep your figures up to date, free of charge

This section, and most of the other information in this Price Book, is brought up to date every three months, until the next annual edition, in the *Price Book Update*.

The *Update* is available free to all Price Book purchasers.

To ensure you receive your copy, simply complete the reply card from the centre of the book and return it to us.

Y:MECHANICAL AND ELECTRICAL SERVICES

Item	Net Price £	Material £	Labour hours	Labour £	Unit	Total Rate £
Y23 : STORAGE CYLINDERS/CALORIFIERS : STEEL STORAGE CALORIFIER						
Storage Calorifiers; galvanised mild steel; heater battery capable of raising temperature of contents from 10 degree C to 65 degree C in one hour; static head not exceeding 1.35 bar; (Includes fixing in position on cradles or legs)						
Horizontal; Primary LPHW at 82 degree C (on)/71 degree C (off)						
400 litres capacity	925.00	1017.50	6.71	76.24	nr	**1093.74**
1000 litres capacity	1360.00	1496.00	8.00	90.88	nr	**1586.88**
2000 litres capacity	2200.00	2420.00	14.08	160.00	nr	**2580.00**
3000 litres capacity	3500.00	3850.00	25.00	284.00	nr	**4134.00**
4000 litres capacity	4050.00	4455.00	40.00	454.40	nr	**4909.40**
4500 litres capacity	4200.00	4620.00	35.71	405.71	nr	**5025.71**
Horizontal; Primary steam at 3.2 bar						
400 litres capacity	815.00	896.50	6.71	76.24	nr	**972.74**
1000 litres capacity	1200.00	1320.00	8.00	90.88	nr	**1410.88**
2000 litres capacity	1950.00	2145.00	14.08	160.00	nr	**2305.00**
3000 litres capacity	3080.00	3388.00	25.00	284.00	nr	**3672.00**
4000 litres capacity	3550.00	3905.00	40.00	454.40	nr	**4359.40**
4500 litres capacity	3700.00	4070.00	50.00	568.00	nr	**4638.00**
Vertical; Primary LPHW at 82 degree C (on)/71 degree C (off)						
400 litres capacity	900.00	990.00	7.04	80.00	nr	**1070.00**
1000 litres capacity	1335.00	1468.50	8.00	90.88	nr	**1559.38**
2000 litres capacity	2175.00	2392.50	14.08	160.00	nr	**2552.50**
3000 litres capacity	3475.00	3822.50	25.00	284.00	nr	**4106.50**
4000 litres capacity	4025.00	4427.50	40.00	454.40	nr	**4881.90**
4500 litres capacity	4175.00	4592.50	50.00	568.00	nr	**5160.50**
Vertical; Primary steam at 3.2 bar						
400 litres capacity	790.00	869.00	7.04	80.00	nr	**949.00**
1000 litres capacity	1175.00	1292.50	8.00	90.88	nr	**1383.38**
2000 litres capacity	1925.00	2117.50	14.08	160.00	nr	**2277.50**
3000 litres capacity	3055.00	3360.50	25.00	284.00	nr	**3644.50**
4000 litres capacity	3525.00	3877.50	40.00	454.40	nr	**4331.90**
4500 litres capacity	3675.00	4042.50	50.00	568.00	nr	**4610.50**

Y:MECHANICAL AND ELECTRICAL SERVICES

Item	Net Price £	Material £	Labour hours	Labour £	Unit	Total Rate £
Y23 : STORAGE CYLINDERS/CALORIFIERS : COPPER STORAGE CALORIFIER						
Storage Calorifiers; copper; heater battery capable of raising temperature; of contents from 10 degree C to 65 degree C in one hour; static head not exceeding 1.35 bar; BS 853; (Includes fixing in position on cradles or legs.)						
Horizontal; Primary LPHW at 82 degree C (on)/71 degree C (off)						
400 litres capacity	1000.00	1100.00	7.04	80.00	nr	1180.00
1000 litres capacity	1600.00	1760.00	8.00	90.88	nr	1850.88
2000 litres capacity	3200.00	3520.00	14.08	160.00	nr	3680.00
3000 litres capacity	3950.00	4345.00	25.00	284.00	nr	4629.00
4000 litres capacity	4800.00	5280.00	40.00	454.40	nr	5734.40
4500 litres capacity	5400.00	5940.00	50.00	568.00	nr	6508.00
Horizontal; Primary steam at 3.2 bar						
400 litres capacity	920.00	1012.00	7.04	80.00	nr	1092.00
1000 litres capacity	1350.00	1485.00	8.00	90.88	nr	1575.88
2000 litres capacity	3000.00	3300.00	14.08	160.00	nr	3460.00
3000 litres capacity	3710.00	4081.00	25.00	284.00	nr	4365.00
4000 litres capacity	4600.00	5060.00	40.00	454.40	nr	5514.40
4500 litres capacity	5200.00	5720.00	50.00	568.00	nr	6288.00
Vertical; Primary LPHW at 82 degree C (on)/71 degree C (off)						
400 litres capacity	980.00	1078.00	7.04	80.00	nr	1158.00
1000 litres capacity	1575.00	1732.50	8.00	90.88	nr	1823.38
2000 litres capacity	3000.00	3300.00	14.08	160.00	nr	3460.00
3000 litres capacity	3750.00	4125.00	25.00	284.00	nr	4409.00
4000 litres capacity	4600.00	5060.00	40.00	454.40	nr	5514.40
4500 litres capacity	5200.00	5720.00	50.00	568.00	nr	6288.00
Vertical; Primary steam at 3.2 bar						
400 litres capacity	900.00	990.00	7.04	80.00	nr	1070.00
1000 litres capacity	1325.00	1457.50	8.00	90.88	nr	1548.38
2000 litres capacity	2800.00	3080.00	14.08	160.00	nr	3240.00
3000 litres capacity	3500.00	3850.00	25.00	284.00	nr	4134.00
4000 litres capacity	4400.00	4840.00	40.00	454.40	nr	5294.40
4500 litres capacity	5000.00	5500.00	50.00	568.00	nr	6068.00

Y:MECHANICAL AND ELECTRICAL SERVICES

Item	Net Price £	Material £	Labour hours	Labour £	Unit	Total Rate £
Y30 : AIR DUCTLINES : MILD STEEL DUCTWORK						
Galvanised Sheet Metal DW142 Class A rectangular Section Ductwork; (including all necessary stiffeners, joints, couplers in the running length and duct support)						
Straight Duct						
Sum of two sides 200mm	4.24	4.67	1.03	11.74	m	**16.41**
Sum of two sides 250mm	4.82	5.30	1.03	11.74	m	**17.04**
Sum of two sides 300mm	5.49	6.04	1.03	11.74	m	**17.78**
Sum of two sides 350mm	6.16	6.78	1.03	11.74	m	**18.52**
Sum of two sides 400mm	6.79	7.46	1.00	11.39	m	**18.85**
Sum of two sides 450mm	7.40	8.14	1.00	11.39	m	**19.53**
Sum of two sides 500mm	8.03	8.83	1.00	11.39	m	**20.22**
Sum of two sides 550mm	8.66	9.53	1.10	12.53	m	**22.06**
Sum of two sides 600mm	9.29	10.22	1.10	12.53	m	**22.75**
Sum of two sides 650mm	9.92	10.91	1.10	12.53	m	**23.44**
Sum of two sides 700mm	10.53	11.58	1.10	12.53	m	**24.11**
Sum of two sides 750mm	11.18	12.30	1.10	12.53	m	**24.83**
Sum of two sides 800mm	11.79	12.97	1.10	12.53	m	**25.50**
Sum of two sides 850mm	12.44	13.69	1.10	12.53	m	**26.22**
Sum of two sides 900mm	13.06	14.36	1.10	12.53	m	**26.89**
Sum of two sides 950mm	13.81	15.19	1.19	13.56	m	**28.75**
Sum of two sides 1000mm	14.31	15.74	1.19	13.56	m	**29.30**
Sum of two sides 1100mm	15.55	17.10	1.19	13.56	m	**30.66**
Sum of two sides 1200mm	16.83	18.51	1.19	13.56	m	**32.07**
Sum of two sides 1300mm	18.07	19.88	1.21	13.79	m	**33.67**
Sum of two sides 1400mm	19.33	21.26	1.21	13.79	m	**35.05**
Sum of two sides 1500mm	20.59	22.65	1.28	14.59	m	**37.24**
Sum of two sides 1600mm	21.86	24.05	1.34	15.27	m	**39.32**
Sum of two sides 1700mm	25.40	27.94	1.34	15.27	m	**43.21**
Sum of two sides 1800mm	26.76	29.44	1.39	15.84	m	**45.28**
Sum of two sides 1900mm	28.11	30.92	1.39	15.84	m	**46.76**
Sum of two sides 2000mm	29.46	32.40	1.39	15.84	m	**48.24**
Sum of two sides 2100mm	38.89	42.78	1.88	21.45	m	**64.23**
Sum of two sides 2200mm	40.59	44.65	1.90	21.65	m	**66.30**
Sum of two sides 2300mm	42.34	46.58	1.90	21.65	m	**68.23**
Sum of two sides 2400mm	44.08	48.48	2.06	23.49	m	**71.97**
Sum of two sides 2500mm	45.79	50.37	2.06	23.49	m	**73.86**
Sum of two sides 2600mm	50.97	56.07	2.29	26.12	m	**82.19**
Sum of two sides 2700mm	52.85	58.13	2.30	26.24	m	**84.37**
Sum of two sides 2800mm	54.68	60.15	2.55	29.06	m	**89.21**
Sum of two sides 2900mm	56.55	62.20	2.56	29.21	m	**91.41**
Sum of two sides 3000mm	58.40	64.24	2.73	31.12	m	**95.36**
Sum of two sides 3100mm	60.25	66.27	2.73	31.12	m	**97.39**
Sum of two sides 3200mm	62.11	68.33	2.75	31.38	m	**99.71**
Extra Over Galvanised Sheet DW 142 Class A Rectangular Section Ductwork; (including all necessary stiffeners, joints etc.)						
Stopped End						
Sum of two sides 200mm	0.09	0.10	0.33	3.76	nr	**3.86**
Sum of two sides 250mm	0.14	0.16	0.33	3.76	nr	**3.92**
Sum of two sides 300mm	0.14	0.16	0.33	3.76	nr	**3.92**

Y:MECHANICAL AND ELECTRICAL SERVICES

Item	Net Price £	Material £	Labour hours	Labour £	Unit	Total Rate £
Y30 : AIR DUCTLINES : MILD STEEL DUCTWORK (contd)						
Extra Over Galvanised Sheet DW 142 Class A Rectangular Section Ductwork; (including all necessary stiffeners, joints etc.) (contd)						
Stopped End (contd)						
Sum of two sides 350mm	0.19	0.21	0.33	3.76	nr	3.97
Sum of two sides 400mm	0.19	0.21	0.33	3.76	nr	3.97
Sum of two sides 450mm	0.24	0.26	0.33	3.76	nr	4.02
Sum of two sides 500mm	0.24	0.26	0.33	3.76	nr	4.02
Sum of two sides 550mm	0.24	0.26	0.33	3.76	nr	4.02
Sum of two sides 600mm	0.29	0.32	0.33	3.76	nr	4.08
Sum of two sides 650mm	0.29	0.32	0.33	3.76	nr	4.08
Sum of two sides 700mm	0.33	0.36	0.33	3.76	nr	4.12
Sum of two sides 750mm	0.39	0.43	0.33	3.76	nr	4.19
Sum of two sides 800mm	0.39	0.43	0.33	3.76	nr	4.19
Sum of two sides 850mm	0.45	0.50	0.50	5.69	nr	6.19
Sum of two sides 900mm	0.47	0.52	0.50	5.69	nr	6.21
Sum of two sides 950mm	0.50	0.55	0.50	5.69	nr	6.24
Sum of two sides 1000mm	0.53	0.59	0.50	5.69	nr	6.28
Sum of two sides 1100mm	0.56	0.62	0.50	5.69	nr	6.31
Sum of two sides 1200mm	0.60	0.67	0.50	5.69	nr	6.36
Sum of two sides 1300mm	0.65	0.71	0.50	5.69	nr	6.40
Sum of two sides 1400mm	0.68	0.74	0.50	5.69	nr	6.43
Sum of two sides 1500mm	0.72	0.79	0.50	5.69	nr	6.48
Sum of two sides 1600mm	0.76	0.83	0.50	5.69	nr	6.52
Sum of two sides 1700mm	0.82	0.90	0.50	5.69	nr	6.59
Sum of two sides 1800mm	0.86	0.95	0.50	5.69	nr	6.64
Sum of two sides 1900mm	0.91	1.00	0.50	5.69	nr	6.69
Sum of two sides 2000mm	0.95	1.05	0.50	5.69	nr	6.74
Sum of two sides 2100mm	1.26	1.39	0.75	8.54	nr	9.93
Sum of two sides 2200mm	1.31	1.44	0.75	8.54	nr	9.98
Sum of two sides 2300mm	1.37	1.51	0.75	8.54	nr	10.05
Sum of two sides 2400mm	1.42	1.57	0.75	8.54	nr	10.11
Sum of two sides 2500mm	1.50	1.65	0.75	8.54	nr	10.19
Sum of two sides 2600mm	1.57	1.73	0.75	8.54	nr	10.27
Sum of two sides 2700mm	1.62	1.78	0.75	8.54	nr	10.32
Sum of two sides 2800mm	1.69	1.86	1.00	11.39	nr	13.25
Sum of two sides 2900mm	1.73	1.91	1.00	11.39	nr	13.30
Sum of two sides 3000mm	1.80	1.98	1.50	17.08	nr	19.06
Sum of two sides 3100mm	1.87	2.05	1.56	17.77	nr	19.82
Sum of two sides 3200mm	1.91	2.10	1.56	17.77	nr	19.87
Taper; Large End Measured						
Sum of two sides 200mm	3.12	3.43	1.23	14.01	nr	17.44
Sum of two sides 250mm	3.22	3.54	1.23	14.01	nr	17.55
Sum of two sides 300mm	3.94	4.34	1.23	14.01	nr	18.35
Sum of two sides 350mm	4.39	4.83	1.21	13.79	nr	18.62
Sum of two sides 400mm	4.79	5.27	1.21	13.79	nr	19.06
Sum of two sides 450mm	5.21	5.73	1.21	13.79	nr	19.52
Sum of two sides 500mm	5.65	6.21	1.21	13.79	nr	20.00
Sum of two sides 550mm	6.06	6.66	1.46	16.65	nr	23.31
Sum of two sides 600mm	6.46	7.10	1.46	16.65	nr	23.75
Sum of two sides 650mm	6.89	7.58	1.46	16.65	nr	24.23
Sum of two sides 700mm	7.32	8.05	1.46	16.65	nr	24.70

Y:MECHANICAL AND ELECTRICAL SERVICES

Item	Net Price £	Material £	Labour hours	Labour £	Unit	Total Rate £
Sum of two sides 750mm	7.74	8.51	1.66	18.92	nr	**27.43**
Sum of two sides 800mm	8.15	8.96	1.66	18.92	nr	**27.88**
Sum of two sides 850mm	8.57	9.43	1.66	18.92	nr	**28.35**
Sum of two sides 900mm	8.98	9.88	1.66	18.92	nr	**28.80**
Sum of two sides 950mm	9.41	10.35	1.69	19.24	nr	**29.59**
Sum of two sides 1000mm	9.83	10.81	1.89	21.53	nr	**32.34**
Sum of two sides 1100mm	10.67	11.74	1.89	21.53	nr	**33.27**
Sum of two sides 1200mm	11.50	12.65	1.89	21.53	nr	**34.18**
Sum of two sides 1300mm	12.36	13.60	1.99	22.69	nr	**36.29**
Sum of two sides 1400mm	13.17	14.49	1.99	22.69	nr	**37.18**
Sum of two sides 1500mm	14.03	15.44	2.14	24.39	nr	**39.83**
Sum of two sides 1600mm	14.86	16.35	2.14	24.39	nr	**40.74**
Sum of two sides 1700mm	18.57	20.43	2.14	24.39	nr	**44.82**
Sum of two sides 1800mm	19.52	21.47	2.24	25.54	nr	**47.01**
Sum of two sides 1900mm	20.48	22.53	2.35	26.80	nr	**49.33**
Sum of two sides 2000mm	27.57	30.33	2.24	25.54	nr	**55.87**
Sum of two sides 2100mm	24.91	27.40	2.53	28.83	nr	**56.23**
Sum of two sides 2200mm	25.88	28.47	2.53	28.83	nr	**57.30**
Sum of two sides 2300mm	27.05	29.75	2.53	28.83	nr	**58.58**
Sum of two sides 2400mm	28.12	30.93	2.70	30.78	nr	**61.71**
Sum of two sides 2500mm	29.20	32.12	2.70	30.78	nr	**62.90**
Sum of two sides 2600mm	30.28	33.31	2.70	30.78	nr	**64.09**
Sum of two sides 2700mm	31.37	34.50	2.70	30.78	nr	**65.28**
Sum of two sides 2800mm	32.42	35.66	3.42	39.01	nr	**74.67**
Sum of two sides 2900mm	33.52	36.87	3.44	39.14	nr	**76.01**
Sum of two sides 3000mm	34.56	38.02	3.91	44.49	nr	**82.51**
Sum of two sides 3100mm	35.60	39.16	3.91	44.49	nr	**83.65**
Sum of two sides 3200mm	36.74	40.41	3.91	44.38	nr	**84.79**
Set Piece						
Sum of two sides 200mm	2.73	3.00	1.43	16.29	nr	**19.29**
Sum of two sides 250mm	3.10	3.41	1.43	16.29	nr	**19.70**
Sum of two sides 300mm	3.52	3.87	1.43	16.29	nr	**20.16**
Sum of two sides 350mm	3.95	4.34	1.41	16.06	nr	**20.40**
Sum of two sides 400mm	4.40	4.84	1.41	16.06	nr	**20.90**
Sum of two sides 450mm	4.87	5.36	1.41	16.06	nr	**21.42**
Sum of two sides 500mm	5.38	5.92	1.41	16.06	nr	**21.98**
Sum of two sides 550mm	5.89	6.48	1.66	18.92	nr	**25.40**
Sum of two sides 600mm	6.46	7.10	1.66	18.92	nr	**26.02**
Sum of two sides 650mm	7.04	7.75	1.66	18.92	nr	**26.67**
Sum of two sides 700mm	7.66	8.42	1.66	18.92	nr	**27.34**
Sum of two sides 750mm	8.26	9.09	1.66	18.92	nr	**28.01**
Sum of two sides 800mm	8.91	9.80	1.66	18.92	nr	**28.72**
Sum of two sides 850mm	9.57	10.53	1.66	18.92	nr	**29.45**
Sum of two sides 900mm	10.56	11.61	1.66	18.92	nr	**30.53**
Sum of two sides 950mm	10.99	12.09	1.89	21.53	nr	**33.62**
Sum of two sides 1000mm	11.74	12.91	1.89	21.53	nr	**34.44**
Sum of two sides 1100mm	13.31	14.65	1.89	21.53	nr	**36.18**
Sum of two sides 1200mm	14.95	16.45	1.89	21.53	nr	**37.98**
Sum of two sides 1300mm	16.72	18.39	1.99	22.69	nr	**41.08**
Sum of two sides 1400mm	18.55	20.41	1.99	22.69	nr	**43.10**
Sum of two sides 1500mm	20.49	22.54	2.14	24.39	nr	**46.93**
Sum of two sides 1600mm	22.54	24.79	2.14	24.39	nr	**49.18**
Sum of two sides 1700mm	27.52	30.27	2.24	25.54	nr	**55.81**
Sum of two sides 1800mm	29.88	32.87	2.26	25.77	nr	**58.64**
Sum of two sides 1900mm	32.31	35.54	2.35	26.80	nr	**62.34**
Sum of two sides 2000mm	34.86	38.35	2.35	26.80	nr	**65.15**
Sum of two sides 2100mm	43.78	48.16	2.53	28.83	nr	**76.99**
Sum of two sides 2200mm	51.27	56.40	2.82	32.18	nr	**88.58**

Y:MECHANICAL AND ELECTRICAL SERVICES

Item	Net Price £	Material £	Labour hours	Labour £	Unit	Total Rate £
Y30 : AIR DUCTLINES : MILD STEEL DUCTWORK (contd)						
Extra Over Galvanised Sheet DW 142 Class A Rectangular Section Ductwork; (including all necessary stiffeners, joints etc.) (contd)						
Set Piece (contd)						
Sum of two sides 2300mm	54.87	60.36	2.82	32.18	nr	**92.54**
Sum of two sides 2400mm	58.59	64.45	3.00	34.20	nr	**98.65**
Sum of two sides 2500mm	62.42	68.66	3.01	34.31	nr	**102.97**
Sum of two sides 2600mm	66.41	73.05	3.02	34.41	nr	**107.46**
Sum of two sides 2700mm	70.49	77.54	3.03	34.52	nr	**112.06**
Sum of two sides 2800mm	74.67	82.14	3.76	42.82	nr	**124.96**
Sum of two sides 2900mm	84.53	92.98	4.12	46.87	nr	**139.85**
Sum of two sides 3000mm	89.16	98.08	4.61	52.49	nr	**150.57**
Sum of two sides 3100mm	93.93	103.32	4.63	52.73	nr	**156.05**
Sum of two sides 3200mm	98.80	108.68	4.63	52.73	nr	**161.41**
90 degree Medium Bend						
Sum of two sides 200mm	2.73	3.00	1.08	12.31	nr	**15.31**
Sum of two sides 250mm	3.10	3.41	1.08	12.31	nr	**15.72**
Sum of two sides 300mm	3.52	3.87	1.08	12.31	nr	**16.18**
Sum of two sides 350mm	3.95	4.34	1.06	12.08	nr	**16.42**
Sum of two sides 400mm	4.40	4.84	1.06	12.08	nr	**16.92**
Sum of two sides 450mm	4.87	5.36	1.06	12.08	nr	**17.44**
Sum of two sides 500mm	5.38	5.92	1.06	12.08	nr	**18.00**
Sum of two sides 550mm	5.89	6.48	1.15	13.11	nr	**19.59**
Sum of two sides 600mm	6.46	7.10	1.15	13.11	nr	**20.21**
Sum of two sides 650mm	7.04	7.75	1.15	13.11	nr	**20.86**
Sum of two sides 700mm	7.66	8.42	1.15	13.11	nr	**21.53**
Sum of two sides 750mm	8.26	9.09	1.21	13.79	nr	**22.88**
Sum of two sides 800mm	8.91	9.80	1.21	13.79	nr	**23.59**
Sum of two sides 850mm	9.57	10.53	1.21	13.79	nr	**24.32**
Sum of two sides 900mm	10.28	11.31	1.21	13.79	nr	**25.10**
Sum of two sides 950mm	10.99	12.09	1.58	18.02	nr	**30.11**
Sum of two sides 1000mm	11.74	12.91	1.61	18.28	nr	**31.19**
Sum of two sides 1100mm	13.31	14.65	1.58	17.97	nr	**32.62**
Sum of two sides 1200mm	14.95	16.45	1.58	18.02	nr	**34.47**
Sum of two sides 1300mm	16.72	18.39	1.86	21.21	nr	**39.60**
Sum of two sides 1400mm	18.55	20.41	1.86	21.21	nr	**41.62**
Sum of two sides 1500mm	20.49	22.54	2.06	23.49	nr	**46.03**
Sum of two sides 1600mm	22.54	24.79	2.06	23.49	nr	**48.28**
Sum of two sides 1700mm	27.52	30.27	2.21	25.20	nr	**55.47**
Sum of two sides 1800mm	29.88	32.87	2.21	25.20	nr	**58.07**
Sum of two sides 1900mm	32.31	35.54	2.45	27.92	nr	**63.46**
Sum of two sides 2000mm	34.86	38.35	2.45	27.92	nr	**66.27**
Sum of two sides 2100mm	43.78	48.16	3.33	37.97	nr	**86.13**
Sum of two sides 2200mm	51.27	56.40	3.62	41.27	nr	**97.67**
Sum of two sides 2300mm	54.87	60.36	3.65	41.57	nr	**101.93**
Sum of two sides 2400mm	58.59	64.45	4.05	46.11	nr	**110.56**
Sum of two sides 2500mm	62.42	68.66	4.07	46.30	nr	**114.96**
Sum of two sides 2600mm	66.41	73.05	4.07	46.30	nr	**119.35**
Sum of two sides 2700mm	70.49	77.54	4.08	46.49	nr	**124.03**
Sum of two sides 2800mm	74.67	82.14	7.09	80.78	nr	**162.92**
Sum of two sides 2900mm	84.53	92.98	7.46	85.00	nr	**177.98**
Sum of two sides 3000mm	89.16	98.08	7.58	86.29	nr	**184.37**

Y:MECHANICAL AND ELECTRICAL SERVICES

Item	Net Price £	Material £	Labour hours	Labour £	Unit	Total Rate £
Sum of two sides 3100mm	93.93	103.32	7.58	86.29	nr	**189.61**
Sum of two sides 3200mm	98.80	108.68	7.58	86.29	nr	**194.97**
45 degree Medium Bend						
Sum of two sides 200mm	2.36	2.59	0.77	8.76	nr	**11.35**
Sum of two sides 250mm	2.63	2.90	0.97	11.06	nr	**13.96**
Sum of two sides 300mm	2.95	3.25	0.97	11.06	nr	**14.31**
Sum of two sides 350mm	3.25	3.57	0.97	11.06	nr	**14.63**
Sum of two sides 400mm	3.64	4.00	0.95	10.85	nr	**14.85**
Sum of two sides 450mm	4.02	4.42	0.95	10.85	nr	**15.27**
Sum of two sides 500mm	4.43	4.87	0.95	10.85	nr	**15.72**
Sum of two sides 550mm	4.85	5.33	1.00	11.39	nr	**16.72**
Sum of two sides 600mm	4.70	5.18	1.00	11.39	nr	**16.57**
Sum of two sides 650mm	5.80	6.38	1.00	11.39	nr	**17.77**
Sum of two sides 700mm	6.29	6.92	1.00	11.39	nr	**18.31**
Sum of two sides 750mm	6.79	7.46	1.06	12.08	nr	**19.54**
Sum of two sides 800mm	7.39	8.13	1.06	12.08	nr	**20.21**
Sum of two sides 850mm	7.95	8.75	1.06	12.08	nr	**20.83**
Sum of two sides 900mm	8.57	9.43	1.06	12.08	nr	**21.51**
Sum of two sides 950mm	9.17	10.09	1.21	13.79	nr	**23.88**
Sum of two sides 1000mm	9.83	10.81	1.21	13.79	nr	**24.60**
Sum of two sides 1100mm	11.19	12.31	1.21	13.79	nr	**26.10**
Sum of two sides 1200mm	12.66	13.92	1.21	13.81	nr	**27.73**
Sum of two sides 1300mm	14.21	15.63	1.65	18.80	nr	**34.43**
Sum of two sides 1400mm	15.86	17.44	1.65	18.80	nr	**36.24**
Sum of two sides 1500mm	17.63	19.39	1.83	20.86	nr	**40.25**
Sum of two sides 1600mm	19.46	21.41	1.83	20.86	nr	**42.27**
Sum of two sides 1700mm	24.26	26.69	1.96	22.33	nr	**49.02**
Sum of two sides 1800mm	26.41	29.06	1.96	22.33	nr	**51.39**
Sum of two sides 1900mm	28.67	31.54	2.15	24.49	nr	**56.03**
Sum of two sides 2000mm	31.03	34.13	2.15	24.49	nr	**58.62**
Sum of two sides 2100mm	38.72	42.60	1.90	21.65	nr	**64.25**
Sum of two sides 2200mm	41.79	45.97	1.90	21.65	nr	**67.62**
Sum of two sides 2300mm	44.98	49.47	1.90	21.65	nr	**71.12**
Sum of two sides 2400mm	48.28	53.11	3.38	38.48	nr	**91.59**
Sum of two sides 2500mm	51.67	56.84	3.38	38.48	nr	**95.32**
Sum of two sides 2600mm	60.17	66.18	3.69	42.03	nr	**108.21**
Sum of two sides 2700mm	64.02	70.42	3.94	44.84	nr	**115.26**
Sum of two sides 2800mm	67.97	74.76	3.94	44.84	nr	**119.60**
Sum of two sides 2900mm	72.05	79.25	5.95	67.80	nr	**147.05**
Sum of two sides 3000mm	76.25	83.87	6.06	69.03	nr	**152.90**
Sum of two sides 3100mm	80.58	88.63	6.06	69.03	nr	**157.66**
Sum of two sides 3200mm	85.01	93.51	6.06	69.03	nr	**162.54**
90 degree Bend with Turning Vane						
Sum of two sides 200mm	3.00	3.30	1.12	12.77	nr	**16.07**
Sum of two sides 250mm	3.75	4.13	1.12	12.77	nr	**16.90**
Sum of two sides 300mm	4.20	4.62	1.12	12.77	nr	**17.39**
Sum of two sides 350mm	4.93	5.42	1.12	12.77	nr	**18.19**
Sum of two sides 400mm	5.81	6.39	1.12	12.77	nr	**19.16**
Sum of two sides 450mm	6.56	7.22	1.12	12.77	nr	**19.99**
Sum of two sides 500mm	7.38	8.12	1.12	12.77	nr	**20.89**
Sum of two sides 550mm	8.49	9.34	1.20	13.67	nr	**23.01**
Sum of two sides 600mm	9.53	10.49	1.20	13.67	nr	**24.16**
Sum of two sides 650mm	10.68	11.75	1.20	13.67	nr	**25.42**
Sum of two sides 700mm	11.87	13.06	1.20	13.67	nr	**26.73**
Sum of two sides 750mm	13.16	14.48	1.26	14.36	nr	**28.84**
Sum of two sides 800mm	14.48	15.93	1.26	14.36	nr	**30.29**

Y:MECHANICAL AND ELECTRICAL SERVICES

Item	Net Price £	Material £	Labour hours	Labour £	Unit	Total Rate £
Y30 : AIR DUCTLINES : MILD STEEL DUCTWORK (contd)						
Extra Over Galvanised Sheet DW 142 Class A Rectangular Section Ductwork; (including all necessary stiffiners, joints etc.) (contd)						
90 degree Bend with Turning Vane						
Sum of two sides 850mm	15.90	17.49	1.26	14.36	nr	**31.85**
Sum of two sides 900mm	17.38	19.12	1.26	14.36	nr	**33.48**
Sum of two sides 950mm	18.96	20.86	1.63	18.58	nr	**39.44**
Sum of two sides 1000mm	20.57	22.63	1.63	18.58	nr	**41.21**
Sum of two sides 1100mm	24.04	26.44	1.63	18.58	nr	**45.02**
Sum of two sides 1200mm	27.76	30.53	1.63	18.58	nr	**49.11**
Sum of two sides 1300mm	31.79	34.96	1.96	22.33	nr	**57.29**
Sum of two sides 1400mm	36.10	39.71	1.96	22.33	nr	**62.04**
Sum of two sides 1500mm	40.68	44.75	2.16	24.60	nr	**69.35**
Sum of two sides 1600mm	45.54	50.09	2.16	24.60	nr	**74.69**
Sum of two sides 1700mm	53.58	58.93	2.31	26.37	nr	**85.30**
Sum of two sides 1800mm	59.13	65.05	2.31	26.37	nr	**91.42**
Sum of two sides 1900mm	64.96	71.46	2.65	30.21	nr	**101.67**
Sum of two sides 2000mm	71.07	78.18	2.65	30.21	nr	**108.39**
Sum of two sides 2100mm	83.44	91.78	3.51	39.97	nr	**131.75**
Sum of two sides 2200mm	90.73	99.81	3.51	39.97	nr	**139.78**
Sum of two sides 2300mm	102.64	112.91	3.80	43.31	nr	**156.22**
Sum of two sides 2400mm	110.67	121.74	4.20	47.86	nr	**169.60**
Sum of two sides 2500mm	119.01	130.91	4.20	47.86	nr	**178.77**
Sum of two sides 2600mm	127.66	140.43	4.22	48.06	nr	**188.49**
Sum of two sides 2700mm	136.60	150.26	4.22	48.06	nr	**198.32**
Sum of two sides 2800mm	145.86	160.44	7.63	86.95	nr	**247.39**
Sum of two sides 2900mm	155.42	170.96	7.63	86.95	nr	**257.91**
Sum of two sides 3000mm	171.02	188.12	8.06	91.85	nr	**279.97**
Sum of two sides 3100mm	181.38	199.52	8.06	91.85	nr	**291.37**
Sum of two sides 3200mm	192.03	211.24	8.13	92.60	nr	**303.84**
Equal Twin Square Bend with Turning Vanes						
Sum of two sides 200mm	4.46	4.90	1.81	20.63	nr	**25.53**
Sum of two sides 250mm	5.33	5.86	1.81	20.63	nr	**26.49**
Sum of two sides 300mm	6.37	7.00	1.81	20.63	nr	**27.63**
Sum of two sides 350mm	7.51	8.26	1.77	20.16	nr	**28.42**
Sum of two sides 400mm	8.82	9.70	1.77	20.16	nr	**29.86**
Sum of two sides 450mm	10.24	11.26	1.77	20.16	nr	**31.42**
Sum of two sides 500mm	11.75	12.92	1.77	20.16	nr	**33.08**
Sum of two sides 550mm	13.41	14.75	1.86	21.21	nr	**35.96**
Sum of two sides 600mm	15.26	16.79	1.86	21.21	nr	**38.00**
Sum of two sides 650mm	17.17	18.89	1.87	21.25	nr	**40.14**
Sum of two sides 700mm	19.26	21.19	1.87	21.25	nr	**42.44**
Sum of two sides 750mm	21.42	23.56	1.96	22.33	nr	**45.89**
Sum of two sides 800mm	23.75	26.12	1.96	22.33	nr	**48.45**
Sum of two sides 850mm	26.17	28.78	1.96	22.33	nr	**51.11**
Sum of two sides 900mm	28.74	31.62	1.96	22.33	nr	**53.95**
Sum of two sides 950mm	31.44	34.58	2.61	29.74	nr	**64.32**
Sum of two sides 1000mm	34.29	37.71	2.61	29.74	nr	**67.45**
Sum of two sides 1100mm	40.32	44.36	2.61	29.74	nr	**74.10**
Sum of two sides 1200mm	46.89	51.58	2.61	29.74	nr	**81.32**
Sum of two sides 1300mm	54.00	59.40	3.18	36.27	nr	**95.67**

Y:MECHANICAL AND ELECTRICAL SERVICES

Item	Net Price £	Material £	Labour hours	Labour £	Unit	Total Rate £
Sum of two sides 1400mm	61.58	67.74	3.18	36.27	nr	**104.01**
Sum of two sides 1500mm	69.66	76.62	3.52	40.11	nr	**116.73**
Sum of two sides 1600mm	78.32	86.15	3.52	40.11	nr	**126.26**
Sum of two sides 1700mm	91.76	100.93	2.42	27.51	nr	**128.44**
Sum of two sides 1800mm	101.58	111.74	2.31	26.37	nr	**138.11**
Sum of two sides 1900mm	111.95	123.15	2.55	29.06	nr	**152.21**
Sum of two sides 2000mm	122.80	135.07	2.55	29.06	nr	**164.13**
Sum of two sides 2100mm	146.60	161.25	3.51	39.97	nr	**201.22**
Sum of two sides 2200mm	159.50	175.45	3.51	39.97	nr	**215.42**
Sum of two sides 2300mm	172.94	190.23	3.80	43.31	nr	**233.54**
Sum of two sides 2400mm	186.89	205.58	4.20	47.86	nr	**253.44**
Sum of two sides 2500mm	201.48	221.63	4.20	47.86	nr	**269.49**
Sum of two sides 2600mm	221.51	243.66	4.22	48.06	nr	**291.72**
Sum of two sides 2700mm	237.36	261.10	4.22	48.06	nr	**309.16**
Sum of two sides 2800mm	253.71	279.08	7.63	86.95	nr	**366.03**
Sum of two sides 2900mm	270.61	297.67	12.82	146.03	nr	**443.70**
Sum of two sides 3000mm	288.13	316.94	13.16	149.87	nr	**466.81**
Sum of two sides 3100mm	312.05	343.26	13.51	153.92	nr	**497.18**
Sum of two sides 3200mm	330.87	363.96	13.51	153.92	nr	**517.88**
Spigot Branch						
Sum of two sides 200mm	0.19	0.21	0.82	9.34	nr	**9.55**
Sum of two sides 250mm	0.24	0.26	0.82	9.34	nr	**9.60**
Sum of two sides 300mm	0.29	0.32	0.82	9.34	nr	**9.66**
Sum of two sides 350mm	0.33	0.36	0.80	9.11	nr	**9.47**
Sum of two sides 400mm	0.39	0.43	0.80	9.11	nr	**9.54**
Sum of two sides 450mm	0.43	0.47	0.80	9.11	nr	**9.58**
Sum of two sides 500mm	0.46	0.51	0.80	9.11	nr	**9.62**
Sum of two sides 550mm	0.52	0.58	0.89	10.14	nr	**10.72**
Sum of two sides 600mm	0.56	0.62	0.89	10.14	nr	**10.76**
Sum of two sides 650mm	0.63	0.69	0.89	10.14	nr	**10.83**
Sum of two sides 700mm	0.67	0.73	0.89	10.14	nr	**10.87**
Sum of two sides 750mm	0.72	0.79	0.89	10.14	nr	**10.93**
Sum of two sides 800mm	0.78	0.86	0.89	10.14	nr	**11.00**
Sum of two sides 850mm	0.82	0.90	0.89	10.17	nr	**11.07**
Sum of two sides 900mm	0.86	0.95	0.89	10.17	nr	**11.12**
Sum of two sides 950mm	0.91	1.00	1.12	12.77	nr	**13.77**
Sum of two sides 1000mm	0.95	1.05	1.12	12.77	nr	**13.82**
Sum of two sides 1100mm	1.06	1.16	1.12	12.77	nr	**13.93**
Sum of two sides 1200mm	1.15	1.26	1.12	12.77	nr	**14.03**
Sum of two sides 1300mm	1.25	1.38	1.20	13.67	nr	**15.05**
Sum of two sides 1400mm	1.34	1.48	1.20	13.67	nr	**15.15**
Sum of two sides 1500mm	1.42	1.57	1.42	16.18	nr	**17.75**
Sum of two sides 1600mm	1.53	1.68	1.42	16.18	nr	**17.86**
Sum of two sides 1700mm	1.62	1.78	1.46	16.65	nr	**18.43**
Sum of two sides 1800mm	1.71	1.88	1.46	16.65	nr	**18.53**
Sum of two sides 1900mm	1.81	2.00	1.60	18.22	nr	**20.22**
Sum of two sides 2000mm	1.91	2.10	1.60	18.22	nr	**20.32**
Sum of two sides 2100mm	2.52	2.77	2.26	25.77	nr	**28.54**
Sum of two sides 2200mm	2.64	2.91	2.26	25.77	nr	**28.68**
Sum of two sides 2300mm	2.74	3.01	2.26	25.77	nr	**28.78**
Sum of two sides 2400mm	2.86	3.15	2.49	28.40	nr	**31.55**
Sum of two sides 2500mm	2.99	3.29	2.49	28.40	nr	**31.69**
Sum of two sides 2600mm	2.99	3.29	2.49	28.40	nr	**31.69**
Sum of two sides 2700mm	3.25	3.57	2.49	28.40	nr	**31.97**
Sum of two sides 2800mm	3.36	3.70	3.41	38.87	nr	**42.57**
Sum of two sides 2900mm	3.47	3.82	3.41	38.87	nr	**42.69**
Sum of two sides 3000mm	3.60	3.96	4.18	47.66	nr	**51.62**
Sum of two sides 3100mm	3.70	4.07	4.18	47.66	nr	**51.73**
Sum of two sides 3200mm	3.83	4.22	4.18	47.66	nr	**51.88**

Y:MECHANICAL AND ELECTRICAL SERVICES

Item	Net Price £	Material £	Labour hours	Labour £	Unit	Total Rate £
Y30 : AIR DUCTLINES : MILD STEEL DUCTWORK (contd)						
Extra Over Galvanised Sheet DW 142 Class A Rectangular Section Ductwork; (including all necessary stiffeners, joints etc.) (contd)						
Branch; Based on Branch Size						
Sum of two sides 200mm	1.65	1.82	0.95	10.85	nr	**12.67**
Sum of two sides 250mm	1.84	2.03	0.95	10.85	nr	**12.88**
Sum of two sides 300mm	2.12	2.33	0.95	10.85	nr	**13.18**
Sum of two sides 350mm	2.36	2.59	1.00	11.39	nr	**13.98**
Sum of two sides 400mm	2.58	2.84	1.00	11.39	nr	**14.23**
Sum of two sides 450mm	2.82	3.10	1.00	11.36	nr	**14.46**
Sum of two sides 500mm	3.04	3.35	1.00	11.39	nr	**14.74**
Sum of two sides 550mm	3.29	3.62	1.02	11.62	nr	**15.24**
Sum of two sides 600mm	3.52	3.87	1.02	11.62	nr	**15.49**
Sum of two sides 650mm	3.78	4.16	1.02	11.62	nr	**15.78**
Sum of two sides 700mm	4.00	4.40	1.02	11.62	nr	**16.02**
Sum of two sides 750mm	4.22	4.65	1.02	11.62	nr	**16.27**
Sum of two sides 800mm	4.46	4.90	1.02	11.62	nr	**16.52**
Sum of two sides 850mm	4.70	5.18	1.02	11.62	nr	**16.80**
Sum of two sides 900mm	4.93	5.42	1.02	11.62	nr	**17.04**
Sum of two sides 950mm	5.17	5.68	1.25	14.24	nr	**19.92**
Sum of two sides 1000mm	5.38	5.92	1.25	14.24	nr	**20.16**
Sum of two sides 1100mm	5.86	6.45	1.25	14.24	nr	**20.69**
Sum of two sides 1200mm	6.33	6.97	1.25	14.24	nr	**21.21**
Sum of two sides 1300mm	6.80	7.48	1.46	16.65	nr	**24.13**
Sum of two sides 1400mm	7.27	7.99	1.46	16.65	nr	**24.64**
Sum of two sides 1500mm	7.74	8.51	1.55	17.66	nr	**26.17**
Sum of two sides 1600mm	8.19	9.01	1.55	17.66	nr	**26.67**
Sum of two sides 1700mm	10.09	11.09	1.61	18.34	nr	**29.43**
Sum of two sides 1800mm	10.63	11.69	1.75	19.95	nr	**31.64**
Sum of two sides 1900mm	11.14	12.26	1.75	19.95	nr	**32.21**
Sum of two sides 2000mm	11.66	12.83	1.75	19.95	nr	**32.78**
Sum of two sides 2100mm	13.70	15.07	2.26	25.77	nr	**40.84**
Sum of two sides 2200mm	14.31	15.74	2.26	25.77	nr	**41.51**
Sum of two sides 2300mm	14.90	16.39	2.26	25.77	nr	**42.16**
Sum of two sides 2400mm	15.50	17.05	2.49	28.40	nr	**45.45**
Sum of two sides 2500mm	16.09	17.70	2.49	28.40	nr	**46.10**
Sum of two sides 2600mm	16.70	18.37	2.49	28.40	nr	**46.77**
Sum of two sides 2700mm	17.28	19.01	2.49	28.40	nr	**47.41**
Sum of two sides 2800mm	17.91	19.70	3.41	38.87	nr	**58.57**
Sum of two sides 2900mm	18.04	20.34	3.57	40.68	nr	**61.02**
Sum of two sides 3000mm	18.63	21.00	4.33	49.31	nr	**70.31**
Sum of two sides 3100mm	19.20	21.65	4.33	49.31	nr	**70.96**
Sum of two sides 3200mm	19.78	22.30	4.33	49.31	nr	**71.61**

Y:MECHANICAL AND ELECTRICAL SERVICES

Item	Net Price £	Material £	Labour hours	Labour £	Unit	Total Rate £
Y30 : AIR DUCTLINES : MILD STEEL DUCTWORK (contd)						
Galvanised Sheet Metal DW 142 Class C rectangular Section Ductwork;(including all necessary stiffeners, joints, couplers in the running the running length and duct supports.)						
Straight Duct:						
Sum of two sides 200mm	4.25	4.67	1.03	11.74	m	**16.41**
Sum of two sides 250mm	4.90	5.39	1.03	11.74	m	**17.13**
Sum of two sides 300mm	5.49	6.04	1.03	11.74	m	**17.78**
Sum of two sides 350mm	6.16	6.77	1.00	11.39	m	**18.16**
Sum of two sides 400mm	6.78	7.46	1.00	11.39	m	**18.85**
Sum of two sides 450mm	7.40	8.14	1.00	11.39	m	**19.53**
Sum of two sides 500mm	8.03	8.83	1.00	11.39	m	**20.22**
Sum of two sides 550mm	8.66	9.53	1.01	11.50	m	**21.03**
Sum of two sides 600mm	9.29	10.22	1.01	11.50	m	**21.72**
Sum of two sides 650mm	9.92	10.92	1.01	11.50	m	**22.42**
Sum of two sides 700mm	10.53	11.59	1.01	11.50	m	**23.09**
Sum of two sides 750mm	11.18	12.30	1.01	11.50	m	**23.80**
Sum of two sides 800mm	11.79	12.97	1.01	11.50	m	**24.47**
Sum of two sides 850mm	13.65	15.01	1.09	12.42	m	**27.43**
Sum of two sides 900mm	14.34	15.77	1.10	12.53	m	**28.30**
Sum of two sides 950mm	15.03	16.53	1.28	14.58	m	**31.11**
Sum of two sides 1000mm	15.71	17.28	1.28	14.58	m	**31.86**
Sum of two sides 1100mm	17.08	18.79	1.29	14.70	m	**33.49**
Sum of two sides 1200mm	18.48	20.32	1.29	14.70	m	**35.02**
Sum of two sides 1300mm	19.85	21.83	1.31	14.93	m	**36.76**
Sum of two sides 1400mm	21.24	23.37	1.34	15.25	m	**38.62**
Sum of two sides 1500mm	22.61	24.88	1.39	15.84	m	**40.72**
Sum of two sides 1600mm	24.01	26.41	1.40	15.95	m	**42.36**
Sum of two sides 1700mm	27.94	30.74	1.51	17.21	m	**47.95**
Sum of two sides 1800mm	29.42	32.37	1.52	17.34	m	**49.71**
Sum of two sides 1900mm	30.90	34.00	1.57	17.91	m	**51.91**
Sum of two sides 2000mm	32.39	35.62	1.58	18.02	m	**53.64**
Sum of two sides 2100mm	43.50	47.85	2.19	24.98	m	**72.83**
Sum of two sides 2200mm	45.42	49.97	2.21	25.20	m	**75.17**
Sum of two sides 2300mm	47.37	52.11	2.22	25.31	m	**77.42**
Sum of two sides 2400mm	49.33	54.26	2.39	27.25	m	**81.51**
Sum of two sides 2500mm	51.28	56.40	2.40	27.38	m	**83.78**
Sum of two sides 2600mm	65.08	71.58	2.57	29.28	m	**100.86**
Sum of two sides 2700mm	67.35	74.08	2.59	29.51	m	**103.59**
Sum of two sides 2800mm	69.62	76.58	2.86	32.54	m	**109.12**
Sum of two sides 2900mm	71.98	79.18	2.87	32.64	m	**111.82**
Sum of two sides 3000mm	74.35	81.78	3.06	34.83	m	**116.61**
Sum of two sides 3100mm	76.70	84.37	3.07	34.94	m	**119.31**
Sum of two sides 3200mm	79.07	86.97	3.08	35.05	m	**122.02**
Extra Over Galvanised Sheet Metal DW 14 Class C Rectangular Section Ductwork; fittings (including all necessary stiffener, joints etc.)						
Stopped End						
Sum of two sides 200mm	0.10	0.11	0.33	3.80	nr	**3.91**
Sum of two sides 250mm	0.14	0.15	0.33	3.80	nr	**3.95**

Y:MECHANICAL AND ELECTRICAL SERVICES

Item	Net Price £	Material £	Labour hours	Labour £	Unit	Total Rate £
Y30 : AIR DUCTLINES : MILD STEEL DUCTWORK (contd)						
Extra Over Galvanised Sheet Metal DW 14 Class C Rectangular Section Ductwork; fittings (including all necessary stiffener, joints etc.) (contd)						
Stopped End (contd)						
Sum of two sides 300mm	0.14	0.15	0.33	3.80	nr	**3.95**
Sum of two sides 350mm	0.19	0.21	0.33	3.80	nr	**4.01**
Sum of two sides 400mm	0.19	0.21	0.33	3.80	nr	**4.01**
Sum of two sides 450mm	0.24	0.26	0.33	3.80	nr	**4.06**
Sum of two sides 500mm	0.24	0.26	0.33	3.80	nr	**4.06**
Sum of two sides 550mm	0.24	0.26	0.33	3.80	nr	**4.06**
Sum of two sides 600mm	0.29	0.32	0.33	3.80	nr	**4.12**
Sum of two sides 650mm	0.29	0.32	0.33	3.80	nr	**4.12**
Sum of two sides 700mm	0.33	0.37	0.33	3.80	nr	**4.17**
Sum of two sides 750mm	0.39	0.43	0.33	3.80	nr	**4.23**
Sum of two sides 800mm	0.39	0.43	0.33	3.80	nr	**4.23**
Sum of two sides 850mm	0.43	0.47	0.33	3.76	nr	**4.23**
Sum of two sides 900mm	0.43	0.47	0.33	3.76	nr	**4.23**
Sum of two sides 950mm	0.43	0.47	0.33	3.76	nr	**4.23**
Sum of two sides 1000mm	0.47	0.52	0.33	3.76	nr	**4.28**
Sum of two sides 1100mm	0.53	0.58	0.33	3.76	nr	**4.34**
Sum of two sides 1200mm	0.57	0.62	0.33	3.76	nr	**4.38**
Sum of two sides 1300mm	0.62	0.69	0.33	3.76	nr	**4.45**
Sum of two sides 1400mm	0.66	0.73	0.33	3.76	nr	**4.49**
Sum of two sides 1500mm	0.72	0.79	0.33	3.76	nr	**4.55**
Sum of two sides 1600mm	0.76	0.84	0.33	3.76	nr	**4.60**
Sum of two sides 1700mm	0.82	0.90	0.50	5.69	nr	**6.59**
Sum of two sides 1800mm	0.86	0.94	0.50	5.69	nr	**6.63**
Sum of two sides 1900mm	0.91	1.00	0.50	5.69	nr	**6.69**
Sum of two sides 2000mm	0.95	1.05	0.50	5.69	nr	**6.74**
Sum of two sides 2100mm	1.26	1.39	0.75	8.54	nr	**9.93**
Sum of two sides 2200mm	1.31	1.45	0.75	8.54	nr	**9.99**
Sum of two sides 2300mm	1.37	1.51	0.75	8.54	nr	**10.05**
Sum of two sides 2400mm	1.43	1.57	0.75	8.54	nr	**10.11**
Sum of two sides 2500mm	1.49	1.64	0.75	8.54	nr	**10.18**
Sum of two sides 2600mm	1.56	1.72	0.75	8.54	nr	**10.26**
Sum of two sides 2700mm	1.62	1.78	0.75	8.54	nr	**10.32**
Sum of two sides 2800mm	1.69	1.86	1.00	11.39	nr	**13.25**
Sum of two sides 2900mm	2.34	2.57	1.00	11.39	nr	**13.96**
Sum of two sides 3000mm	2.44	2.68	1.50	17.10	nr	**19.78**
Sum of two sides 3100mm	2.52	2.77	1.50	17.10	nr	**19.87**
Sum of two sides 3200mm	2.57	2.83	1.50	17.10	nr	**19.93**
Taper; Large End Measured						
Sum of two sides 200mm	3.11	3.43	1.23	14.01	nr	**17.44**
Sum of two sides 250mm	3.53	3.88	1.23	14.01	nr	**17.89**
Sum of two sides 300mm	3.94	4.34	1.23	14.01	nr	**18.35**
Sum of two sides 350mm	4.39	4.83	1.21	13.79	nr	**18.62**
Sum of two sides 400mm	4.79	5.27	1.21	13.79	nr	**19.06**
Sum of two sides 450mm	5.20	5.72	1.21	13.79	nr	**19.51**
Sum of two sides 500mm	5.65	6.21	1.21	13.79	nr	**20.00**
Sum of two sides 550mm	6.06	6.67	1.46	16.65	nr	**23.32**
Sum of two sides 600mm	6.46	7.11	1.46	16.65	nr	**23.76**
Sum of two sides 650mm	6.89	7.58	1.46	16.65	nr	**24.23**

Y:MECHANICAL AND ELECTRICAL SERVICES

Item	Net Price £	Material £	Labour hours	Labour £	Unit	Total Rate £
Sum of two sides 700mm	7.32	8.05	1.46	16.65	nr	**24.70**
Sum of two sides 750mm	7.74	8.51	1.66	18.92	nr	**27.43**
Sum of two sides 800mm	8.15	8.97	1.66	18.92	nr	**27.89**
Sum of two sides 850mm	8.57	9.42	1.66	18.92	nr	**28.34**
Sum of two sides 900mm	9.00	9.90	1.66	18.92	nr	**28.82**
Sum of two sides 950mm	9.41	10.35	1.89	21.53	nr	**31.88**
Sum of two sides 1000mm	9.83	10.81	1.89	21.53	nr	**32.34**
Sum of two sides 1100mm	10.67	11.74	1.89	21.53	nr	**33.27**
Sum of two sides 1200mm	11.50	12.65	1.89	21.53	nr	**34.18**
Sum of two sides 1300mm	12.36	13.60	1.99	22.69	nr	**36.29**
Sum of two sides 1400mm	13.18	14.49	1.99	22.69	nr	**37.18**
Sum of two sides 1500mm	14.03	15.44	2.14	24.39	nr	**39.83**
Sum of two sides 1600mm	14.86	16.35	2.14	24.39	nr	**40.74**
Sum of two sides 1700mm	18.57	20.43	2.24	25.54	nr	**45.97**
Sum of two sides 1800mm	19.51	21.47	2.24	25.54	nr	**47.01**
Sum of two sides 1900mm	20.48	22.53	2.35	26.80	nr	**49.33**
Sum of two sides 2000mm	21.42	23.57	2.35	26.80	nr	**50.37**
Sum of two sides 2100mm	24.91	27.40	2.53	28.83	nr	**56.23**
Sum of two sides 2200mm	25.98	28.58	2.53	28.83	nr	**57.41**
Sum of two sides 2300mm	27.06	29.76	2.53	28.83	nr	**58.59**
Sum of two sides 2400mm	28.12	30.94	2.70	30.78	nr	**61.72**
Sum of two sides 2500mm	29.20	32.12	2.70	30.78	nr	**62.90**
Sum of two sides 2600mm	44.80	49.28	2.74	31.21	nr	**80.49**
Sum of two sides 2700mm	46.25	50.88	2.74	31.21	nr	**82.09**
Sum of two sides 2800mm	47.69	52.46	3.46	39.41	nr	**91.87**
Sum of two sides 2900mm	49.30	54.23	3.47	39.55	nr	**93.78**
Sum of two sides 3000mm	50.85	55.93	3.95	45.02	nr	**100.95**
Sum of two sides 3100mm	52.43	57.67	3.95	45.02	nr	**102.69**
Sum of two sides 3200mm	54.05	59.45	3.95	45.02	nr	**104.47**
Set Piece						
Sum of two sides 200mm	2.73	3.00	1.43	16.29	nr	**19.29**
Sum of two sides 250mm	3.10	3.41	1.43	16.29	nr	**19.70**
Sum of two sides 300mm	3.52	3.87	1.43	16.29	nr	**20.16**
Sum of two sides 350mm	3.94	4.34	1.41	16.06	nr	**20.40**
Sum of two sides 400mm	4.40	4.84	1.41	16.06	nr	**20.90**
Sum of two sides 450mm	4.87	5.36	1.41	16.06	nr	**21.42**
Sum of two sides 500mm	5.38	5.92	1.41	16.06	nr	**21.98**
Sum of two sides 550mm	5.90	6.49	1.66	18.92	nr	**25.41**
Sum of two sides 600mm	6.46	7.11	1.66	18.92	nr	**26.03**
Sum of two sides 650mm	7.04	7.75	1.66	18.92	nr	**26.67**
Sum of two sides 700mm	7.65	8.42	1.66	18.92	nr	**27.34**
Sum of two sides 750mm	8.26	9.09	1.66	18.92	nr	**28.01**
Sum of two sides 800mm	8.91	9.80	1.66	18.92	nr	**28.72**
Sum of two sides 850mm	9.58	10.53	1.66	18.92	nr	**29.45**
Sum of two sides 900mm	10.28	11.31	1.66	18.92	nr	**30.23**
Sum of two sides 950mm	10.99	12.09	1.89	21.53	nr	**33.62**
Sum of two sides 1000mm	11.74	12.91	1.89	21.53	nr	**34.44**
Sum of two sides 1.100mm	13.31	14.65	1.89	21.53	nr	**36.18**
Sum of two sides 1200mm	14.96	16.46	1.89	21.53	nr	**37.99**
Sum of two sides 1300mm	18.88	20.77	2.14	24.39	nr	**45.16**
Sum of two sides 1400mm	20.88	22.97	2.14	24.39	nr	**47.36**
Sum of two sides 1500mm	22.99	25.29	2.14	24.39	nr	**49.68**
Sum of two sides 1600mm	25.20	27.72	2.31	26.37	nr	**54.09**
Sum of two sides 1700mm	30.78	33.86	2.51	28.62	nr	**62.48**
Sum of two sides 1800mm	33.31	36.64	2.51	28.62	nr	**65.26**
Sum of two sides 1900mm	35.94	39.54	2.61	29.74	nr	**69.28**
Sum of two sides 2000mm	38.68	42.55	2.51	28.62	nr	**71.17**
Sum of two sides 2100mm	49.26	54.18	2.89	32.92	nr	**87.10**

Y:MECHANICAL AND ELECTRICAL SERVICES

Item	Net Price £	Material £	Labour hours	Labour £	Unit	Total Rate £
Y30 : AIR DUCTLINES : MILD STEEL DUCTWORK (contd)						
Extra Over Galvanised Sheet Metal DW 14 Class C Rectangular Section Ductwork; fittings (including all necessary stiffener, joints etc.) (contd)						
Set Piece (contd)						
Sum of two sides 2200mm	52.18	57.39	2.89	32.92	nr	**90.31**
Sum of two sides 2300mm	56.47	62.11	2.53	28.83	nr	**90.94**
Sum of two sides 2400mm	60.26	66.29	2.70	30.78	nr	**97.07**
Sum of two sides 2500mm	64.16	70.58	2.70	30.78	nr	**101.36**
Sum of two sides 2600mm	83.95	92.35	2.74	31.21	nr	**123.56**
Sum of two sides 2700mm	88.51	97.36	2.74	31.21	nr	**128.57**
Sum of two sides 2800mm	93.21	102.53	3.46	39.41	nr	**141.94**
Sum of two sides 2900mm	105.52	116.07	4.00	45.56	nr	**161.63**
Sum of two sides 3000mm	111.32	122.45	4.20	47.86	nr	**170.31**
Sum of two sides 3100mm	117.25	128.98	4.50	51.31	nr	**180.29**
Sum of two sides 3200mm	123.34	135.68	5.41	61.57	nr	**197.25**
90 degree Medium Bend						
Sum of two sides 200mm	2.73	3.00	1.08	12.31	nr	**15.31**
Sum of two sides 250mm	3.10	3.41	1.08	12.31	nr	**15.72**
Sum of two sides 300mm	3.52	3.87	1.08	12.31	nr	**16.18**
Sum of two sides 350mm	3.94	4.34	1.06	12.08	nr	**16.42**
Sum of two sides 400mm	4.40	4.84	1.06	12.08	nr	**16.92**
Sum of two sides 450mm	4.87	5.36	1.06	12.08	nr	**17.44**
Sum of two sides 500mm	5.38	5.92	1.06	12.08	nr	**18.00**
Sum of two sides 550mm	5.90	6.49	1.15	13.11	nr	**19.60**
Sum of two sides 600mm	6.46	7.11	1.15	13.11	nr	**20.22**
Sum of two sides 650mm	7.04	7.75	1.15	13.11	nr	**20.86**
Sum of two sides 700mm	7.65	8.42	1.15	13.11	nr	**21.53**
Sum of two sides 750mm	8.26	9.09	1.21	13.79	nr	**22.88**
Sum of two sides 800mm	8.91	9.80	1.21	13.79	nr	**23.59**
Sum of two sides 850mm	9.58	10.53	1.21	13.79	nr	**24.32**
Sum of two sides 900mm	10.28	11.31	1.21	13.79	nr	**25.10**
Sum of two sides 950mm	10.99	12.09	1.58	18.02	nr	**30.11**
Sum of two sides 1000mm	11.74	12.91	1.58	18.02	nr	**30.93**
Sum of two sides 1100mm	13.31	14.65	1.58	18.02	nr	**32.67**
Sum of two sides 1200mm	14.96	16.46	1.58	18.02	nr	**34.48**
Sum of two sides 1300mm	18.88	20.77	2.01	22.92	nr	**43.69**
Sum of two sides 1400mm	20.88	22.97	2.01	22.92	nr	**45.89**
Sum of two sides 1500mm	22.99	25.29	2.22	25.31	nr	**50.60**
Sum of two sides 1600mm	25.20	27.72	2.23	25.42	nr	**53.14**
Sum of two sides 1700mm	30.78	33.86	2.46	28.05	nr	**61.91**
Sum of two sides 1800mm	33.31	36.64	2.46	28.05	nr	**64.69**
Sum of two sides 1900mm	35.94	39.54	2.71	30.87.	nr	**70.41**
Sum of two sides 2000mm	38.68	42.55	2.71	30.87	nr	**73.42**
Sum of two sides 2100mm	49.26	54.18	3.69	42.03	nr	**96.21**
Sum of two sides 2200mm	52.80	58.08	3.70	42.19	nr	**100.27**
Sum of two sides 2300mm	56.47	62.11	3.72	42.34	nr	**104.45**
Sum of two sides 2400mm	60.26	66.29	4.13	47.07	nr	**113.36**
Sum of two sides 2500mm	64.16	70.58	4.15	47.26	nr	**117.84**
Sum of two sides 2600mm	83.95	92.35	4.29	48.88	nr	**141.23**
Sum of two sides 2700mm	88.51	97.36	4.29	48.88	nr	**146.24**
Sum of two sides 2800mm	93.21	102.53	7.35	83.75	nr	**186.28**
Sum of two sides 2900mm	105.52	116.07	7.69	87.62	nr	**203.69**

Y:MECHANICAL AND ELECTRICAL SERVICES

Item	Net Price £	Material £	Labour hours	Labour £	Unit	Total Rate £
Sum of two sides 3000mm	111.32	122.45	7.81	88.98	nr	**211.43**
Sum of two sides 3100mm	117.25	128.98	7.81	88.98	nr	**217.96**
Sum of two sides 3200mm	166.50	183.14	7.87	89.68	nr	**272.82**
45 degree Medium Bend						
Sum of two sides 200mm	2.35	2.59	0.97	11.06	nr	**13.65**
Sum of two sides 250mm	2.63	2.89	0.97	11.06	nr	**13.95**
Sum of two sides 300mm	2.95	3.24	0.97	11.06	nr	**14.30**
Sum of two sides 350mm	3.27	3.59	0.95	10.85	nr	**14.44**
Sum of two sides 400mm	3.64	4.00	0.95	10.85	nr	**14.85**
Sum of two sides 450mm	4.01	4.42	0.95	10.85	nr	**15.27**
Sum of two sides 500mm	4.43	4.87	0.95	10.85	nr	**15.72**
Sum of two sides 550mm	4.84	5.33	1.00	11.39	nr	**16.72**
Sum of two sides 600mm	5.31	5.85	1.00	11.39	nr	**17.24**
Sum of two sides 650mm	5.80	6.38	1.00	11.39	nr	**17.77**
Sum of two sides 700mm	6.30	6.93	1.00	11.39	nr	**18.32**
Sum of two sides 750mm	6.81	7.49	1.00	11.39	nr	**18.88**
Sum of two sides 800mm	7.39	8.13	1.06	12.08	nr	**20.21**
Sum of two sides 850mm	7.96	8.75	1.06	12.08	nr	**20.83**
Sum of two sides 900mm	8.57	9.42	1.06	12.08	nr	**21.50**
Sum of two sides 950mm	9.18	10.09	1.21	13.79	nr	**23.88**
Sum of two sides 1000mm	9.83	10.81	1.21	13.79	nr	**24.60**
Sum of two sides 1100mm	11.20	12.32	1.21	13.79	nr	**26.11**
Sum of two sides 1200mm	12.66	13.93	1.21	13.79	nr	**27.72**
Sum of two sides 1300mm	14.21	15.64	1.65	18.80	nr	**34.44**
Sum of two sides 1400mm	15.86	17.45	1.65	18.80	nr	**36.25**
Sum of two sides 1500mm	17.63	19.40	1.83	20.86	nr	**40.26**
Sum of two sides 1600mm	19.47	21.42	1.83	20.86	nr	**42.28**
Sum of two sides 1700mm	27.51	30.27	2.21	25.20	nr	**55.47**
Sum of two sides 1800mm	29.85	32.84	2.21	25.20	nr	**58.04**
Sum of two sides 1900mm	32.30	35.53	2.42	27.51	nr	**63.04**
Sum of two sides 2000mm	34.85	38.33	2.42	27.51	nr	**65.84**
Sum of two sides 2100mm	44.20	48.63	2.26	25.77	nr	**74.40**
Sum of two sides 2200mm	47.53	52.28	2.27	25.89	nr	**78.17**
Sum of two sides 2300mm	50.96	56.05	2.28	26.00	nr	**82.05**
Sum of two sides 2400mm	54.52	59.97	3.76	42.82	nr	**102.79**
Sum of two sides 2500mm	58.18	64.00	3.77	42.98	nr	**106.98**
Sum of two sides 2600mm	77.71	85.48	3.92	44.67	nr	**130.15**
Sum of two sides 2700mm	82.04	90.25	3.92	44.67	nr	**134.92**
Sum of two sides 2800mm	86.51	95.17	6.17	70.31	nr	**165.48**
Sum of two sides 2900mm	91.70	100.87	6.21	70.75	nr	**171.62**
Sum of two sides 3000mm	97.05	106.75	6.29	71.64	nr	**178.39**
Sum of two sides 3100mm	102.55	112.81	6.29	71.64	nr	**184.45**
Sum of two sides 3200mm	108.21	119.04	6.33	72.09	nr	**191.13**
90 degree Square Bend with Turning Vane						
Sum of two sides 200mm	3.00	3.30	1.12	12.77	nr	**16.07**
Sum of two sides 250mm	3.58	3.94	1.12	12.77	nr	**16.71**
Sum of two sides 300mm	4.28	4.70	1.12	12.77	nr	**17.47**
Sum of two sides 350mm	4.93	5.42	1.12	12.77	nr	**18.19**
Sum of two sides 400mm	5.70	6.27	1.12	12.77	nr	**19.04**
Sum of two sides 450mm	6.56	7.22	1.12	12.77	nr	**19.99**
Sum of two sides 500mm	7.47	8.22	1.12	12.77	nr	**20.99**
Sum of two sides 550mm	8.48	9.33	1.20	13.67	nr	**23.00**
Sum of two sides 600mm	9.54	10.49	1.20	13.67	nr	**24.16**
Sum of two sides 650mm	10.68	11.75	1.20	13.67	nr	**25.42**
Sum of two sides 700mm	11.87	13.06	1.20	13.67	nr	**26.73**
Sum of two sides 750mm	13.16	14.48	1.26	14.36	nr	**28.84**

Y:MECHANICAL AND ELECTRICAL SERVICES

Item	Net Price £	Material £	Labour hours	Labour £	Unit	Total Rate £
Y30 : AIR DUCTLINES : MILD STEEL DUCTWORK (contd)						
Extra Over Galvanised Sheet Metal DW 14 Class C Rectangular Section Ductwork; fittings (including all necessary stiffener, joints etc.) (contd)						
90 degree Square Bend with Turning Vane (contd)						
Sum of two sides 800mm	14.49	15.94	1.26	14.36	nr	**30.30**
Sum of two sides 850mm	15.90	17.49	1.26	14.36	nr	**31.85**
Sum of two sides 900mm	17.38	19.12	1.26	14.36	nr	**33.48**
Sum of two sides 950mm	18.96	20.86	1.63	18.58	nr	**39.44**
Sum of two sides 1000mm	20.58	22.64	1.63	18.58	nr	**41.22**
Sum of two sides 1100mm	24.04	26.44	1.63	18.58	nr	**45.02**
Sum of two sides 1200mm	27.76	30.54	1.63	18.58	nr	**49.12**
Sum of two sides 1300mm	31.79	34.97	1.96	22.33	nr	**57.30**
Sum of two sides 1400mm	38.43	42.28	2.11	24.08	nr	**66.36**
Sum of two sides 1500mm	43.18	47.50	2.32	26.43	nr	**73.93**
Sum of two sides 1600mm	48.20	53.03	2.33	26.55	nr	**79.58**
Sum of two sides 1700mm	56.84	62.53	2.56	29.21	nr	**91.74**
Sum of two sides 1800mm	62.57	68.83	2.56	29.21	nr	**98.04**
Sum of two sides 1900mm	68.36	75.19	2.82	32.08	nr	**107.27**
Sum of two sides 2000mm	74.96	82.45	2.82	32.08	nr	**114.53**
Sum of two sides 2100mm	88.98	97.88	6.49	73.96	nr	**171.84**
Sum of two sides 2200mm	96.49	106.14	6.49	73.96	nr	**180.10**
Sum of two sides 2300mm	104.26	114.68	6.54	74.44	nr	**189.12**
Sum of two sides 2400mm	112.37	123.60	7.04	80.21	nr	**203.81**
Sum of two sides 2500mm	120.77	132.84	7.19	81.94	nr	**214.78**
Sum of two sides 2600mm	145.22	159.75	7.41	84.37	nr	**244.12**
Sum of two sides 2700mm	154.63	170.10	7.46	85.00	nr	**255.10**
Sum of two sides 2800mm	164.42	180.86	7.87	89.68	nr	**270.54**
Sum of two sides 2900mm	175.19	192.71	7.87	89.68	nr	**282.39**
Sum of two sides 3000mm	192.78	212.06	8.33	94.92	nr	**306.98**
Sum of two sides 3100mm	204.46	224.90	8.33	94.92	nr	**319.82**
Sum of two sides 3200mm	216.47	238.12	8.33	94.92	nr	**333.04**
Equal Twin Square Bend with Turning Vanes						
Sum of two sides 200mm	4.46	4.90	1.81	20.63	nr	**25.53**
Sum of two sides 250mm	5.33	5.86	1.81	20.63	nr	**26.49**
Sum of two sides 300mm	6.37	7.00	1.81	20.63	nr	**27.63**
Sum of two sides 350mm	7.52	8.27	1.77	20.16	nr	**28.43**
Sum of two sides 400mm	8.82	9.70	1.77	20.16	nr	**29.86**
Sum of two sides 450mm	10.24	11.27	1.77	20.16	nr	**31.43**
Sum of two sides 500mm	11.76	12.94	1.77	20.16	nr	**33.10**
Sum of two sides 550mm	13.41	14.75	1.86	21.21	nr	**35.96**
Sum of two sides 600mm	15.64	17.20	1.86	21.21	nr	**38.41**
Sum of two sides 650mm	17.18	18.89	1.86	21.21	nr	**40.10**
Sum of two sides 700mm	19.27	21.19	1.86	21.21	nr	**42.40**
Sum of two sides 750mm	21.42	23.57	1.96	22.33	nr	**45.90**
Sum of two sides 800mm	23.75	26.12	1.96	22.33	nr	**48.45**
Sum of two sides 850mm	26.17	28.79	1.96	22.33	nr	**51.12**
Sum of two sides 900mm	28.75	31.62	1.96	22.33	nr	**53.95**
Sum of two sides 950mm	31.46	34.60	2.61	29.74	nr	**64.34**
Sum of two sides 1000mm	34.30	37.73	2.61	29.74	nr	**67.47**
Sum of two sides 1100mm	40.36	44.39	2.61	29.74	nr	**74.13**

Y:MECHANICAL AND ELECTRICAL SERVICES

Item	Net Price £	Material £	Labour hours	Labour £	Unit	Total Rate £
Sum of two sides 1200mm	46.90	51.59	2.61	29.74	nr	81.33
Sum of two sides 1300mm	56.16	61.78	3.33	37.97	nr	99.75
Sum of two sides 1400mm	63.91	70.30	3.33	37.97	nr	108.27
Sum of two sides 1500mm	72.16	79.38	3.68	41.88	nr	121.26
Sum of two sides 1600mm	80.99	89.09	3.69	42.03	nr	131.12
Sum of two sides 1700mm	95.03	104.53	4.05	46.11	nr	150.64
Sum of two sides 1800mm	105.03	115.54	4.05	46.11	nr	161.65
Sum of two sides 1900mm	115.59	127.15	4.22	48.06	nr	175.21
Sum of two sides 2000mm	126.64	139.30	4.22	48.06	nr	187.36
Sum of two sides 2100mm	153.54	168.90	6.49	73.96	nr	242.86
Sum of two sides 2200mm	166.77	183.45	6.49	73.96	nr	257.41
Sum of two sides 2300mm	180.56	198.61	6.54	74.44	nr	273.05
Sum of two sides 2400mm	194.83	214.31	7.04	80.21	nr	294.52
Sum of two sides 2500mm	209.76	230.73	7.19	81.94	nr	312.67
Sum of two sides 2600mm	249.36	274.29	7.41	84.37	nr	358.66
Sum of two sides 2700mm	266.02	292.62	7.46	85.00	nr	377.62
Sum of two sides 2800mm	283.14	311.45	13.16	149.87	nr	461.32
Sum of two sides 2900mm	301.99	332.19	13.16	149.87	nr	482.06
Sum of two sides 3000mm	321.53	353.68	13.51	153.92	nr	507.60
Sum of two sides 3100mm	348.24	383.07	13.89	158.19	nr	541.26
Sum of two sides 3200mm	369.24	406.16	13.89	158.19	nr	564.35
Spigot Branch						
Sum of two sides 200mm	0.19	0.21	0.82	9.34	nr	9.55
Sum of two sides 250mm	0.24	0.26	0.82	9.34	nr	9.60
Sum of two sides 300mm	0.29	0.32	0.82	9.34	nr	9.66
Sum of two sides 350mm	0.33	0.37	0.80	9.11	nr	9.48
Sum of two sides 400mm	0.39	0.43	0.80	9.11	nr	9.54
Sum of two sides 450mm	0.43	0.47	0.80	9.11	nr	9.58
Sum of two sides 500mm	0.47	0.52	0.80	9.11	nr	9.63
Sum of two sides 550mm	0.53	0.58	0.89	10.14	nr	10.72
Sum of two sides 600mm	0.57	0.62	0.89	10.14	nr	10.76
Sum of two sides 650mm	0.62	0.69	0.89	10.14	nr	10.83
Sum of two sides 700mm	0.66	0.73	0.89	10.14	nr	10.87
Sum of two sides 750mm	0.72	0.79	0.89	10.14	nr	10.93
Sum of two sides 800mm	0.76	0.84	0.89	10.14	nr	10.98
Sum of two sides 850mm	0.82	0.90	0.89	10.14	nr	11.04
Sum of two sides 900mm	0.86	0.94	0.89	10.14	nr	11.08
Sum of two sides 950mm	0.91	1.00	1.12	12.77	nr	13.77
Sum of two sides 1000mm	0.95	1.05	1.12	12.77	nr	13.82
Sum of two sides 1100mm	1.05	1.16	1.12	12.77	nr	13.93
Sum of two sides 1200mm	1.15	1.26	1.12	12.77	nr	14.03
Sum of two sides 1300mm	1.25	1.37	1.20	13.67	nr	15.04
Sum of two sides 1400mm	1.34	1.48	1.20	13.67	nr	15.15
Sum of two sides 1500mm	1.43	1.57	1.42	16.18	nr	17.75
Sum of two sides 1600mm	1.52	1.67	1.42	16.18	nr	17.85
Sum of two sides 1700mm	1.62	1.78	1.46	16.65	nr	18.43
Sum of two sides 1800mm	1.72	1.89	1.46	16.65	nr	18.54
Sum of two sides 1900mm	1.81	1.99	1.60	18.22	nr	20.21
Sum of two sides 2000mm	1.91	2.10	1.60	18.22	nr	20.32
Sum of two sides 2100mm	2.52	2.77	2.26	25.77	nr	28.54
Sum of two sides 2200mm	2.64	2.91	2.26	25.77	nr	28.68
Sum of two sides 2300mm	2.75	3.03	2.26	25.77	nr	28.80
Sum of two sides 2400mm	2.86	3.15	2.49	28.40	nr	31.55
Sum of two sides 2500mm	2.99	3.29	2.49	28.40	nr	31.69
Sum of two sides 2600mm	3.11	3.43	2.49	28.40	nr	31.83
Sum of two sides 2700mm	3.25	3.58	2.49	28.40	nr	31.98
Sum of two sides 2800mm	3.36	3.70	3.41	38.87	nr	42.57
Sum of two sides 2900mm	3.47	3.82	3.41	38.87	nr	42.69

Y:MECHANICAL AND ELECTRICAL SERVICES

Item	Net Price £	Material £	Labour hours	Labour £	Unit	Total Rate £
Y30 : AIR DUCTLINES : MILD STEEL DUCTWORK (contd)						
Extra Over Galvanised Sheet Metal DW 14 Class C Rectangular Section Ductwork; fittings (including all necessary stiffener, joints etc.) (contd)						
Spigot Branch (contd)						
Sum of two sides 3000mm	3.60	3.96	4.18	47.66	nr	51.62
Sum of two sides 3100mm	3.71	4.08	4.18	47.66	nr	51.74
Sum of two sides 3200mm	3.83	4.22	4.18	47.66	nr	51.88
Branch; Based on Branch Size						
Sum of two sides 200mm	0.26	0.29	0.95	10.85	nr	11.14
Sum of two sides 250mm	1.88	2.07	0.95	10.83	nr	12.90
Sum of two sides 300mm	2.12	2.33	0.95	10.83	nr	13.16
Sum of two sides 350mm	2.35	2.59	1.00	11.39	nr	13.98
Sum of two sides 400mm	2.59	2.85	1.00	11.39	nr	14.24
Sum of two sides 450mm	2.82	3.11	1.00	11.39	nr	14.50
Sum of two sides 500mm	3.04	3.35	1.00	11.39	nr	14.74
Sum of two sides 550mm	3.27	3.59	1.02	11.62	nr	15.21
Sum of two sides 600mm	3.52	3.87	1.02	11.62	nr	15.49
Sum of two sides 650mm	3.78	4.16	1.02	11.62	nr	15.78
Sum of two sides 700mm	4.00	4.40	1.02	11.62	nr	16.02
Sum of two sides 750mm	4.22	4.64	1.02	11.62	nr	16.26
Sum of two sides 800mm	4.46	4.90	1.02	11.62	nr	16.52
Sum of two sides 850mm	4.71	5.18	1.02	11.62	nr	16.80
Sum of two sides 900mm	4.93	5.42	1.02	11.62	nr	17.04
Sum of two sides 950mm	5.16	5.68	1.25	14.24	nr	19.92
Sum of two sides 1000mm	5.38	5.92	1.25	14.24	nr	20.16
Sum of two sides 1100mm	5.87	6.46	1.25	14.24	nr	20.70
Sum of two sides 1200mm	6.34	6.97	1.25	14.24	nr	21.21
Sum of two sides 1300mm	6.80	7.47	1.46	16.65	nr	24.12
Sum of two sides 1400mm	7.27	7.99	1.46	16.65	nr	24.64
Sum of two sides 1500mm	7.74	8.51	1.55	17.66	nr	26.17
Sum of two sides 1600mm	8.19	9.01	1.55	17.66	nr	26.67
Sum of two sides 1700mm	10.09	11.10	1.61	18.34	nr	29.44
Sum of two sides 1800mm	10.63	11.69	1.61	18.34	nr	30.03
Sum of two sides 1900mm	11.14	12.26	1.75	19.95	nr	32.21
Sum of two sides 2000mm	11.67	12.83	1.75	19.95	nr	32.78
Sum of two sides 2100mm	13.70	15.07	2.42	27.51	nr	42.58
Sum of two sides 2200mm	14.31	15.74	2.42	27.51	nr	43.25
Sum of two sides 2300mm	14.91	16.40	2.42	27.51	nr	43.91
Sum of two sides 2400mm	15.50	17.05	2.65	30.13	nr	47.18
Sum of two sides 2500mm	16.10	17.71	2.65	30.13	nr	47.84
Sum of two sides 2600mm	23.97	26.37	2.67	30.37	nr	56.74
Sum of two sides 2700mm	24.73	27.21	2.67	30.37	nr	57.58
Sum of two sides 2800mm	25.53	28.09	3.58	40.82	nr	68.91
Sum of two sides 2900mm	26.35	28.99	3.58	40.82	nr	69.81
Sum of two sides 3000mm	27.24	29.96	4.35	49.52	nr	79.48
Sum of two sides 3100mm	28.05	30.86	4.35	49.52	nr	80.38
Sum of two sides 3200mm	28.91	31.80	4.35	49.52	nr	81.32

Y:MECHANICAL AND ELECTRICAL SERVICES

Item	Net Price £	Material £	Labour hours	Labour £	Unit	Total Rate £
Galvanised Sheet Metal DW 142 Class A Spirally Wound Circular Section Ductwork (including all necessary stiffeners, joints coplers in the running length and duct supports.)						
Straight Duct						
80mm dia.	4.41	4.86	0.13	1.48	m	**6.34**
100mm dia.	4.41	4.86	0.75	8.54	m	**13.40**
150mm dia.	6.67	7.34	0.75	8.54	m	**15.88**
200mm dia.	11.53	12.68	0.75	8.54	m	**21.22**
250mm dia.	13.95	15.35	1.05	11.96	m	**27.31**
300mm dia.	13.95	15.35	1.05	11.96	m	**27.31**
355mm dia.	19.78	21.76	1.05	11.96	m	**33.72**
400mm dia.	19.78	21.76	1.05	11.96	m	**33.72**
450mm dia.	22.41	24.65	1.05	11.96	m	**36.61**
500mm dia.	24.52	26.98	1.05	11.96	m	**38.94**
630mm dia.	33.88	37.27	1.20	13.67	m	**50.94**
710mm dia.	37.08	40.79	1.20	13.67	m	**54.46**
800mm dia.	49.59	54.55	1.25	14.24	m	**68.79**
900mm dia.	55.40	60.94	1.26	14.36	m	**75.30**
1000mm dia.	61.56	67.72	1.43	16.29	m	**84.01**
1120mm dia.	79.32	87.25	2.10	23.93	m	**111.18**
1250mm dia.	88.29	97.11	2.10	23.93	m	**121.04**
1400mm dia.	100.66	110.72	2.40	27.38	m	**138.10**
1600mm dia.	113.18	124.50	2.65	30.21	m	**154.71**
Extra Over Galvanised Sheet Metal DW 142 Class A Spirally Wound Circular Section Ductwork; fittings; (including all necesary stiffeners, joints, etc)						
Stopped End						
80mm dia.	1.74	1.92	0.13	1.48	nr	**3.40**
100mm dia.	1.74	1.92	0.13	1.48	nr	**3.40**
150mm dia.	3.28	3.61	0.13	1.48	nr	**5.09**
200mm dia.	4.08	4.49	0.17	1.94	nr	**6.43**
250mm dia.	6.42	7.06	0.25	2.85	nr	**9.91**
300mm dia.	6.42	7.06	0.25	2.85	nr	**9.91**
355mm dia.	8.64	9.50	0.38	4.33	nr	**13.83**
400mm dia.	8.64	9.50	0.38	4.33	nr	**13.83**
450mm dia.	10.50	11.56	0.38	4.33	nr	**15.89**
500mm dia.	11.56	12.71	0.38	4.33	nr	**17.04**
630mm dia.	17.74	19.52	0.50	5.69	nr	**25.21**
710mm dia.	51.84	57.03	0.60	6.84	nr	**63.87**
800mm dia.	54.06	59.46	0.70	7.98	nr	**67.44**
900mm dia.	62.11	68.33	0.80	9.11	nr	**77.44**
1000mm dia.	62.11	68.33	0.90	10.25	nr	**78.58**
1120mm dia.	92.48	101.73	1.00	11.39	nr	**113.12**
1250mm dia.	92.48	101.73	1.00	11.39	nr	**113.12**
1400mm dia.	103.80	114.18	1.00	11.39	nr	**125.57**
1600mm dia.	117.64	129.40	1.00	11.39	nr	**140.79**

Y:MECHANICAL AND ELECTRICAL SERVICES

Item	Net Price £	Material £	Labour hours	Labour £	Unit	Total Rate £
Y30 : AIR DUCTLINES : MILD STEEL DUCTWORK (contd)						
Extra Over Galvanised Sheet Metal DW 142 Class A Spirally Wound Circular Section Ductwork; fittings; (including all necesary stiffeners, joints, etc) (contd)						
Taper, Large End Measured						
100mm dia.	7.00	7.70	0.25	2.85	nr	**10.55**
150mm dia.	7.93	8.72	0.25	2.85	nr	**11.57**
200mm dia.	14.39	15.83	0.38	4.33	nr	**20.16**
250mm dia.	15.64	17.20	0.50	5.69	nr	**22.89**
300mm dia.	21.95	24.15	0.50	5.69	nr	**29.84**
355mm dia.	31.29	34.42	0.75	8.54	nr	**42.96**
400mm dia.	31.29	34.42	0.75	8.54	nr	**42.96**
450mm dia.	33.85	37.24	0.75	8.54	nr	**45.78**
500mm dia.	34.90	38.40	0.75	8.54	nr	**46.94**
630mm dia.	46.00	50.60	0.75	8.54	nr	**59.14**
710mm dia.	77.88	85.67	0.83	9.49	nr	**95.16**
800mm dia.	86.40	95.04	0.92	10.48	nr	**105.52**
900mm dia.	124.95	137.44	1.00	11.39	nr	**148.83**
1000mm dia.	124.95	137.44	1.08	12.31	nr	**149.75**
1120mm dia.	158.11	173.92	3.00	34.20	nr	**208.12**
1250mm dia.	158.11	173.92	3.00	34.20	nr	**208.12**
1400mm dia.	173.00	190.30	3.51	39.97	nr	**230.27**
1600mm dia.	186.84	205.52	4.00	45.56	nr	**251.08**
90 degree Bend Segmented; Medium						
80mm dia.	6.19	6.81	0.25	2.85	nr	**9.66**
100mm dia.	11.32	12.45	0.25	2.85	nr	**15.30**
150mm dia.	11.32	12.45	0.25	2.85	nr	**15.30**
200mm dia.	15.72	17.29	0.38	4.33	nr	**21.62**
250mm dia.	21.71	23.89	0.50	5.69	nr	**29.58**
300mm dia.	23.22	25.55	0.50	5.69	nr	**31.24**
355mm dia.	34.45	37.89	0.75	8.54	nr	**46.43**
400mm dia.	34.45	37.89	0.75	8.54	nr	**46.43**
450mm dia.	41.10	45.22	0.75	8.54	nr	**53.76**
500mm dia.	50.56	55.61	0.75	8.54	nr	**64.15**
630mm dia.	72.16	79.38	0.75	8.54	nr	**87.92**
710mm dia.	155.42	170.97	0.83	9.46	nr	**180.43**
800mm dia.	176.21	193.83	0.92	10.48	nr	**204.31**
900mm dia.	207.61	228.38	1.00	11.39	nr	**239.77**
1000mm dia.	227.58	250.34	1.08	12.31	nr	**262.65**
1120mm dia.	296.84	326.52	3.00	34.20	nr	**360.72**
1250mm dia.	329.78	362.76	3.00	34.20	nr	**396.96**
1400mm dia.	573.71	631.08	3.51	39.97	nr	**671.05**
1600mm dia.	786.02	864.62	4.00	45.56	nr	**910.18**
45 degree Medium Bend						
80mm dia.	4.90	5.39	0.25	2.85	nr	**8.24**
100mm dia.	4.90	5.39	0.25	2.85	nr	**8.24**
150mm dia.	8.17	8.98	0.25	2.85	nr	**11.83**
200mm dia.	14.71	16.18	0.35	3.99	nr	**20.17**
250mm dia.	15.40	16.94	0.50	5.69	nr	**22.63**
300mm dia.	20.50	22.55	0.50	5.69	nr	**28.24**
355mm dia.	23.24	25.56	0.75	8.54	nr	**34.10**

Y:MECHANICAL AND ELECTRICAL SERVICES

Item	Net Price £	Material £	Labour hours	Labour £	Unit	Total Rate £
400mm dia.	23.24	25.56	0.75	8.54	nr	**34.10**
450mm dia.	26.74	29.41	0.75	8.54	nr	**37.95**
500mm dia.	32.69	35.96	0.75	8.54	nr	**44.50**
630mm dia.	47.53	52.28	0.75	8.54	nr	**60.82**
710mm dia.	51.84	57.03	0.83	9.46	nr	**66.49**
800mm dia.	112.80	124.08	0.92	10.48	nr	**134.56**
900mm dia.	147.34	162.07	1.00	11.39	nr	**173.46**
1000mm dia.	164.06	180.47	1.08	12.31	nr	**192.78**
1120mm dia.	213.34	234.68	3.00	34.20	nr	**268.88**
1250mm dia.	229.11	252.02	3.00	34.20	nr	**286.22**
1400mm dia.	326.03	358.63	3.51	39.97	nr	**398.60**
1600mm dia.	786.02	864.62	4.00	45.56	nr	**910.18**

90 degree Equal Twin Bend

Item	Net Price £	Material £	Labour hours	Labour £	Unit	Total Rate £
80mm dia.	18.91	20.80	0.50	5.69	nr	**26.49**
100mm dia.	18.91	20.80	0.50	5.69	nr	**26.49**
150mm dia.	31.76	34.94	0.50	5.69	nr	**40.63**
200mm dia.	70.17	77.19	0.75	8.54	nr	**85.73**
250mm dia.	97.85	107.63	1.00	11.39	nr	**119.02**
300mm dia.	101.35	111.49	1.00	11.39	nr	**122.88**
355mm dia.	123.77	136.15	1.50	17.10	nr	**153.25**
400mm dia.	123.77	136.15	1.50	17.10	nr	**153.25**
450mm dia.	145.96	160.55	1.50	17.10	nr	**177.65**
500mm dia.	164.64	181.10	1.50	17.10	nr	**198.20**
630mm dia.	252.58	277.84	1.50	17.10	nr	**294.94**
710mm dia.	475.39	522.93	1.58	18.02	nr	**540.95**
800mm dia.	572.19	629.41	1.67	19.05	nr	**648.46**
900mm dia.	656.03	721.63	1.75	19.95	nr	**741.58**
1000mm dia.	700.41	770.46	1.83	20.86	nr	**791.32**
1120mm dia.	961.87	1058.05	4.00	45.56	nr	**1103.61**
1250mm dia.	1047.70	1152.47	4.00	45.56	nr	**1198.03**

Conical Branch

Item	Net Price £	Material £	Labour hours	Labour £	Unit	Total Rate £
80mm dia.	12.72	13.99	0.50	5.69	nr	**19.68**
100mm dia.	12.72	13.99	0.50	5.69	nr	**19.68**
150mm dia.	17.16	18.88	0.50	5.69	nr	**24.57**
200mm dia.	28.48	31.33	0.75	8.54	nr	**39.87**
250mm dia.	37.48	41.23	1.00	11.39	nr	**52.62**
300mm dia.	37.48	41.23	1.00	11.39	nr	**52.62**
355mm dia.	51.26	56.39	1.50	17.10	nr	**73.49**
400mm dia.	51.26	56.39	1.50	17.10	nr	**73.49**
450mm dia.	72.40	79.64	1.50	17.10	nr	**96.74**
500mm dia.	72.40	79.64	1.50	17.10	nr	**96.74**
630mm dia.	104.27	114.70	1.50	17.10	nr	**131.80**
710mm dia.	108.01	118.81	1.58	18.02	nr	**136.83**
800mm dia.	108.01	118.81	1.67	19.05	nr	**137.86**
900mm dia.	176.56	194.21	1.75	19.95	nr	**214.16**
1000mm dia.	176.56	194.21	1.83	20.86	nr	**215.07**
1120mm dia.	246.97	271.67	4.00	45.56	nr	**317.23**
1250mm dia.	246.97	271.67	4.50	51.31	nr	**322.98**

45 degree Branch

Item	Net Price £	Material £	Labour hours	Labour £	Unit	Total Rate £
80mm dia.	16.58	18.24	0.50	5.69	nr	**23.93**
100mm dia.	16.58	18.24	0.50	5.69	nr	**23.93**
150mm dia.	20.19	22.21	0.50	5.69	nr	**27.90**
200mm dia.	29.77	32.75	1.00	11.39	nr	**44.14**
250mm dia.	70.53	77.58	1.00	11.39	nr	**88.97**

Y:MECHANICAL AND ELECTRICAL SERVICES

Item	Net Price £	Material £	Labour hours	Labour £	Unit	Total Rate £
Y30 : AIR DUCTLINES : MILD STEEL DUCTWORK (contd)						
Extra Over Galvanised Sheet Metal DW 142 Class A Spirally Wound Circular Section Ductwork; fittings; (including all necesary stiffeners, joints, etc) (contd)						
45 degree Branch						
300mm dia.	70.64	77.70	1.00	11.39	nr	89.09
355mm dia.	70.64	77.70	1.50	17.10	nr	94.80
400mm dia.	70.64	77.70	1.50	17.10	nr	94.80
450mm dia.	103.80	114.18	1.50	17.10	nr	131.28
500mm dia.	103.80	114.18	1.50	17.10	nr	131.28
630mm dia.	103.80	114.18	1.50	17.10	nr	131.28
710mm dia.	176.56	194.21	1.58	18.02	nr	212.23
800mm dia.	246.97	271.67	1.67	19.05	nr	290.72
900mm dia.	246.97	271.67	2.00	22.78	nr	294.45
1000mm dia.	246.97	271.67	2.00	22.78	nr	294.45
1120mm dia.	246.97	271.67	4.00	45.56	nr	317.23
1250mm dia.	246.97	271.67	4.00	45.56	nr	317.23

Y:MECHANICAL AND ELECTRICAL SERVICES

Item	Net Price £	Material £	Labour hours	Labour £	Unit	Total Rate £
Galvanised Sheet Metal High DW 142 Class C Spirally Wound Circular Section Ductwork; taped joints (including all necessary stiffeners, joints, couples in in the running length and duct supports)						
Straight Duct						
80mm dia.	4.63	5.09	0.75	8.54	m	13.63
100mm dia.	4.63	5.09	0.75	8.54	m	13.63
150mm dia.	6.99	7.69	0.75	8.54	m	16.23
200mm dia.	12.08	13.29	0.75	8.54	m	21.83
250mm dia.	14.62	16.08	1.05	11.96	m	28.04
300mm dia.	16.28	17.91	1.05	11.96	m	29.87
355mm dia.	20.72	22.79	1.05	11.96	m	34.75
400mm dia.	20.72	22.79	1.05	11.96	m	34.75
450mm dia.	23.48	25.82	1.05	11.96	m	37.78
500mm dia.	25.69	28.26	1.05	11.96	m	40.22
630mm dia.	35.50	39.05	1.20	13.67	m	52.72
710mm dia.	38.85	42.73	1.20	13.67	m	56.40
800mm dia.	51.95	57.15	1.25	14.24	m	71.39
900mm dia.	58.04	63.85	1.26	14.36	m	78.21
1000mm dia.	64.50	70.95	1.43	16.29	m	87.24
1120mm dia.	83.10	91.41	2.10	23.93	m	115.34
1250mm dia.	92.50	101.74	2.10	23.93	m	125.67
1400mm dia.	105.46	116.00	2.40	27.38	m	143.38
1600mm dia.	118.58	130.44	2.65	30.21	m	160.65
Extra Over Galvanised Sheet Metal DW 14 Class C Spirally Wound Circular Section Ductwork; fittings; taped joints (including all necessary stiffeners, joints, etc)						
Stopped End						
80mm dia.	2.09	2.30	0.13	1.48	nr	3.78
100mm dia.	2.09	2.30	0.13	1.48	nr	3.78
150mm dia.	3.93	4.33	0.13	1.48	nr	5.81
200mm dia.	4.90	5.39	0.17	1.94	nr	7.33
250mm dia.	4.90	5.39	0.25	2.85	nr	8.24
300mm dia.	7.70	8.47	0.25	2.85	nr	11.32
355mm dia.	10.36	11.39	0.38	4.33	nr	15.72
400mm dia.	10.36	11.39	0.38	4.33	nr	15.72
450mm dia.	12.60	13.86	0.38	4.33	nr	18.19
500mm dia.	13.86	15.25	0.38	4.33	nr	19.58
630mm dia.	21.28	23.41	0.50	5.69	nr	29.10
710mm dia.	62.18	68.40	0.60	6.84	nr	75.24
800mm dia.	64.84	71.32	0.70	7.98	nr	79.30
900mm dia.	74.50	81.95	0.80	9.11	nr	91.06
1000mm dia.	74.50	81.95	0.90	10.25	nr	92.20
1120mm dia.	110.92	122.01	1.00	11.39	nr	133.40
1250mm dia.	110.92	122.01	1.00	11.39	nr	133.40
1400mm dia.	124.50	136.95	1.00	11.39	nr	148.34
1600mm dia.	124.50	136.95	1.00	11.39	nr	148.34

Y:MECHANICAL AND ELECTRICAL SERVICES

Item	Net Price £	Material £	Labour hours	Labour £	Unit	Total Rate £
Y30 : AIR DUCTLINES : MILD STEEL DUCTWORK (contd)						
Extra Over Galvanised Sheet Metal DW 14 Class C Spirally Wound Circular Section Ductwork; fittings; taped joints (including all necessary stiffeners, joints, etc)						
Taper, Large End Measured						
100mm dia.	7.39	8.13	0.25	2.85	nr	**10.98**
150mm dia.	8.37	9.20	0.25	2.85	nr	**12.05**
200mm dia.	14.66	16.12	0.38	4.33	nr	**20.45**
250mm dia.	16.50	18.15	0.50	5.69	nr	**23.84**
300mm dia.	23.16	25.47	0.50	5.69	nr	**31.16**
355mm dia.	33.01	36.31	0.75	8.54	nr	**44.85**
400mm dia.	33.01	36.31	0.75	8.54	nr	**44.85**
450mm dia.	35.71	39.28	0.75	8.54	nr	**47.82**
500mm dia.	36.82	40.50	0.75	8.54	nr	**49.04**
630mm dia.	48.53	53.38	0.75	8.54	nr	**61.92**
710mm dia.	82.15	90.37	0.83	9.46	nr	**99.83**
800mm dia.	91.15	100.26	0.92	10.48	nr	**110.74**
900mm dia.	131.81	144.99	1.00	11.39	nr	**156.38**
1000mm dia.	131.81	144.99	1.08	12.31	nr	**157.30**
1120mm dia.	166.79	183.47	3.00	34.20	nr	**217.67**
1250mm dia.	166.79	183.47	3.00	34.20	nr	**217.67**
1400mm dia.	182.50	200.75	3.51	39.97	nr	**240.72**
1600mm dia.	197.10	216.81	4.00	45.56	nr	**262.37**
90 degree Medium Bend						
80mm dia.	6.66	7.33	0.25	2.85	nr	**10.18**
100mm dia.	6.66	7.33	0.25	2.85	nr	**10.18**
150mm dia.	12.19	13.41	0.25	2.85	nr	**16.26**
200mm dia.	16.93	18.62	0.38	4.33	nr	**22.95**
250mm dia.	23.38	25.72	0.50	5.69	nr	**31.41**
300mm dia.	25.00	27.50	0.50	5.69	nr	**33.19**
355mm dia.	37.09	40.79	0.75	8.54	nr	**49.33**
400mm dia.	37.09	40.79	0.75	8.54	nr	**49.33**
450mm dia.	44.25	48.68	0.75	8.54	nr	**57.22**
500mm dia.	54.43	59.87	0.75	8.54	nr	**68.41**
630mm dia.	77.69	85.46	0.75	8.54	nr	**94.00**
710mm dia.	167.33	184.06	0.83	9.46	nr	**193.52**
800mm dia.	189.71	208.68	0.92	10.48	nr	**219.16**
900mm dia.	223.51	245.87	1.00	11.39	nr	**257.26**
1000mm dia.	245.02	269.52	1.08	12.31	nr	**281.83**
1120mm dia.	319.58	351.53	3.00	34.20	nr	**385.73**
1250mm dia.	355.04	390.54	3.00	34.20	nr	**424.74**
1400mm dia.	617.65	679.41	3.51	39.97	nr	**719.38**
1600mm dia.	846.22	930.84	4.00	45.56	nr	**976.40**
45 degree Medium Bend						
80mm dia.	5.42	5.96	0.25	2.85	nr	**8.81**
100mm dia.	5.42	5.96	0.25	2.85	nr	**8.81**
150mm dia.	9.03	9.93	0.25	2.85	nr	**12.78**
200mm dia.	16.26	17.89	0.38	4.33	nr	**22.22**
250mm dia.	17.03	18.73	0.50	5.69	nr	**24.42**
300mm dia.	22.66	24.93	0.50	5.69	nr	**30.62**

Y:MECHANICAL AND ELECTRICAL SERVICES

Item	Net Price £	Material £	Labour hours	Labour £	Unit	Total Rate £
355mm dia.	25.69	28.26	0.75	8.54	nr	**36.80**
400mm dia.	25.69	28.26	0.75	8.54	nr	**36.80**
450mm dia.	29.56	32.52	0.75	8.54	nr	**41.06**
500mm dia.	36.14	39.75	0.75	8.54	nr	**48.29**
630mm dia.	52.54	57.79	0.75	8.54	nr	**66.33**
710mm dia.	97.46	107.21	0.83	9.46	nr	**116.67**
800mm dia.	124.70	137.16	0.92	10.48	nr	**147.64**
900mm dia.	162.88	179.17	1.00	11.39	nr	**190.56**
1000mm dia.	181.37	199.50	1.08	12.31	nr	**211.81**
1120mm dia.	235.85	259.43	3.00	34.20	nr	**293.63**
1250mm dia.	253.28	278.60	3.00	34.20	nr	**312.80**
1400mm dia.	360.42	396.46	4.50	51.31	nr	**447.77**
1600mm dia.	473.90	521.29	4.00	45.56	nr	**566.85**
90 degree Equal Twin Bend						
80mm dia.	19.40	21.34	0.50	5.69	nr	**27.03**
100mm dia.	19.40	21.34	0.50	5.69	nr	**27.03**
150mm dia.	32.59	35.85	0.50	5.69	nr	**41.54**
200mm dia.	71.99	79.19	0.75	8.54	nr	**87.73**
250mm dia.	100.39	110.43	1.00	11.39	nr	**121.82**
300mm dia.	103.99	114.39	1.00	11.39	nr	**125.78**
355mm dia.	126.99	139.69	1.50	17.10	nr	**156.79**
400mm dia.	126.99	139.69	1.50	17.10	nr	**156.79**
450mm dia.	149.75	164.73	1.50	17.10	nr	**181.83**
500mm dia.	168.92	185.82	1.50	17.10	nr	**202.92**
630mm dia.	259.15	285.07	1.50	17.10	nr	**302.17**
710mm dia.	487.76	536.53	1.58	17.97	nr	**554.50**
800mm dia.	587.07	645.78	1.75	19.95	nr	**665.73**
900mm dia.	673.09	740.40	1.75	19.95	nr	**760.35**
1000mm dia.	718.63	790.50	1.83	20.86	nr	**811.36**
1120mm dia.	986.89	1085.57	4.00	45.56	nr	**1131.13**
1250mm dia.	1074.95	1182.45	4.00	45.56	nr	**1228.01**
Conical Branch						
80mm dia.	12.96	14.25	0.50	5.69	nr	**19.94**
100mm dia.	12.96	14.25	0.50	5.69	nr	**19.94**
150mm dia.	17.48	19.23	0.50	5.69	nr	**24.92**
200mm dia.	29.02	31.92	0.75	8.54	nr	**40.46**
250mm dia.	38.18	42.00	1.00	11.39	nr	**53.39**
300mm dia.	38.18	42.00	1.00	11.39	nr	**53.39**
355mm dia.	52.23	57.45	1.50	17.10	nr	**74.55**
400mm dia.	52.23	57.45	1.50	17.10	nr	**74.55**
450mm dia.	73.76	81.13	1.50	17.10	nr	**98.23**
500mm dia.	73.76	81.13	1.50	17.10	nr	**98.23**
630mm dia.	105.75	116.32	1.50	17.10	nr	**133.42**
710mm dia.	110.04	121.04	1.58	18.02	nr	**139.06**
800mm dia.	110.04	121.04	1.67	19.05	nr	**140.09**
900mm dia.	179.87	197.86	1.75	19.95	nr	**217.81**
1000mm dia.	179.87	197.86	1.83	20.86	nr	**218.72**
1120mm dia.	195.45	215.00	4.00	45.56	nr	**260.56**
1250mm dia.	251.61	276.78	4.00	45.56	nr	**322.34**
1400mm dia.	307.24	337.96	4.50	51.31	nr	**389.27**
1600mm dia.	374.50	411.95	5.00	56.95	nr	**468.90**
45 degree Branch						
80mm dia.	20.61	22.67	0.50	5.69	nr	**28.36**
100mm dia.	20.61	22.67	0.50	5.69	nr	**28.36**
150mm dia.	25.09	27.60	0.50	5.69	nr	**33.29**

Y:MECHANICAL AND ELECTRICAL SERVICES

Item	Net Price £	Material £	Labour hours	Labour £	Unit	Total Rate £
Y30 : AIR DUCTLINES : MILD STEEL DUCTWORK (contd)						
Extra Over Galvanised Sheet Metal DW 14 Class C Spirally Wound Circular Section Ductwork; fittings; taped joints (including all necessary stiffeners, joints, etc) (contd)						
45 degree Branch (contd)						
200mm dia.	37.00	40.70	0.75	8.56	nr	49.26
250mm dia.	87.65	96.42	1.00	11.39	nr	107.81
300mm dia.	87.79	96.57	1.00	11.39	nr	107.96
355mm dia.	87.79	96.57	1.50	17.10	nr	113.67
400mm dia.	87.79	96.57	1.50	17.10	nr	113.67
450mm dia.	129.00	141.90	1.50	17.10	nr	159.00
500mm dia.	129.00	141.90	1.50	17.10	nr	159.00
630mm dia.	129.00	141.90	1.50	17.10	nr	159.00
710mm dia.	219.42	241.36	1.58	18.02	nr	259.38
800mm dia.	306.93	337.63	1.84	20.98	nr	358.61
900mm dia.	306.93	337.63	2.00	22.78	nr	360.41
1000mm dia.	306.93	337.63	2.00	22.78	nr	360.41
1120mm dia.	306.93	337.63	4.00	45.56	nr	383.19
1250mm dia.	306.93	337.63	4.00	45.56	nr	383.19
1400mm dia.	394.95	434.44	4.50	51.31	nr	485.75
1600mm dia.	394.95	434.44	5.00	56.95	nr	491.39

Keep your figures up to date, free of charge

This section, and most of the other information in this Price Book, is brought up to date every three months, until the next annual edition, in the *Price Book Update*.

The *Update* is available free to all Price Book purchasers.

To ensure you receive your copy, simply complete the reply card from the centre of the book and return it to us.

Y:MECHANICAL AND ELECTRICAL SERVICES

Item	Net Price £	Material £	Labour hours	Labour £	Unit	Total Rate £
Galvanised Sheet Metal Spirally Wound DW 142 Class C Flat Oval Section Ductwork; DW 142; (including all necessary stiffeners, joints, couplers in the running length and duct supports.)						
Straight Duct						
345 x 102mm	15.94	17.53	2.35	26.80	m	**44.33**
427 x 102mm	18.88	20.77	2.59	29.51	m	**50.28**
508 x 102mm	21.17	23.28	2.72	31.04	m	**54.32**
559 x 152mm	21.17	23.28	2.97	33.80	m	**57.08**
531 x 203mm	29.34	32.27	2.97	33.80	m	**66.07**
851 x 203mm	38.82	42.70	4.95	56.39	m	**99.09**
582 x 254mm	32.42	35.66	3.13	35.59	m	**71.25**
1303 x 254mm	81.53	89.68	7.04	80.21	m	**169.89**
632 x 305mm	35.98	39.58	3.40	38.74	m	**78.32**
1275 x 305mm	73.46	80.81	7.04	80.21	m	**161.02**
765 x 356mm	42.57	46.83	4.95	56.39	m	**103.22**
1247 x 356mm	74.55	82.00	7.04	80.21	m	**162.21**
1727 x 356mm	104.56	132.55	9.01	102.61	m	**235.16**
737 x 406mm	43.12	47.43	4.95	56.39	m	**103.82**
818 x 406mm	45.56	50.12	5.38	61.24	m	**111.36**
978 x 406mm	50.60	55.66	5.99	68.20	m	**123.86**
1379 x 406mm	82.74	91.01	7.58	86.29	m	**177.30**
1699 x 406mm	104.56	115.02	9.01	102.61	m	**217.63**
709 x 457mm	43.12	47.43	4.95	56.39	m	**103.82**
1671 x 457mm	104.56	115.02	8.93	101.70	m	**216.72**
678 x 508mm	43.12	47.43	4.95	56.39	m	**103.82**
Extra Over Galvanised Sheet Metal Spirally Wound DW 142 Class C Flat Oval Section Ductwork; fittings (including all necessary joints etc.)						
Plug End						
345 x 102mm	21.46	23.61	0.17	1.94	nr	**25.55**
427 x 102mm	21.83	24.02	0.17	1.94	nr	**25.96**
508 x 102mm	22.22	24.44	0.25	2.85	nr	**27.29**
559 x 152mm	27.67	30.44	0.25	2.85	nr	**33.29**
531 x 203mm	28.07	30.88	0.38	4.33	nr	**35.21**
851 x 203mm	30.37	33.41	0.38	4.33	nr	**37.74**
582 x 254mm	28.76	31.64	0.38	4.33	nr	**35.97**
1303 x 254mm	58.20	64.02	0.60	6.84	nr	**70.86**
632 x 305mm	28.87	31.76	0.38	4.33	nr	**36.09**
1275 x 305mm	48.22	53.04	0.60	6.84	nr	**59.88**
765 x 356mm	36.95	40.65	0.60	6.84	nr	**47.49**
1727 x 356mm	78.22	86.04	0.90	10.25	nr	**96.29**
737 x 406mm	31.14	34.25	0.38	4.36	nr	**38.61**
818 x 406mm	31.68	58.46	0.38	4.33	nr	**62.79**
978 x 406mm	33.09	36.40	0.38	4.33	nr	**40.73**
1379 x 406mm	76.77	84.45	0.60	6.84	nr	**91.29**
1699 x 406mm	79.05	86.95	0.90	10.25	nr	**97.20**
709 x 457mm	30.33	33.36	0.38	4.33	nr	**37.69**
1671 x 457mm	78.93	86.82	1.80	20.48	nr	**107.30**
678 x 508mm	41.27	45.39	0.38	4.33	nr	**49.72**

Y:MECHANICAL AND ELECTRICAL SERVICES

Item	Net Price £	Material £	Labour hours	Labour £	Unit	Total Rate £
Y30 : AIR DUCTLINES : MILD STEEL DUCTWORK (contd)						
Extra Over Galvanised Sheet Metal Spirally Wound DW 142 Class C Flat Oval Section Ductwork; fittings (including all necessary joints, etc.) (contd)						
Taper, Large End Measured						
345 x 102mm	43.76	48.13	0.82	9.34	nr	**57.47**
427 x 102mm	45.46	50.00	0.92	10.48	nr	**60.48**
508 x 102mm	47.43	52.17	0.98	11.17	nr	**63.34**
559 x 152mm	57.58	63.34	1.09	12.42	nr	**75.76**
531 x 203mm	56.17	61.79	1.09	12.42	nr	**74.21**
851 x 203mm	79.49	87.44	1.00	11.39	nr	**98.83**
582 x 254mm	60.50	66.55	1.16	13.21	nr	**79.76**
1303 x 254mm	133.64	147.00	1.00	11.39	nr	**158.39**
632 x 305mm	70.64	77.71	0.61	6.95	nr	**84.66**
1275 x 305mm	135.71	149.28	1.00	11.39	nr	**160.67**
765 x 356mm	79.64	87.61	1.00	11.39	nr	**99.00**
1247 x 356mm	137.89	151.68	1.00	11.39	nr	**163.07**
1727 x 356mm	175.94	193.54	1.08	12.31	nr	**205.85**
737 x 406mm	86.24	94.87	1.00	11.39	nr	**106.26**
818 x 406mm	89.82	98.80	1.10	12.53	nr	**111.33**
978 x 406mm	102.01	112.21	1.25	14.24	nr	**126.45**
1379 x 406mm	150.03	165.03	1.25	14.24	nr	**179.27**
1699 x 406mm	173.31	190.64	1.25	14.24	nr	**204.88**
709 x 457mm	86.45	95.09	1.00	11.39	nr	**106.48**
1671 x 457mm	173.79	191.17	1.25	14.24	nr	**205.41**
678 x 508mm	90.18	99.20	1.00	11.39	nr	**110.59**
90 degree Easy Bend						
345 x 102mm	69.92	76.91	0.25	2.85	nr	**79.76**
427 x 102mm	70.66	77.73	0.50	5.69	nr	**83.42**
508 x 102mm	71.41	78.55	0.50	5.69	nr	**84.24**
559 x 152mm	70.51	77.56	0.50	5.69	nr	**83.25**
531 x 203mm	72.13	79.34	0.75	8.54	nr	**87.88**
851 x 203mm	81.21	89.33	0.75	8.54	nr	**97.87**
582 x 254mm	77.95	85.75	0.75	8.54	nr	**94.29**
1303 x 254mm	107.27	117.99	0.83	9.46	nr	**127.45**
632 x 305mm	88.27	97.10	0.75	8.54	nr	**105.64**
1275 x 305mm	122.41	134.65	0.83	9.46	nr	**144.11**
765 x 356mm	100.97	111.07	0.75	8.54	nr	**119.61**
1247 x 356mm	139.20	153.11	0.83	9.46	nr	**162.57**
1727 x 356mm	173.52	190.87	1.08	12.31	nr	**203.18**
737 x 406mm	115.11	126.62	0.83	9.46	nr	**136.08**
818 x 406mm	117.16	128.87	0.75	8.54	nr	**137.41**
978 x 406mm	140.40	154.44	0.83	9.46	nr	**163.90**
1379 x 406mm	166.16	182.78	1.00	11.39	nr	**194.17**
1699 x 406mm	179.88	197.86	1.08	12.31	nr	**210.17**
709 x 457mm	125.62	138.19	0.75	8.54	nr	**146.73**
1671 x 457mm	207.50	228.25	1.08	12.31	nr	**240.56**
678 x 508mm	131.18	144.30	0.75	8.54	nr	**152.84**
45 degree Easy Bend						
345 x 102mm	63.26	69.59	0.89	10.14	nr	**79.73**
427 x 102mm	63.86	70.24	0.82	9.34	nr	**79.58**
508 x 102mm	64.41	70.85	0.71	8.09	nr	**78.94**

Y:MECHANICAL AND ELECTRICAL SERVICES

Item	Net Price £	Material £	Labour hours	Labour £	Unit	Total Rate £
559 x 152mm	59.63	65.60	0.68	7.75	nr	**73.35**
531 x 203mm	60.66	66.73	0.74	8.43	nr	**75.16**
851 x 203mm	97.96	107.76	0.16	1.82	nr	**109.58**
582 x 254mm	62.40	68.64	0.66	7.52	nr	**76.16**
1303 x 254mm	91.12	100.23	1.00	11.39	nr	**111.62**
632 x 305mm	63.70	70.07	0.50	5.69	nr	**75.76**
1275 x 305mm	102.38	112.61	1.00	11.39	nr	**124.00**
765 x 356mm	78.90	86.79	0.75	8.56	nr	**95.35**
1247 x 356mm	106.78	117.46	1.00	11.39	nr	**128.85**
1727 x 356mm	132.45	145.70	1.09	12.37	nr	**158.07**
737 x 406mm	88.72	97.60	0.60	6.84	nr	**104.44**
818 x 406mm	89.95	98.95	0.35	3.99	nr	**102.94**
978 x 406mm	108.01	118.82	0.75	8.54	nr	**127.36**
1379 x 406mm	126.79	139.47	1.00	11.39	nr	**150.86**
1699 x 406mm	136.40	150.04	1.10	12.49	nr	**162.53**
709 x 457mm	96.42	106.06	0.70	7.98	nr	**114.04**
1671 x 457mm	157.32	173.05	1.09	12.37	nr	**185.42**
678 x 508mm	101.42	111.57	0.80	9.11	nr	**120.68**

90 degree Hard Bend with Turning Vanes

Item	Net Price £	Material £	Labour hours	Labour £	Unit	Total Rate £
345 x 102mm	64.68	71.15	0.48	5.47	nr	**76.62**
427 x 102mm	67.00	73.70	0.34	3.87	nr	**77.57**
508 x 102mm	69.56	76.51	0.15	1.71	nr	**78.22**
559 x 152mm	86.03	94.63	0.03	0.34	nr	**94.97**
531 x 203mm	86.15	94.77	0.11	1.25	nr	**96.02**
851 x 203mm	136.66	150.33	1.50	17.10	nr	**167.43**
582 x 254mm	88.82	97.70	1.50	17.10	nr	**114.80**
1303 x 254mm	324.99	357.49	1.58	18.02	nr	**375.51**
632 x 305mm	99.67	109.64	1.50	17.10	nr	**126.74**
1275 x 305mm	357.01	392.71	1.58	18.02	nr	**410.73**
765 x 356mm	114.48	125.93	1.50	17.10	nr	**143.03**
1247 x 356mm	356.37	392.01	1.58	18.02	nr	**410.03**
1727 x 356mm	565.82	622.40	1.58	18.02	nr	**640.42**
737 x 406mm	118.71	130.58	1.50	17.10	nr	**147.68**
818 x 406mm	156.15	171.76	1.50	17.10	nr	**188.86**
978 x 406mm	166.45	183.09	1.50	17.10	nr	**200.19**
1379 x 406mm	369.64	406.60	1.58	18.02	nr	**424.62**
1699 x 406mm	564.54	620.99	1.83	20.86	nr	**641.85**
709 x 457mm	126.41	139.05	1.50	17.10	nr	**156.15**
1671 x 457mm	628.72	691.60	1.83	20.90	nr	**712.50**
678 x 508mm	121.29	133.42	1.58	18.02	nr	**151.44**

45 degree Branch

Item	Net Price £	Material £	Labour hours	Labour £	Unit	Total Rate £
345 x 102mm	46.14	50.75	0.50	5.69	nr	**56.44**
427 x 102mm	48.44	53.29	0.50	5.69	nr	**58.98**
508 x 102mm	50.97	56.07	1.00	11.39	nr	**67.46**
599 x 152mm	61.04	67.14	1.50	17.10	nr	**84.24**
531 x 203mm	61.77	67.94	1.50	17.10	nr	**85.04**
851 x 203mm	91.16	100.28	1.50	17.10	nr	**117.38**
582 x 254mm	66.42	73.06	1.50	17.10	nr	**90.16**
1303 x 254mm	166.75	183.42	0.83	9.46	nr	**192.88**
632 x 305mm	74.75	82.23	1.50	17.10	nr	**99.33**
1275 x 305mm	163.86	180.24	1.58	18.02	nr	**198.26**
765 x 356mm	87.41	96.15	1.50	17.10	nr	**113.25**
1247 x 356mm	164.58	181.04	1.58	18.02	nr	**199.06**
1727 x 356mm	237.47	261.22	1.58	18.02	nr	**279.24**
737 x 406mm	95.29	104.82	1.50	17.10	nr	**121.92**
818 x 406mm	103.30	113.63	1.50	17.10	nr	**130.73**

Y:MECHANICAL AND ELECTRICAL SERVICES

Item	Net Price £	Material £	Labour hours	Labour £	Unit	Total Rate £
Y30 : AIR DUCTLINES : MILD STEEL DUCTWORK (contd)						
Extra Over Galvanised Sheet Metal Spirally Wound DW 142 Class C Flat Oval Section Ductwork; fittings (including all necessary joints, etc.) (contd)						
45 degree Branch (contd)						
978 x 406mm	119.17	131.08	1.50	17.10	nr	**148.18**
1379 x 406mm	184.66	203.13	1.67	19.05	nr	**222.18**
1699 x 406mm	238.32	262.16	1.90	21.65	nr	**283.81**
709 x 457mm	95.59	105.15	1.50	17.10	nr	**122.25**
1671 x 457mm	220.30	242.33	1.83	20.86	nr	**263.19**
678 x 508mm	99.55	109.50	1.50	17.10	nr	**126.60**
90 degree Branch						
345 x 102mm	27.46	30.21	0.50	5.69	nr	**35.90**
427 x 102mm	28.00	30.80	0.50	5.69	nr	**36.49**
508 x 102mm	28.52	31.37	1.00	11.39	nr	**42.76**
559 x 152mm	34.82	38.30	1.00	11.39	nr	**49.69**
531 x 203mm	35.55	39.10	1.00	11.39	nr	**50.49**
851 x 203mm	45.25	49.78	1.50	17.10	nr	**66.88**
582 x 254mm	40.18	44.20	1.50	17.10	nr	**61.30**
1303 x 254mm	74.96	82.45	1.58	18.02	nr	**100.47**
632 x 305mm	42.93	47.22	1.50	17.10	nr	**64.32**
1275 x 305mm	75.82	83.40	1.58	18.02	nr	**101.42**
765 x 356mm	49.94	54.93	1.50	17.10	nr	**72.03**
1247 x 356mm	80.28	88.31	1.58	18.02	nr	**106.33**
1727 x 356mm	100.73	110.80	1.83	20.86	nr	**131.66**
737 x 406mm	52.22	57.45	1.50	17.10	nr	**74.55**
818 x 406mm	54.58	60.04	1.50	17.10	nr	**77.14**
978 x 406mm	59.19	65.11	0.83	9.46	nr	**74.57**
1379 x 406mm	97.81	107.59	1.67	19.05	nr	**126.64**
1699 x 406mm	97.81	107.59	1.83	20.86	nr	**128.45**
709 x 457mm	54.40	59.84	1.50	17.10	nr	**76.94**
1671 x 457mm	102.28	112.51	1.83	20.86	nr	**133.37**
678 x 508mm	58.32	64.15	1.50	17.10	nr	**81.25**

Y:MECHANICAL AND ELECTRICAL SERVICES

Item	Net Price £	Material £	Labour hours	Labour £	Unit	Total Rate £
Y31 : AIR DUCTLINE ANCILLARIES : DAMPERS						
Volume Control Dampers for Rectangular Section ductwork (including all necessary joints.)						
Damper, Aerofoil Flanged Ends						
Sum of two sides 200mm	17.60	19.36	1.60	18.22	nr	**37.58**
Sum of two sides 250mm	19.68	21.65	1.60	18.22	nr	**39.87**
Sum of two sides 300mm	20.14	22.16	1.60	18.22	nr	**40.38**
Sum of two sides 350mm	20.53	22.58	1.60	18.22	nr	**40.80**
Sum of two sides 400mm	22.75	25.03	1.60	18.22	nr	**43.25**
Sum of two sides 450mm	24.98	27.48	1.60	18.22	nr	**45.70**
Sum of two sides 500mm	25.52	28.07	1.60	18.22	nr	**46.29**
Sum of two sides 550mm	27.75	30.53	1.60	18.22	nr	**48.75**
Sum of two sides 600mm	28.37	31.20	1.70	19.37	nr	**50.57**
Sum of two sides 650mm	30.75	33.82	2.00	22.78	nr	**56.60**
Sum of two sides 700mm	31.37	34.50	2.10	23.93	nr	**58.43**
Sum of two sides 750mm	33.75	37.12	2.10	23.93	nr	**61.05**
Sum of two sides 800mm	34.52	37.97	2.10	23.93	nr	**61.90**
Sum of two sides 850mm	36.98	40.67	2.10	23.93	nr	**64.60**
Sum of two sides 900mm	37.75	41.52	2.20	25.09	nr	**66.61**
Sum of two sides 950mm	40.28	44.31	2.30	26.24	nr	**70.55**
Sum of two sides 1000mm	41.13	45.24	2.30	26.24	nr	**71.48**
Sum of two sides 1100mm	44.59	49.05	2.40	27.38	nr	**76.43**
Sum of two sides 1200mm	48.20	53.02	2.60	29.66	nr	**82.68**
Sum of two sides 1300mm	51.89	57.08	2.80	31.90	nr	**88.98**
Sum of two sides 1400mm	55.73	61.31	3.11	35.37	nr	**96.68**
Sum of two sides 1500mm	59.66	65.62	3.26	37.10	nr	**102.72**
Sum of two sides 1600mm	63.73	70.10	3.46	39.41	nr	**109.51**
Sum of two sides 1700mm	67.88	74.67	3.61	41.12	nr	**115.79**
Sum of two sides 1800mm	72.19	79.40	3.91	44.49	nr	**123.89**
Sum of two sides 1900mm	76.64	84.31	4.20	47.86	nr	**132.17**
Sum of two sides 2000mm	81.18	89.30	4.35	49.52	nr	**138.82**
Sum of two sides 2100mm	196.18	215.80	4.41	50.18	nr	**265.98**
Sum of two sides 2200mm	196.18	215.80	4.41	50.18	nr	**265.98**
Sum of two sides 2300mm	212.10	233.31	4.72	53.73	nr	**287.04**
Sum of two sides 2400mm	212.10	233.31	4.90	55.83	nr	**289.14**
Sum of two sides 2500mm	228.32	251.15	5.00	56.95	nr	**308.10**
Sum of two sides 2600mm	228.32	251.15	5.10	58.11	nr	**309.26**
Sum of two sides 2700mm	245.23	269.75	5.32	60.58	nr	**330.33**
Sum of two sides 2800mm	245.23	269.75	5.52	62.93	nr	**332.68**
Sum of two sides 2900mm	251.62	276.78	5.62	63.99	nr	**340.77**
Sum of two sides 3000mm	254.40	279.84	5.78	65.84	nr	**345.68**
Sum of two sides 3100mm	256.30	281.93	6.02	68.61	nr	**350.54**
Sum of two sides 3200mm	270.65	297.72	6.17	70.31	nr	**368.03**
Fire damper, for Rectangular Section Ductwork (including all necessary joints)						
Damper						
Sum of two sides 200mm	30.60	33.66	1.60	18.22	nr	**51.88**
Sum of two sides 250mm	30.83	33.91	1.60	18.22	nr	**52.13**
Sum of two sides 300mm	30.83	33.91	1.60	18.22	nr	**52.13**
Sum of two sides 350mm	31.06	34.16	1.60	18.22	nr	**52.38**
Sum of two sides 400mm	31.29	34.42	1.60	18.22	nr	**52.64**
Sum of two sides 450mm	32.36	35.60	1.60	18.22	nr	**53.82**

Y:MECHANICAL AND ELECTRICAL SERVICES

Item	Net Price £	Material £	Labour hours	Labour £	Unit	Total Rate £
Y31 : AIR DUCTLINE ANCILLARIES : DAMPERS (contd)						
Fire damper, for Rectangular Section Ductwork (including all necessary joints) (contd)						
Damper (contd)						
Sum of two sides 500mm	33.67	37.04	1.60	18.22	nr	**55.26**
Sum of two sides 550mm	36.13	39.74	1.60	18.22	nr	**57.96**
Sum of two sides 600mm	37.67	41.44	1.70	19.37	nr	**60.81**
Sum of two sides 650mm	39.44	43.38	2.00	22.78	nr	**66.16**
Sum of two sides 700mm	41.05	45.16	2.10	23.93	nr	**69.09**
Sum of two sides 750mm	42.82	47.10	2.10	23.93	nr	**71.03**
Sum of two sides 800mm	44.43	48.88	2.15	24.49	nr	**73.37**
Sum of two sides 850mm	46.28	50.91	2.20	25.09	nr	**76.00**
Sum of two sides 900mm	47.97	52.77	2.30	26.24	nr	**79.01**
Sum of two sides 950mm	49.89	54.88	2.30	26.24	nr	**81.12**
Sum of two sides 1000mm	51.58	56.74	2.40	27.38	nr	**84.12**
Sum of two sides 1100mm	55.27	60.80	2.60	29.66	nr	**90.46**
Sum of two sides 1200mm	59.04	64.94	2.80	31.90	nr	**96.84**
Sum of two sides 1300mm	62.96	69.26	3.11	35.37	nr	**104.63**
Sum of two sides 1400mm	66.88	73.57	3.26	37.10	nr	**110.67**
Sum of two sides 1500mm	70.96	78.05	3.40	38.74	nr	**116.79**
Sum of two sides 1600mm	75.18	82.70	3.46	39.41	nr	**122.11**
Sum of two sides 1700mm	79.41	87.35	3.61	41.12	nr	**128.47**
Sum of two sides 1800mm	83.72	92.09	3.91	44.49	nr	**136.58**
Sum of two sides 1900mm	88.10	96.91	4.20	47.86	nr	**144.77**
Sum of two sides 2000mm	92.63	101.90	4.35	49.52	nr	**151.42**
Sum of two sides 2100mm	279.90	307.89	4.44	50.62	nr	**358.51**
Sum of two sides 2200mm	279.90	307.89	4.57	52.01	nr	**359.90**
Sum of two sides 2300mm	351.55	386.70	4.72	53.73	nr	**440.43**
Sum of two sides 2400mm	351.55	386.70	4.90	55.83	nr	**442.53**
Sum of two sides 2500mm	371.92	409.11	4.95	56.39	nr	**465.50**
Sum of two sides 2600mm	371.92	409.11	5.10	58.11	nr	**467.22**
Sum of two sides 2700mm	392.75	432.03	5.32	60.58	nr	**492.61**
Sum of two sides 2800mm	392.75	432.03	5.52	62.93	nr	**494.96**
Sum of two sides 2900mm	413.43	454.78	5.52	62.93	nr	**517.71**
Sum of two sides 3000mm	413.43	454.78	5.78	65.84	nr	**520.62**
Sum of two sides 3100mm	434.73	478.20	6.02	68.61	nr	**546.81**
Sum of two sides 3200mm	434.73	478.20	6.17	70.31	nr	**548.51**
Dampers, for Circular Section Ductwork; (including all necessary joints.)						
Volume Control Damper (Balancing)						
100mm dia.	31.44	34.59	0.80	9.11	nr	**43.70**
150mm dia.	32.98	36.28	1.11	12.66	nr	**48.94**
200mm dia.	35.13	38.65	1.05	11.96	nr	**50.61**
250mm dia.	37.90	41.69	1.20	13.67	nr	**55.36**
300mm dia.	41.20	45.33	1.35	15.39	nr	**60.72**
355mm dia.	49.51	54.46	1.65	18.80	nr	**73.26**
400mm dia.	49.51	54.46	1.90	21.65	nr	**76.11**
450mm dia.	54.50	59.95	2.10	23.93	nr	**83.88**
500mm dia.	60.12	66.13	2.95	33.60	nr	**99.73**
630mm dia.	80.33	88.37	4.57	52.01	nr	**140.38**

Y:MECHANICAL AND ELECTRICAL SERVICES

Item	Net Price £	Material £	Labour hours	Labour £	Unit	Total Rate £
Fire damper, fusible link (wall frame)						
100mm dia.	29.75	32.73	0.80	9.11	nr	**41.84**
150mm dia.	29.98	32.98	0.80	9.11	nr	**42.09**
200mm dia.	31.44	34.59	0.90	10.25	nr	**44.84**
250mm dia.	34.59	38.05	1.20	13.67	nr	**51.72**
300mm dia.	38.21	42.03	1.35	15.39	nr	**57.42**
355mm dia.	45.59	50.15	1.65	18.80	nr	**68.95**
400mm dia.	45.82	50.40	1.90	21.65	nr	**72.05**
450mm dia.	49.74	54.71	2.10	23.93	nr	**78.64**
500mm dia.	53.81	59.19	2.95	33.60	nr	**92.79**
630mm dia.	66.50	73.15	4.57	52.01	nr	**125.16**
710mm dia.	75.41	82.96	5.21	59.32	nr	**142.28**
800mm dia.	80.33	88.37	5.81	66.22	nr	**154.59**
900mm dia.	89.94	98.94	6.41	73.01	nr	**171.95**
1000mm dia.	99.94	109.93	7.04	80.21	nr	**190.14**
Dampers, for flat oval section ductwork; (including all necessary joints)						
Damper, multi leaf spigot end						
345 x 102mm	45.20	49.72	1.20	13.67	nr	**63.39**
427 x 102mm	47.20	51.92	1.35	15.39	nr	**67.31**
508 x 102mm	49.20	54.12	1.60	18.22	nr	**72.34**
559 x 152mm	55.04	60.55	1.90	21.65	nr	**82.20**
531 x 203mm	58.73	64.61	1.90	21.65	nr	**86.26**
851 x 203mm	67.04	73.74	4.57	52.01	nr	**125.75**
582 x 254mm	64.81	71.29	2.10	23.93	nr	**95.22**
1303 x 254mm	156.52	172.17	6.41	73.01	nr	**245.18**
632 x 305mm	71.11	78.22	2.95	33.60	nr	**111.82**
1275 x 305mm	171.89	189.08	6.41	73.01	nr	**262.09**
765 x 356mm	80.49	88.54	4.57	52.01	nr	**140.55**
1247 x 356mm	197.65	217.41	6.41	73.01	nr	**290.42**
1727 x 365mm	216.86	238.55	8.20	93.36	nr	**331.91**
737 x 406mm	84.18	92.60	4.57	52.01	nr	**144.61**
818 x 406mm	87.25	95.98	5.21	59.32	nr	**155.30**
978 x 406mm	91.94	101.14	5.52	62.93	nr	**164.07**
1379 x 406mm	214.64	236.10	7.04	80.21	nr	**316.31**
1699 x 406mm	226.86	249.54	8.26	94.13	nr	**343.67**
709 x 457mm	89.33	98.26	4.50	51.31	nr	**149.57**
1671 x 457mm	240.85	264.93	8.13	92.60	nr	**357.53**
678 x 508mm	87.64	96.40	4.57	52.01	nr	**148.41**
Fire damper, shutter type with frame						
345 x 102mm	33.21	36.53	1.20	13.67	nr	**50.20**
427 x 102mm	35.90	39.49	1.35	15.39	nr	**54.88**
508 x 102mm	38.51	142.62	-	-	nr	**142.62**
559 x 152mm	41.51	45.66	1.90	21.65	nr	**67.31**
531 x 203mm	41.74	45.92	1.90	21.65	nr	**67.57**
831 x 203mm	51.35	56.49	4.57	52.01	nr	**108.50**
582 x 254mm	44.74	49.22	2.10	23.93	nr	**73.15**
1303 x 254mm	116.93	128.62	6.41	73.01	nr	**201.63**
632 x 305mm	50.05	55.05	2.95	33.60	nr	**88.65**
1275 x 305mm	128.77	141.64	6.41	73.01	nr	**214.65**
765 x 356mm	56.96	62.66	4.57	52.01	nr	**114.67**
1247 x 356mm	134.53	147.98	6.41	73.01	nr	**220.99**
1727 x 356mm	155.59	171.15	8.20	93.36	nr	**264.51**
737 x 406mm	57.50	63.25	4.57	52.01	nr	**115.26**
818 x 406mm	60.81	66.89	5.21	59.32	nr	**126.21**

Y:MECHANICAL AND ELECTRICAL SERVICES

Item	Net Price £	Material £	Labour hours	Labour £	Unit	Total Rate £
Y31 : AIR DUCTLINE ANCILLARIES : DAMPERS (contd)						
Dampers, for flat oval section ductwork; (including all necessary joints) (contd)						
Fire damper, shutter type with frame (contd)						
978 x 406mm	65.73	72.30	5.52	62.93	nr	**135.23**
1379 x 406mm	147.83	162.61	7.04	80.21	nr	**242.82**
1699 x 406mm	157.59	173.35	8.20	93.36	nr	**266.71**
709 x 457mm	59.81	65.79	4.50	51.31	nr	**117.10**
1671 x 457mm	164.05	180.46	8.13	92.60	nr	**273.06**
678 x 508mm	60.27	66.30	4.57	52.01	nr	**118.31**

Y:MECHANICAL AND ELECTRICAL SERVICES

Item	Net Price £	Material £	Labour hours	Labour £	Unit	Total Rate £
Y31 : AIR DUCTLINE ANCILLARIES : ACCESS OPENINGS, DOORS AND COVERS						
Access Doors, Cam type						
Rectangular Duct						
150 x 150mm	8.07	8.88	1.25	14.24	nr	**23.12**
200 x 200mm	8.92	9.81	1.25	14.24	nr	**24.05**
300 x 150mm	9.15	10.06	1.25	14.24	nr	**24.30**
300 x 300mm	10.46	11.50	1.25	14.24	nr	**25.74**
400 x 400mm	12.15	13.36	1.35	15.39	nr	**28.75**
450 x 300mm	11.92	8.74	1.50	17.10	nr	**25.84**
450 x 450mm	13.15	14.46	1.50	17.10	nr	**31.56**
Access Doors, Hinge/Cam type						
Rectangular Duct						
150 x 150mm	8.07	8.88	1.25	14.24	nr	**23.12**
200 x 200mm	8.92	9.81	1.25	14.24	nr	**24.05**
300 x 150mm	9.15	10.06	1.25	14.24	nr	**24.30**
300 x 300mm	10.46	11.50	1.25	14.24	nr	**25.74**
400 x 400mm	12.15	13.36	1.35	15.39	nr	**28.75**
450 x 300mm	11.92	13.11	1.50	17.10	nr	**30.21**
450 x 450mm	13.15	14.46	1.50	17.10	nr	**31.56**
Test Holes; Sealed; Rubber Grommet						
Any Size	3.15	3.46	0.50	5.69	nr	**9.15**

Y:MECHANICAL AND ELECTRICAL SERVICES

Item	Net Price £	Material £	Labour hours	Labour £	Unit	Total Rate £
Y46 : GRILLES/DIFFUSERS/LOUVRES : GRILLES						
Single Deflection Grille; Aluminium; Adjustable horizontal blade; screwed fittings						
Extract or Supply; Rectangular Without Volume Control						
150 x 150mm	9.23	10.15	0.88	10.02	nr	20.17
200 x 150mm	10.15	11.16	0.88	10.00	nr	21.16
200 x 200mm	11.38	12.52	1.05	11.93	nr	24.45
300 x 100mm	10.15	11.16	1.05	11.93	nr	23.09
300 x 150mm	11.99	13.19	1.50	17.06	nr	30.25
300 x 200mm	13.76	15.14	1.75	19.89	nr	35.03
300 x 300mm	17.37	19.11	2.10	23.87	nr	42.98
400 x 100mm	11.38	12.52	2.10	23.87	nr	36.39
400 x 150mm	13.76	15.14	2.10	23.87	nr	39.01
400 x 200mm	16.22	17.84	2.20	25.02	nr	42.86
400 x 300mm	20.99	23.09	2.70	30.70	nr	53.79
600 x 200mm	20.99	23.09	3.51	39.86	nr	62.95
600 x 300mm	28.29	31.12	3.51	39.86	nr	70.98
600 x 400mm	35.52	39.07	3.70	42.07	nr	81.14
600 x 500mm	42.74	47.02	3.80	43.19	nr	90.21
600 x 600mm	50.05	55.05	4.00	45.44	nr	100.49
800 x 400mm	45.20	49.72	4.41	50.04	nr	99.76
800 x 600mm	64.50	70.95	4.50	51.17	nr	122.12
1000 x 300mm	42.74	47.02	5.00	56.80	nr	103.82
1000 x 400mm	54.89	60.38	5.32	60.42	nr	120.80
1000 x 600mm	79.03	86.93	5.62	63.82	nr	150.75
1000 x 800mm	103.17	113.48	5.92	67.22	nr	180.70
1200 x 600mm	94.86	104.35	6.33	71.90	nr	176.25
1200 x 800mm	123.77	136.15	6.62	75.23	nr	211.38
1200 x 1000mm	152.83	168.11	7.04	80.00	nr	248.11
With Volume Control						
150 x 150mm	17.07	18.77	0.72	8.18	nr	26.95
200 x 150mm	18.68	20.55	0.83	9.44	nr	29.99
200 x 200mm	20.83	22.92	0.90	10.23	nr	33.15
300 x 100mm	18.68	20.55	0.90	10.23	nr	30.78
300 x 200mm	21.91	24.10	1.06	12.05	nr	36.15
300 x 300mm	31.52	34.67	1.20	13.64	nr	48.31
400 x 100mm	20.83	22.92	1.06	12.05	nr	34.97
400 x 150mm	25.06	27.57	1.13	12.84	nr	40.41
400 x 200mm	29.44	32.39	1.20	13.64	nr	46.03
400 x 300mm	37.90	41.69	1.34	15.23	nr	56.92
600 x 200mm	37.90	41.69	1.50	17.06	nr	58.75
600 x 300mm	50.81	55.90	1.66	18.87	nr	74.77
600 x 400mm	63.65	70.02	1.80	20.47	nr	90.49
600 x 500mm	76.49	84.14	2.00	22.72	nr	106.86
600 x 600mm	89.41	98.35	2.60	29.58	nr	127.93
800 x 300mm	63.65	70.02	2.00	22.72	nr	92.74
800 x 400mm	80.80	88.88	2.60	29.58	nr	118.46
800 x 600mm	115.08	126.59	3.61	41.01	nr	167.60
1000 x 300mm	76.49	84.14	3.00	34.11	nr	118.25
1000 x 400mm	97.94	107.73	3.61	41.01	nr	148.74
1000 x 600mm	140.76	154.83	4.61	52.35	nr	207.18
1000 x 800mm	183.58	201.94	5.62	63.82	nr	265.76

Y:MECHANICAL AND ELECTRICAL SERVICES

Item	Net Price £	Material £	Labour hours	Labour £	Unit	Total Rate £
1200 x 600mm	168.97	185.87	3.51	39.86	nr	**225.73**
1200 x 800mm	220.25	242.27	3.51	39.86	nr	**282.13**
1200 x 1000mm	271.83	299.01	9.01	102.34	nr	**401.35**
Double Deflection Grille; Aluminium; Adjustable horizontal and vertical blade; screwed fittings						
Extract or Supply; Rectangular Without Volume Control						
150 x 150mm	10.92	12.01	0.62	7.05	nr	**19.06**
200 x 150mm	12.15	13.36	0.69	7.84	nr	**21.20**
200 x 200mm	13.84	15.22	0.72	8.18	nr	**23.40**
300 x 150mm	14.68	16.15	0.72	8.18	nr	**24.33**
300 x 200mm	17.14	18.86	0.82	9.32	nr	**28.18**
300 x 300mm	22.14	24.35	0.88	10.00	nr	**34.35**
400 x 100mm	13.84	15.22	1.00	11.36	nr	**26.58**
400 x 150mm	17.14	18.86	0.88	10.00	nr	**28.86**
400 x 200mm	20.45	22.49	0.94	10.69	nr	**33.18**
400 x 300mm	27.06	29.77	1.12	12.74	nr	**42.51**
600 x 200mm	27.06	29.77	1.26	14.32	nr	**44.09**
600 x 300mm	36.98	40.67	1.40	15.91	nr	**56.58**
600 x 400mm	46.97	51.67	1.50	17.06	nr	**68.73**
600 x 500mm	56.89	62.58	1.76	20.00	nr	**82.58**
600 x 600mm	66.80	73.48	1.10	12.50	nr	**85.98**
800 x 300mm	46.97	51.67	1.76	20.00	nr	**71.67**
800 x 400mm	60.19	66.21	1.10	12.50	nr	**78.71**
800 x 600mm	86.64	95.30	3.00	34.11	nr	**129.41**
1000 x 300mm	56.89	62.58	2.40	27.31	nr	**89.89**
1000 x 400mm	73.42	80.76	3.00	34.11	nr	**114.87**
1000 x 600mm	106.55	117.20	3.80	43.19	nr	**160.39**
1000 x 800mm	139.61	153.57	4.61	52.35	nr	**205.92**
1200 x 600mm	127.84	140.63	4.61	52.35	nr	**192.98**
1200 x 800mm	167.51	184.26	6.02	68.43	nr	**252.69**
1200 x 1000mm	207.26	227.98	8.00	90.88	nr	**318.86**
With Volume Control						
150 x 150mm	18.76	20.63	0.72	8.18	nr	**28.81**
200 x 150mm	20.68	22.75	0.83	9.44	nr	**32.19**
200 x 200mm	23.29	25.62	0.90	10.23	nr	**35.85**
300 x 100mm	20.68	22.75	0.90	10.23	nr	**32.98**
300 x 150mm	24.60	27.06	0.98	11.14	nr	**38.20**
300 x 200mm	28.44	31.29	1.06	12.05	nr	**43.34**
300 x 300mm	36.29	39.91	1.20	13.64	nr	**53.55**
400 x 100mm	23.29	25.62	1.06	12.06	nr	**37.68**
400 x 150mm	28.44	31.29	1.13	12.84	nr	**44.13**
400 x 200mm	33.67	37.04	1.20	13.64	nr	**50.68**
400 x 300mm	43.97	48.37	1.34	15.23	nr	**63.60**
600 x 200mm	43.97	48.37	1.50	17.06	nr	**65.43**
600 x 300mm	59.50	65.45	1.66	18.87	nr	**84.32**
600 x 400mm	75.11	82.62	1.80	20.47	nr	**103.09**
600 x 500mm	90.64	99.70	2.00	22.72	nr	**122.42**
600 x 600mm	106.16	116.78	2.60	29.58	nr	**146.36**
800 x 300mm	75.11	82.62	2.00	22.72	nr	**105.34**
800 x 400mm	95.79	105.36	2.60	29.58	nr	**134.94**
800 x 600mm	137.22	150.94	3.61	41.01	nr	**191.95**
1000 x 300mm	90.64	99.70	3.00	34.11	nr	**133.81**
1000 x 400mm	116.47	128.11	3.61	41.01	nr	**169.12**
1000 x 600mm	168.28	185.11	4.61	52.35	nr	**237.46**

Y:MECHANICAL AND ELECTRICAL SERVICES

Item	Net Price £	Material £	Labour hours	Labour £	Unit	Total Rate £
Y46 : GRILLES/DIFFUSERS/LOUVRES : GRILLES (contd)						
Double Deflection Grille; Aluminium; Adjustable horizontal and vertical blade; screwed fittings (contd)						
Extract or Supply; Rectangular With Volume Control (contd)						
1000 x 800mm	220.02	242.02	5.62	63.82	nr	**305.84**
1200 x 600mm	201.95	222.15	5.62	63.82	nr	**285.97**
1200 x 800mm	263.99	290.39	3.51	39.86	nr	**330.25**
1200 x 1000mm	326.18	358.80	9.01	102.34	nr	**461.14**
UPVC Plastic Grilles						
Supply or Extract Eggcrate Grille; fixed blade						
150 x 150mm	5.54	6.09	0.62	7.05	nr	**13.14**
200 x 200mm	7.30	8.03	0.72	8.18	nr	**16.21**
250 x 250mm	8.61	9.47	0.80	9.09	nr	**18.56**
300 x 300mm	11.15	12.26	1.00	11.36	nr	**23.62**
315 x 315mm	8.15	8.96	1.05	11.93	nr	**20.89**
Supply or Extact Grille; 45 Degree Fixed Blade						
150 x 150mm	5.15	5.67	0.62	7.05	nr	**12.72**
200 x 200mm	6.84	7.53	0.72	8.18	nr	**15.71**
250 x 250mm	7.23	7.95	0.80	9.09	nr	**17.04**
300 x 300mm	9.46	10.40	1.00	11.36	nr	**21.76**
315 x 315mm	7.23	7.95	1.05	11.93	nr	**19.88**
Supply or Extract Grille; 0 Degree fixed blade						
150 x 150mm	5.15	5.67	0.62	7.05	nr	**12.72**
200 x 200mm	6.84	7.53	0.72	8.18	nr	**15.71**
250 x 250mm	7.23	7.95	0.80	9.09	nr	**17.04**
300 x 300mm	9.46	10.40	1.00	11.36	nr	**21.76**
315 x 315mm	7.23	7.95	1.05	11.93	nr	**19.88**
Supply or Extract Deflection Grille: Single Adjustable Blade						
150 x 150mm	5.15	5.67	0.62	7.05	nr	**12.72**
200 x 200mm	6.84	7.53	0.72	8.18	nr	**15.71**
250 x 250mm	7.23	7.95	0.80	9.09	nr	**17.04**
300 x 300mm	9.46	10.40	1.00	11.36	nr	**21.76**
315 x 315mm	7.23	7.95	1.05	11.93	nr	**19.88**
Supply or Extract Deflection Grille; Double Adjustable Blades						
150 x 150mm	6.00	6.60	0.62	7.05	nr	**13.65**
200 x 200mm	7.61	8.37	0.72	8.18	nr	**16.55**
250 x 250mm	10.30	11.33	0.80	9.09	nr	**20.42**
300 x 300mm	14.15	15.56	1.00	11.36	nr	**26.92**
315 x 315mm	10.30	11.33	1.05	11.93	nr	**23.26**

Y:MECHANICAL AND ELECTRICAL SERVICES

Item	Net Price £	Material £	Labour hours	Labour £	Unit	Total Rate £
Supply or Extract Door Transfer Grille						
150 x 150mm	10.22	11.25	0.62	7.05	nr	**18.30**
200 x 200mm	12.76	14.04	0.72	8.18	nr	**22.22**
250 x 250mm	14.53	15.98	0.80	9.09	nr	**25.07**
300 x 300mm	18.83	20.72	1.00	11.36	nr	**32.08**
315 x 315mm	13.68	15.05	1.05	11.93	nr	**26.98**
Extra Over Supply or Extract Upvc Plastic Grilles						
Grille Damper						
150 x 150mm	3.92	4.31	0.30	3.41	nr	**7.72**
200 x 200mm	5.07	5.58	0.30	3.41	nr	**8.99**
250 x 250mm	6.00	6.60	0.30	3.41	nr	**10.01**
300 x 300mm	7.30	8.03	0.50	5.68	nr	**13.71**
315 x 315mm	6.00	6.60	0.50	5.68	nr	**12.28**
Grille Neck Reducer						
150 x 150mm	4.69	5.16	0.80	9.09	nr	**14.25**
200 x 200mm	5.15	5.67	0.80	9.09	nr	**14.76**
250 x 250mm	6.00	6.60	0.80	9.09	nr	**15.69**
300 x 300mm	7.30	8.03	0.80	9.09	nr	**17.12**
315 x 315mm	6.00	6.60	0.80	9.09	nr	**15.69**

Y:MECHANICAL AND ELECTRICAL SERVICES

Item	Net Price £	Material £	Labour hours	Labour £	Unit	Total Rate £
Y46 : GRILLES/DIFFUSERS/LOUVRES : DIFFUSERS						
Ceiling Mounted Diffusers; circular multi-core aluminium diffuser fitted to supply or extract ductwork						
Without Volume Control						
150mm dia.	58.27	64.10	0.80	9.09	nr	73.19
203mm dia.	69.73	76.70	1.10	12.50	nr	89.20
305mm dia.	98.55	108.41	1.40	15.91	nr	124.32
381mm dia.	124.38	136.82	1.50	17.06	nr	153.88
457mm dia.	213.87	235.25	2.00	22.72	nr	257.97
With Volume Control						
150mm dia.	78.10	85.92	1.00	11.36	nr	97.28
203mm dia.	88.87	97.75	1.20	13.64	nr	111.39
305mm dia.	120.46	132.51	1.60	18.18	nr	150.69
381mm dia.	149.45	164.39	1.90	21.60	nr	185.99
457mm dia.	241.39	265.53	2.40	27.31	nr	292.84
Ceiling Mounted Diffusers; rectangular / square flush mounted aluminium multi cone diffuser, four way flow to supply or extract ductwork						
Without Volume Control						
150 x 150 nominal neck size	28.60	31.46	1.80	20.47	nr	51.93
300 x 150 nominal neck size	32.83	36.11	2.35	26.73	nr	62.84
300 x 300 nominal neck size	41.44	45.58	2.80	31.82	nr	77.40
450 x 150 nominal neck size	37.13	40.84	2.80	31.82	nr	72.66
450 x 300 nominal neck size	50.05	55.05	3.11	35.28	nr	90.33
450 x 450 nominal neck size	62.88	69.17	3.40	38.64	nr	107.81
600 x 150 nominal neck size	41.44	45.58	3.11	35.28	nr	80.86
600 x 300 nominal neck size	58.58	64.44	3.40	38.64	nr	103.08
600 x 600 nominal neck size	92.87	102.15	4.00	45.44	nr	147.59
900 x 300 nominal neck size	75.72	83.29	4.00	45.44	nr	128.73
With Volume Control						
150 x 150 nominal neck size	36.44	40.08	1.80	20.47	nr	60.55
300 x 150 nominal neck size	42.74	47.02	2.35	26.73	nr	73.75
300 x 300 nominal neck size	55.58	61.14	2.80	31.82	nr	92.96
450 x 150 nominal neck size	49.20	54.12	2.80	31.82	nr	85.94
450 x 300 nominal neck size	68.42	75.26	3.11	35.28	nr	110.54
450 x 450 nominal neck size	87.56	96.32	3.51	39.86	nr	136.18
600 x 150 nominal neck size	55.58	61.14	3.11	35.28	nr	96.42
600 x 300 nominal neck size	81.10	89.21	3.51	39.86	nr	129.07
600 x 600 nominal neck size	132.23	145.45	5.62	63.82	nr	209.27
900 x 300 nominal neck size	106.70	117.37	5.62	63.82	nr	181.19

Y:MECHANICAL AND ELECTRICAL SERVICES

Item	Net Price £	Material £	Labour hours	Labour £	Unit	Total Rate £
Slot Diffusers; continuous aluminium slot diffuser with flanged frame in 1.5m lengths fitted to plenum chamber						
Diffuser						
1 slot	37.28	41.01	3.76	42.71	m	**83.72**
2 slot	50.89	55.98	3.76	42.71	m	**98.69**
3 slot	64.57	71.03	3.76	42.71	m	**113.74**
4 slot	78.18	86.00	4.50	51.17	m	**137.17**
6 slot	105.47	116.02	4.50	51.17	m	**167.19**
8 slot	132.69	145.95	4.50	51.17	m	**197.12**
Diffuser with equalizing deflectors						
1 slot	52.35	57.59	5.26	59.79	m	**117.38**
2 slot	65.96	72.55	5.26	59.79	m	**132.34**
3 slot	79.64	87.61	5.26	59.79	m	**147.40**
4 slot	93.25	102.57	6.33	71.90	m	**174.47**
6 slot	120.54	132.59	6.33	71.90	m	**204.49**
8 slot	147.75	162.53	6.33	71.90	m	**234.43**
Ends						
1 slot	5.15	5.67	1.00	11.36	Pair	**17.03**
2 slot	5.46	6.00	1.00	11.36	Pair	**17.36**
3 slot	5.77	6.34	1.00	11.36	Pair	**17.70**
4 slot	6.15	6.76	1.30	14.77	Pair	**21.53**
6 slot	6.77	7.44	1.40	15.91	Pair	**23.35**
8 slot	7.46	8.20	1.50	17.06	Pair	**25.26**
Plenum boxes 1.0m long						
1 slot	34.98	38.48	2.75	31.29	m	**69.77**
2 slot	34.98	38.48	2.75	31.29	m	**69.77**
3 slot	38.28	42.11	2.75	31.29	m	**73.40**
4 slot	38.28	42.11	3.51	39.86	m	**81.97**
6 slot	41.90	46.09	3.51	39.86	m	**85.95**
8 slot	41.90	46.09	3.51	39.86	m	**85.95**
Plenum boxes 2.0m long						
1 slot	34.98	38.48	3.26	37.00	m	**75.48**
2 slot	42.43	46.68	3.26	37.00	m	**83.68**
3 slot	47.43	52.18	3.26	37.00	m	**89.18**
4 slot	47.43	52.18	3.76	42.71	m	**94.89**
6 slot	51.05	56.15	3.76	42.71	m	**98.86**
8 slot	51.05	56.15	3.76	42.71	m	**98.86**
Perforated Diffusers; aluminium perforated diffuser; quick release face plate; fitted to supply or extract ductwork						
Circular; without volume control						
150mm dia.	46.66	51.33	0.80	9.09	nr	**60.42**
300mm dia.	86.33	94.96	1.40	15.91	nr	**110.87**

Y:MECHANICAL AND ELECTRICAL SERVICES

Item	Net Price £	Material £	Labour hours	Labour £	Unit	Total Rate £
Y46 : GRILLES/DIFFUSERS/LOUVRES : DIFFUSERS (contd)						
Perforated Diffusers; aluminium perforated diffuser; quick release face plate; fitted to supply or extract ductwork (contd)						
Circular; with volume control						
150mm dia.	65.27	71.79	1.00	11.36	nr	**83.15**
300mm dia.	105.24	115.77	1.60	18.18	nr	**133.95**
Rectangular; without volume control						
150 x 150mm dia.; 300 x 300mm face size	49.12	54.04	1.00	11.36	nr	**65.40**
300 x 150mm dia.; 600 x 300mm face size	56.81	62.49	1.00	11.36	nr	**73.85**
Rectangular; with volume control						
150 x 150mm dia.; 300 x 300mm face size	63.65	70.02	1.40	15.91	nr	**85.93**
300 x 150mm dia.; 600 x 300mm face size	67.04	73.74	1.40	15.91	nr	**89.65**
300 x 300mm dia.; 600 x 600mm face size	101.78	111.96	1.40	15.91	nr	**127.87**
Upvc Diffusers						
Cellular						
300 x 300mm	15.61	17.17	2.80	31.82	nr	**48.99**
600 x 600mm	33.82	37.21	4.00	45.44	nr	**82.65**
Multicone						
300 x 300mm	15.61	17.17	2.80	31.82	nr	**48.99**
450 x 450mm	23.37	25.71	3.40	38.64	nr	**64.35**
500 x 500mm	23.37	25.71	3.80	43.19	nr	**68.90**
600 x 600mm	33.82	37.21	4.00	45.44	nr	**82.65**
625 x 625mm	33.82	37.21	4.26	48.34	nr	**85.55**
Extra Over Upvc Diffusers						
Opposed Blade Damper						
300 x 300mm	4.38	4.82	1.20	13.64	nr	**18.46**
450 x 450mm	6.92	7.61	1.50	17.06	nr	**24.67**
600 x 600mm	13.91	15.31	2.60	29.58	nr	**44.89**
Plenum Box						
300mm	6.92	7.61	2.80	31.82	nr	**39.43**
450mm	10.38	11.42	3.40	38.64	nr	**50.06**
600mm	13.91	15.31	4.00	45.44	nr	**60.75**
Plenum Spigot Reducer						
600mm	5.23	5.75	1.00	11.36	nr	**17.11**
Cellular Diffuser; Blanking						
300mm	3.46	3.81	0.88	10.00	nr	**13.81**
600mm	5.23	5.75	1.10	12.50	nr	**18.25**

Y:MECHANICAL AND ELECTRICAL SERVICES

Item	Net Price £	Material £	Labour hours	Labour £	Unit	Total Rate £
Multicone Diffuser; Blanking						
300mm	3.46	3.81	0.88	10.00	nr	**13.81**
450mm	4.38	4.82	0.90	10.23	nr	**15.05**
600mm	5.23	5.75	1.10	12.50	nr	**18.25**

Y:MECHANICAL AND ELECTRICAL SERVICES

Item	Net Price £	Material £	Labour hours	Labour £	Unit	Total Rate £
Y46 : GRILLES/DIFFUSERS/LOUVRES : EXTERNAL LOUVRES						
Weather Louvres; opening mounted; 300mm deep galvanised steel louvres; (screw fixing in positiion.)						
Louvre Units; Louvre, 12mm galvanised mesh birdscreen						
900 x 600	157.67	173.44	2.25	25.59	nr	**199.03**
900 x 900	213.33	234.66	2.25	25.59	nr	**260.25**
900 x 1200	264.37	290.81	2.50	28.40	nr	**319.21**
900 x 1500	312.42	343.66	2.50	28.40	nr	**372.06**
900 x 1800	358.08	393.89	2.50	28.40	nr	**422.29**
900 x 2100	493.31	542.64	2.50	28.40	nr	**571.04**
900 x 2400	528.67	581.54	2.76	31.38	nr	**612.92**
900 x 2700	591.17	650.29	2.76	31.38	nr	**681.67**
900 x 3000	624.84	687.32	2.76	31.38	nr	**718.70**
1200 x 600	191.96	211.15	2.25	25.59	nr	**236.74**
1200 x 900	259.91	285.91	2.50	28.40	nr	**314.31**
1200 x 1200	322.34	354.57	2.50	28.40	nr	**382.97**
1200 x 1500	381.15	419.26	2.50	28.40	nr	**447.66**
1200 x 1800	437.11	480.82	2.76	31.38	nr	**512.20**
1200 x 2100	601.16	661.28	2.76	31.38	nr	**692.66**
1200 x 2400	644.67	709.14	2.76	31.38	nr	**740.52**
1500 x 600	223.78	246.16	2.25	25.59	nr	**271.75**
1500 x 900	303.12	333.43	2.50	28.40	nr	**361.83**
1500 x 1200	376.15	413.76	2.50	28.40	nr	**442.16**
1500 x 1500	444.95	489.45	2.76	31.38	nr	**520.83**
1500 x 1800	510.68	561.75	2.76	31.38	nr	**593.13**
1500 x 2100	701.10	771.21	3.00	34.11	nr	**805.32**
1800 x 600	253.69	279.06	2.50	28.40	nr	**307.46**
1800 x 900	343.86	378.25	2.50	28.40	nr	**406.65**
1800 x 1200	426.96	469.66	2.76	31.38	nr	**501.04**
1800 x 1500	505.38	555.91	3.00	34.11	nr	**590.02**

Y:MECHANICAL AND ELECTRICAL SERVICES

Item	Net Price £	Material £	Labour hours	Labour £	Unit	Total Rate £
Y50 : THERMAL INSULATION :PREFORMED RIGID SECTIONS AND SLABS						
Rigid Mineral Fibre, Preformed sections, wire fixings, aluminium sheeting fastened with screws in the running length. Horizontal joints sealed with water resistant sealant						
Plant Rooms; Pipework dia. 3/4 inch insulation thickness						
25mm	2.43	2.68	0.45	5.12	m	**7.80**
40mm	3.83	4.22	0.46	5.23	m	**9.45**
50mm	5.51	6.06	0.54	6.14	m	**12.20**
80mm	12.71	13.99	0.54	6.14	m	**20.13**
Plant Rooms; Pipework dia. 1 inch insulation thickness						
25mm	2.65	2.91	0.45	5.12	m	**8.03**
40mm	4.15	4.56	0.46	5.23	m	**9.79**
50mm	5.85	6.44	0.54	6.14	m	**12.58**
80mm	12.35	13.59	0.54	6.14	m	**19.73**
Plant Rooms; Pipework dia. 1 1/2 inch insulation thickness						
25mm	3.10	3.41	0.46	5.23	m	**8.64**
40mm	4.64	5.10	0.46	5.23	m	**10.33**
50mm	6.45	7.10	0.46	5.23	m	**12.33**
80mm	13.36	14.70	0.54	6.14	m	**20.84**
100mm	15.61	17.17	0.54	6.14	m	**23.31**
Plant Rooms; Pipework dia. 2 inch insulation thickness						
25mm	3.57	3.93	0.46	5.23	m	**9.16**
40mm	5.27	5.80	0.46	5.23	m	**11.03**
50mm	7.24	7.96	0.46	5.23	m	**13.19**
80mm	14.38	15.81	0.54	6.14	m	**21.95**
100mm	17.82	19.61	0.54	6.14	m	**25.75**
Plant Rooms; Pipework dia. 2 1/2 inch insulation thickness						
25mm	4.09	4.50	0.48	5.45	m	**9.95**
40mm	5.94	6.54	0.49	5.57	m	**12.11**
50mm	8.52	9.37	0.49	5.57	m	**14.94**
80mm	15.57	17.13	0.58	6.59	m	**23.72**
100mm	20.95	23.04	0.59	6.71	m	**29.75**
Plant Rooms; Pipework dia. 3 inch insulation thickness						
25mm	4.58	5.04	0.48	5.45	m	**10.49**
40mm	6.45	7.10	0.49	5.57	m	**12.67**
50mm	8.52	9.37	0.49	5.57	m	**14.94**
80mm	16.63	18.29	0.58	6.59	m	**24.88**
100mm	23.50	25.85	0.59	6.71	m	**32.56**
120mm	31.19	34.31	0.63	7.16	m	**41.47**

Y:MECHANICAL AND ELECTRICAL SERVICES

Item	Net Price £	Material £	Labour hours	Labour £	Unit	Total Rate £
Y50 : THERMAL INSULATION :PREFORMED RIGID SECTIONS AND SLABS (contd)						
Rigid Mineral Fibre, Preformed sections, wire fixings, aluminium sheeting fastened with screws in the running length. Horizontal joints sealed with water resistant sealant (contd)						
Plant Rooms; Pipework dia. 4 inch insulation thickness						
25mm	5.95	6.55	0.48	5.45	m	**12.00**
40mm	8.38	9.22	0.49	5.57	m	**14.79**
50mm	10.95	12.04	0.49	5.57	m	**17.61**
80mm	20.69	22.75	0.58	6.59	m	**29.34**
100mm	27.61	30.37	0.59	6.71	m	**37.08**
120mm	34.67	38.13	0.63	7.16	m	**45.29**
Plant Rooms; Pipework dia. 6 inch insulation thickness						
25mm	8.35	9.19	0.55	6.25	m	**15.44**
40mm	11.25	12.38	0.55	6.25	m	**18.63**
50mm	14.30	15.73	0.66	7.50	m	**23.23**
80mm	24.99	27.49	0.66	7.50	m	**34.99**
100mm	33.75	37.12	0.67	7.61	m	**44.73**
120mm	40.49	44.54	0.69	7.84	m	**52.38**
(Fittings to include, Elbows, Tees, Reducers and the like)						
Plant Rooms; fittings dia. 3/4 inch insulation thickness						
25mm	0.24	0.26	0.63	7.16	nr	**7.42**
40mm	0.34	0.37	0.66	7.50	nr	**7.87**
50mm	0.43	0.48	0.66	7.50	nr	**7.98**
80mm	0.46	0.51	0.91	10.35	nr	**10.86**
Plant Rooms; fittings dia. 1 inch insulation thickness						
25mm	0.26	0.28	0.63	7.16	nr	**7.44**
40mm	0.36	0.40	0.66	7.50	nr	**7.90**
50mm	0.45	0.50	0.68	7.73	nr	**8.23**
80mm	0.46	0.51	0.91	10.35	nr	**10.86**
Plant Rooms; fittings dia. 1 1/2 inch insulation thickness						
25mm	0.30	0.33	0.64	7.27	nr	**7.60**
40mm	0.43	0.48	0.66	7.50	nr	**7.98**
50mm	0.53	0.58	0.67	7.61	nr	**8.19**
80mm	0.50	0.56	0.93	10.57	nr	**11.13**
100mm	0.60	0.66	0.99	11.25	nr	**11.91**

Y:MECHANICAL AND ELECTRICAL SERVICES

Item	Net Price £	Material £	Labour hours	Labour £	Unit	Total Rate £
Plant Rooms; fittings dia. 2 inch insulation thickness						
25mm	0.38	0.42	0.66	7.50	nr	**7.92**
40mm	0.52	0.57	0.69	7.84	nr	**8.41**
50mm	0.62	0.68	0.71	8.07	nr	**8.75**
80mm	0.56	0.61	0.97	11.03	nr	**11.64**
100mm	0.71	0.78	1.04	11.82	nr	**12.60**
Plant Rooms; fittings dia. 2 1/2 inch insulation thickness						
25mm	0.26	0.28	0.81	9.21	nr	**9.49**
40mm	0.36	0.40	0.82	9.32	nr	**9.72**
50mm	0.40	0.44	0.85	9.66	nr	**10.10**
80mm	0.44	0.49	1.10	12.50	nr	**12.99**
100mm	0.56	0.61	1.17	13.30	nr	**13.91**
Plant Rooms; fittings dia. 3 inch insulation thickness						
25mm	0.30	0.33	0.81	9.21	nr	**9.54**
40mm	0.37	0.41	0.84	9.55	nr	**9.96**
50mm	0.43	0.48	0.86	9.78	nr	**10.26**
80mm	0.47	0.52	1.13	12.84	nr	**13.36**
100mm	0.59	0.65	1.20	13.64	nr	**14.29**
120mm	0.45	0.50	1.53	17.40	nr	**17.90**
Plant Rooms; fittings dia. 4 inch insulation thickness						
25mm	0.35	0.39	0.84	9.55	nr	**9.94**
40mm	0.42	0.46	0.87	9.89	nr	**10.35**
50mm	0.50	0.56	0.89	10.12	nr	**10.68**
80mm	0.53	0.58	1.16	13.18	nr	**13.76**
100mm	0.65	0.71	1.25	14.20	nr	**14.91**
120mm	0.49	0.54	1.60	18.18	nr	**18.72**
Plant Rooms; fittings dia. 6 inch insulation thickness						
25mm	0.28	0.31	1.06	12.05	nr	**12.36**
40mm	0.32	0.35	1.10	12.50	nr	**12.85**
50mm	0.35	0.39	1.34	15.23	nr	**15.62**
80mm	0.37	0.41	1.44	16.37	nr	**16.78**
100mm	0.43	0.48	1.53	17.40	nr	**17.88**
120mm	0.38	0.42	1.92	21.85	nr	**22.27**

Y:MECHANICAL AND ELECTRICAL SERVICES

Item	Net Price £	Material £	Labour hours	Labour £	Unit	Total Rate £
Y50 : THERMAL INSULATION :PREFORMED RIGID SECTIONS AND SLABS (contd)						
Rigid Mineral Fibre, Preformed sections, wire fixings, aluminium sheeting fastened with screws in the running length. Horizontal joints sealed with water resistant sealant (contd)						
Non-Plant Room Areas; Pipework dia. 3/4 inch insulation thickness						
25mm	2.43	2.68	0.39	4.43	m	**7.11**
40mm	3.83	4.22	0.40	4.54	m	**8.76**
50mm	5.51	6.06	0.40	4.54	m	**10.60**
80mm	12.71	13.99	0.47	5.34	m	**19.33**
Non-Plant Rooms; Pipework dia. 1 inch insulation thickness						
25mm	2.65	2.91	0.39	4.43	m	**7.34**
40mm	4.15	4.56	0.40	4.54	m	**9.10**
50mm	5.85	6.44	0.40	4.54	m	**10.98**
80mm	12.35	13.59	0.47	5.34	m	**18.93**
Non-Plant Rooms; Pipework dia. 1 1/2 inch insulation thickness						
25mm	3.10	3.41	0.40	4.54	m	**7.95**
40mm	4.64	5.10	0.40	4.54	m	**9.64**
50mm	6.45	7.10	0.40	4.54	m	**11.64**
80mm	13.36	14.70	0.47	5.34	m	**20.04**
100mm	15.61	17.17	0.47	5.34	m	**22.51**
Non-Plant Rooms; Pipework dia. 2 inch insulation thickness						
25mm	3.57	3.93	0.40	4.54	m	**8.47**
40mm	5.27	5.80	0.40	4.54	m	**10.34**
50mm	7.24	7.96	0.40	4.54	m	**12.50**
80mm	14.38	15.81	0.47	5.34	m	**21.15**
100mm	17.82	19.61	0.47	5.34	m	**24.95**
Non-Plant Rooms; Pipework dia. 2 1/2 inch insulation thickness						
25mm	4.09	4.50	0.42	4.77	m	**9.27**
40mm	5.94	6.54	0.43	4.89	m	**11.43**
50mm	8.52	9.37	0.43	4.89	m	**14.26**
80mm	15.57	17.13	0.50	5.68	m	**22.81**
100mm	20.95	23.04	0.51	5.80	m	**28.84**
Non-Plant Rooms; Pipework dia. 3 inch insulation thickness						
25mm	4.58	5.04	0.42	4.77	m	**9.81**
40mm	6.45	7.10	0.43	4.89	m	**11.99**
50mm	8.52	9.37	0.43	4.89	m	**14.26**
80mm	16.63	18.29	0.50	5.68	m	**23.97**
100mm	23.50	25.85	0.51	5.80	m	**31.65**
120mm	31.19	34.31	0.55	6.25	m	**40.56**

Y:MECHANICAL AND ELECTRICAL SERVICES

Item	Net Price £	Material £	Labour hours	Labour £	Unit	Total Rate £
Non-Plant Rooms; Pipework dia. 4 inch insulation thickness						
25mm	5.95	6.55	0.42	4.77	m	**11.32**
40mm	8.38	9.22	0.43	4.89	m	**14.11**
50mm	10.95	12.04	0.43	4.89	m	**16.93**
80mm	20.69	22.75	0.50	5.68	m	**28.43**
100mm	27.61	30.37	0.50	5.68	m	**36.05**
120mm	34.67	38.13	0.55	6.25	m	**44.38**
Non-Plant Rooms; Pipework dia. 6 inch insulation thickness						
25mm	8.35	9.19	0.48	5.45	m	**14.64**
40mm	11.25	12.38	0.48	5.45	m	**17.83**
50mm	14.30	15.73	0.49	5.57	m	**21.30**
80mm	24.99	27.49	0.57	6.48	m	**33.97**
100mm	33.75	37.12	0.58	6.59	m	**43.71**
120mm	40.49	44.54	0.60	6.82	m	**51.36**
(Fittings to include, Elbows, Tees, Reducers and the like)						
Non-Plant Rooms; fittings dia. 3/4 inch insulation thickness						
25mm	0.24	0.26	0.55	6.25	nr	**6.51**
40mm	0.34	0.37	0.57	6.48	nr	**6.85**
50mm	0.43	0.48	0.58	6.59	nr	**7.07**
80mm	0.46	0.51.	0.79	8.98	nr	**9.49**
Non-Plant Rooms; fittings dia. 1 inch insulation thickness						
25mm	0.26	0.28	0.55	6.25	nr	**6.53**
40mm	0.36	0.40	0.57	6.48	nr	**6.88**
50mm	0.45	0.50	0.59	6.71	nr	**7.21**
80mm	0.46	0.51	0.79	8.98	nr	**9.49**
Non-Plant Rooms; fittings dia. 1 1/2 inch insulation thickness						
25mm	0.30	0.33	0.56	6.36	nr	**6.69**
40mm	0.43	0.48	0.58	6.59	nr	**7.07**
50mm	0.53	0.58	0.59	6.71	nr	**7.29**
80mm	0.50	0.56	0.81	9.21	nr	**9.77**
100mm	0.60	0.66	0.86	9.78	nr	**10.44**
Non-Plant Rooms; fittings dia. 2 inch insulation thickness						
25mm	0.38	0.42	0.58	6.59	nr	**7.01**
40mm	0.52	0.57	0.60	6.82	nr	**7.39**
50mm	0.62	0.68	0.62	7.05	nr	**7.73**
80mm	0.56	0.61	0.84	9.55	nr	**10.16**
100mm	0.71	0.78	9.01	102.34	nr	**103.12**

Y:MECHANICAL AND ELECTRICAL SERVICES

Item	Net Price £	Material £	Labour hours	Labour £	Unit	Total Rate £
Y50 : THERMAL INSULATION :PREFORMED RIGID SECTIONS AND SLABS (contd)						
Rigid Mineral Fibre, Preformed sections, wire fixings, aluminium sheeting fastened with screws in the running length. Horizontal joints sealed with water resistant sealant (contd)						
(Fittings to include, Elbows, Tees, Reducers and the like) (contd)						
Non-Plant Rooms; fittings dia. 2 1/2 inch insulation thickness						
25mm	0.26	0.28	0.70	7.96	nr	**8.24**
40mm	0.36	0.40	0.72	8.18	nr	**8.58**
50mm	0.40	0.44	0.74	8.41	nr	**8.85**
80mm	0.44	0.49	0.96	10.91	nr	**11.40**
100mm	0.56	0.61	1.02	11.59	nr	**12.20**
Non-Plant Rooms; fittings dia. 3 inch insulation thickness						
25mm	0.30	0.33	0.70	7.96	nr	**8.29**
40mm	0.37	0.41	0.73	8.30	nr	**8.71**
50mm	0.43	0.48	0.75	8.52	nr	**9.00**
80mm	0.47	0.52	0.98	11.14	nr	**11.66**
100mm	0.59	0.65	1.04	11.82	nr	**12.47**
120mm	0.45	0.50	1.33	15.13	nr	**15.63**
Non-Plant Rooms; fittings dia. 4 inch insulation thickness						
25mm	0.35	0.39	0.73	8.30	nr	**8.69**
40mm	0.42	0.46	0.76	8.64	nr	**9.10**
50mm	0.50	0.56	0.78	8.86	nr	**9.42**
80mm	0.53	0.58	1.01	11.47	nr	**12.05**
100mm	0.65	0.71	1.09	12.39	nr	**13.10**
120mm	0.49	0.54	1.39	15.80	nr	**16.34**
Non-Plant Rooms; fittings dia. 6 inch insulation thickness						
25mm	0.28	0.31	0.93	10.57	nr	**10.88**
40mm	0.32	0.35	0.96	10.91	nr	**11.26**
50mm	0.35	0.39	0.99	11.25	nr	**11.64**
80mm	0.37	0.41	1.25	14.20	nr	**14.61**
100mm	0.43	0.48	1.33	15.13	nr	**15.61**
120mm	0.38	0.42	1.67	19.00	nr	**19.42**

Y:MECHANICAL AND ELECTRICAL SERVICES

Item	Net Price £	Material £	Labour hours	Labour £	Unit	Total Rate £
Y50 : THERMAL INSULATION : VALVE BOXES						
Rigid Mineral Fibre, Preformed sections, wire fixings, aluminium sheeting fastened with screws in the running length. Horizontal joints sealed with water resistant sealant.						
Plant Rooms; Valve box dia. 3/4 inch insulation thickness						
25mm	8.58	9.44	0.60	6.82	nr	**16.26**
40mm	11.22	12.34	0.62	7.05	nr	**19.39**
50mm	14.10	15.51	0.64	7.27	nr	**22.78**
80mm	19.42	21.36	0.77	8.75	nr	**30.11**
Plant Rooms; Valve box dia. 1 inch insulation thickness						
25mm	9.16	10.07	0.60	6.82	nr	**16.89**
40mm	11.96	13.15	0.62	7.05	nr	**20.20**
50mm	14.82	16.30	0.64	7.27	nr	**23.57**
80mm	18.88	20.77	0.77	8.75	nr	**29.52**
Plant Rooms; Valve box dia. 1 1/2 inch insulation thickness						
25mm	10.68	11.75	0.61	6.93	nr	**18.68**
40mm	13.51	14.87	0.63	7.16	nr	**22.03**
50mm	16.59	18.25	0.66	7.50	nr	**25.75**
80mm	20.58	22.64	0.79	8.98	nr	**31.62**
100mm	24.05	26.46	0.87	9.89	nr	**36.35**
Plant Rooms; Valve box dia. 2 inch insulation thickness						
25mm	12.83	14.12	0.63	7.16	nr	**21.28**
40mm	16.12	17.73	0.66	7.50	nr	**25.23**
50mm	19.39	21.33	0.69	7.84	nr	**29.17**
80mm	23.14	25.46	0.84	9.55	nr	**35.01**
100mm	28.28	31.11	0.92	10.45	nr	**41.56**
Plant Rooms; Valve box dia. 2 1/2 inch insulation thickness						
25mm	13.40	14.74	0.54	6.14	nr	**20.88**
40mm	16.20	17.82	0.58	6.59	nr	**24.41**
50mm	19.08	20.98	0.60	6.82	nr	**27.80**
80mm	26.20	28.82	0.82	9.32	nr	**38.14**
100mm	33.12	36.44	0.91	10.35	nr	**46.79**
Plant Rooms; Valve box dia. 3 inch insulation thickness						
25mm	14.48	15.93	0.54	6.14	nr	**22.07**
40mm	17.37	19.10	0.58	6.59	nr	**25.69**
50mm	20.50	22.55	0.60	6.82	nr	**29.37**
80mm	27.87	30.66	0.84	9.55	nr	**40.21**
100mm	34.90	38.39	0.93	10.57	nr	**48.96**
120mm	38.50	42.35	1.31	14.89	nr	**57.24**

Y:MECHANICAL AND ELECTRICAL SERVICES

Item	Net Price £	Material £	Labour hours	Labour £	Unit	Total Rate £
Y50 : THERMAL INSULATION : VALVE BOXES (contd)						
Rigid Mineral Fibre, Preformed sections, wire fixings, aluminium sheeting fastened with screws in the running length. Horizontal joints sealed with water resistant sealant.(contd)						
Plant Rooms; Valve box dia. 4 inch insulation thickness						
25mm	17.48	19.23	0.58	6.59	nr	**25.82**
40mm	21.18	23.29	0.62	7.05	nr	**30.34**
50mm	24.59	27.04	0.66	7.50	nr	**34.54**
80mm	32.24	35.46	0.88	10.00	nr	**45.46**
100mm	40.16	44.18	1.00	11.36	nr	**55.54**
120mm	42.89	47.18	1.39	15.80	nr	**62.98**
Plant Rooms; Valve box dia. 6 inch insulation thickness						
25mm	21.09	23.20	0.79	8.98	nr	**32.18**
40mm	24.98	27.48	0.85	9.66	nr	**37.14**
50mm	28.28	31.11	0.88	10.00	nr	**41.11**
80mm	36.31	39.94	1.07	12.16	nr	**52.10**
100mm	44.28	48.71	1.20	13.64	nr	**62.35**
120mm	49.71	54.68	1.73	19.65	nr	**74.33**
Non-Plant Rooms; Valve box dia. 3/4 inch insulation thickness						
25mm	8.58	9.44	0.52	5.91	nr	**15.35**
40mm	11.22	12.34	0.54	6.14	nr	**18.48**
50mm	14.10	15.51	0.56	6.36	nr	**21.87**
80mm	19.42	21.36	0.67	7.61	nr	**28.97**
Non-Plant Rooms; Valve box dia. 1 inch insulation thickness						
25mm	9.16	10.07	0.52	5.91	nr	**15.98**
40mm	11.96	13.15	0.55	6.25	nr	**19.40**
50mm	14.82	16.30	0.57	6.48	nr	**22.78**
80mm	18.88	20.77	0.67	7.61	nr	**28.38**
Non-Plant Rooms; Valve box dia. 1 1/2 inch insulation thickness						
25mm	10.68	11.75	0.53	6.02	nr	**17.77**
40mm	13.51	14.87	0.55	6.25	nr	**21.12**
50mm	16.59	18.25	0.57	6.48	nr	**24.73**
80mm	20.58	22.64	0.69	7.84	nr	**30.48**
100mm	24.05	26.46	0.76	8.64	nr	**35.10**
Non-Plant Rooms; Valve box dia. 2 inch insulation thickness						
25mm	12.83	14.12	0.55	6.25	nr	**20.37**
40mm	16.12	17.73	0.58	6.59	nr	**24.32**
50mm	19.39	21.33	0.60	6.82	nr	**28.15**

Y:MECHANICAL AND ELECTRICAL SERVICES

Item	Net Price £	Material £	Labour hours	Labour £	Unit	Total Rate £
80mm	23.14	25.46	0.73	8.30	nr	**33.76**
100mm	28.28	31.11	0.80	9.09	nr	**40.20**
Non-Plant Rooms; Valve box dia. 2 1/2 inch insulation thickness						
25mm	13.40	14.74	0.47	5.34	nr	**20.08**
40mm	16.20	17.82	0.50	5.68	nr	**23.50**
50mm	19.08	20.98	0.52	5.91	nr	**26.89**
80mm	26.20	28.82	0.71	8.07	nr	**36.89**
100mm	33.12	36.44	0.79	8.98	nr	**45.42**
Non-Plant Rooms; Valve box dia. 3 inch insulation thickness						
25mm	14.48	15.93	0.47	5.34	nr	**21.27**
40mm	17.37	19.10	0.50	5.68	nr	**24.78**
50mm	20.50	22.55	0.53	6.02	nr	**28.57**
80mm	27.87	30.66	0.73	8.30	nr	**38.96**
100mm	34.90	38.39	0.81	9.21	nr	**47.60**
120mm	38.50	42.35	1.14	12.95	nr	**55.30**
Non-Plant Rooms; Valve box dia. 4 inch insulation thickness						
25mm	17.48	19.23	0.50	5.68	nr	**24.91**
40mm	21.18	23.29	0.54	6.14	nr	**29.43**
50mm	24.59	27.04	0.57	6.48	nr	**33.52**
80mm	32.24	35.46	0.77	8.75	nr	**44.21**
100mm	40.16	44.18	0.87	9.89	nr	**54.07**
120mm	42.89	47.18	1.21	13.75	nr	**60.93**
Non-Plant Rooms; Valve box dia. 6 inch insulation thickness						
25mm	21.09	23.20	0.69	7.84	nr	**31.04**
40mm	24.98	27.48	0.74	8.41	nr	**35.89**
50mm	28.28	31.11	0.77	8.75	nr	**39.86**
80mm	36.31	39.94	0.93	10.57	nr	**50.51**
100mm	44.28	48.71	1.04	11.82	nr	**60.53**
120mm	49.71	54.68	1.50	17.06	nr	**71.74**

Y:MECHANICAL AND ELECTRICAL SERVICES

Item	Net Price £	Material £	Labour hours	Labour £	Unit	Total Rate £
Y50 : THERMAL INSULATION : INSULATION OF DUCTWORK						
Rigid Mineral Fibre Slabs; foil faced; adhesive and tape fixings; with wire netting; plain sheet aluminium cladding; coloured identification markings						
Plant Rooms; Concealed ductwork; plain sheet aluminium cladding						
40mm thick	7.71	8.49	1.21	13.75	m²	**22.24**
50mm thick	8.48	9.32	1.21	13.75	m²	**23.07**
Non-Plant Rooms; Concealed ductwork; plain sheet aluminium cladding						
40mm thick	7.20	7.92	1.10	12.50	m²	**20.42**
50mm thick	8.24	9.06	1.10	12.50	m²	**21.56**

Keep your figures up to date, free of charge

This section, and most of the other information in this Price Book, is brought up to date every three months, until the next annual edition, in the *Price Book Update*.

The *Update* is available free to all Price Book purchasers.

To ensure you receive your copy, simply complete the reply card from the centre of the book and return it to us.

Y:MECHANICAL AND ELECTRICAL SERVICES

Item	Net Price £	Material £	Labour hours	Labour £	Unit	Total Rate £
Y52 : VIBRATION ISOLATION MOUNTINGS : VIBRATION						
Anti-Vibration Mountings; per set of four; installed with fans or other items of plant						
Sets of mountings for fan 305-601mm dia.						
up to 23 kg per mounting	23.32	25.65	2.00	22.72	Set	**48.37**
23-41 kg per mounting	23.32	25.65	2.00	22.72	Set	**48.37**
Sets of mountings for fan 610-1219mm dia.						
35-115 kg per mounting	77.38	85.12	2.00	22.72	Set	**107.84**
Sets of mountings for fan 762-1219mm dia.						
115-220 kg	93.28	102.61	2.00	22.72	Set	**125.33**
Butterfly Type Damper; (installed for fans of the diameter shown)						
Dampers; Hand Operated						
305mm (dia. of fan)	160.06	176.07	2.00	22.72	nr	**198.79**
483mm (dia. of fan)	202.46	222.71	2.00	22.72	nr	**245.43**
762mm (dia. of fan)	283.02	311.32	2.00	22.72	nr	**334.04**
1219mm (dia. of fan)	579.82	637.80	2.00	22.72	nr	**660.52**
Dampers; Air Operated (Horizontal)						
305mm (dia. of fan)	110.24	121.26	2.00	22.72	nr	**143.98**
483mm (dia. of fan)	140.98	155.08	2.00	22.72	nr	**177.80**
762mm (dia. of fan)	217.30	239.03	2.00	22.72	nr	**261.75**
1219mm (dia. of fan)	478.06	525.87	2.00	22.72	nr	**548.59**

Y:MECHANICAL AND ELECTRICAL SERVICES

Item	Net Price £	Material £	Labour hours	Labour £	Unit	Total Rate £
Y53 : CONTROL COMPONENTS - MECHANICAL : ANCILLARIES						
Room Thermostats; light and medium duty; installed and connected to provide system control						
Range 3 to 27 degree C; 240 Volt						
1 amp; on/off type	16.67	18.33	0.30	3.41	nr	**21.74**
Range 0 to +15 degree C; 240 Volt						
6 amp; frost thermostat	10.94	12.03	0.30	3.41	nr	**15.44**
Range 3 to 27 degree C; 250 Volt						
2 amp; changeover type; dead zone	25.73	28.30	0.30	3.41	nr	**31.71**
2 amp; changeover type	12.95	14.25	0.30	3.41	nr	**17.66**
2 amp; changeover type; concealed setting	15.41	16.95	0.30	3.41	nr	**20.36**
6 amp; on/off type	24.14	26.55	0.30	3.41	nr	**29.96**
6 amp; temperature set-back	19.25	21.17	0.30	3.41	nr	**24.58**
16 amp; on/off type	14.31	15.75	0.30	3.41	nr	**19.16**
16 amp; on/off type; concealed setting	15.61	17.17	0.30	3.41	nr	**20.58**
20 amp; on/off type; concealed setting; industrial non-ventilated cover	17.04	18.74	0.30	3.70	nr	**22.44**
20 amp; indicated "off" position	16.21	17.83	0.30	3.70	nr	**21.53**
20 amp; manual; double pole on/off and neon indicator	31.43	34.57	0.30	3.41	nr	**37.98**
20 amp; indicated "off" position	16.76	18.43	0.30	3.41	nr	**21.84**
Range 10 to 40 degree C; 240 Volt						
20 amp; changeover contacts	18.18	20.00	0.30	3.41	nr	**23.41**
2 amp; 'Heating-Cooling' switch	24.14	26.55	0.30	3.41	nr	**29.96**
Surface Thermostats						
Cylinder Thermostat						
6 amp; changeover type; with cable	11.63	13.11	0.25	2.84	nr	**15.95**
Electrical Thermostats; installed and connected to provide system control						
Range -20 to +70 degree C; 240 Volt						
8 amp; metal base; plastic cover	62.29	68.52	0.30	3.41	nr	**71.93**
16 amp; waterproof; aluminium cover	74.21	81.63	0.30	3.41	nr	**85.04**
Range O to 40 degree C; 240 Volt						
2 stage; internal sensor	73.15	80.47	0.30	3.41	nr	**83.88**
4 stage; heating and/or cooling; (for us with sensor)	101.54	111.70	0.30	3.41	nr	**115.11**
Range 30 to 150 degree C; 240 Volt						
16 amp; waterproof; aluminium cover	74.21	81.63	0.30	3.41	nr	**85.04**

Y:MECHANICAL AND ELECTRICAL SERVICES

Item	Net Price £	Material £	Labour hours	Labour £	Unit	Total Rate £
Sensors; for use with electronic thermostats						
Sensors; for electronic thermostats						
-20 to 70 degree C; nickel plated copper tube	11.99	13.19	0.50	5.68	nr	**18.87**
5 to 45 degree C; wall mounted plastic box	10.59	11.65	0.50	5.68	nr	**17.33**
Radiator Thermostats						
Angled Valve Body; thermostatic head; built in sensor						
15mm; liquid filled	8.80	9.68	0.84	9.55	nr	**19.23**
15mm; wax filled	8.80	9.68	0.84	9.55	nr	**19.23**
Immersion Thermostats; stem type; domestic water boilers; fitted; (Electrical work elsewhere)						
Temperature range 0 to 40 degree C						
Non standard; 280mm stem	6.27	6.90	0.25	2.84	nr	**9.74**
Temperature range 18 to 88 degree C						
13 amp; 178mm stem	4.28	4.71	0.25	2.84	nr	**7.55**
20 amp; 178mm stem	6.46	7.10	0.25	2.84	nr	**9.94**
Non standard; pocket clip; 280mm stem	5.99	6.59	0.25	2.84	nr	**9.43**
Temperature range 40 to 80 degree C						
13 amp; 178mm stem	2.94	3.24	0.25	2.84	nr	**6.08**
20 amp; 178mm stem	4.44	4.89	0.25	2.84	nr	**7.73**
Non standard; pocket clip; 280mm stem	6.73	7.40	0.25	2.84	nr	**10.24**
13 amp; 457mm stem	3.25	3.58	0.25	2.84	nr	**6.42**
20 amp; 457mm stem	4.87	5.35	0.25	2.84	nr	**8.19**
Temperature Range 50 to 100 degree C						
Non standard; 1780mm stem	5.65	6.22	0.25	2.84	nr	**9.06**
Non standard; 280mm stem	5.88	6.47	0.25	2.84	nr	**9.31**
Pockets for thermostats						
For 178mm stem	7.39	8.13	0.25	2.84	nr	**10.97**
For 280mm stem	7.54	8.30	0.25	2.84	nr	**11.14**
Immersion Thermostats; stem type; industrial installations; fitted; (Electrical work elsewhere)						
Temperature range 5 to 105 degree C						
For 305mm stem	95.62	105.19	0.50	5.68	nr	**110.87**

DAVIS LANGDON & EVEREST
CHARTERED QUANTITY SURVEYORS
CONSTRUCTION COST CONSULTANTS

LONDON
Princes House
39 Kingsway
London
WC2B 6TP
Tel : (0171) 497 9000
Fax : (0171) 497 8858

BRISTOL
St Lawrence House
29/31 Broad Street
Bristol BS1 2HF
Tel : (0117) 927 7832
Fax : (0117) 925 1350

CAMBRIDGE
36 Storey's Way
Cambridge CB3 ODT
Tel : (01223) 351258
Fax : (01223) 321002

CARDIFF
3 Raleigh Walk
Brigantine Place
Atlantic Wharf
Cardiff CF1 5LN
Tel : (01222) 471306
Fax : (01222) 471465

CHESTER
Ford Lane Farm
Lower Lane
Aldford
Chester CH3 6HP
Tel : (01244) 620222
Fax : (01244) 620303

EDINBURGH
74 Great King Street
Edinburgh
EH3 6QU
Tel : (0131) 557 5306
Fax : (0131) 557 5704

GATESHEAD
11 Regent Terrace
Gateshead
Tyne and Wear
NE8 1LU
Tel : (0191) 477 3844
Fax : (0191) 490 1742

GLASGOW
Cumbrae House
15 Carlton Court
Glasgow G5 9JP
Tel : (0141) 429 6677
Fax : (0141) 429 2255

IPSWICH
17 St Helens Street
Ipswich IP4 1HE
Tel : (01473) 253405
Fax : (01473) 231215

LEEDS
Duncan House
14 Duncan Street
Leeds
LS1 6DL
Tel : (0113) 243 2481
Fax : (0113) 2424601

LIVERPOOL
Cunard Building
Water Street
Liverpool
L3 1JR
Tel : (0151) 236 1992
Fax : (0151) 227 5401

MANCHESTER
Boulton House
Chorlton Street
Manchester M1 3HY
Tel : (0161) 228 2011
Fax : (0161) 228 6317

MILTON KEYNES
6 Bassett Court
Newport Pagnell
Buckinghamshire
MK16 OJN
Tel : (01908) 613777
Fax : (01908) 210642

NEWPORT
34 Godfrey Road
Newport
Gwent NP9 4PE
Tel : (01633) 259712
Fax : (01633) 215694

NORWICH
63 Thorpe Road
Norwich NR1 1UD
Tel : (01603) 628194
Fax : (01603) 615928

OXFORD
Avalon House
Marcham Road
Abingdon
Oxford OX14 1TZ
Tel : (01235) 555025
Fax : (01235) 554909

PLYMOUTH
3 Russell Court
St Andrews Street
Plymouth
Devon PL6 2AX
Tel : (01752) 668372
Fax : (01752) 221219

PORTSMOUTH
Kings House
4 Kings Road
Portsmouth PO5 3BQ
Tel : (01705) 815218
Fax : (01705) 827156

SOUTHAMPTON
Clifford House
New Road
Southampton
SO14 OAB
Tel : (01703) 333438
Fax : (01703) 226099

**DAVIS LANGDON
CONSULTANCY**
Princes House
39 Kingsway
London WC2B 6TP
Tel : (0171) 379 3322
Fax : (0171) 379 3030

A MEMBER OF
DAVIS LANGDON & SEAH INTERNATIONAL

Electrical Installations
MATERIAL COSTS/MEASURED WORK PRICES

DIRECTIONS

The following explanations are given for each of the column headings and letter codes. It should be noted that not only are full material costs per item declared but also the published list price, the latest published price increase (update).

Unit — Prices for each unit are given as singular (1 metre, 1 nr).

Net price — Manufacturer's latest issued material/component price list, plus nominal allowance for fixings, plus any percentage uplift advised by manufacturer, plus where appropriate waste less percentage discount. The net price also reflects the applicable quantity for the minimum purchase of each batch of material and against which discounts etc. apply.

Material cost — Net price plus percentage allowance for overheads and profit.

Labour constant — Gang norm (in manhours) for each operation.

Labour cost — Labour constant multiplied by the appropriate all-in manhour cost.(See also relevant Rates of Wages Section)

Measured work price — Material cost plus Labour cost.

MATERIAL COSTS

The Material Costs given includes for delivery to sites in the London area at March/April 1995 with an allowance for waste, overheads and profit, and represents the price paid by contractors after the deduction of all trade discounts but excludes any charges in respect of VAT.

MEASURED WORK PRICES

These prices are intended to apply to new work in the London area and include allowances for all charges, preliminary items and profit. The prices are for reasonable quantities of work and the user should make suitable adjustments if the quantities are especially small or especially large. Adjustments may also be required for locality (eg outside London) and for the market conditions (eg volume of work on hand or on offer) at the time of use.

DIRECTIONS

LABOUR RATE

The labour rate on which these prices have been based is £9.85 per man hour which is the London Standard Rate at December 1994 plus allowances for all other emoluments and expenses. To this rate has been added 25% to cover site and head office overheads and preliminary items together with a further 2½% for profit, resulting in an inclusive rate of **£12.62** per man hour. The rate of £9.85 per man hour has been calculated on a working year of 1732.50 hours; a detailed build-up of the rate is given at the end of these Directions.

In calculating the 'Measured Work Prices' the following assumptions have been made:
- (a)   That the work is carried out as a sub-contract under the Standard Form of Building Contract and that such facilities as are usual would be afforded by the main contractor.
- (b)   That, unless otherwise stated, the work is being carried out in open areas at a height which would not require more than simple scaffolding.
- (c)   That the building in which the work is being carried out is no more than six storey's high.

Where these assumptions are not valid, as for example where work is carried out in ducts and similar confined spaces or in multi-storey structures when additional time is needed to get to and from upper floors, then an appropriate adjustment must be made to the prices. Such adjustment will normally be to the labour element only.

No allowance has been made in the prices for any cash discount to the main contractor.

DIRECTIONS

LABOUR RATE - ELECTRICAL

The following detail shows how the labour rate of £9.85 per man hour has been calculated.

The annual cost of notional eleven man gang.

	TECHNICIAN	APPROVED ELECTRICIANS	ELECTRICIANS	SUB-TOTALS
	1 NR	4 NR	6 NR	
Hourly Rate from 03/12/1994	7.58	6.60	6.12	
Working hours/annum	1,732.50	6,930.00	10,395.00	
x Hourly rate = £/annum	13,132.35	45,738.00	63,617.40	122,487.75
Daily travel rate/day	6.02	5.24	4.86	
Days/annum	231	924	1386	
£/annum	1,390.62	4,841.76	6,735.96	12,968.34
Daily travel fare	4.00	4.00	4.00	
Days/annum	226	904	1356	
£/annum	904.00	3,616.00	5,424.00	9,944.00
JIB Combined Benefits Scheme(nr of weeks)	48	192	288	
Stamp value	26.35	23.57	22.20	
£/annum	1,264.80	4,525.44	6,393.60	12,183.84
National insurance contributions				
Weekly gross pay EACH	303.83	264.54	245.30	
% Contributions	0.102	0.102	0.102	
£ Contributions EACH	30.99	26.98	25.02	
x Nr of men = £ contributions/week	30.99	107.92	150.12	
£ Contributions/annum	1,481.32	5,158.58	7,175.74	13,815.64

	SUB-TOTALS
SUB-TOTAL	171,399.57
TRAINING (INCLUDING ANY TRADE REGISTRATIONS) - SAY 1%	1,714.00
SEVERANCE PAY AND SUNDRY COSTS - SAY 1.5%	2,596.70
EMPLOYER'S LIABILITY AND THIRD PARTY INSURANCE - SAY 2%	3,514.21
ANNUAL COST OF NOTIONAL 11 MAN GANG (10.5 MEN ACTUALLY WORKING)	179,224.48
ANNUAL COST PER PRODUCTIVE MAN (ie+10.5)	17,069.00
ALL IN MAN HOUR (BASED 0N 1732.6 HOURS)	9.85

Notes:
(1)   The following assumptions have been made in the above calculations:-
    (a)   The working week of 38 hours is made up of 7.5 hours Monday to Friday.
    (b)   The actual hours worked are five days of 8.5 hours each.
    (c)   Five days in the year are lost through sickness or similar reasons.
    (d)   A working year of 1732.50 hours.
(2)   National insurance contributions are those effective from March 1994.
(3)   Weekly Holiday Credit/Welfare Stamp values are those effective from 31 January 1994.

V:ELECTRICAL SUPPLY/POWER/LIGHTING SYSTEMS

Item	Net Price £	Material £	Labour hours	Labour £	Unit	Total Rate £
V10 : ELECTRICITY GENERATION PLANT : GENERATING PLANT						
Standby Diesel Generating Sets; Supply and installation fixing to base; all supports and fixings (Rawlbolts or equal); all necessary connections to equipment (Provision and fixing of plates, discs etc. for identification is included						
Three phase, 440 Volt, four wire 50 Hz packaged standby diesel generating set, complete with radio and television suppressors, daily service fuel tank and associated piping, 4 metres of exhaust pipe and primary exhaust silencer, control panel, mains failure relay, starting battery with charger, all internal wiring, interconnections, earthing and labels rated for standby duty; delivery, installation and commissioning is included						
50 kVA	11055.14	12160.65	-	-	nr	**12160.65**
100 kVA	13662.48	13662.48	-	-	nr	**15028.73**
150 kVA	17625.64	17625.64	-	-	nr	**19388.20**
300 kVA	24248.30	26673.13	-	-	nr	**26673.13**
500 kVA	40127.02	44139.72	-	-	nr	**44139.72**
750 kVA	58274.13	64101.54	-	-	nr	**64101.54**
1000 kVA	72275.57	79503.13	-	-	nr	**79503.13**
EXTRA FOR:						
Residential Silencer						
300 kVA	716.50	788.15	-	-	nr	**788.15**
500 kVA	895.88	985.47	-	-	nr	**985.47**
750 kVA	1013.73	1115.10	-	-	nr	**1115.10**
Synchronisation Panel						
2 x 100 kVA	7296.39	8026.03	-	-	nr	**8026.03**
2 x 300 kVA	7904.42	8694.86	-	-	nr	**8694.86**
2 x 500 kVA	8260.07	9086.08	-	-	nr	**9086.08**
Note: Prices exclude costs for bulk storage tank, accoustic treatment or engine room ventilation and have been priced as a total package, ie, supplied, delivered, installed and commissioned by the manufacturer.						

V:ELECTRICAL SUPPLY/POWER/LIGHTING SYSTEMS

Item	Net Price £	Material £	Labour hours	Labour £	Unit	Total Rate £
V11 : HV SUPPLY/INSTALLATION/PUBLIC UTILITY : TRANSFORMERS						
Step Down Transformers; mounted on skids; provision for inserting axles and wheels; all necessary connections to equipment. (Provision and fixing of plates, discs, etc., for identification is included).						
500 kVA;						
Three Phase 11 kV/415 Volt 50 Hz and LV cable boxes, with glands and silica gel breathers.						
Naturally cooled, oil filled type ONAN; offload tap changer excluding conservator, dial thermometer or Bucholz relay.	4100.00	4510.00	38.46	485.38	nr	**4995.38**
Naturally cooled, silicon filled, type LNAN in ventilated steel tank.	5326.00	5858.60	38.46	485.38	nr	**6343.98**
Dry type; class C insulated air cooled; type AN in ventilated steel tank; temperature indicator.	10400.00	11440.00	38.46	485.38	nr	**11925.38**
Dry type; cast resin filled, type AN in ventilated steel tank; temperature indicator.	11100.00	12210.00	38.46	485.38	nr	**12695.38**
Extra For:						
Conservator	176.00	193.60	8.00	100.96	nr	**294.56**
Dial thermometer	286.00	314.60	2.00	25.24	nr	**339.84**
Bucholz relay	148.00	162.80	3.00	37.90	nr	**200.70**
Pressure Relief Valve	268.00	294.80	2.00	25.24	Unit	**320.04**
800 kVA;						
Three Phase 11 kV/415 Volt 50 Hz and LV cable boxes, with glands and silica gel breathers.						
Naturally cooled, oil filled type ONAN; offloead tap changer excluding conservator, dial thermometer or Bucholz relay	4900.00	5390.00	45.45	573.64	nr	**5963.64**
Naturally cooled, silicon filled, type LNAN in ventilated steel tank.	6229.00	6851.90	45.45	573.64	nr	**7425.54**
Dry type; class C insulated air cooled; type AN in ventilated steel tank; temperature indicator.	11800.00	12980.00	45.45	573.64	nr	**13553.64**
Dry type; cast resin filled, type AN in ventilated steel tank; temperature indicator.	13050.00	14355.00	45.45	573.64	nr	**14928.64**

V:ELECTRICAL SUPPLY/POWER/LIGHTING SYSTEMS

Item	Net Price £	Material £	Labour hours	Labour £	Unit	Total Rate £
V11 : HV SUPPLY/INSTALLATION/PUBLIC UTILITY : TRANSFORMERS (contd)						
Step Down Transformers; mounted on skids; provision for inserting axles and wheels; all necessary connections to equipment. (Provision and fixing of plates, discs, etc., for identification is included) (contd)						
800 kVA; Three Phase 11 kV/415 Volt 50 Hz and LV cable boxes, with glands and silica gel breathers (contd)						
Extra For:						
Conservator	178.00	195.80	10.00	126.20	nr	**322.00**
Dial thermometer	286.00	314.60	2.00	25.24	nr	**339.84**
Bucholz relay	148.00	162.80	3.00	37.90	nr	**200.70**
Pressure Relief Valve	268.00	294.80	2.00	25.24	nr	**320.04**
1000 kVA						
Three Phase 11KV/415 volt 50 Hz; HV and LV cable boxes, with glands and silica gel breathers.						
Naturally cooled, oil filled type ONAN; offload tap changer excluding conservator, dial thermometer or Bucholz relay	5383.00	5921.30	83.33	1051.67	nr	**6972.97**
Naturally cooled, silicon filled, type LNAN in ventilated steel tank.	7052.00	7757.20	83.33	1051.67	nr	**8808.87**
Dry type; class C insulated air cooled; type AN in ventilated steel tank; temperature indicator.	13200.00	14520.00	83.33	1051.67	nr	**15571.67**
Dry type; cast resin filled, type AN in ventilated steel tank; temperature indicator.	15300.00	16830.00	83.33	1051.67	nr	**17881.67**
Extra For:						
Conservator	181.00	199.10	10.00	126.20	nr	**325.30**
Dial thermometer	286.00	314.60	2.00	25.24	nr	**339.84**
Bucholz relay	148.00	162.80	3.00	37.90	nr	**200.70**
Pressure Relief Valve	268.00	294.80	2.00	25.24	nr	**320.04**
1500 kVA;						
Three Phase 11KV/415 Volt 50 Hz; HV and LV cable boxes, with glands and silica gel breathers.						
Naturally cooled, oil filled type ONAN; offload tap changer excluding conservator, dial thermometer or Bucholz relay	7989.00	8787.90	100.00	1262.00	nr	**10049.90**

V:ELECTRICAL SUPPLY/POWER/LIGHTING SYSTEMS

Item	Net Price £	Material £	Labour hours	Labour £	Unit	Total Rate £
Naturally cooled, silicon filled, type LNAN in ventilated steel tank.	10168.00	11184.80	100.00	1262.00	nr	**12446.80**
Dry type; class C insulated air cooled; type AN in ventilated steel tank; temperature indicator.	17600.00	19360.00	100.00	1262.00	nr	**20622.00**
Dry type; cast resin filled; type AN in ventilated steel tank; temperature indicator.	19200.00	21120.00	100.00	1262.00	nr	**22382.00**
Extra For:						
Conservator	184.00	202.40	12.05	152.05	nr	**354.45**
Dial thermometer	286.00	314.60	2.00	25.24	nr	**339.84**
Bucholz relay	148.00	159.10	3.00	37.90	nr	**197.00**
Pressure Relief Valve	268.00	294.80	2.00	25.24	nr	**320.04**
2000 kVA;						
Three Phase 11KV/415 volt 50Hz; HV and LV cable boxes, with glands and silica gel breathers.						
Naturally cooled, oil filled type ONAN; offload tap changer excluding conservator, dial thermometer or Bucholz relay	11500.00	12650.00	125.00	1577.50	nr	**14227.50**
Naturally cooled, silicon filled, type LNAN in ventilated steel tank.	14472.00	15919.20	125.00	1577.50	nr	**17496.70**
Dry type; class C insulated air cooled; type AN in ventilated steel tank; temperature indicator	23000.00	25300.00	125.00	1577.50	nr	**26877.50**
Dry type; cast resin filled, type AN in ventilated steel tank; temperature indicator	24500.00	26950.00	125.00	1577.50	nr	**28527.50**
Extra For:						
Conservator	186.00	204.60	12.05	152.05	nr	**356.65**
Dial thermometer	286.00	314.60	2.00	25.24	nr	**339.84**
Bucholz relay	148.00	162.80	3.00	37.90	nr	**200.70**
Pressure Relief Valve	268.00	294.80	2.00	25.24	nr	**320.04**

V:ELECTRICAL SUPPLY/POWER/LIGHTING SYSTEMS

Item	Net Price £	Material £	Labour hours	Labour £	Unit	Total Rate £
V32 : UNINTERRUPTED POWER SUPPLY : U.P.S SYSTEMS/INVERTERS/BATTERIES/ CHARGERS						
Uninterruptable Power Supply; 240 Volt AC input and output; standard 13 Amp socket outlet connection; sheet steel enclosure; installed in office environs; self contained battery pack. (Includes postioning on site and commissioning).						
Single Phase; I/P and O/P						
1.2kVA (7 minute supply)	1293.00	1422.30	0.31	3.94	nr	**1426.24**
1.2kVA (18 minute supply)	1479.00	1626.90	3.21	40.45	nr	**1667.35**
1.6kVA (7 minute supply)	1376.00	1513.60	3.40	42.93	nr	**1556.53**
1.6kVA (15 minute supply)	1596.00	1755.60	3.40	42.93	nr	**1798.53**
2.2kVA (10 minute supply)	1760.00	1936.00	3.61	45.56	nr	**1981.56**
2.2kVA (15 minute supply)	1923.00	2115.30	3.61	45.56	nr	**2160.86**
3.0kVA (7 minute supply)	1950.00	2145.00	4.00	50.48	nr	**2195.48**
3.0kVA (24 minute supply)	2340.00	2574.00	4.00	50.48	nr	**2624.48**
5.0kVA (5 minute supply)	2654.00	2919.40	4.50	56.85	nr	**2976.25**
5.0kVA (15 minute supply)	3156.00	3471.60	4.50	56.85	nr	**3528.45**
7.5kVA (6 minute supply)	3168.00	3484.80	4.90	61.86	nr	**3546.66**
7.5kVA (18 minute supply)	4119.00	4530.90	4.90	61.86	nr	**4592.76**
10.0kVA (5 minute supply)	4519.00	4970.90	5.00	63.10	nr	**5034.00**
10.0kVA (12 minute supply)	5688.00	6256.80	5.00	63.10	nr	**6319.90**
Uninterruptable Power Supply; positioned on site; final connections and commissioning. (Includes delivery up to 75 miles to a road/level site).						
Medium size static; three phase input, single phase output; integral sealed battery.						
7.5 kVA (32 minutes supply)	5333.00	5866.30	30.30	382.42	nr	**6248.72**
10.0 kVA (30 minutes supply)	5609.00	6169.90	30.30	382.42	nr	**6552.32**
15.0 kVA (20 minutes supply)	6344.00	6978.40	35.71	450.71	nr	**7429.11**
20.0 kVA (12 minutes supply)	6741.00	7415.10	40.00	504.80	nr	**7919.90**
25.0 kVA (6 minutes supply)	7295.00	8024.50	40.00	504.80	nr	**8529.30**
Medium size static; three phase input, three phase output; integral sealed battery;						
10.0 kVA (30 minutes supply)	6959.00	7654.90	35.71	450.71	nr	**8105.61**
15.0 kVA (20 minutes supply)	7072.00	7779.20	35.71	450.71	nr	**8229.91**
20.0 kVA (12 minutes supply)	7463.00	8209.30	40.00	504.80	nr	**8714.10**
30.0 kVA (6 minutes supply)	7738.00	8511.80	30.30	382.42	nr	**8894.22**

V:ELECTRICAL SUPPLY/POWER/LIGHTING SYSTEMS

Item	Net Price £	Material £	Labour hours	Labour £	Unit	Total Rate £
Large size static; three phase input/output; lead acid batteries on separate open racks.						
40 kVA (16 minutes supply)	13033.00	14336.30	43.48	548.70	nr	**14885.00**
60 kVA (9 minutes supply)	14358.00	15793.80	45.45	573.64	nr	**16367.44**
80 kVA (16 minutes supply)	18630.00	20493.00	45.45	573.64	nr	**21066.64**
100 kVA (12 minutes supply)	19522.00	21474.20	50.00	631.00	nr	**22105.20**
200 kVA (10 minutes supply)	31901.00	35091.10	55.56	701.11	nr	**35792.21**
300 kVA (10 minutes supply)	39189.00	43107.90	62.50	788.75	nr	**43896.65**
Integral diesel rotary; three phase input/output; no break supply including ventilation and accoustic attenuation						
75 kVA	70140.00	77154.00	-	-	nr	**77154.00**
100 kVA	72240.00	79464.00	-	-	nr	**79464.00**
150 kVA	81900.00	90090.00	-	-	nr	**90090.00**
200 kVA	140079.00	154086.90	-	-	nr	**154086.90**
250 kVA	162033.00	178236.30	-	-	nr	**178236.30**
330 kVA	180294.00	198323.40	-	-	nr	**198323.40**
400 kVA	207552.00	228307.20	-	-	nr	**228307.20**
500 kVA	216267.00	237893.70	-	-	nr	**237893.70**
630 kVA	252387.00	277625.70	-	-	nr	**277625.70**
800 kVA	288969.00	317865.90	-	-	nr	**317865.90**
1000 kVA	341070.00	375177.00	-	-	nr	**375177.00**

Note: The integral and integrated Diesel Rotary Systems are each priced as a total package, ie. supplied, delivered, installed and commissioned by the manufacturer.

W:COMMUNICATIONS/SECURITY/CONTROL SYSTEMS

Item	Net Price £	Material £	Labour hours	Labour £	Unit	Total Rate £
W10 : TELECOMMUNICATIONS : TELECOMMUNICATION WIRING						
Multipair Internal Telephone cable; BS 6746; loose laid on tray or drawn in conduit or trunking.(Including cable sleeves).						
0.5 millimeter diameter conductor p.v.c. insulated and sheathed multipair cables; BT specification CW 1308						
3 pair	0.12	0.13	0.05	0.63	m	**0.76**
4 pair	0.14	0.16	0.06	0.76	m	**0.92**
6 pair	0.20	0.23	0.06	0.76	m	**0.99**
10 pair	0.39	0.43	0.07	0.88	m	**1.31**
12 pair	0.41	0.48	0.07	0.88	m	**1.36**
15 pair	0.52	0.60	0.07	0.88	m	**1.48**
20 pair + 1 wire	0.70	0.81	0.09	1.14	m	**1.95**
25 pair	0.83	0.96	0.10	1.26	m	**2.22**
40 pair + earth	1.25	1.37	0.13	1.64	m	**3.01**
60 pair + earth	2.01	2.32	0.15	1.89	m	**4.21**
80 pair + earth	2.21	2.56	0.20	2.52	m	**5.08**
Telephone Undercarpet Cable; Low Profile; laid loose						
0.5 millimetre; PVC Insulated; PVC Sheathed multicore cable; BT CW 1316						
6 Core	0.32	0.35	0.05	0.63	m	**0.98**
Telephone Drop Wire Cable; drawn in conduit or trunking						
0.5 millimetre conductor; PVC insulate twisted pairs;Polyehylene sheathed; BT CW 1378						
Drop Wire 10	0.19	0.21	0.06	0.76	m	**0.97**

W:COMMUNICATIONS/SECURITY/CONTROL SYSTEMS

Item	Net Price £	Material £	Labour hours	Labour £	Unit	Total Rate £
W20 : RADIO/TV/CCTV : COAXIAL CABLES						
Television Aerial Cable; coaxial; PVC sheathed; fixed to backgrounds						
General purpose TV aerial downlead; copper stranded inner conductor; cellular polythene insulation; copper braid outer conductor; 75 ohm impedance						
7/0.25mm	0.18	0.20	0.06	0.76	m	**0.96**
Low loss TV aerial downlead; solid copper inner conductor; cellular polythene insulation; copper braid outer; conductor; 75 ohm impedance						
1/1.12mm	0.28	0.32	0.06	0.76	m	**1.08**
Low loss air spaced; solid copper inner conductor; air spaced polythene insulation; copper braid outer conductor; 75 ohm impedance						
1/1.00mm	0.23	0.27	0.06	0.76	m	**1.03**
Satelite aerial downlead; solid copper inner conductor; air spaced polythene insulation; copper tape and braid outer conductor; 75 ohm impedance						
1/1.00mm	0.35	0.40	0.07	0.88	m	**1.28**
Satelite TV coaxial; solid copper inner conductor; semi air spaced polyethylene dielectric insulation; plain annealed copper foil and copper braid screen in outer conductor; PVC sheath; 75 ohm impedance						
1/1.25mm	0.47	0.54	0.08	1.01	m	**1.55**
Satelite TV coaxial; solid copper inner conductor; air spaced polyethylene dielectric insulation; plain annealed copper foil and copper braid screen in outer conductor; PVC sheath; 75 ohm impedance						
1/1.67mm	0.78	0.90	0.09	1.14	m	**2.04**
Radio Frequency Cable; BS 2316 ; PVC sheathed; laid loose						
7/0.41mm tinned copper inner conductor; solid polyethylene dielectric insulation; bare copper wire braid; PVC sheath; 75 ohm impedance						
Cable	0.66	0.76	0.05	0.63	m	**1.39**

W:COMMUNICATIONS/SECURITY/CONTROL SYSTEMS

Item	Net Price £	Material £	Labour hours	Labour £	Unit	Total Rate £
W20 : RADIO/TV/CCTV : COAXIAL CABLES (contd)						
Radio Frequency Cable; BS 2316; PVC sheathed; laid loose (contd)						
Twin 1/0.58mm copper covered steel solid core wire conductor; solid polyethylene dielectric insulation; bare copper wire braid; PVC sheath; 75 ohm impedance						
Cable	1.01	1.11	0.05	0.63	m	**1.74**
Video Cable; PVC Flame Retardant sheath; laid loose						
7/0.1mm silver coated copper covered annealed steel wire conductor; polyethylene dielectric insulation with tin coated copper wire braid; 75 ohms impedance						
Cable	0.68	0.76	0.05	0.63	m	**1.39**

W:COMMUNICATIONS/SECURITY/CONTROL SYSTEMS

Item	Net Price £	Material £	Labour hours	Labour £	Unit	Total Rate £
W23 : CLOCKS : CLOCK SYSTEMS						
Clock Timing Systems; Master and Slave Units; fixed to background; having battery standby power reserve; 24 DC; (Including supports and fixings).						
Quartz master clock with solid state digital readout for parallel loop operation; one minute, half minute and one second pulse; maximum of 80 clocks per loop						
Two outputs of 1 Amp per loop	461.25	507.38	4.00	50.48	nr	**557.86**
Four outputs of 1 Amp per loop	507.38	558.11	6.02	76.02	nr	**634.13**
Power supplies for above, giving 24 hours power reserve						
2 6 Amp hour batteries	230.63	253.69	3.00	37.90	nr	**291.59**
2 15 Amp hour batteries	333.12	366.44	5.00	63.10	nr	**429.54**
Larger quartz master clock with solid state digital readout for parallel loop operation, one minute, half minute and one second pulse maximum of 500 impulse clocks over 6 circuits.						
Facility for BST & GMT programmed changeover	1383.75	1522.12	8.00	100.96	nr	**1623.08**
Power supply for larger master clock with 24 hours reserve power, with 2 Nr 38A hour batteries	538.12	591.94	7.04	88.87	nr	**680.81**
Radio receiver to accept BBC Rugby Transmitter MSF signal						
To synchronise time of above larger master clock	281.87	310.06	7.04	88.87	nr	**398.93**
Wall clocks for slave (impulse) systems; 24V DC, white dial with black numerals fitted with axxis polycarbonate disc; BS 467.7 Class O						
227mm diameter 1 minute impulse	46.13	50.74	0.50	6.31	nr	**57.05**
305mm diameter 1 minute impulse	48.69	53.56	2.00	25.24	nr	**78.80**
227mm diameter 1/2 minute impulse	48.69	53.56	2.00	25.24	nr	**78.80**
305mm diameter 1/2 minute impulse	48.69	53.56	2.00	25.24	nr	**78.80**
227mm diameter 1 second impulse	69.19	76.11	2.00	25.24	nr	**101.35**
305mm diameter 1 second impulse	71.70	78.87	2.00	25.24	nr	**104.11**
Quartz Battery Movement; BS 467.7 Class O; white dial with black numerals and sweep second hand; fitted with axxis polycarbonate disc; stove enamel case						
227mm diameter	21.53	23.68	0.80	10.10	nr	**33.78**
305mm diameter	28.70	31.57	0.80	10.10	nr	**41.67**

W:COMMUNICATIONS/SECURITY/CONTROL SYSTEMS

Item	Net Price £	Material £	Labour hours	Labour £	Unit	Total Rate £
W23 : CLOCKS : CLOCK SYSTEMS (contd)						
Clock Timing Systems (contd)						
Large wall clock; BS 467.7 Class O; synchronous mains movement; white dial with black numerals; fitted with axxis polycarbonate disc; stove enamel case						
458mm diameter	128.13	140.94	1.50	18.95	nr	**159.89**
610mm diameter	143.50	157.85	1.50	18.95	nr	**176.80**
Weather resistant clock; synchronous mains movement; white dial with black numerals; fitted with axxis polycarbonate disc; stove enamel case and heavy duty hanging bracket and extended handset						
610mm diameter	256.25	281.87	2.00	25.24	nr	**307.11**
Elapsed time clock; BS 467.7 Class O; 240V AC, 50 Hz mains supply; 12 hour duration; dial with 0-55 and 1-12 duration; remote control facility; IP66; axxis polycarbonate disc; spun metal movement cover; 6 point fixing bezel						
227mm diameter	563.75	620.13	0.50	6.31	nr	**626.44**
Matching clock; BS 467.7 Class O; 240V AC, 50/60 Hz mains supply; 12 hour duration; dial with 1-12; dial with 1-12 duration; IP 66; axxis polycarbonate disc; spun metal movement cover; semi flush mount on 6 point fixing bezel						
227mm diameter	153.75	169.12	0.50	6.31	nr	**175.43**
Remote Control Unit; BS 1363; integral Stop/Start and Reset; 2 gang; flush or surface mounted						
Remote Control Unit	66.63	73.29	0.50	6.31	nr	**79.60**
Digital clocks; 240V, 50 hz supply; with/without synchronisation from Master clock;12/24 hour display;stand alone operation; mutifunctional diplay; o/a size 144 x 72 x 200mm						
Type hours/minutes/seconds or minutes/seconds/10th seconds	189.62	208.59	0.40	5.05	nr	**213.64**
Type hours/minutes or minutes/seconds	169.12	186.04	0.40	5.05	nr	**191.09**

W:COMMUNICATIONS/SECURITY/CONTROL SYSTEMS

Item	Net Price £	Material £	Labour hours	Labour £	Unit	Total Rate £
Digital clocks; 240V, 50 hz supply; with/without synchronisation from Master clock;12/24 hour display;stand alone operation; mutifunctional diplay; o/a size 240 x 90 x 56mm						
Type hours/minutes or minutes/seconds	138.38	152.21	0.60	7.57	nr	159.78

Keep your figures up to date, free of charge

This section, and most of the other information in this Price Book, is brought up to date every three months, until the next annual edition, in the *Price Book Update*.

The *Update* is available free to all Price Book purchasers.

To ensure you receive your copy, simply complete the reply card from the centre of the book and return it to us.

W:COMMUNICATIONS/SECURITY/CONTROL SYSTEMS

W30 : DATA TRANSMISSION : GUIDELINES FOR NETWORK DESIGN

IEEE 802.3 ETHERNET systems

These systems are laid out in a BUS topology - devices (workstations/printers/signal repeaters) are connected to a single long main cable. This main cable is called a trunk. The extreme ends of the trunk are fitted with terminators, at least one of which should be grounded. A long trunk is made up of several individual trunks joined with signal repeaters.

Thin Ethernet systems use network interface cards with the transceivers built in. Connection to the network is via a T-connector joined directly to the trunk cable segment. Thick Ethernet systems use transceivers which are permanently fixed to the trunk cable. Connection between the workstation NIC and the transceiver is via a patch cable.

	10 Base 5 'Thick-net'	10 Base 2 'Thin-net'	1 Base 5	10 Base T	100 Base T	100 Base VG
Length of individual trunk segments	500m	185m	500m	100m	100m	500m
Number of trunk segments:	5nr	5nr				
Devices / segment (includes repeaters):	100nr	30nr				1,024nr
Minimum separation between devices:	2.4m	0.5m				
Maximum length of patch cables:	50m	n/a				200m

IEEE 802.5 TOKEN RING systems

These systems are connected in a RING topology electrically, but are star-wired physically. Each device is connected to a wiring hub/ concentrator with data travelling from the sending device by way of the hub to the next device, where it is re-transmitted on to the next device. Each device contains a network interface card (NIC) and a 2.4/3.0m adapter cable (fly-lead). The fly lead is connected directly, or via extension cables (patch cables) to a wiring concentrator / hub (multiple access unit / MAU).
The length of cable between the fly-lead and the MAU is called the lobe. The MAUs are wired together to form the ring. The original IEEE 802.5 standard applies to passive hubs, but modern hardware contains signal repeaters and re-timers which allow transmission over greater distances and/or through lower quality media.

SUMMARY:

	passive MAU (IEEE 802.5)	passive MAU			active MAU ***		
		Type 3 UTP 4Mbs (16Mbs)	Cat 5 UTP 4Mbs (16Mbs)	Type 1 STP 4Mbs (16Mbs)	Type 3 UTP 4Mbs (16Mbs)	Cat 5 UTP 4Mbs (16Mbs)	Type 1 STP 4Mbs (16Mbs)
Devices per ring:	96Nr	72Nr	72Nr	260Nr *	260Nr	260Nr	260Nr
Devices per MAU:	8Nr	8/16Nr	8/16Nr	8/16Nr	10/16Nr	10/16Nr	10/16Nr
MAUs per ring:	12Nr	9/5Nr	9/5Nr	33/17Nr *	26/17Nr	26/17Nr	26/17Nr
Lobe length:	45m	150m (60m)	150m (60m)	385m (173m)	300m (150m)	400m (300m)	600m (400m)
Length between MAUs:	45m	300m (100m)	300m (100m)	770m (346m)	300m (150m)	400m (300m)	600m (400m)
Drive Distance **:	120m	300m (100m)	300m (100m)	770m (346m)	7.5Km(3.75Km)	10Km (7.5Km)	15Km (10Km)

NB   The distances quoted above assume the simplest ring - two widely spaced MAUs with 16 separate devices. More MAUs and devices will increase signal attenuation and will reduce the distances unless signal repeaters/lobe extenders are fitted.
   *   Use of discrete signal re-timers required for enhanced connection parameters shown for Type 1 STP.
   **   Maximum Drive Distance consists of longest lobe length + (main ring length - shortest ring segment) .
     Device fly leads and inter-MAU patch cables up to 3m long do not form part of the maximum distance calculations.
   *** Assumes use of signal boosting/re-timing hardware within MAUs but not discrete signal repeaters/lobe extenders which would increase these figures further.

Unassisted 16Mbs networks have very short transmission distances - typically 160m maximum, even assuming high-quality cable such as Type 1 STP

Cable qualities:

Level 1	voice-grade	Capable of 20Kbs (Kbits/sec), an example is telephone cable. Rarely used for networks but modern hardware can make use of it.
Level 2	data-grade	Capable of 4Mbs. Token Ring (4Mbs) / 1Base5 supported. Unshielded twisted-pair cable.
IBM Type 3	data-grade	Capable of 4Mbs (16Mbs with certain hardware). Token Ring (4Mbs/16Mbs) Shielded twisted-pair
Level 3	high-speed data-grade	Capable of 10Mbs and 16Mbs. Token Ring (4Mbs/16Mbs) / 10BaseT supported. Unshielded twisted-pair cable.
IBM Type 9	high-speed data-grade	Capable of 16Mbs. Length limited to 100m. Token Ring (4Mbs/16Mbs) Shielded twisted-pair
IBM Type 6	high-speed data-grade	Capable of 16Mbs. Token Ring (4Mbs/16Mbs) Shielded twisted-pair (extension/patch cables rather than trunks))
Category 4	extended-distance high-speed data-grade	Capable of 20Mbs. Token Ring (4Mbs/16Mbs) / 10BaseT supported. UTP/STP
Category 5	low-loss extended-distance high-speed data-grade	Capable of 100Mbs. Token Ring (4Mbs/16Mbs)/ Ethernet / Fast Ethernet supported. UTP/STP
IBM Type 1	high-speed data-grade	Capable of 16Mbs. Token Ring (4Mbs/16Mbs) Shielded twisted-pair.
IEEE 802.3	Thin Ethernet	Capable of 10Mbs over 200m. Shielded coaxial cable.
IEEE 802.3	Thick Ethernet	Capable of 10Mbs over 500m. Shielded coaxial cable.

Column groups: **IEEE 802.3 ETHERNET-based systems** (10Base5 … 100BaseT) — **IEE802.12** (100BaseVG) — **IEEE 802.5 IBM TOKEN RING systems** (4Mbs systems: 4Mbs/100m … 4Mbs; 16Mbs systems: 16Mbs/100m … 16Mbs).

10Base5 Ethernet Thicknet 10Mbs/500m Thick coax (Trunk)	10Base5 Ethernet Thicknet TP (Drops)	10Base2 Thin Ethernet Thinnet 10Mbs/185m Thin coax	1Base5 Starlan 1Mbs/500m TP	10BaseT 10Mbs/100m TP	100BaseT Fast Ethernet 100Mbs/100m TP	100BaseVG Compatible: Fast Ethernet Token Ring 100Mbs/500m TP	4Mbs/100m TP	4Mbs/300m TP	4Mbs/600m TP	4Mbs TP (Drops)	16Mbs/100m TP	16Mbs/300m TP	16Mbs TP (Drops)	Cable	Ω
Thick coax	-	-	-	-	-	-	-	-	-	-	-	-	-	FES 10Base5-Trunk COAX	50 Ω
Thick coax	-	-	-	-	-	-	-	-	-	-	-	-	-	Belden 9880 COAX	50 Ω
Thick coax	-	-	-	-	-	-	-	-	-	-	-	-	-	Belden 9880NH COAX	50 Ω
-	STP	-	-	-	-	-	-	-	-	-	-	-	-	FES 10Base5, 4pr FTP	78 Ω
-	STP	-	-	-	-	-	-	-	-	-	-	-	-	FES 10Base5, 4pr LSF FTP	78 Ω
-	STP	-	-	-	-	-	-	-	-	-	-	-	-	FES 10Base5, miniature 4pr F1	78 Ω
-	STP	-	-	-	-	-	-	-	-	-	-	-	-	Belden 9901 4pr FTP	78 Ω
-	STP	-	-	-	-	-	-	-	-	-	-	-	-	Belden 9901NH 4pr FTP	78 Ω
-	STP	-	-	-	-	-	-	-	-	-	-	-	-	Belden 9902 5pr FTP	78 Ω
-	STP	-	-	-	-	-	-	-	-	-	-	-	-	Belden 9903 4pr FTP	78 Ω
-	-	Thin coax	-	-	-	-	-	-	-	-	-	-	-	FES 10Base2 COAX	50 Ω
-	-	Thin coax	-	-	-	-	-	-	-	-	-	-	-	FES 10Base2 LSF COAX	50 Ω
-	-	Thin coax	-	-	-	-	-	-	-	-	-	-	-	FES 10Base2 Dual COAX	50 Ω
-	-	Thin coax	-	-	-	-	-	-	-	-	-	-	-	Belden 9907 COAX	50 Ω
-	-	Thin coax	-	-	-	-	-	-	-	-	-	-	-	Belden 9907 COAX	50 Ω
-	-	Thin coax	-	-	-	-	-	-	-	-	-	-	-	Belden 9907NH COAX	50 Ω
-	-	-	UTP	UTP	-	UTP	UTP	UTP	-	-	UTP	-	-	AT&T 1024 UTP	100 Ω
-	-	-	UTP	UTP	-	UTP	UTP	UTP	-	-	UTP	-	-	IBM Type 3 UTP	105 Ω
-	-	-	UTP	UTP	-	UTP	UTP	UTP	-	-	UTP	-	-	Belden 1154A UTP	105 Ω
-	-	-	UTP	UTP	-	UTP	UTP	UTP	-	-	UTP	-	-	AT&T 1010 UTP	100 Ω
-	-	-	UTP	UTP	-	UTP	UTP	UTP	-	-	UTP	-	-	FES UTP-2PR-LEV3	100 Ω
-	-	-	UTP	UTP	-	UTP	UTP	UTP	-	-	UTP	-	-	FES UTP-4PR-LEV3	100 Ω
-	-	-	UTP	UTP	-	UTP	UTP	UTP	-	-	UTP	-	-	FES UTP-12PR-LEV3	100 Ω
-	-	-	UTP	UTP	-	UTP	UTP	UTP	-	-	UTP	-	-	FES UTP-25PR-LEV3	100 Ω
-	-	-	UTP	UTP	-	UTP	UTP	UTP	-	-	UTP	-	-	Belden 1455A UTP	100 Ω
-	-	-	UTP	UTP	-	UTP	UTP	UTP	-	-	UTP	-	-	Belden 1456A UTP	100 Ω
-	-	-	UTP	UTP	UTP	UTP	UTP	UTP	-	-	UTP	UTP	-	AT&T 1061 UTP	100 Ω
-	-	-	UTP	UTP	UTP	UTP	UTP	UTP	-	-	UTP	UTP	-	FES UTP-4PR-CAT5	100 Ω
-	-	-	UTP	UTP	UTP	UTP	UTP	UTP	-	-	UTP	UTP	-	FES UTP-4PR-CAT5 LSF	100 Ω
-	-	-	UTP	UTP	UTP	UTP	UTP	UTP	-	-	UTP	UTP	-	Belden 1583A UTP	100 Ω
-	-	-	STP	STP	STP	STP	STP	STP	-	-	STP	STP	-	FES FTP-4PR-CAT5	100 Ω
-	-	-	STP	STP	STP	STP	STP	STP	-	-	STP	STP	-	Belden 1584A FTP	100 Ω
-	-	-	-	-	-	-	-	-	-	UTP	-	-	UTP	FES UTP-PATCH26-CAT5	100 Ω
-	-	-	-	-	-	-	-	-	-	UTP	-	-	UTP	FES UTP-PATCH24-CAT5	100 Ω
-	-	-	-	-	-	-	-	-	-	STP	-	-	STP	FES FTP-PATCH-CAT5	100 Ω
-	-	-	UTP	UTP	UTP	UTP	UTP	UTP	-	-	UTP	UTP	-	Belden Datatwist 350 UTP	100 Ω
-	-	-	UTP	UTP	UTP	UTP	UTP	UTP	-	-	UTP	UTP	-	BICC H9682 UTP	100 Ω
-	-	-	-	-	-	-	STP	-	-	-	STP	-	-	IBM Type 9 FTP	150Ω
-	-	-	-	-	-	-	STP	STP	STP	-	STP	STP	-	IBM Type 1 FTP	(150Ω)
-	-	-	-	-	-	-	STP	STP	STP	-	STP	STP	-	IBM Type 1A FTP	(150Ω)
-	-	-	-	-	-	-	STP	STP	STP	-	STP	STP	-	IBM Type 1A LSF FTP	(150Ω)
-	-	-	-	-	-	-	STP	STP	STP	-	STP	STP	-	Belden Type 1A FTP	150Ω
-	-	-	-	-	-	-	-	-	-	STP	-	-	STP	IBM Type 6A FTP	(150Ω)
-	-	-	-	-	-	-	-	-	-	STP	-	-	STP	IBM Type 6A LSF FTP	(150Ω)
-	-	-	-	-	-	-	-	-	-	STP	-	-	STP	Belden Type 6A FTP	150Ω

W:COMMUNICATIONS/SECURITY/CONTROL SYSTEMS

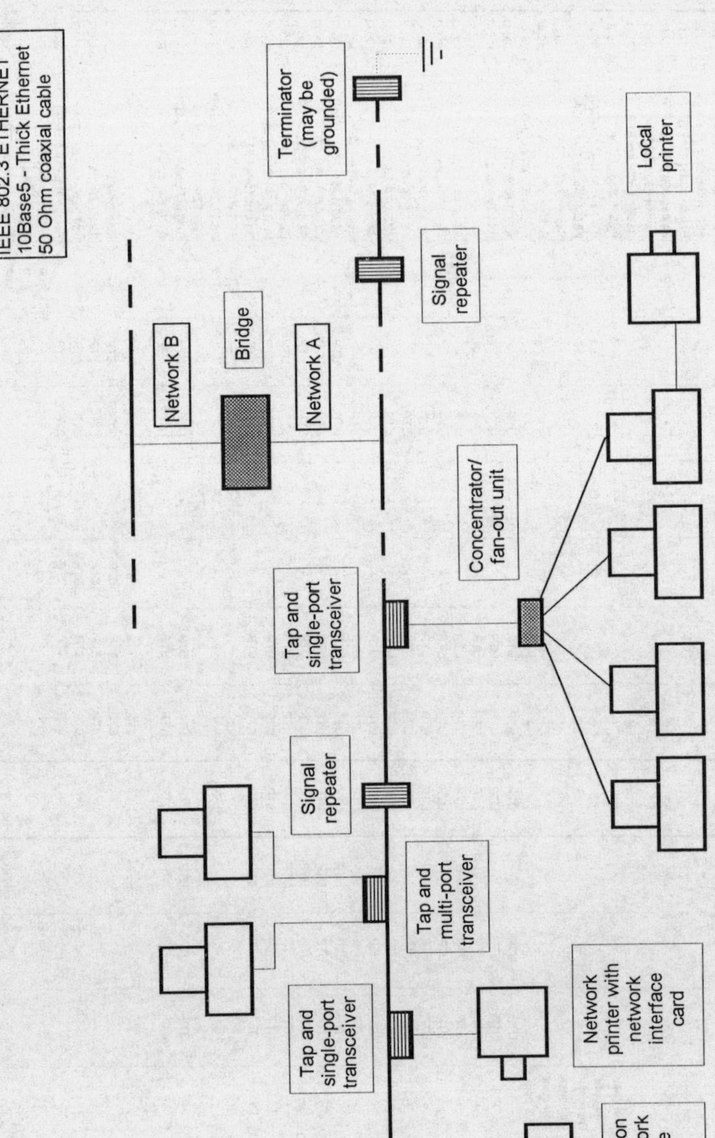

Rules: Up to 500m of trunk cable may be treated as a single trunk segment.
Up to 5 trunk segments may be joined together using signal repeaters.
Up to 100 devices (including repeaters) can be connected to each segment.
Up to 50m of patch cable/fly lead can be used between the device and it's transceiver.
Each transceiver must be separated from the next by at least 2.4m of trunk cable.

W:COMMUNICATIONS/SECURITY/CONTROL SYSTEMS

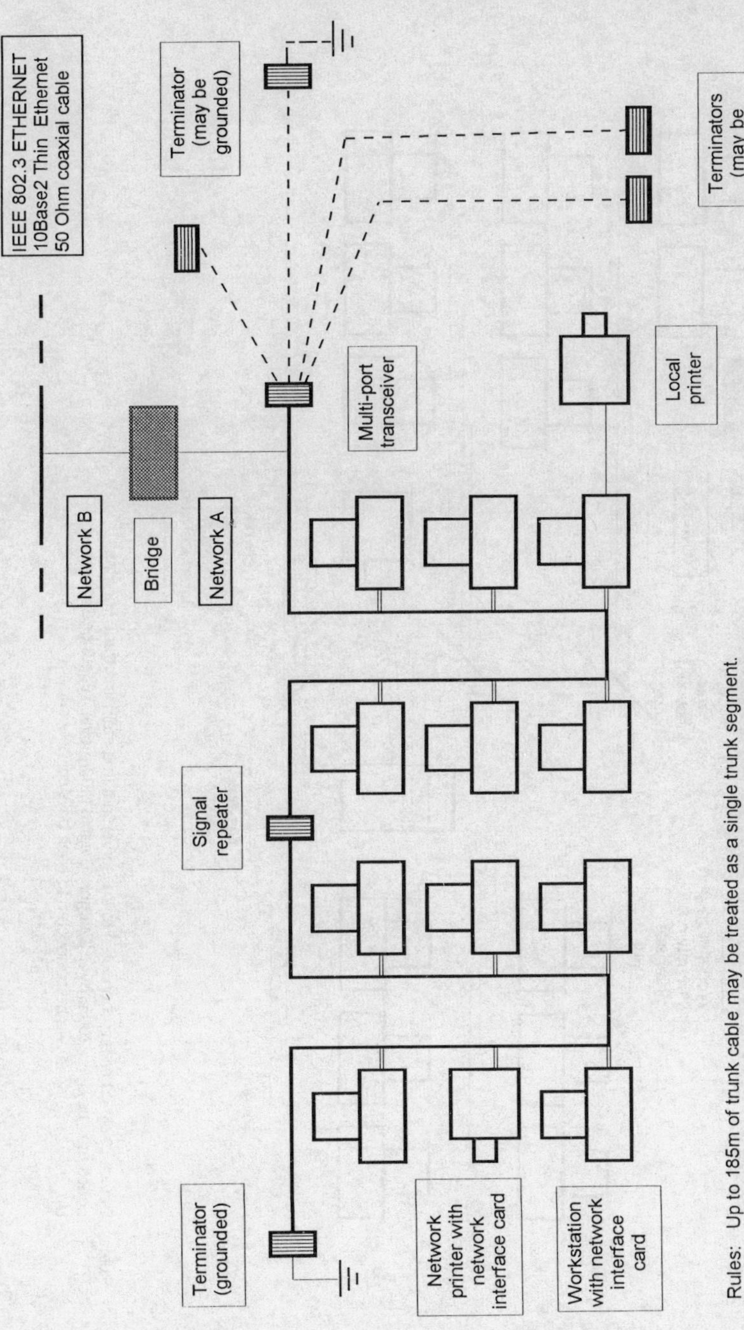

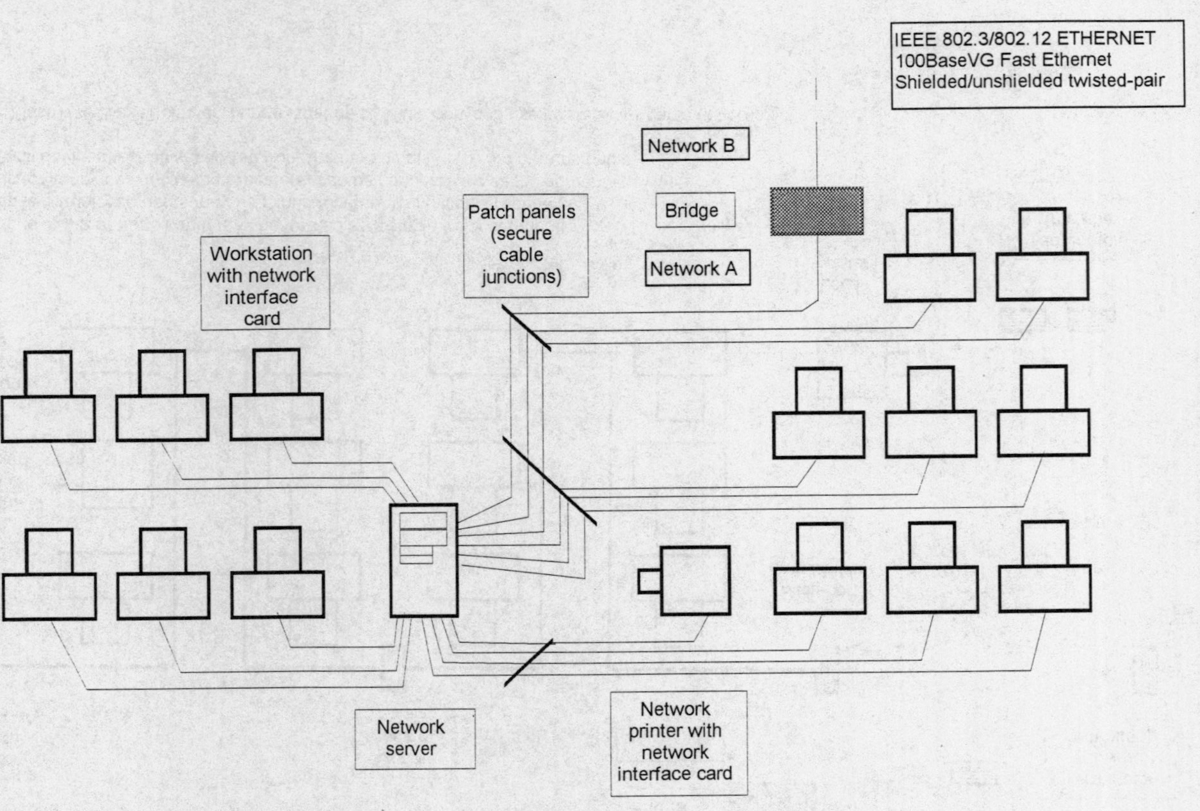

Rules: Device cable lobes are connected to the server, not to a main line/trunk line/ring.
Lobes may be up to 200m over Category 5 cable (100m over Type 3 or Category 4).
Up to 1024 devices can be connected to each physical network.

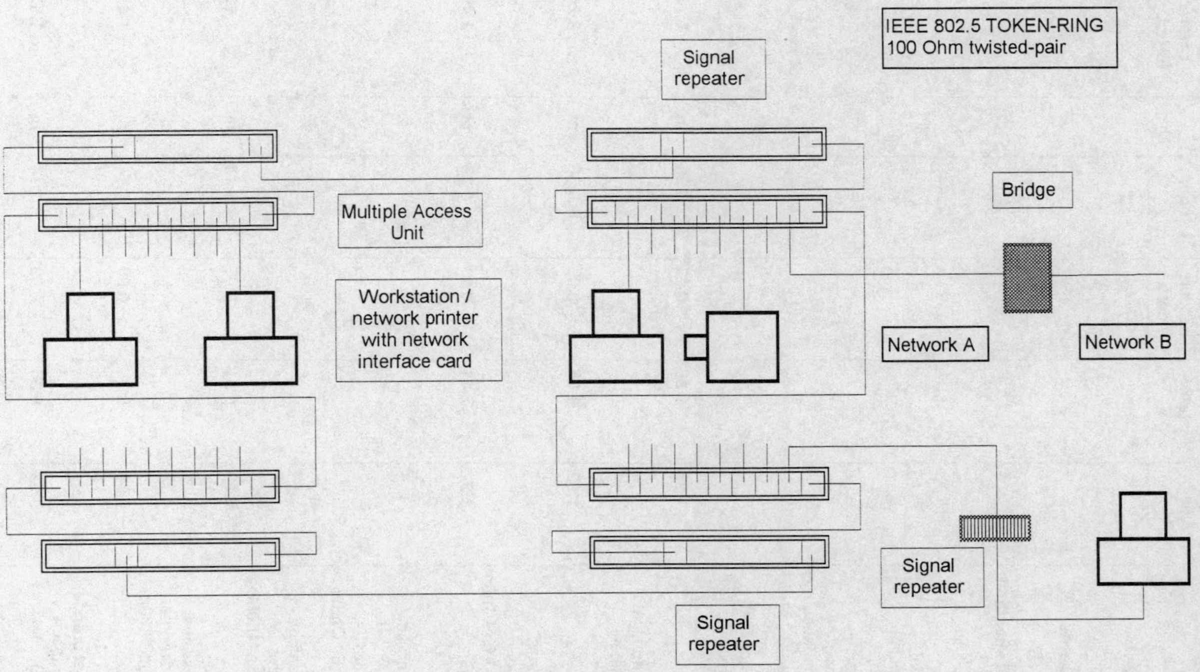

Rules: Maximum transmission distance (longest lobe + total ring - shortest ring segment) is 770m (346m). Self-repeating MAUs increase this to 15Km (10Km). Devices can be up to 150m (60m) from a MAU. If repeater/retiming equipment is used this maximum lobe length is extended to 600m (400m). MAUs can normally be up to 300m (100m) from the next. This distance is extended to 600m (400m) with the extra equipment. Normally up to 72 devices (including repeaters) can be connected to each ring,. Repeaters/retimers increase this to 260 devices per ring.

N.B. Figures are for 4Mbit/sec networks (16Mbs in brackets) over Type 1 shielded twisted-pair. Type 3/Category 5 cables are capable of 40-45% of these distances.

W:COMMUNICATIONS/SECURITY/CONTROL SYSTEMS

Item	Net Price £	Material £	Labour hours	Labour £	Unit	Total Rate £
W30 : DATA TRANSMISSION : DATA COMMUNICATIONS						
Local area networking; IEEE Ethernet 802.3 Systems						
Data Transmission Cables; PVC sheathed; fixed to backgrounds or laid in trunking.						
Trunk/backbone data cable; coaxial; tinned copper inner conductor; cellular polyethene insulation;polyester / aluminium foil shield; tinned copper braid screen; second foil shield and second braid screen; nominal impedance 50 Ohms						
1 core 2.17mm; PVC sheathed; nominal outside diameter 10.3mm; nominal attenuation 13.8dB/100m at 100Mhz (10Base5 100BaseVG)	1.23	1.36	0.03	0.38	m	1.74
1 core 2.47mm; PVC sheathed; nominal outside diameter 10.27mm; nominal attenuation 13.8dB/100m at 100Mhz (10Base5 100BaseVG)	1.69	1.86	0.03	0.38	m	2.24
1 core 2.47mm; flame resistant PVC sheathed; nominal outside diameter 10.27mm; nominal attenuation 13.8dB/100m at 100Mhz (10Base5 100BaseVG)	1.92	2.11	0.03	0.38	m	2.49
Segment data cable; coaxial; tinned copper inner conductors; cellular polyethene insulation; polyester / aluminium foil shield; tinned copper braid screen; nominal impedance 50 Ohms						
19 32AWG strand core; PVC sheathed; nominal outside diameter 4.62mm; nominal attenuation 4.3dB/100m at 100Mhz (10Base2)	0.29	0.31	0.09	1.14	m	1.45
19 32AWG strand core; flame resistant PVC sheathed; nominal outside diameter 4.62mm; nominal attenuation 4.3dB/100m at 100Mhz (10Base2)	0.31	0.34	0.09	1.14	m	1.48
Multicore; PVC sheathed; nominal outside diameter 4.62mm; nominal attenuation 4.3dB/100m at 10Mhz (10Base2)	0.32	0.36	0.09	1.14	m	1.50
Multicore; coloured PVC sheathed; nominal outside diameter 4.62mm; nominal attenuation 4.3dB/100m at 10Mhz (10Base2)	0.54	0.60	0.09	1.14	m	1.74

W:COMMUNICATIONS/SECURITY/CONTROL SYSTEMS

Item	Net Price £	Material £	Labour hours	Labour £	Unit	Total Rate £
Multicore; flame resistant PVC sheathed; nominal outside diameter 4.62mm; nominal attenuation 4.3dB/100m at 10Mhz (10Base2)	0.55	0.61	0.09	1.14	m	**1.75**
Outlet drop data cable; dual coaxial; parallel tinned copper inner conductors; cellular polyethene insulation; polyester / aluminium foil shield; tinned copper braid screen; nominal impedance 50 Ohms						
2 x 19 core 32AWG; PVC sheathed; nominal outside dimensions 4.62 x 9.74mm; nominal attenuation 4.3dB/100m at 10Mhz. (10Base2)	0.63	0.69	0.06	0.76	m	**1.45**
1.0m length; PVC sheathed; terminated with BNC and make-before-break safety tap connectors. (10Base2)	18.04	19.84	0.06	0.76	m	**20.60**
2.0m length; PVC sheathed; terminated with BNC and make-before-break safety tap connectors. (10Base2)	20.19	22.21	0.06	0.76	m	**22.97**
2.5m length; PVC sheathed; terminated with BNC and make-before-break safety tap connectors. (10Base2)	19.94	21.93	0.06	0.76	m	**22.69**
3.0m length; PVC sheathed; terminated with BNC and make-before-break safety tap connectors. (10Base2)	22.12	24.33	0.06	0.76	m	**25.09**
5.0m length; PVC sheathed; terminated with BNC and make-before-break safety tap connectors. (10Base2)	29.33	32.26	0.06	0.76	m	**33.02**
Outlet drop data cable; shielded twisted pair; tinned copper inner conductors with polyolefin insulation and polyester / aluminium foil screen to each; tinned copperbraid shield; nominal impedance 78 Ohm						
4 pair 28AWG; PVC sheathed; nominal outside diameter 6.6mm (10Base5)	1.07	1.18	0.08	1.01	m	**2.19**
4 pair 3,28AWG and 1,24AWG; PVC sheathed; nominal outside diameter 6.6mm (10Base5)	0.91	1.00	0.08	1.01	m	**2.01**
4 pair 20AWG; PVC sheathed; nominal outside diameter 10.54mm (10Base5)	2.16	2.37	0.08	1.01	m	**3.38**
4 pair 20AWG; flame resistant; PVC sheathed; nominal outside diameter 10.54mm. (10Base5)	1.49	1.64	0.08	1.01	m	**2.65**
5 pair 20AWG; PVC sheathed; nominal outside diameter 13.16mm (10Base5)	2.96	3.26	0.08	1.01	m	**4.27**

W:COMMUNICATIONS/SECURITY/CONTROL SYSTEMS

Item	Net Price £	Material £	Labour hours	Labour £	Unit	Total Rate £
W30 : DATA TRANSMISSION : DATA COMMUNICATIONS (contd)						
Local area networking; IEEE Ethernet 802.3 Systems (contd)						
Data Transmission Cables; PVC sheathed; fixed to backgrounds or laid in trunking (contd)						
Data cable; unshielded twisted pair; solid copper conductors; PVC insulation; nomimal impedance 100 Ohm; Level 2 media/Data grade/IBM type 3						
4 pair 24AWG; PVC sheathed; nominal outside diameter 4.8mm. (1Base5 10BaseT 100BaseVG TR4/300 TR 16/100)	0.12	0.14	0.08	1.01	m	**1.15**
Data cable; unshielded twisted pair; solid copper conductors; PVC insulation; nomimal impedance 105 Ohm; Level 2 media/Data grade/IBM type 3						
4 pair 24AWG; PVC sheathed; nominal outside diameter 4.7mm; nominal attenuation 2.1dB/100m at 1 Mhz. (1Base5 10BaseT 100BaseVG TR4/300 TR 16/100)	0.17	0.19	0.08	1.01	m	**1.20**
Data cable; unshielded twisted pair; solid copper conductors; PVC insulation; nomimal impedance 100 Ohm; Level 3 media/High Speed/Data grade/IBM type 3						
4 pair 24AWG; PVC sheathed; nominal outside diameter 4.6mm; nominal attenuation 10dB/100m at 10 Mhz. (1Base5 10BaseT 100BaseVG TR4/300 TR 16/100)	0.14	0.16	0.08	1.01	m	**1.17**
2 pair 24AWG; PVC sheathed; nominal outside diameter 3.9mm; nominal attenuation 10dB/100m at 10 Mhz. (1Base5 10BaseT 100BaseVG TR4/300 TR 16/100)	0.13	0.15	0.06	0.76	m	**0.91**
4 pair 24AWG; PVC sheathed; nominal outside diameter 4.9mm; nominal attenuation 10dB/100m at 10 Mhz. (1Base5 10BaseT 100BaseVG TR4/300 TR 16/100)	0.18	0.20	0.08	1.01	m	**1.21**
12 pair 24AWG; PVC sheathed; nominal outside diameter 7.6mm; nominal attenuation 10dB/100m at 10 Mhz. (1Base5 10BaseT 100BaseVG TR4/300 TR 16/100)	0.59	0.65	0.08	1.01	m	**1.66**

W:COMMUNICATIONS/SECURITY/CONTROL SYSTEMS

Item	Net Price £	Material £	Labour hours	Labour £	Unit	Total Rate £
25 pair 24AWG; PVC sheathed; nominal outside diameter 10.4mm; nominal attenuation 10dB/100m at 10 Mhz. (1Base5 10BaseT 100BaseVG TR4/300 TR 16/100)	1.11	1.22	0.10	1.26	m	**2.48**
Data cable; unshielded twisted pair; solid copper conductors; polyolefin insulation; nomimal impedance 100 Ohm; Category 4 media/Extended Distance High-speed Data-grade						
4 pair 24AWG; PVC sheathed; nominal outside diameter 4.93mm; nominal attenuation 7.2dB/100m at 16 Mhz. (1Base5 10BaseT 100BaseVG TR4/300 TR 16/100)	0.31	0.34	0.08	1.01	m	**1.35**
Data cable; shielded twisted pair; solid copper conductors; polyolefin insulation; nomimal impedance 100 Ohm; Category 4 media/Extended Distance High-speed Data-grade						
4 pair 24AWG; PVC sheathed; nominal outside diameter 5.25mm; nominal attenuation 9.8dB/100m at 16 Mhz. (1Base5 10BaseT 100BaseVG TR4/300 TR 16/100)	0.38	0.42	0.08	1.01	m	**1.43**
Data cable; unshielded twisted pair; solid copper conductors; polythene insulation; nomimal impedance 100 Ohm; Category 5 media/Low loss Extended-Distance High-speed Data-grade						
4 pair 24AWG; PVC sheathed; nominal outside diameter 4.8mm; (1Base5 10BaseT 100BaseVG TR4/300 TR 16/300)	0.46	0.50	0.08	1.01	m	**1.51**
4 pair 24AWG; PVC sheathed; nominal outside diameter 5.51mm; nominal attenuation 8.2dB/100m at 16Mhz (1Base5 10BaseT 100BaseVG TR4/300 TR 16/300)	0.20	0.22	0.08	1.01	m	**1.23**
4 pair 24AWG; PVC sheathed; nominal outside diameter 5.2mm; nominal attenuation 19.1dB/100m at 100Mhz (1Base5 10BaseT 100BaseT 100BaseVG TR4/300 TR 16/300)	0.25	0.27	0.08	1.01	m	**1.28**
4 pair 24AWG; PVC sheathed; nominal outside diameter 4.8mm; nominal attenuation 21.9dB/100m at 100Mhz (1Base5 10BaseT 100BaseT 100BaseVG TR4/300 TR 16/300)	0.19	0.21	0.08	1.01	m	**1.22**

W:COMMUNICATIONS/SECURITY/CONTROL SYSTEMS

Item	Net Price £	Material £	Labour hours	Labour £	Unit	Total Rate £
W30 : DATA TRANSMISSION : DATA COMMUNICATIONS (contd)						
Local area networking; IEEE Ethernet 802.3 Systems (contd)						
Data Transmission Cables; PVC sheathed; fixed to backgrounds or laid in trunking (contd)						
Data cable; shielded twisted pairs; solid copper conductors; polethene insulation; aluminium foil shield nomimal impedance 100 Ohm; Category 5 media/Low loss Extended-Distance High-speed Data-grade						
4 pair 24AWG; PVC sheathed; nominal outside diameter 5.3mm; nominal attenuation 21.98dB/100m at 100Mhz (1Base5 10BaseT 100BaseT 100BaseVG TR4/300 TR 16/300)	0.32	0.36	0.08	1.01	m	1.37
4 pair 24AWG; PVC sheathed; nominal outside diameter 6.1mm; nominal attenuation 8.2dB/100m at 16Mhz (1Base5 10BaseT 100BaseT 100BaseVG TR4/300 TR 16/300)	0.30	0.33	0.08	1.01	m	1.34
Accessories for Ethernet Systems; Fixed to backgrounds						
Surge Protectors; shielded; BNC connectors						
10KA surge capacity (10Base2)	51.67	56.83	0.42	5.30	nr	62.13
Line terminator; nominal impedance 50Ohm						
fitted to make-before-break safety line tap (10Base2)	3.20	3.52	0.18	2.31	nr	5.83
fitted to make-before-break safety line tap; including ground connector (10Base2)	3.97	4.37	0.18	2.31	nr	6.68
fitted to BNC jack (10Base2)	1.91	2.10	0.17	2.09	nr	4.19
Connectors; BNC push-connector; cable secured by twist-on barrel; nominal impedance 50 Ohm						
end socket; straight (10Base2)	1.55	1.70	0.17	2.09	nr	3.79
end socket; right-angled (10Base2)	3.80	4.18	0.17	2.09	nr	6.27
in-line jack (10Base2)	1.42	1.56	0.17	2.09	nr	3.65
bulkhead jack (for fascia plates) (10Base2)	2.40	2.63	0.17	2.09	nr	4.72

W:COMMUNICATIONS/SECURITY/CONTROL SYSTEMS

Item	Net Price £	Material £	Labour hours	Labour £	Unit	Total Rate £
Connectors; N series BNC twist-connector; cable secured by twist-on barrel; nominal impedance 50 Ohm						
end socket; straight (10Base5)	1.68	1.85	0.17	2.09	nr	**3.94**
end socket; right-angled (10Base5)	3.44	3.78	0.17	2.09	nr	**5.87**
bulkhead jack (for fascia plates) (10Base5)	2.77	3.04	0.17	2.09	nr	**5.13**
Line tap; passive; insulation-displacement connection through sheath of cable; plate phosphor-bronze contacts						
make-before-break safety tap connection; switching time 500ms (10Base2)	8.75	9.63	0.10	1.26	nr	**10.89**
Line tap; for fixing into transceiver body						
insulation-displacment connection (vampire/bee-sting); low-profile; accepts cable diameters between 9.53mm and 10.41mm (10Base5)	16.24	17.87	0.17	2.09	nr	**19.96**
N-series BNC twist-connector; single (10Base5)	12.70	13.97	0.10	1.26	nr	**15.23**
BNC twist-connector; single (10Base2)	10.03	11.03	0.10	1.26	nr	**12.29**
BNC push-connector; tee (10Base2)	9.40	10.34	0.10	1.26	nr	**11.60**
Transceiver body; including indicator LEDs for power, signal quality error/SEQ, transmission/reception/collision errors						
1-port; for connection to separate line tap/adaptor (10Base5 10Base2)	82.77	91.05	0.17	2.09	nr	**93.14**
2-port; for connection to separate line tap/adaptor (10Base5 10Base2)	171.38	188.52	0.25	3.15	nr	**191.67**
4-port; for connection to separate line tap/adaptor (10Base5 10Base2)	242.46	266.71	0.60	7.57	nr	**274.28**
1-port; including line tap; insulation-displacement connection (vampire/bee-sting); accepts cable diameters between 9.53mm and 10.41mm (10Base5)	106.14	116.75	0.17	2.09	nr	**118.84**
2-port; including line tap; insulation-displacement connection (vampire/bee-sting) (10Base5)	239.95	263.95	0.33	4.20	nr	**268.15**
1-port; including N-series BNC-type connector (1OBase5)	106.14	116.75	0.17	2.09	nr	**118.84**
2-port; including N-series BNC-type connector (1OBase5)	239.95	263.95	0.33	4.20	nr	**268.15**

W:COMMUNICATIONS/SECURITY/CONTROL SYSTEMS

Item	Net Price £	Material £	Labour hours	Labour £	Unit	Total Rate £
W30 : DATA TRANSMISSION : DATA COMMUNICATIONS (contd)						
Local area networking; IEEE Ethernet 802.3 Systems (contd)						
Accessories for Ethernet Systems; Fixed to backgrounds (contd)						
Transceiver body; including indicator LEDs for power, signal quality error/SEQ, transmission/reception/collision errors (contd)						
1-port; including BNC-type connector (1OBase5)	131.07	144.17	0.17	2.09	nr	**146.26**
2-port; including BNC-type connector (1OBase5)	239.95	263.95	0.33	4.20	nr	**268.15**
Transceiver body; no status indicators						
1-port; including integral line tap; insulation-displacement connection (vampire/bee-sting); accepts cable diameters between 9.53mm and 10.41mm (10Base5)	97.37	107.10	0.17	2.09	nr	**109.19**
1-port; including N-series BNC-type connector (1OBase5)	97.37	107.10	0.17	2.09	nr	**109.19**
Repeaters/retimers; fixing between extended segments or at junctions between media; including status indicators						
2, AUI connectors (10Base5 10Base2)	339.84	373.82	0.83	10.48	nr	**384.30**
1, AUI/1 BNC connectors (10Base5 10Base2)	339.84	373.82	0.67	8.41	nr	**382.23**
1, AUI/1 RJ45 connectors (10Base5 10Base2)	242.46	266.71	0.67	8.41	nr	**275.12**
2, BNC connectors (10Base5 10Base2)	388.53	427.38	0.50	6.31	nr	**433.69**
1, BNC/1, RJ45 connectors (10Base5 10Base2)	242.46	266.71	0.50	6.31	nr	**273.02**
Multiport transceiver; unpowered AUI connection to LAN; status indicators						
2-port; AUI connectors (10Base5 10Base2)	174.30	191.73	0.50	6.31	nr	**198.04**
4-port; AUI connectors (10Base5 10Base2)	242.46	266.71	1.17	14.73	nr	**281.44**
Multiport transceiver; mains powered; single AUI connection to LAN; status indicators						
8-port; AUI connectors (10Base5 10Base2)	563.80	620.18	1.50	18.95	nr	**639.13**
Multiport transceiver; mains powered; electrically stackable; single AUI connection to LAN; status indicators						
8-port; AUI connectors (10Base5 10Base2)	587.15	645.87	1.50	18.95	nr	**664.82**

W:COMMUNICATIONS/SECURITY/CONTROL SYSTEMS

Item	Net Price £	Material £	Labour hours	Labour £	Unit	Total Rate £
Multiport repeater/hubs; mains powered; electrically stackable; single AUI connection to LAN; fixing between segments and/or connected to several segments; including status indicators						
8-port; AUI connectors (10BaseT)	437.21	480.94	1.50	18.95	nr	499.89
2-port; BNC connectors (10Base2)	583.28	641.60	0.50	6.31	nr	647.91
4-port; BNC connectors (10Base2)	924.09	1016.50	0.83	10.52	nr	1027.02
4-port; BNC connectors; slim line case (10Base2)	680.65	748.72	0.83	10.52	nr	759.24
8-port; BNC connectors (10Base2)	1167.53	1284.28	1.50	18.95	nr	1303.23
8-port; BNC connectors; slim line case (10Base2)	875.40	962.94	1.50	18.95	nr	981.89
8-port; BNC connectors (10BaseT)	875.40	962.94	1.50	18.95	nr	981.89
4-port; RJ45 connectors (10BaseT)	167.48	184.23	0.83	10.52	nr	194.75
8-port; RJ45 connectors (10BaseT)	177.22	194.94	1.50	18.95	nr	213.89
12-port; RJ45 connectors (10BaseT)	485.90	534.49	1.67	21.03	nr	555.52
Multiport repeater/hubs; mains powered; electrically stackable; single AUI connection to LAN; fixing between segments and/or connected to several segments; including status indicators						
8-port; RJ45 connectors (10BaseT)	222.99	245.29	1.50	18.95	nr	264.24
12-port; RJ45 connectors/1, RS-232 connector/1, Inter-bus connector; SNMP software management (10BaseT)	680.65	748.72	2.17	27.38	nr	776.10
Multiport repeater/hubs; unpowered; electrically stackable; single AUI connection to LAN; fixing between segments and/or connected to several segments; including status indicators						
8-port; RJ45 connectors (10BaseT)	213.25	234.58	1.50	18.95	nr	253.53
Local bridges/filters; mains powered; spanning tree-type bridges; fixing between adjacent LANs; BNC and AUI connectors; filtering speed 25000 packets per second (pps); forwarding at 12500pps; storage for 2000 MAC addresses ;includes status indicators						
enhanced unit; SNMP software management (10Base5 10Base2)	1459.65	1605.62	0.33	4.20	nr	1609.82
standard unit (10Base5 10Base2)	1264.90	1391.39	0.33	4.20	nr	1395.59

W:COMMUNICATIONS/SECURITY/CONTROL SYSTEMS

Item	Net Price £	Material £	Labour hours	Labour £	Unit	Total Rate £
W30 : DATA TRANSMISSION : DATA COMMUNICATIONS (contd)						
Local area networking; IEEE Ethernet 802.3 Systems (contd)						
Accessories for Ethernet Systems; Fixed to backgrounds (contd)						
Patch panel; passive multi-port connectors; mechanical connections						
16-port, unkeyed unshielded RJ45 connectors (1Base5 10Base5 10Base2 10BaseT 100BaseT 100BaseVG)	116.30	127.94	0.50	6.31	nr	**134.25**
16-port, unkeyed shielded RJ45 connectors (1Base5 10Base5 10Base2 10BaseT 100BaseT 100BaseVG)	202.54	222.79	0.50	6.31	nr	**229.10**
48-port, unkeyed unshielded RJ45 connectors (1Base5 10Base5 10Base2 10BaseT 100BaseT 100BaseVG)	302.84	333.12	1.37	17.24	nr	**350.36**
48-port, unkeyed shielded RJ45 connectors (1Base5 10Base5 10Base2 10BaseT 100BaseT 100BaseVG)	405.08	445.59	1.37	17.24	nr	**462.83**
Patch panel; passive multi-port connectors; insulation-displacement connections						
16-port, unkeyed unshielded RJ45 connectors (1Base5 10Base5 10Base2 10BaseT 100BaseT 100BaseVG)	78.87	86.76	0.33	4.20	nr	**90.96**
16-port, keyed unshielded RJ45 connectors (1Base5 10Base5 10Base2 10BaseT 100BaseT 100BaseVG)	107.11	117.82	0.33	4.20	nr	**122.02**
16-port, keyed shielded RJ45 connectors (1Base5 10Base5 10Base2 10BaseT 100BaseT 100BaseVG)	121.72	133.89	0.33	4.20	nr	**138.09**
24-port, unkeyed RJ45 connectors (1Base5 10Base5 10Base2 10BaseT 100BaseT 100BaseVG)	122.69	134.96	0.57	7.15	nr	**142.11**
48-port, unkeyed unshielded RJ45 connectors (1Base5 10Base5 10Base2 10BaseT 100BaseT 100BaseVG)	180.14	198.16	1.13	14.31	nr	**212.47**
Data outlet face plate; to suit BS 4662 wall, pattress boxes or service trunking; fitted with keyed RJ45 shuttered data outlet; Category 5 twisted pair system						
single gang; 1 outlet; shielded	6.52	7.18	0.45	5.68	nr	**12.86**
single gang; 2 outlet; shielded	10.03	11.03	0.45	5.68	nr	**16.71**

W:COMMUNICATIONS/SECURITY/CONTROL SYSTEMS

Item	Net Price £	Material £	Labour hours	Labour £	Unit	Total Rate £
Data outlet face plate; to suit BS 4662 wall, pattress boxes or service trunking; fitted with keyed RJ45 shuttered data outlet; twisted pair system						
single gang; 1 outlet; shielded	6.87	7.56	0.45	5.68	nr	**13.24**
single gang; 2 outlet; shielded	10.13	11.14	0.45	5.68	nr	**16.82**
Data outlet face plate; to suit BS 4662 wall, pattress boxes or service trunking; fitted with unkeyed RJ45 in shuttered modules; including blanking modules where required; Category 5 twisted pair system						
single gang; 1 outlet; unshielded	8.87	9.76	0.45	5.68	nr	**15.44**
single gang; 2 outlet; unshielded	15.41	16.95	0.45	5.68	nr	**22.63**
double gang; 1 outlet; unshielded	10.24	11.26	0.45	5.68	nr	**16.94**
double gang; 2 outlet; unshielded	16.38	18.02	0.45	5.68	nr	**23.70**
double gang; 3 outlet; unshielded	23.13	25.44	0.45	5.68	nr	**31.12**
double gang; 4 outlet; unshielded	33.93	37.32	0.45	5.68	nr	**43.00**
Data outlet face plate; modular; to suit BS 4662 wall, pattress boxes or service trunking; fitted with unkeyed RJ45 in shuttered modules; including blanking modules where required; twisted pair system						
single gang; 1 outlet; unshielded	6.15	6.76	0.45	5.68	nr	**12.44**
single gang; 2 outlet; unshielded	9.19	10.90	0.45	5.68	nr	**16.58**
double gang; 1 outlet; unshielded	7.51	8.26	0.45	5.68	nr	**13.94**
double gang; 2 outlet; unshielded	11.27	12.40	0.45	5.68	nr	**18.08**
double gang; 3 outlet; unshielded	15.66	17.23	0.45	5.68	nr	**22.91**
double gang; 4 outlet; unshielded	19.42	21.36	0.45	5.68	nr	**27.04**
Data outlet face plate; modular; to suit BS 4662 wall, pattress boxes or service trunking; fitted with 25 pin D-type data outlet; twisted pair system						
single gang; 1 outlet; shielded	3.32	3.65	0.45	5.68	nr	**9.33**
single gang; 2 outlet; shielded	3.42	3.76	0.45	5.68	nr	**9.44**
double gang; 1 outlet; shielded	3.77	4.15	0.45	5.68	nr	**9.83**
double gang; 2 outlet; shielded	5.23	5.75	0.45	5.68	nr	**11.43**
double gang; 3 outlet; shielded	6.38	7.02	0.45	5.68	nr	**12.70**
double gang; 4 outlet; shielded	6.93	7.62	0.45	5.68	nr	**13.30**
Data outlet face plate; to suit BS 4662 wall, pattress boxes or service trunking; fitted with 25 pin D-type data outlet; twisted pair system						
single gang; 1 outlet; shielded	7.43	8.17	0.45	5.68	nr	**13.85**
single gang; 2 outlet; shielded	13.27	14.60	0.45	5.68	nr	**20.28**

W:COMMUNICATIONS/SECURITY/CONTROL SYSTEMS

Item	Net Price £	Material £	Labour hours	Labour £	Unit	Total Rate £
W30 : DATA TRANSMISSION : DATA COMMUNICATIONS (contd)						
Local area networking; IEEE Ethernet 802.3 Systems (contd)						
Accessories for Ethernet Systems; Fixed to backgrounds (contd)						
Data outlet face plate; to suit BS 4662 wall, pattress boxes or service trunking; fitted with 25 pin D-type data outlet; coaxial cable system						
single gang; 1 outlet; unshielded	6.25	6.88	0.45	5.68	nr	**12.56**
single gang; 2 outlet; unshielded	9.47	10.42	0.45	5.68	nr	**16.10**
Data outlet face plate; modular; to suit BS 4662 wall, pattress boxes or service trunking; fitted with BNC data outlets; including blanking modules where required; nominal impedance 50 Ohm; coaxial cable system						
single gang; 1 outlet	3.50	3.85	0.45	5.68	nr	**9.53**
single gang; 2 outlet	4.62	5.08	0.45	5.68	nr	**10.76**
double gang; 1 outlet	3.95	4.34	0.45	5.68	nr	**10.02**
double gang; 2 outlet	5.98	6.58	0.45	5.68	nr	**12.26**
double gang; 3 outlet	7.71	8.48	0.45	5.68	nr	**14.16**
double gang; 4 outlet	8.82	9.70	0.45	5.68	nr	**15.38**
Data outlet face plate; modular; to suit BS 4662 wall, pattress boxes or service trunking; fitted with make-before-break safety tap data outlets; including blanking modules where required; 10Base2 coaxial cable system						
single gang; 1 outlet	1.73	1.91	0.45	5.68	nr	**7.59**
double gang; 1 outlet	2.94	3.23	0.45	5.68	nr	**8.91**
double gang; 2 outlet	3.05	3.36	0.45	5.68	nr	**9.04**
Data outlet face plate; to suit BS 4662 wall, pattress boxes or service trunking; fitted with make-before-break safety tap data outlets; 10Base2 coaxial cable system						
single gang; 2 outlet	27.46	30.21	0.45	5.68	nr	**35.89**
Data outlet box; surface mounted; fitted with hinged face plate						
make-before-break safety tap BNC push-connector; surface mounted; (10Base2)	19.91	21.90	0.45	5.68	nr	**27.58**
make-before-break safety tap BNC push-connector; flush mounted; (10Base2)	19.91	21.90	0.45	5.68	nr	**27.58**

W:COMMUNICATIONS/SECURITY/CONTROL SYSTEMS

Item	Net Price £	Material £	Labour hours	Labour £	Unit	Total Rate £
Local area networking; IEEE Token Ring Systems						
Data Transmission Cables; PVC sheathed; fixed to backgrounds or laid in trunking.						
Data cable; unshielded twisted pair; solid copper conductors; PVC insulation; nomimal impedance 100 Ohm; Level 2 media/Data grade/IBM type 3						
4 pair 24AWG; PVC sheathed; nominal outside diameter 4.8mm. (1Base5 10BaseT 100BaseVG TR4/300 TR 16/100)	0.12	0.14	0.08	1.01	m	**1.15**
Data cable; unshielded twisted pair; solid copper conductors; PVC insulation; nomimal impedance 105 Ohm; Level 2 media/Data grade/IBM type 3						
4 pair 24AWG; PVC sheathed; nominal outside diameter 4.7mm; nominal attenuation 2.1dB/100m at 1 Mhz. (1Base5 10BaseT 100BaseVG TR4/300 TR 16/100)	0.17	0.19	0.08	1.01	m	**1.20**
Data cable; unshielded twisted pair; solid copper conductors; PVC insulation; nomimal impedance 100 Ohm; Level 3 media/High Speed/Data grade/IBM type 3						
4 pair 24AWG; PVC sheathed; nominal outside diameter 4.6mm; nominal attenuation 10dB/100m at 10 Mhz. (1Base5 10BaseT 100BaseVG TR4/300 TR 16/100)	0.14	0.16	0.08	1.01	m	**1.17**
2 pair 24AWG; PVC sheathed; nominal outside diameter 3.9mm; nominal attenuation 10dB/100m at 10 Mhz. (1Base5 10BaseT 100BaseVG TR4/300 TR 16/100)	0.13	0.15	0.06	0.76	m	**0.91**
4 pair 24AWG; PVC sheathed; nominal outside diameter 4.9mm; nominal attenuation 10dB/100m at 10 Mhz. (1Base5 10BaseT 100BaseVG TR4/300 TR 16/100)	0.18	0.20	0.08	1.01	m	**1.21**
12 pair 24AWG; PVC sheathed; nominal outside diameter 7.6mm; nominal attenuation 10dB/100m at 10 Mhz. (1Base5 10BaseT 100BaseVG TR4/300 TR 16/100)	0.59	0.65	0.08	1.01	m	**1.66**

W:COMMUNICATIONS/SECURITY/CONTROL SYSTEMS

Item	Net Price £	Material £	Labour hours	Labour £	Unit	Total Rate £
W30 : DATA TRANSMISSION : DATA COMMUNICATIONS (contd)						
Local area networking; IEEE Token Ring Systems (contd)						
Data Transmission Cables; PVC sheathed; fixed to backgrounds or laid in trunking (contd)						
25 pair 24AWG; PVC sheathed; nominal outside diameter 10.4mm; nominal attenuation 10dB/100m at 10 Mhz. (1Base5 10BaseT 100BaseVG TR4/300 TR 16/100)	1.11	1.22	0.10	1.26	m	2.48
Data cable; unshielded twisted pair; solid copper conductors; polyolefin insulation; nomimal impedance 100 Ohm; Category 4 media/Extended Distance High-speed Data-grade						
4 pair 24AWG; PVC sheathed; nominal outside diameter 4.93mm; nominal attenuation 7.2dB/100m at 16 Mhz. (1Base5 10BaseT 100BaseVG TR4/300 TR 16/100)	0.31	0.34	0.08	1.01	m	1.35
Data cable; shielded twisted pair; solid copper conductors; polyolefin insulation; nomimal impedance 100 Ohm; Category 4 media/Extended Distance High-speed Data-grade						
4 pair 24AWG; PVC sheathed; nominal outside diameter 5.25mm; nominal attenuation 9.8dB/100m at 16 Mhz. (1Base5 10BaseT 100BaseT 100BaseVG TR4/300 TR 16/100)	0.38	0.42	0.08	1.01	m	1.43
Data cable; unshielded twisted pair; solid copper conductors; polythene insulation; nomimal impedance 100 Ohm; Category 5 media/Low loss Extended-Distance High-speed Data-grade						
4 pair 24AWG; PVC sheathed; nominal outside diameter 4.8mm; ('1Base5 10BaseT 100BaseVG TR4/300 TR 16/300)	0.46	0.50	0.08	1.01	m	1.51
4 pair 24AWG; PVC sheathed; nominal outside diameter 5.51mm; nominal attenuation 8.2dB/100m at 16Mhz (1Base5 10BaseT 100BaseVG TR4/300 TR 16/300)	0.20	0.22	0.08	1.01	m	1.23
4 pair 24AWG; PVC sheathed; nominal outside diameter 5.2mm; nominal attenuation 19.1dB/100m at 100Mhz (1Base5 10BaseT 100BaseT 100BaseVG TR4/300 TR 16/300)	0.25	0.27	0.08	1.01	m	1.28

W:COMMUNICATIONS/SECURITY/CONTROL SYSTEMS

Item	Net Price £	Material £	Labour hours	Labour £	Unit	Total Rate £
4 pair 24AWG; PVC sheathed; nominal outside diameter 4.8mm; nominal attenuation 21.9dB/100m at 100Mhz (1Base5 10BaseT 100BaseT 100BaseVG TR4/300 TR 16/300)	0.19	0.21	0.08	1.01	m	**1.22**
4 pair 24AWG; PVC sheathed; nominal attenuation 21.6dB/100m at 100Mhz (1Base5 10BaseT 100BaseT 100BaseVG TR4/300 TR16/300)	0.25	0.27	0.08	1.01	m	**1.28**
Data cable; shielded twisted pairs; solid copper conductors; polethene insulation; aluminium foil shield nomimal impedance 100 Ohm; Category 5 media/Low loss Extended-Distance High-speed Data-grade						
4 pair 24AWG; PVC sheathed; nominal outside diameter 5.3mm; nominal attenuation 21.98dB/100m at 100Mhz (1Base5 10BaseT 100BaseT 100BaseVG TR4/300 TR 16/300)	0.32	0.36	0.08	1.01	m	**1.37**
4 pair 24AWG; PVC sheathed; nominal outside diameter 6.1mm; nominal attenuation 8.2dB/100m at 16Mhz (1Base5 10BaseT 100BaseT 100BaseVG TR4/300 TR 16/300)	0.31	0.34	0.08	1.01	m	**1.35**
4 pair 24AWG; PVC sheathed; nominal attenuation 21.6dB/100m at 100Mhz (1Base5 10BaseT 100BaseT 100BaseVG TR4/300 TR16/300)	0.29	0.31	0.08	1.01	m	**1.32**
Data cable; shielded twisted pairs; solid copper conductors; polythene insulation; aluminium foil shield; overall copper braid screen; nomimal impedance 150 Ohm; IBM Type 9 Indoor						
2 pair 26AWG; PVC sheathed; nominal outside diameter 6.6mm;nominal attenuation 6.6dB/100m at 16Mhz (TR4/100 TR16/100)	0.44	0.48	0.06	0.76	m	**1.24**
Data cable; shielded twisted pairs; solid copper conductors; polythene insulation; double layer aluminium foil shield; copper braid screen; nomimal impedance 150 Ohm; IBM Type 1A Indoor						
2 pair 22AWG; PVC sheathed; nominal outside diameter 10.9mm;nominal attenuation 4.0dB/100m at 16Mhz (TR4/600 TR16/300)	0.28	0.30	0.06	0.76	m	**1.06**
2 pair 22AWG; flame resistant PVC sheathed; nominal outside diameter 10.9mm ; nominal attenuation 4.0dB/100m at 16 Mhz (TR4/600 TR16/300)	0.48	0.52	0.06	0.76	m	**1.28**

W:COMMUNICATIONS/SECURITY/CONTROL SYSTEMS

Item	Net Price £	Material £	Labour hours	Labour £	Unit	Total Rate £
W30 : DATA TRANSMISSION : DATA COMMUNICATIONS (contd)						
Local area networking; IEEE Token Ring Systems (contd)						
Data Transmission Cables; PVC sheathed; fixed to backgrounds or laid in trunking (contd)						
Data cable; shielded twisted pairs; solid copper conductors; polythene insulation; double layer aluminium foil shield; copper braid screen; nomimal impedance 150 Ohm; IBM Type 1A Indoor (contd)						
2 pair 22AWG; flame resistant PVC sheathed; nominal outside diamensions 7.87 x 11.55mm ; nominal attenuation 4.0dB/100m at 16 Mhz (TR4/600 TR16/300)	0.67	0.73	0.06	0.76	m	**1.49**
Drop cable; shielded twisted pair; solid copper conductors; polythene insulation; double layer aluminium foil shield; overall copper braid screen; nominal impedance 150 Ohm; IBM Type 6 fire resistant/IBM type 6A						
2 pair 26AWG; PVC sheathed; nominal outside diameter 8.8mm; (TR4 TR16)	3.10	3.41	0.06	0.76	m	**4.17**
2 pair 26AWG; PVC sheathed; nominal outside diameter 9.41mm; (TR4 TR16)	0.74	0.82	0.06	0.76	m	**1.58**
2 pair 26AWG; fire resistant PVC sheathed; nominal outside diameter 8.8mm; (TR4 TR16)	0.54	0.60	0.06	0.76	m	**1.36**
Patch cable; shielded twisted-pair; solid copper conductors; polythene insulation; aluminium foil shield ; nominal impedance 100 Ohm; Category 5 media/Low-loss Extended-distance High speed Data grade						
4 pair 26AWG; PVC sheathed; nominal outside diameter 4.8mm (1Base5 10BaseT 100BaseT 100BaseVG)	0.37	0.41	0.08	1.01	m	**1.42**
Patch cable; unshielded twisted-pair; solid copper conductors; polythene insulation; nominal impedance 100 Ohm; Category 5 media/Low-loss Extended-distance High-speed Data grade						
4 pair 24AWG; PVC sheathed; nominal outside diameter 5.25mm (1Base5 10BaseT 100BaseT 100BaseVG)	0.32	0.36	0.08	1.01	m	**1.37**
4 pair 26AWG; PVC sheathed; nominal outside diameter 4.5mm (1Base5 10BaseT 100BaseT 100BaseVG)	0.30	0.33	0.08	1.01	m	**1.34**

W:COMMUNICATIONS/SECURITY/CONTROL SYSTEMS

Item	Net Price £	Material £	Labour hours	Labour £	Unit	Total Rate £
Outside cable; shielded twisted-pair; solid copper conductors; polythene insulation; double layer corrugated metallic shield; nominal impedance 150 Ohm; IBM Type 1 Outdoor						
2 pair 22AWG; PVC sheathed; nominal outside diameter 14.9mm (TR4/600 TR16/300)	3.10	3.41	0.06	0.76	m	4.17
Accessories for Token Ring Systems; fixed to outlet boxes or trunking						
Connectors; unshielded; Category 5 twisted pairs						
RJ45 jack; 8-way; 24-28AWG	0.84	0.92	0.17	2.09	nr	3.01
RJ45 identification boot; PVC coloured	0.18	0.19	0.02	0.20	nr	0.39
Connectors; shielded; Category 5 twisted pairs						
RJ45 jack; 8-way; 24-28AWG	0.84	0.92	0.17	2.09	nr	3.01
RJ45 identification boot; PVC coloured	0.18	0.19	0.02	0.20	nr	0.39
IBM Universal Data Connector	2.29	2.52	0.17	2.09	nr	4.61
Multiple/multi-station access units (MAU/MSAU); mains powered; electronic relays; status indicators						
8-lobe; Ring-in, Ring-out;IBM data connectors (TR4 TR16)	278.92	306.81	1.50	18.95	nr	325.76
8-lobe; Ring-in, Ring-out; shielded RJ45 connectors (TR4 TR16)	291.15	320.27	1.50	18.95	nr	339.22
Multiple/multi-station access units (MAU/MSAU); mains powered; power failure protected; electronic relays; self-monitoring and diagostics; status indicators						
8-lobe; Ring-in, Ring-out;IBM data connectors (TR4 TR16)	631.50	694.65	1.50	18.95	nr	713.60
Patch panels; passive multi-port connectors; mechanical operation						
16-port, unkeyed unshielded RJ45 connectors (TR4 TR16)	116.30	127.94	2.83	35.75	nr	163.69
16-port, unkeyed shielded RJ45 connectors (TR4 TR16)	202.54	222.79	2.83	35.75	nr	258.54
48-port, unkeyed unshielded RJ45 connectors (TR4 TR16)	302.84	333.12	8.20	103.44	nr	436.56
48-port, keyed unshielded RJ45 connectors (TR4 TR16)	405.08	445.59	8.20	103.44	nr	549.03

W:COMMUNICATIONS/SECURITY/CONTROL SYSTEMS

Item	Net Price £	Material £	Labour hours	Labour £	Unit	Total Rate £
W30 : DATA TRANSMISSION : DATA COMMUNICATIONS (contd)						
Local area networking; IEEE Token Ring Systems (contd)						
Accessories for Token Ring Systems; fixed to outlet boxes or trunking (contd)						
Patch panels; passive multi-port connectors; insulation-displacment operation						
16-port, unkeyed unshielded RJ45 connectors (TR4 TR16)	78.87	86.76	2.83	35.75	nr	**122.51**
16-port, keyed unshielded RJ45 connectors (TR4 TR16)	107.11	117.82	2.83	35.75	nr	**153.57**
16-port, keyed shielded RJ45 connectors (TR4 TR16)	121.72	133.89	2.83	35.75	nr	**169.64**
24-port, unkeyed unshielded RJ45 connectors (TR4 TR16)	122.69	134.96	4.17	52.58	nr	**187.54**
48-port, keyed unshielded RJ45 connectors (TR4 TR16)	180.14	198.16	8.20	103.44	nr	**301.60**
Data outlet face plate; to suit BS 4662 wall, pattress boxes or service trunking; fitted with keyed RJ45 shuttered data outlet; Category 5 twisted pair system						
single gang; 1 outlet; shielded	6.52	7.18	0.45	5.68	nr	**12.86**
single gang; 2 outlet; shielded	10.03	11.03	0.45	5.68	nr	**16.71**
Data outlet face plate; to suit BS 4662 wall, pattress boxes or service trunking; fitted with keyed RJ45 shuttered data outlet; twisted pair system						
single gang; 1 outlet; unshielded	6.87	7.56	0.45	5.68	nr	**13.24**
single gang; 2 outlet; unshielded	10.13	11.14	0.45	5.68	nr	**16.82**
Data outlet face plate; to suit BS 4662 wall, pattress boxes or service trunking; fitted with unkeyed RJ45 in shuttered modules; including blanking modules where required; Category 5 twisted pair system						
single gang; 1 outlet; unshielded	8.87	9.76	0.45	5.68	nr	**15.44**
single gang; 2 outlet; unshielded	15.41	16.95	0.45	5.68	nr	**22.63**
double gang; 1 outlet; unshielded	10.24	11.26	0.45	5.68	nr	**16.94**
double gang; 2 outlet; unshielded	16.38	18.02	0.45	5.68	nr	**23.70**
double gang; 3 outlet; unshielded	23.13	25.44	0.45	5.68	nr	**31.12**
double gang; 4 outlet; unshielded	29.25	32.18	0.45	5.68	nr	**37.86**

W:COMMUNICATIONS/SECURITY/CONTROL SYSTEMS

Item	Net Price £	Material £	Labour hours	Labour £	Unit	Total Rate £
Data outlet face plate; modular; to suit BS 4662 wall, pattress boxes or service trunking; fitted with keyed RJ45 in shuttered modules; including blanking modules where required; twisted pair system						
single gang; 1 outlet; unshielded	6.15	6.76	0.45	5.68	nr	**12.44**
single gang; 2 outlet; unshielded	9.91	10.90	0.45	5.68	nr	**16.58**
double gang; 1 outlet; unshielded	7.51	8.26	0.45	5.68	nr	**13.94**
double gang; 2 outlet; unshielded	11.27	12.40	0.45	5.68	nr	**18.08**
double gang; 3 outlet; unshielded	15.66	17.23	0.45	5.68	nr	**22.91**
double gang; 4 outlet; unshielded	19.42	21.36	0.45	5.68	nr	**27.04**
Data outlet face plate; to suit BS 4662 wall, pattress boxes or service trunking; fitted with IBM Universal Data connector; twisted pair system						
single gang; 1 outlet; unshielded	6.33	6.96	0.45	5.68	nr	**12.64**
Data outlet face plate; modular; to suit BS 4662 wall, pattress boxes or service trunking; fitted with IBM Universal Data Connector and hinged fascia plate; including blanking modules where required; twisted pair systems						
single gang; 1 outlet; unshielded	4.02	4.42	0.45	5.68	nr	**10.10**
double gang; 1 outlet; unshielded	5.23	5.75	0.45	5.68	nr	**11.43**
double gang; 2 outlet; unshielded	7.64	8.40	0.45	5.68	nr	**14.08**
Data outlet face plate; to suit BS 4662 wall, pattress boxes or service trunking; fitted with 25 pin D-type data outlet; twisted pair systems						
single gang; 1 outlet; shielded	3.32	3.65	0.45	5.68	nr	**9.33**
single gang; 2 outlet; shielded	3.42	3.76	0.45	5.68	nr	**9.44**
double gang; 1 outlet; shielded	3.77	4.15	0.45	5.68	nr	**9.83**
double gang; 2 outlet; shielded	5.23	5.75	0.45	5.68	nr	**11.43**
double gang; 3 outlet; shielded	6.38	7.02	0.45	5.68	nr	**12.70**
double gang; 4 outlet; shielded	6.93	7.62	0.45	5.68	nr	**13.30**
Data outlet face plate; to suit BS 4662 wall, pattress boxes or service trunking; fitted with 25 pin D-type data outlet; twisted pair systems						
single gang; 1 outlet; unshielded	7.43	8.17	0.45	5.68	nr	**13.85**
single gang; 2 outlet; unshielded	13.27	14.60	0.45	5.68	nr	**20.28**

W:COMMUNICATIONS/SECURITY/CONTROL SYSTEMS

Item	Net Price £	Material £	Labour hours	Labour £	Unit	Total Rate £
W50 : FIRE DETECTION AND ALARM : FIRE DETECTION SYSTEMS						
Fire Detection Equipment; fixed to backgrounds. (Including supports and fixings.)						
Zone control panel - small installation, complte with Batteries and charger; mild steel case; flush or surface mounting						
1 zone unit	117.67	129.44	3.00	37.90	nr	**167.34**
2 zone unit	162.86	179.15	3.51	44.28	nr	**223.43**
4 zone unit	289.45	318.39	4.00	50.48	nr	**368.87**
8 zone unit	417.23	458.95	5.00	63.10	nr	**522.05**
8 zone repeat indicator	125.76	138.33	4.00	50.48	nr	**188.81**
5 zone unit complete with batteries and charger and 5 alarm sectors	1042.52	1146.77	4.00	50.48	nr	**1197.25**
10 zone unit complete with batteries and charger and 10 alarm sectors	1320.50	1452.55	5.00	63.10	nr	**1515.65**
Multi-zone addressable Fire Alarm Panel; BS 5839 Part 4; single loop arangement; incorporating 16 detection circuits and 2 master alarm circuits; integral battery standby providing 72 hour back-up; mild steel case; surface fixed						
16 zone unit	523.21	575.53	7.04	88.87	nr	**664.40**
Repeat indicator with LCD display and integral batteries and charger	441.26	485.39	6.02	76.02	nr	**561.41**
Zone module including end-of-line device	56.37	62.01	5.00	63.10	nr	**125.11**
Power supply/charger; 24 volt; free standing						
1.5 Amp Lead Acid	165.49	182.03	6.02	76.02	nr	**258.05**
3.0 Amp Lead Acid	357.47	393.22	6.02	76.02	nr	**469.24**
6.0 Amp Lead Acid	514.29	565.72	6.02	76.02	nr	**641.74**
Battery Packs; 24 volt lead acid; sealed						
6.0 Amp/hr	66.61	73.28	8.00	100.96	nr	**174.24**
12.0 Amp/hr	134.26	147.69	8.00	100.96	nr	**248.65**
24.0 Amp/hr	201.62	221.78	8.00	100.96	nr	**322.74**
Accessories						
Manual call point units: surface mounted						
Plastic covered push to break glass unit	9.07	9.98	0.50	6.31	nr	**16.29**
Plastic covered push to break glass unit; Waterproof	14.15	15.56	0.80	10.10	nr	**25.66**

W:COMMUNICATIONS/SECURITY/CONTROL SYSTEMS

Item	Net Price £	Material £	Labour hours	Labour £	Unit	Total Rate £
Manual call point units: flush mounted						
Plastic covered push to break glass unit	8.51	9.36	0.56	7.07	nr	16.43
Plastic covered push to break glass unit; Waterproof	13.58	14.94	0.86	10.86	nr	25.80
Addressable Manual Call Point						
Plastic covered push to break glass unit	39.97	43.97	1.00	12.62	nr	56.59
Plastic covered push to break glass unit; Weather resistant	47.71	52.49	1.25	15.77	nr	68.26
Detectors						
Smoke, ionisation type with mounting base	39.97	43.97	0.75	9.47	nr	53.44
Smoke, optical type with mounting base	37.67	41.44	0.75	9.47	nr	50.91
Fixed temperature heat detector with mounting base (60 degrees)	17.37	19.11	0.75	9.47	nr	28.58
Rate of Rise heat detector with mounting base (90 degrees)	17.37	19.11	0.75	9.47	nr	28.58
Addressable Duct Detector including optical smoke detector and addressable base	214.23	235.65	2.00	25.24	nr	260.89
Sundries for detection system						
Xenon flasher, 24 volt, conduit box	33.62	36.98	0.50	6.31	nr	43.29
6" bell, conduit box	19.22	21.14	0.75	9.47	nr	30.61
6" bell, conduit box; weather resistant	30.49	33.54	1.00	12.62	nr	46.16
Siren; 24V polarised	38.47	42.32	1.00	12.62	nr	54.94
Siren; 240V	72.97	80.27	1.25	15.77	nr	96.04
Magnetic Door Holder; 240V ; surface fixed	48.84	53.73	1.50	18.95	nr	72.68

Keep your figures up to date, free of charge

This section, and most of the other information in this Price Book, is brought up to date every three months, until the next annual edition, in the *Price Book Update*.

The *Update* is available free to all Price Book purchasers.

To ensure you receive your copy, simply complete the reply card from the centre of the book and return it to us.

W:COMMUNICATIONS/SECURITY/CONTROL SYSTEMS

Item	Net Price £	Material £	Labour hours	Labour £	Unit	Total Rate £
W52 : LIGHTNING PROTECTION : GENERAL						
Lightning Protection Equipment; fixed to backgrounds. (Including supports and fixings.)						
Multiple point air terminal fixed to structure; copper						
15 mm	38.75	42.63	1.00	12.62	nr	**55.25**
Elevation rod, fixed to structure						
500 x 10 mm	6.56	7.22	0.75	9.47	nr	**16.69**
1000 x 10 mm	9.97	10.97	0.90	11.36	nr	**22.33**
Tapered rod, fixed to structure						
500 x 15 mm	54.87	60.36	0.75	9.47	nr	**69.83**
1000 x 15 mm	77.07	84.78	0.90	11.36	nr	**96.14**
Down conductor, PVC covered copper tape						
12.5 x 1.5 mm	1.45	1.67	0.20	2.52	m	**4.19**
25.0 x 3.0 mm	4.77	5.51	0.20	2.52	m	**8.03**
25.0 x 6.0 mm	8.42	9.72	0.30	3.79	m	**13.51**
Down conductor, bare copper tape						
20.0 x 3 mm	2.60	3.01	0.20	2.52	m	**5.53**
25.0 x 3 mm	3.26	3.76	0.20	2.52	m	**6.28**
31.0 x 3 mm	4.12	4.76	0.30	3.79	m	**8.55**
38.0 x 3 mm	5.07	5.86	0.30	3.79	m	**9.65**
Down conductor, PVC covered aluminium tape						
20.0 x 3 mm	3.10	3.58	0.20	2.52	m	**6.10**
25.0 x 3 mm	3.13	3.61	0.20	2.52	m	**6.13**
Down conductor, bare aluminium tape						
12.5 x 1.5 mm	0.64	0.73	0.20	2.52	m	**3.25**
20.0 x 3.0 mm	1.78	2.06	0.20	2.52	m	**4.58**
25.0 x 3.0 mm	1.92	2.21	0.20	2.52	m	**4.73**
25.0 x 6.0 mm	3.83	4.43	0.20	2.52	m	**6.95**
Rod to tape gunmetal coupling						
15 mm	7.09	7.80	0.20	2.52	nr	**10.32**
Test clamp						
Oblong	4.47	4.92	0.15	1.89	nr	**6.81**
Screwdown	11.07	12.18	0.18	2.27	nr	**14.45**
Bolted	12.66	13.93	0.25	3.15	nr	**17.08**

W:COMMUNICATIONS/SECURITY/CONTROL SYSTEMS

Item	Net Price £	Material £	Labour hours	Labour £	Unit	Total Rate £
Copper lattice earth plate; laid in ground and connected						
600 x 600 x 3 mm	37.12	40.83	0.33	4.16	nr	**44.99**
900 x 900 x 3 mm	66.16	72.77	0.33	4.16	nr	**76.93**
Copper earth electrodes; driven into ground and connected						
1220 x 15 mm	10.44	11.48	1.20	15.15	nr	**26.63**
1220 x 19 mm	18.57	20.42	1.30	16.41	nr	**36.83**
Steel cored copper bonded earth electrodes driven into ground and connected						
1200 x 16 mm	5.22	5.74	1.70	21.46	nr	**27.20**
2400 x 16 mm	10.28	11.31	1.80	22.74	nr	**34.05**
3000 x 16 mm	12.76	14.04	1.90	23.99	nr	**38.03**
Driving stud, coupling and spike fitted to earth rod						
16 mm	0.64	0.70	0.20	2.52	nr	**3.22**
20 mm	1.33	1.47	0.20	2.52	nr	**3.99**

Y:MECHANICAL AND ELECTRICAL SERVICES

Item	Net Price £	Material £	Labour hours	Labour £	Unit	Total Rate £
Y60 : CONDUIT AND CABLE TRUNKING : METAL AND PLASTICS CONDUIT						
Heavy Gauged, Screwed Welded Steel; surface fixed on saddles to backgrounds, with standard pattern boxes and fittings including all fixings and supports. (Forming holes, conduit entry, draw wires etc. and components for earth continuity are included.)						
Black Enamelled						
20 mm dia.	1.46	1.69	0.50	6.31	m	8.00
25 mm dia.	2.00	2.31	0.55	6.94	m	9.25
32 mm dia.	2.77	3.20	0.60	7.57	m	10.77
38 mm dia.	3.54	4.09	0.70	8.84	m	12.93
50 mm dia.	5.74	6.63	1.15	14.52	m	21.15
Galvanised						
20 mm dia.	2.21	2.55	0.50	6.31	m	8.86
25 mm dia.	3.02	3.49	0.55	6.94	m	10.43
32 mm dia.	3.83	4.42	0.60	7.57	m	11.99
38 mm dia.	4.89	5.65	0.70	8.84	m	14.49
50 mm dia.	7.23	8.35	1.15	14.52	m	22.87
High Impact Unscrewed PVC; surface fixed on saddles to backgrounds; with standard pattern boxes and fittings; including all fixings and supports.						
Light Gauge						
16 mm dia.	0.79	0.91	0.25	3.15	m	4.06
20 mm dia.	0.99	1.15	0.25	3.15	m	4.30
25 mm dia.	1.66	1.91	0.30	3.79	m	5.70
32 mm dia.	2.21	2.56	0.30	3.79	m	6.35
38 mm dia.	2.81	3.25	0.35	4.42	m	7.67
50 mm dia.	4.63	5.35	0.35	4.42	m	9.77
Heavy Gauge						
16 mm dia.	1.11	1.28	0.25	3.15	m	4.43
20 mm dia.	1.36	1.57	0.25	3.15	m	4.72
25 mm dia.	1.80	2.08	0.30	3.79	m	5.87
32 mm dia.	2.89	3.34	0.30	3.79	m	7.13
38 mm dia.	3.80	4.39	0.35	4.42	m	8.81
50 mm dia.	6.03	6.96	0.35	4.42	m	11.38

Y: MECHANICAL AND ELECTRICAL SERVICES

Item	Net Price £	Material £	Labour hours	Labour £	Unit	Total Rate £
Flexible Conduits; including adaptors and locknuts. (For connections to equipment.)						
Metallic, PVC covered conduit; not exceeding 1m long; including zinc plated mild steel adaptors, lock nuts and earth conductor						
16 mm dia.	4.10	4.74	0.50	6.31	nr	**11.05**
20 mm dia.	5.01	5.79	0.50	6.31	nr	**12.10**
25 mm dia.	6.60	7.62	0.50	6.31	nr	**13.93**
32 mm dia.	9.41	10.87	0.60	7.57	nr	**18.44**
38 mm dia.	11.98	13.84	0.60	7.57	nr	**21.41**
50 mm dia.	12.91	15.33	1.00	12.62	nr	**27.95**
PVC Conduit; not exceeding 1m long; including nylon adaptors, lock nuts						
16 mm dia.	1.14	1.32	0.50	6.31	nr	**7.63**
20 mm dia.	1.42	1.64	0.50	6.31	nr	**7.95**
25 mm dia.	2.05	2.37	0.50	6.31	nr	**8.68**
32 mm.dia.	3.19	3.68	0.60	7.57	nr	**11.25**
Grey Cast Iron Adaptable Boxes; with heavy covers, fixed to backgrounds; including all supports and fixings. (Cutting and connecting conduit to boxes is included.)						
Square Pattern - Black						
150 x 150 x 50 mm	22.09	24.30	0.92	11.61	nr	**35.91**
150 x 150 x 75 mm	27.65	30.41	0.92	11.61	nr	**42.02**
150 x 150 x 100 mm	34.18	37.60	0.92	11.61	nr	**49.21**
225 x 225 x 75 mm	54.35	59.79	1.00	12.62	nr	**72.41**
300 x 300 x 100 mm	121.34	133.47	1.05	13.26	nr	**146.73**
Square Pattern - Galvanised						
150 x 150 x 50 mm	30.93	34.02	0.92	11.61	nr	**45.63**
150 x 150 x 75 mm	35.67	39.24	0.92	11.61	nr	**50.85**
150 x 150 x 100 mm	47.38	52.12	0.92	11.61	nr	**63.73**
225 x 225 x 75 mm	74.82	82.30	1.00	12.62	nr	**94.92**
300 x 300 x 100 mm	154.72	170.19	1.05	13.26	nr	**183.45**
Rectangular Pattern - Black						
150 x 75 x 50 mm	14.58	16.04	0.92	11.61	nr	**27.65**
150 x 100 x 50 mm	22.56	24.82	0.92	11.61	nr	**36.43**
150 x 100 x 75 mm	24.44	26.88	0.92	11.61	nr	**38.49**
225 x 150 x 75 mm	42.60	46.86	0.98	12.37	nr	**59.23**
225 x 150 x 100 mm	57.54	63.29	1.00	12.62	nr	**75.91**
300 x 150 x 75 mm	79.15	87.07	1.05	13.26	nr	**100.33**

Y:MECHANICAL AND ELECTRICAL SERVICES

Item	Net Price £	Material £	Labour hours	Labour £	Unit	Total Rate £
Y60 : CONDUIT AND CABLE TRUNKING : METAL AND PLASTICS CONDUIT (contd)						
Grey Cast Iron Adaptable Boxes; with heavy covers, fixed to backgrounds; including all supports and fixings. (Cutting and connecting conduit to boxes is included.) (contd)						
Rectangular Pattern - Galvanised						
150 x 75 x 50 mm	19.85	21.84	0.92	11.61	nr	**33.45**
150 x 100 x 50 mm	31.27	34.39	0.92	11.61	nr	**46.00**
150 x 100 x 75 mm	31.06	34.17	0.92	11.61	nr	**45.78**
225 x 150 x 75 mm	59.35	65.28	0.98	12.37	nr	**77.65**
225 x 150 x 100 mm	77.55	85.31	1.00	12.62	nr	**97.93**
300 x 150 x 75 mm	102.74	113.02	1.05	13.26	nr	**126.28**
Sheet Steel Adaptable Boxes; with plain or knockout sides; fixed to backgrounds; including supports and fixings. (Cutting and connecting conduit to boxes is included.)						
Square Pattern - Black						
75 x 75 x 37 mm	4.63	5.09	0.88	11.11	nr	**16.20**
75 x 75 x 50 mm	5.25	5.77	0.88	11.11	nr	**16.88**
75 x 75 x 75 mm	6.29	6.93	0.88	11.11	nr	**18.04**
100 x 100 x 50 mm	5.56	6.12	0.90	11.36	nr	**17.48**
150 x 150 x 50 mm	7.45	8.20	0.90	11.36	nr	**19.56**
150 x 150 x 75 mm	8.27	9.09	0.92	11.61	nr	**20.70**
150 x 150 x 100 mm	10.30	11.34	0.92	11.61	nr	**22.95**
225 x 225 x 50 mm	12.96	14.25	0.98	12.37	nr	**26.62**
225 x 225 x 100 mm	15.70	17.26	1.00	12.62	nr	**29.88**
300 x 300 x 100 mm	25.72	28.30	1.05	13.26	nr	**41.56**
Square Pattern - Galvanised						
75 x 75 x 37 mm	4.99	5.49	0.88	11.11	nr	**16.60**
75 x 75 x 50 mm	5.35	5.88	0.88	11.11	nr	**16.99**
75 x 75 x 75 mm	6.68	7.35	0.90	11.36	nr	**18.71**
100 x 100 x 50 mm	5.95	6.55	0.90	11.36	nr	**17.91**
150 x 150 x 50 mm	6.99	7.69	1.00	12.62	nr	**20.31**
150 x 150 x 75 mm	9.34	10.27	0.92	11.61	nr	**21.88**
150 x 150 x 100 mm	11.83	13.01	0.92	11.61	nr	**24.62**
225 x 225 x 50 mm	16.04	17.64	0.98	12.37	nr	**30.01**
225 x 225 x 100 mm	19.26	21.18	1.00	12.62	nr	**33.80**
300 x 300 x 100 mm	31.68	34.84	1.05	13.26	nr	**48.10**
Rectangular Pattern - Black						
100 x 75 x 50 mm	5.34	5.88	0.88	11.11	nr	**16.99**
150 x 75 x 50 mm	5.68	6.25	0.88	11.11	nr	**17.36**
150 x 75 x 75 mm	6.81	7.49	0.90	11.36	nr	**18.85**
150 x 100 x 75 mm	7.45	8.20	0.90	11.36	nr	**19.56**
225 x 75 x 50 mm	7.20	7.92	0.90	11.36	nr	**19.28**
225 x 150 x 75 mm	11.73	12.90	0.92	11.61	nr	**24.51**
225 x 150 x 100 mm	12.68	13.94	0.92	11.61	nr	**25.55**
300 x 150 x 50 mm	13.42	14.76	0.98	12.37	nr	**27.13**

Y: MECHANICAL AND ELECTRICAL SERVICES

Item	Net Price £	Material £	Labour hours	Labour £	Unit	Total Rate £
300 x 150 x 75 mm	14.32	15.75	1.00	12.62	nr	**28.37**
300 x 150 x 100 mm	16.40	18.04	1.05	13.26	nr	**31.30**
Rectangular Pattern - Galvanised						
100 x 75 x 50 mm	6.93	7.62	0.88	11.11	nr	**18.73**
150 x 75 x 50 mm	6.19	6.81	0.88	11.11	nr	**17.92**
150 x 75 x 75 mm	7.68	8.44	0.90	11.36	nr	**19.80**
150 x 100 x 75 mm	7.58	8.33	0.90	11.36	nr	**19.69**
225 x 75 x 50 mm	7.67	8.43	0.90	11.36	nr	**19.79**
225 x 150 x 75 mm	13.93	15.32	0.92	11.61	nr	**26.93**
225 x 150 x 100 mm	15.28	16.80	0.92	11.61	nr	**28.41**
300 x 150 x 50 mm	13.42	14.76	0.98	12.37	nr	**27.13**
300 x 150 x 75 mm	17.61	19.38	1.00	12.62	nr	**32.00**
300 x 150 x 100 mm	19.85	21.84	1.05	13.26	nr	**35.10**
PVC Adaptable Boxes; fixed to backgrounds; including all supports and fixings. (Cutting and connecting conduit to boxes is included.)						
Square Pattern						
75 x 75 x 53 mm	3.08	3.38	0.88	11.11	nr	**14.49**
100 x 100 x 75 mm	5.18	5.69	0.90	11.36	nr	**17.05**
150 x 150 x 75 mm	6.64	7.30	0.92	11.61	nr	**18.91**
Terminal Strips; (To be fixed in metal or polythene adaptable boxes.)						
20 Amp High Density Polythene						
2 way	0.54	0.59	0.25	3.15	nr	**3.74**
3 way	0.67	0.74	0.25	3.15	nr	**3.89**
4 way	0.82	0.90	0.25	3.15	nr	**4.05**
5 way	0.96	1.06	0.25	3.15	nr	**4.21**
6 way	1.10	1.21	0.25	3.15	nr	**4.36**
7 way	1.26	1.38	0.25	3.15	nr	**4.53**
8 way	1.39	1.53	0.30	3.79	nr	**5.32**
9 way	1.53	1.68	0.30	3.79	nr	**5.47**
10 way	1.67	1.84	0.35	4.42	nr	**6.26**
11 way	1.82	2.00	0.35	4.42	nr	**6.42**
12 way	1.97	2.16	0.35	4.42	nr	**6.58**
13 way	2.10	2.31	0.35	4.42	nr	**6.73**
14 way	2.25	2.47	0.35	4.42	nr	**6.89**
15 way	2.39	2.62	0.40	5.05	nr	**7.67**
16 way	2.53	2.78	0.40	5.05	nr	**7.83**
18 way	2.41	2.65	0.40	5.05	nr	**7.70**

Y:MECHANICAL AND ELECTRICAL SERVICES

Item	Net Price £	Material £	Labour hours	Labour £	Unit	Total Rate £
Y60 : CONDUIT AND CABLE TRUNKING : METAL AND PLASTICS CONDUIT (contd)						
Galvanised Steel Trunking; fixed to backgrounds; jointed with standard connectors (including plates for air gap between trunking and background; earth continuity straps included).						
Single Compartment						
50 x 50 mm	5.46	6.31	0.42	5.30	m	11.61
75 x 50 mm	6.54	7.55	0.50	6.31	m	13.86
75 x 75 mm	7.28	8.41	0.54	6.82	m	15.23
100 x 50 mm	7.48	8.64	0.54	6.82	m	15.46
100 x 75 mm	8.62	9.95	0.67	8.46	m	18.41
100 x 100 mm	9.23	10.65	0.75	9.47	m	20.12
150 x 50 mm	11.04	11.23	1.00	12.62	m	23.85
150 x 100 mm	13.55	15.65	0.96	12.12	m	27.77
150 x 150 mm	16.02	18.51	1.08	13.64	m	32.15
225 x 75 mm	17.34	20.03	1.12	14.15	m	34.18
225 x 150 mm	20.60	26.67	1.00	12.62	m	39.29
225 x 225 mm	26.00	30.03	1.20	15.15	m	45.18
300 x 75 mm	20.60	23.79	1.20	15.15	m	38.94
300 x 100 mm	24.60	28.42	1.25	15.77	m	44.19
300 x 150 mm	26.00	30.03	1.25	15.77	m	45.80
300 x 225 mm	28.75	33.21	1.35	17.05	m	50.26
300 x 300 mm	31.42	36.29	1.40	17.68	m	53.97
Double Compartment						
50 x 50 mm	7.10	8.20	0.45	5.68	m	13.88
75 x 50 mm	7.90	9.12	0.55	6.94	m	16.06
75 x 75 mm	8.75	10.10	0.60	7.57	m	17.67
100 x 50 mm	8.83	10.20	0.60	7.57	m	17.77
100 x 75 mm	9.68	11.18	0.75	9.47	m	20.65
100 x 100 mm	10.68	12.34	0.83	10.48	m	22.82
150 x 50 mm	12.39	14.31	0.83	10.48	m	24.79
150 x 100 mm	15.47	17.87	1.05	13.26	m	31.13
150 x 150 mm	18.86	21.78	1.20	15.15	m	36.93
Triple Compartment						
75 x 50 mm	9.53	11.01	0.65	8.21	m	19.22
75 x 75 mm	10.47	12.09	0.70	8.84	m	20.93
100 x 50 mm	10.20	11.78	0.70	8.84	m	20.62
100 x 75 mm	11.39	13.16	0.85	10.73	m	23.89
100 x 100 mm	12.79	14.77	0.93	11.74	m	26.51
150 x 50 mm	13.79	15.93	0.93	11.74	m	27.67
150 x 100 mm	17.56	20.28	0.87	10.97	m	31.25
150 x 150 mm	23.15	26.74	1.30	16.41	m	43.15
Four Compartment						
100 x 50 mm	11.57	13.36	0.75	9.47	m	22.83
100 x 75 mm	13.13	15.16	0.90	11.36	m	26.52
100 x 100 mm	14.87	17.18	0.98	12.37	m	29.55
150 x 50 mm	15.14	17.48	0.98	12.37	m	29.85
150 x 100 mm	19.65	22.69	1.20	15.15	m	37.84
150 x 150 mm	27.40	31.64	1.35	17.05	m	48.69

Y: MECHANICAL AND ELECTRICAL SERVICES

Item	Net Price £	Material £	Labour hours	Labour £	Unit	Total Rate £
Galvanised Steel Trunking Fittings; (Cutting and jointing trunking to fittings is included.)						
Additional Connector or Stop End						
50 x 50 mm	0.82	0.90	0.16	2.02	nr	2.92
75 x 50 mm	0.89	0.97	0.16	2.02	nr	2.99
75 x 75 mm	0.95	1.05	0.16	2.02	nr	3.07
100 x 50 mm	1.00	1.10	0.20	2.52	nr	3.62
100 x 75 mm	1.07	1.18	0.20	2.52	nr	3.70
100 x 100 mm	1.13	1.24	0.20	2.52	nr	3.76
150 x 50 mm	1.24	1.36	0.20	2.52	nr	3.88
150 x 100 mm	1.33	1.46	0.20	2.52	nr	3.98
150 x 150 mm	1.84	2.02	0.20	2.52	nr	4.54
225 x 75 mm	1.87	2.06	0.25	3.15	nr	5.21
225 x 150 mm	2.71	2.98	0.25	3.15	nr	6.13
225 x 225 mm	3.27	3.59	0.25	3.15	nr	6.74
300 x 75 mm	2.71	2.98	0.30	3.79	nr	6.77
300 x 100 mm	3.27	3.59	0.30	3.79	nr	7.38
300 x 150 mm	3.39	3.72	0.30	3.79	nr	7.51
300 x 225 mm	3.54	3.90	0.30	3.79	nr	7.69
300 x 300 mm	3.69	4.06	0.30	3.79	nr	7.85
Flanged Connector or Stop End						
50 x 50 mm	0.82	0.90	0.16	2.02	nr	2.92
75 x 50 mm	0.89	0.97	0.16	2.02	nr	2.99
75 x 75 mm	0.95	1.05	0.16	2.02	nr	3.07
100 x 50 mm	1.00	1.10	0.20	2.52	nr	3.62
100 x 75 mm	1.07	1.18	0.20	2.52	nr	3.70
100 x 100 mm	1.13	1.24	0.20	2.52	nr	3.76
150 x 50 mm	1.24	1.36	0.20	2.52	nr	3.88
150 x 100 mm	1.33	1.46	0.20	2.52	nr	3.98
150 x 150 mm	1.84	2.02	0.20	2.52	nr	4.54
225 x 75 mm	1.87	2.06	0.25	3.15	nr	5.21
225 x 150 mm	2.71	2.98	0.25	3.15	nr	6.13
225 x 225 mm	3.27	3.59	0.25	3.15	Unit	6.74
300 x 75 mm	2.71	2.98	0.30	3.79	nr	6.77
300 x 100 mm	3.27	3.59	0.30	3.79	nr	7.38
300 x 150 mm	3.39	3.72	0.30	3.79	nr	7.51
300 x 225 mm	3.54	3.90	0.30	3.79	nr	7.69
300 x 300 mm	3.69	4.06	0.30	3.79	nr	7.85
Bends 90 Degree; Single Compartment						
50 x 50 mm	5.05	5.55	0.50	6.31	nr	11.86
75 x 50 mm	5.86	6.44	0.55	6.94	nr	13.38
75 x 75 mm	6.34	6.97	0.60	7.57	nr	14.54
100 x 50 mm	6.53	7.18	0.62	7.83	nr	15.01
100 x 75 mm	6.72	7.39	0.67	8.46	nr	15.85
100 x 100 mm	6.99	7.69	0.70	8.84	nr	16.53
150 x 50 mm	7.48	8.23	0.75	9.47	nr	17.70
150 x 100 mm	10.11	11.12	1.25	15.77	nr	26.89
150 x 150 mm	11.26	12.39	1.18	14.85	nr	27.24
225 x 75 mm	15.65	17.22	0.85	10.73	nr	27.95
225 x 150 mm	20.53	22.58	0.95	12.00	nr	34.58
225 x 225 mm	23.93	26.32	0.95	12.00	nr	38.32

Y:MECHANICAL AND ELECTRICAL SERVICES

Item	Net Price £	Material £	Labour hours	Labour £	Unit	Total Rate £
Y60 : CONDUIT AND CABLE TRUNKING : METAL AND PLASTICS CONDUIT (contd)						
Galvanised Steel Trunking Fittings (contd)						
Bends 90 Degree; Single Compartment (contd)						
300 x 75 mm	22.08	24.28	0.95	12.02	nr	36.30
300 x 100 mm	22.55	24.80	1.05	13.26	nr	38.06
300 x 150 mm	23.91	26.30	1.15	14.52	nr	40.82
300 x 225 mm	27.76	30.53	1.15	14.52	nr	45.05
300 x 300 mm	34.85	38.34	1.25	15.77	nr	54.11
Bends 90 Degree; Double Compartment						
50 x 50 mm	5.97	6.57	0.50	6.31	nr	12.88
75 x 50 mm	6.75	7.43	0.55	6.94	nr	14.37
75 x 75 mm	7.37	8.11	0.60	7.57	nr	15.68
100 x 50 mm	7.48	8.23	0.62	7.83	nr	16.06
100 x 75 mm	7.70	8.47	0.67	8.46	nr	16.93
100 x 100 mm	8.12	8.93	0.70	8.84	nr	17.77
150 x 50 mm	8.59	9.45	0.75	9.47	nr	18.92
150 x 100 mm	11.13	12.25	0.80	10.10	nr	22.35
150 x 150 mm	12.40	13.64	0.85	10.73	nr	24.37
Bends 90 Degree; Triple Compartment						
75 x 50 mm	7.83	8.62	0.55	6.94	nr	15.56
75 x 75 mm	8.51	9.36	0.60	7.57	nr	16.93
100 x 50 mm	8.64	9.51	0.62	7.83	nr	17.34
100 x 75 mm	8.83	9.71	0.67	8.46	nr	18.17
100 x 100 mm	9.36	10.30	0.70	8.84	nr	19.14
150 x 50 mm	9.63	10.59	0.75	9.47	nr	20.06
150 x 100 mm	12.42	13.66	0.80	10.10	nr	23.76
150 x 150 mm	13.77	15.15	0.85	10.73	nr	25.88
Bends 90 Degrees; Four Compartments						
100 x 50 mm	9.54	10.49	0.62	7.83	nr	18.32
100 x 75 mm	9.95	10.95	0.67	8.46	nr	19.41
100 x 100 mm	10.63	11.69	0.70	8.84	nr	20.53
150 x 50 mm	10.67	11.74	0.75	9.47	nr	21.21
150 x 100 mm	13.63	15.00	0.80	10.10	nr	25.10
150 x 150 mm	15.13	16.64	0.85	10.73	nr	27.37
Tees; Single Compartment						
50 x 50 mm	5.84	6.42	0.71	8.96	nr	15.38
75 x 50 mm	6.75	7.43	0.71	8.96	nr	16.39
75 x 75 mm	6.99	7.69	0.75	9.47	nr	17.16
100 x 50 mm	7.88	8.67	0.75	9.47	nr	18.14
100 x 75 mm	8.16	8.98	0.85	10.73	nr	19.71
100 x 100 mm	8.40	9.24	0.85	10.73	nr	19.97
150 x 50 mm	8.65	9.52	0.95	12.00	nr	21.52
150 x 100 mm	12.15	13.36	0.95	12.00	nr	25.36
150 x 150 mm	13.28	14.61	1.05	13.26	nr	27.87

Y: MECHANICAL AND ELECTRICAL SERVICES

Item	Net Price £	Material £	Labour hours	Labour £	Unit	Total Rate £
225 x 75 mm	19.71	21.69	1.05	13.26	nr	**34.95**
225 x 150 mm	28.97	31.86	1.15	14.52	nr	**46.38**
225 x 225 mm	33.74	37.11	1.15	14.52	nr	**51.63**
300 x 75 mm	31.08	34.19	1.25	15.77	nr	**49.96**
300 x 100 mm	32.57	35.83	1.25	15.77	nr	**51.60**
300 x 150 mm	33.90	37.29	1.35	17.05	nr	**54.34**
300 x 225 mm	39.11	43.03	1.40	17.68	nr	**60.71**
300 x 300mm	46.85	51.54	1.50	18.95	nr	**70.49**
Tees; Double Compartment						
50 x 50 mm	7.68	8.45	0.71	8.96	nr	**17.41**
75 x 50 mm	8.58	9.44	0.71	8.96	nr	**18.40**
75 x 75 mm	8.82	9.70	0.75	9.47	nr	**19.17**
100 x 50 mm	9.51	10.46	0.75	9.47	nr	**19.93**
100 x 75 mm	9.95	10.95	0.85	10.73	nr	**21.68**
100 x 100 mm	10.22	11.24	0.85	10.73	nr	**21.97**
150 x 50 mm	10.41	11.45	0.95	12.00	nr	**23.45**
150 x 100 mm	13.86	15.24	0.95	12.00	nr	**27.24**
150 x 150 mm	14.96	16.46	1.05	13.26	nr	**29.72**
Tees; Triple Compartment						
75 x 50 mm	10.57	11.63	0.71	8.96	nr	**20.59**
75 x 75 mm	10.78	11.86	0.75	9.47	nr	**21.33**
100 x 50 mm	11.49	12.63	0.75	9.47	nr	**22.10**
100 x 75 mm	11.95	13.14	0.85	10.73	nr	**23.87**
100 x 100 mm	12.17	13.38	0.85	10.73	nr	**24.11**
150 x 50 mm	12.40	13.64	0.95	12.00	nr	**25.64**
150 x 100 mm	15.81	17.39	0.95	12.00	nr	**29.39**
150 x 150 mm	16.96	18.65	1.05	13.26	nr	**31.91**
Tees; Four Compartment						
100 x 50 mm	13.43	14.77	0.75	9.47	nr	**24.24**
100 x 75 mm	13.87	15.26	0.85	10.73	nr	**25.99**
100 x 100 mm	14.11	15.52	0.85	10.73	nr	**26.25**
150 x 50 mm	14.33	15.76	0.95	12.00	nr	**27.76**
150 x 100 mm	17.74	19.51	0.95	12.00	nr	**31.51**
150 x 150 mm	18.90	20.79	1.05	13.26	nr	**34.05**
Crossovers; Single Compartment						
50 x 50 mm	7.36	8.10	0.81	10.23	nr	**18.33**
75 x 50 mm	9.64	10.60	0.81	10.23	nr	**20.83**
75 x 75 mm	11.16	12.28	0.85	10.73	nr	**23.01**
100 x 50 mm	11.86	13.05	0.85	10.73	nr	**23.78**
100 x 75 mm	12.68	13.94	0.95	12.00	nr	**25.94**
100 x 100 mm	13.48	14.83	0.95	12.00	nr	**26.83**
150 x 50 mm	14.11	15.52	1.05	13.26	nr	**28.78**
150 x 100 mm	16.49	18.13	1.05	13.26	nr	**31.39**
150 x 150 mm	18.82	20.70	1.15	14.52	nr	**35.22**
225 x 75 mm	28.99	31.89	1.15	14.52	nr	**46.41**
225 x 150 mm	38.55	42.41	1.25	15.77	nr	**58.18**
225 x 225 mm	44.22	48.64	1.25	15.77	nr	**64.41**
300 x 75 mm	41.56	45.71	1.30	16.41	nr	**62.12**
300 x 100 mm	42.82	47.10	1.30	16.41	nr	**63.51**
300 x 150 mm	44.22	48.64	1.35	17.05	nr	**65.69**
300 x 225 mm	50.38	55.42	1.35	17.05	nr	**72.47**
300 x 300mm	60.03	66.03	1.50	18.95	nr	**84.98**

Y:MECHANICAL AND ELECTRICAL SERVICES

Item	Net Price £	Material £	Labour hours	Labour £	Unit	Total Rate £
Y60 : CONDUIT AND CABLE TRUNKING : METAL AND PLASTICS CONDUIT (contd)						
Galvanised Steel Trunking Fittings (contd)						
Crossovers; Double Compartment						
50 x 50 mm	9.44	10.38	0.81	10.23	nr	**20.61**
75 x 50 mm	11.69	12.86	0.81	10.23	nr	**23.09**
75 x 75 mm	13.14	14.45	0.85	10.73	nr	**25.18**
100 x 50 mm	14.16	15.58	0.85	10.73	nr	**26.31**
100 x 75 mm	14.63	16.09	0.95	12.00	nr	**28.09**
100 x 100 mm	15.37	16.91	0.95	12.00	nr	**28.91**
150 x 50 mm	16.01	17.61	0.95	12.00	nr	**29.61**
150 x 100 mm	18.36	20.19	1.05	13.26	nr	**33.45**
150 x 150 mm	19.33	21.26	1.15	14.52	nr	**35.78**
Crossovers; Triple Compartment						
75 x 50 mm	13.89	15.28	0.81	10.23	nr	**25.51**
75 x 75 mm	15.37	16.91	0.85	10.73	nr	**27.64**
100 x 50 mm	16.38	18.02	0.85	10.73	nr	**28.75**
100 x 75 mm	16.85	18.54	0.95	12.00	nr	**30.54**
100 x 100 mm	17.61	19.37	0.95	12.00	nr	**31.37**
150 x 50 mm	18.21	20.03	1.05	13.26	nr	**33.29**
150 x 100 mm	20.61	22.67	1.05	13.26	nr	**35.93**
150 x 150 mm	21.59	23.75	1.15	14.52	nr	**38.27**
Crossovers; Four Compartments						
100 x 50 mm	18.59	20.45	0.85	10.73	nr	**31.18**
100 x 75 mm	19.10	21.01	0.95	12.00	nr	**33.01**
100 x 100 mm	19.85	21.84	0.95	12.00	nr	**33.84**
150 x 50 mm	20.46	22.51	1.05	13.26	nr	**35.77**
150 x 100 mm	22.82	25.11	1.05	13.26	nr	**38.37**
150 x 150 mm	23.79	26.17	1.15	14.52	nr	**40.69**
Anodised Aluminium Lightweight Trunking; fixed to brackets; standard coupling joints; earth continuity straps						
Suspended Surface Trunking						
66 x 32mm	4.29	4.95	0.45	5.68	m	**10.63**
Suspended Recessed Trunking						
90 x 65mm	7.03	8.12	0.45	5.68	m	**13.80**

Y: MECHANICAL AND ELECTRICAL SERVICES

Item	Net Price £	Material £	Labour hours	Labour £	Unit	Total Rate £
Anodised Aluminium Lightweight Trunking Fittings; (Cutting and jointing trunking to fittings is included.)						
Extra on 66 x 32mm trunking for:						
Bend	8.55	9.41	0.50	6.31	nr	**15.72**
Tee	8.27	9.09	0.55	6.94	nr	**16.03**
Cross	12.18	13.39	0.60	7.57	nr	**20.96**
End cap	12.63	13.89	0.15	1.89	nr	**15.78**
PVC cover (per 1.8m length)	2.64	2.90	0.12	1.51	nr	**4.41**
T bolt fitting suspension unit	1.58	1.74	0.16	2.02	nr	**3.76**
Extra on 90 x 65mm trunking for:						
Bend	13.68	15.05	0.50	6.31	nr	**21.36**
Tee	17.17	18.88	0.55	6.94	nr	**25.82**
Cross	18.85	20.73	0.60	7.57	nr	**28.30**
End cap	2.11	2.32	0.15	1.89	nr	**4.21**
PVC cover (per 1.8m length)	2.64	2.90	0.12	1.51	nr	**4.41**
T bolt fitting suspension unit	2.84	3.13	0.16	2.02	nr	**5.15**
Galvanised Steel Flush Floor Trunking; fixed to backgrounds; supports and fixings; standard coupling joints; (including plates for air gap between trunking and background; earth continuity straps included.)						
Double Compartment						
265 x 60mm	24.20	27.95	1.20	15.15	m	**43.10**
Triple Compartment						
395 x 60mm	35.24	40.70	1.80	22.74	m	**63.44**
Galvanised Steel Flush Floor Trunking; Fittings (Cutting and jointing trunking to fittings is included.)						
Stop End; Double Compartment						
265 x 60mm	2.72	2.99	0.36	4.54	nr	**7.53**
Stop End; Triple Compartment						
395 x 60mm	4.11	4.52	0.55	6.94	nr	**11.46**
Rising Bend; Double Compartment						
265 x 60mm	17.51	19.26	1.20	15.15	nr	**34.41**
Rising Bend; Triple Compartment						
395 x 60mm	26.59	29.25	1.80	22.74	nr	**51.99**
Junction Box; Double Compartment						
265 x 60mm	28.88	31.77	1.00	12.62	nr	**44.39**

Y:MECHANICAL AND ELECTRICAL SERVICES

Item	Net Price £	Material £	Labour hours	Labour £	Unit	Total Rate £
Y60 : CONDUIT AND CABLE TRUNKING : METAL AND PLASTICS CONDUIT (contd)						
Galvanised Steel Flush Floor Trunking; Fittings (Cutting and jointing trunking to fittings is included.) (contd)						
Junction Box; Triple Compartment						
395 x 60mm	54.17	59.58	1.25	15.77	nr	**75.35**
Single Compartment PVC Trunking; grey finish; clip on lid; fixed to backgrounds; including supports and fixings (standard coupling joints)						
Single Compartment						
50 x 50mm	8.23	9.51	0.25	3.15	m	**12.66**
75 x 50mm	9.45	10.92	0.25	3.15	m	**14.07**
75 x 75mm	11.74	13.56	0.25	3.15	m	**16.71**
100 x 50mm	13.51	15.60	0.30	3.79	m	**19.39**
100 x 75mm	15.00	17.32	0.35	4.42	m	**21.74**
100 x 100mm	17.35	20.04	0.35	4.42	m	**24.46**
150 x 75mm	29.31	33.86	0.40	5.05	m	**38.91**
150 x 100mm	36.72	42.41	0.40	5.05	m	**47.46**
150 x 50mm	38.24	44.17	0.45	5.68	m	**49.85**
Single Compartment PVC Trunking Fittings; (cutting and jointing trunking to fittings is included)						
Crossover						
50 x 50mm	23.21	25.53	0.18	2.27	nr	**27.80**
75 x 50mm	25.44	27.99	0.18	2.27	nr	**30.26**
75 x 75mm	27.64	30.41	0.18	2.27	nr	**32.68**
100 x 50mm	42.97	47.26	0.20	2.52	nr	**49.78**
100 x 75mm	47.83	52.61	0.20	2.52	nr	**55.13**
100 x 100mm	47.83	52.61	0.25	3.15	nr	**55.76**
150 x 75mm	55.75	61.33	0.25	3.15	nr	**64.48**
150 x 100mm	69.93	76.92	0.25	3.15	nr	**80.07**
150 x 150mm	88.26	97.09	0.25	3.15	nr	**100.24**
Stop End						
50 x 50mm	2.35	2.58	0.05	0.63	nr	**3.21**
75 x 50mm	2.62	2.88	0.05	0.63	nr	**3.51**
75 x 75mm	2.62	2.88	0.05	0.63	nr	**3.51**
100 x 50mm	3.49	3.83	0.05	0.63	nr	**4.46**
100 x 75mm	4.30	4.73	0.05	0.63	nr	**5.36**
100 x 100mm	4.30	4.73	0.08	1.01	nr	**5.74**
150 x 75mm	13.95	15.34	0.08	1.01	nr	**16.35**
150 x 100mm	13.95	15.34	0.08	1.01	nr	**16.35**
150 x 150mm	19.39	21.33	0.08	1.01	nr	**22.34**

Y: MECHANICAL AND ELECTRICAL SERVICES

Item	Net Price £	Material £	Labour hours	Labour £	Unit	Total Rate £
Flanged Coupling						
50 x 50mm	6.46	7.11	0.20	2.52	nr	**9.63**
75 x 50mm	7.37	8.11	0.20	2.52	nr	**10.63**
75 x 75mm	8.87	9.76	0.20	2.52	nr	**12.28**
100 x 50mm	9.66	10.63	0.30	3.79	nr	**14.42**
100 x 75mm	11.96	13.16	0.30	3.79	nr	**16.95**
100 x 100mm	11.96	13.16	0.30	3.79	nr	**16.95**
150 x 75mm	13.95	15.34	0.40	5.05	nr	**20.39**
150 x 100mm	17.34	18.99	0.40	5.05	nr	**24.04**
150 x 150mm	21.97	24.16	0.40	5.05	nr	**29.21**
Internal Coupling						
50 x 50mm	2.35	2.58	0.06	0.76	nr	**3.34**
75 x 50mm	2.62	2.88	0.06	0.76	nr	**3.64**
75 x 75mm	2.62	2.88	0.06	0.76	nr	**3.64**
100 x 50mm	3.49	3.83	0.08	1.01	nr	**4.84**
100 x 75mm	4.36	4.80	0.08	1.01	nr	**5.81**
100 x 100mm	4.36	4.80	0.08	1.01	nr	**5.81**
External Coupling						
50 x 50mm	2.90	3.19	0.06	0.76	nr	**3.95**
75 x 50mm	3.62	3.98	0.06	0.76	nr	**4.74**
75 x 75mm	4.23	4.65	0.06	0.76	nr	**5.41**
100 x 50mm	6.68	7.35	0.08	1.01	nr	**8.36**
100 x 75mm	8.40	9.24	0.08	1.01	nr	**10.25**
100 x 100mm	8.40	9.24	0.08	1.01	nr	**10.25**
150 x 75mm	12.32	13.55	0.10	1.26	nr	**14.81**
150 x 100mm	15.21	16.73	0.10	1.26	nr	**17.99**
150 x 150mm	18.85	20.74	0.10	1.26	nr	**22.00**
Angle, Flat Cover						
50 x 50mm	5.21	5.73	0.10	1.26	nr	**6.99**
75 x 50mm	7.04	7.74	0.10	1.26	nr	**9.00**
75 x 75mm	8.65	9.52	0.10	1.26	nr	**10.78**
100 x 50mm	16.04	17.64	0.10	1.26	nr	**18.90**
100 x 75mm	23.02	25.32	0.15	1.89	nr	**27.21**
100 x 100mm	31.41	34.56	0.15	1.89	nr	**36.45**
150 x 75mm	37.93	41.73	0.15	1.89	nr	**43.62**
150 x 100mm	43.25	47.57	0.20	2.52	nr	**50.09**
150 x 150mm	66.04	72.65	0.20	2.52	nr	**75.17**
Angle, Internal or External Cover						
50 x 50mm	5.21	5.73	0.10	1.26	nr	**6.99**
75 x 50mm	7.04	7.74	0.10	1.26	nr	**9.00**
75 x 75mm	15.54	17.09	0.10	1.26	nr	**18.35**
100 x 50mm	17.53	19.28	0.10	1.26	nr	**20.54**
100 x 75mm	23.02	25.32	0.15	1.89	nr	**27.21**
100 x 100mm	31.41	34.56	0.15	1.89	nr	**36.45**
150 x 75mm	36.38	40.01	0.15	1.89	nr	**41.90**
150 x 100mm	43.25	47.57	0.20	2.52	nr	**50.09**
150 x 150mm	66.04	72.65	0.20	2.52	nr	**75.17**

Y:MECHANICAL AND ELECTRICAL SERVICES

Item	Net Price £	Material £	Labour hours	Labour £	Unit	Total Rate £
Y60 : CONDUIT AND CABLE TRUNKING : **METAL AND PLASTICS CONDUIT (contd)**						
Single Compartment PVC Trunking **Fittings; (cutting and jointing trunking** **to fittings is included) (contd)**						
Tee, Flat Cover						
50 x 50mm	8.14	8.96	0.15	1.89	nr	10.85
75 x 50mm	10.36	11.39	0.15	1.89	nr	13.28
75 x 75mm	11.71	12.88	0.15	1.89	nr	14.77
100 x 50mm	18.97	20.86	0.20	2.52	nr	23.38
100 x 75mm	27.19	29.91	0.20	2.52	nr	32.43
100 x 100mm	27.19	29.91	0.20	2.52	nr	32.43
150 x 75mm	46.91	51.60	0.25	3.15	nr	54.75
150 x 100mm	60.20	66.22	0.25	3.15	nr	69.37
150 x 150mm	75.71	83.28	0.25	3.15	nr	86.43
Tee, Internal or External Cover						
50 x 50mm	15.79	17.37	0.15	1.89	nr	19.26
75 x 50mm	17.22	18.94	0.15	1.89	nr	20.83
75 x 75mm	20.52	22.58	0.15	1.89	nr	24.47
100 x 50mm	24.48	26.92	0.20	2.52	nr	29.44
100 x 75mm	32.73	36.01	0.20	2.52	nr	38.53
100 x 100mm	32.73	36.01	0.20	2.52	nr	38.53
150 x 75mm	46.91	51.60	0.25	3.15	nr	54.75
150 x 100mm	60.20	66.22	0.25	3.15	nr	69.37
150 x 150mm	75.71	83.28	0.25	3.15	nr	86.43
Division Strip (1.8m long)						
50mm	6.40	7.04	0.05	0.63	nr	7.67
75mm	8.48	9.33	0.05	0.63	nr	9.96
100mm	10.79	11.87	0.07	0.88	nr	12.75
PVC Minature Trunking; white finish; **fixed to backgrounds; including supports** **and fixing; standard coupling joints**						
Single Compartment						
16 x 16mm	1.80	2.08	0.20	2.52	m	4.60
25 x 16mm	2.24	2.59	0.20	2.52	m	5.11
38 x 16mm	2.81	3.25	0.22	2.78	m	6.03
38 x 25mm	3.41	3.93	0.25	3.15	m	7.08
Compartmented						
38 x 16mm	3.30	3.81	0.22	2.78	m	6.59
38 x 25mm	4.01	4.63	0.25	3.15	m	7.78

Y: MECHANICAL AND ELECTRICAL SERVICES

Item	Net Price £	Material £	Labour hours	Labour £	Unit	Total Rate £
PVC Miniature Trunking Fittings; single compartment; white finish; (cutting and jointing trunking to fittings is included)						
Coupling						
16 x 16mm	0.58	0.63	0.10	1.26	nr	**1.89**
25 x 16mm	0.58	0.63	0.10	1.26	nr	**1.89**
38 x 16mm	0.58	0.63	0.15	1.89	nr	**2.52**
38 x 25mm	1.34	1.48	0.20	2.52	nr	**4.00**
Stop End						
16 x 16mm	0.57	0.62	0.10	1.26	nr	**1.88**
25 x 16mm	0.57	0.62	0.10	1.26	nr	**1.88**
38 x 16mm	0.57	0.62	0.15	1.89	nr	**2.51**
38 x 25mm	0.70	0.77	0.20	2.52	nr	**3.29**
Bend; Flat, Internal or External						
16 x 16mm	0.58	0.63	0.15	1.89	nr	**2.52**
25 x 16mm	0.58	0.63	0.15	1.89	nr	**2.52**
38 x 16mm	0.58	0.63	0.20	2.52	nr	**3.15**
38 x 25mm	1.40	1.54	0.25	3.15	nr	**4.69**
Tee						
16 x 16mm	0.97	1.06	0.20	2.52	nr	**3.58**
25 x 16mm	0.97	1.06	0.20	2.52	nr	**3.58**
38 x 16mm	0.97	1.06	0.25	3.15	nr	**4.21**
38 x 25mm	0.97	1.06	0.30	3.79	nr	**4.85**
PVC Bench Trunking; White or Grey Finish; fixed to backgrounds; including supports and fixings; standard coupling joints						
Trunking						
90 x 90mm	27.01	31.20	0.35	4.42	m	**35.62**
150 x 150mm	98.42	113.67	0.45	5.68	m	**119.35**
PVC Bench Trunking Fittings; White or Grey Finish. (Cutting and jointing trunking to fittings is included.)						
Stop End						
90 x 90mm	6.13	6.75	0.05	0.63	nr	**7.38**
150 x 150mm	31.55	34.71	0.08	1.01	nr	**35.72**
Coupling						
90 x 90mm	10.51	11.56	0.10	1.26	nr	**12.82**
150 x 150mm	31.55	34.71	0.15	1.89	nr	**36.60**
Internal or External Bend						
90 x 90mm	30.84	33.92	0.15	1.89	nr	**35.81**

Y:MECHANICAL AND ELECTRICAL SERVICES

Item	Net Price £	Material £	Labour hours	Labour £	Unit	Total Rate £
Y60 : CONDUIT AND CABLE TRUNKING : METAL AND PLASTICS CONDUIT (contd)						
PVC Bench Trunking Fittings; White or Grey Finish. (Cutting and jointing trunking to fittings is included.) (contd)						
Socket Plate - 1 Gang						
90 x 90mm - 1 gang	13.30	14.63	0.10	1.26	nr	**15.89**
90 x 90mm - 2 gang	15.62	17.18	0.10	1.26	nr	**18.44**
PVC Underfloor Trunking; Single Compartment; fitted in floor screed; standard coupling joints						
Trunking						
90 x 90mm	1.56	1.80	0.25	3.15	m	**4.95**
150 x 150mm	98.42	113.67	0.30	3.79	m	**117.46**
PVC Underfloor Trunking Fittings; Single Compartment; fitted in floor screed; (Cutting and jointing trunking to fittings is included.)						
Jointing Sleeve						
60 x 25mm	0.21	0.23	0.08	1.01	nr	**1.24**
90 x 35mm	0.41	0.46	0.10	1.26	nr	**1.72**
Duct Connector						
90 x 35mm	0.35	0.39	0.18	2.27	nr	**2.66**
Socket Reducer						
90 x 35mm	0.63	0.69	0.10	1.26	nr	**1.95**
Vertical Access Box; 2 compartment						
Shallow	32.62	35.88	0.33	4.16	nr	**40.04**
Duct Bend; Vertical						
60 x 25mm	7.08	7.79	0.30	3.79	nr	**11.58**
90 x 35mm	7.99	8.79	0.40	5.05	nr	**13.84**
Duct Bend; Horizontal						
60 x 25mm	8.34	9.18	0.35	4.42	nr	**13.60**
90 x 35mm	6.60	7.26	0.45	5.68	nr	**12.94**

Y: MECHANICAL AND ELECTRICAL SERVICES

Item	Net Price £	Material £	Labour hours	Labour £	Unit	Total Rate £
Zinc Coated Steel Underfloor Ducting; fixed to backgrounds; standard coupling joints; earth continuity straps; (Including supports and fixing, packing shims where required)						
Double Compartment						
75 x 25mm	4.64	5.35	0.50	6.31	m	**11.66**
75 x 38mm	5.00	5.77	0.50	6.31	m	**12.08**
100 x 25mm	5.47	6.31	0.55	6.94	m	**13.25**
100 x 38mm	5.92	6.83	0.60	7.57	m	**14.40**
150 x 25mm	6.38	7.37	0.60	7.57	m	**14.94**
150 x 38mm	7.34	8.48	0.70	8.84	m	**17.32**
225 x 25mm	9.13	10.54	0.75	9.47	m	**20.01**
225 x 38mm	9.53	11.00	0.75	9.47	m	**20.47**
Triple Compartment						
75 x 25mm	5.00	5.77	0.75	9.47	m	**15.24**
75 x 38mm	5.33	6.16	0.75	9.47	m	**15.63**
100 x 25mm	5.69	6.57	0.90	11.36	m	**17.93**
100 x 38mm	6.24	7.21	0.90	11.36	m	**18.57**
150 x 25mm	7.21	8.32	0.90	11.36	m	**19.68**
150 x 38mm	7.76	8.96	1.00	12.62	m	**21.58**
225 x 25mm	9.41	10.87	1.00	12.62	m	**23.49**
225 x 38mm	9.96	11.50	1.10	13.88	m	**25.38**
Zinc Coated Steel Underfloor Ducting Fittings; (cutting and jointing to fittings is included.)						
Stop End; Double Compartment						
75 x 25mm	0.97	1.07	0.15	1.89	nr	**2.96**
75 x 38mm	1.03	1.13	0.15	1.89	nr	**3.02**
100 x 25mm	1.18	1.30	0.20	2.52	nr	**3.82**
100 x 38mm	1.18	1.30	0.20	2.52	nr	**3.82**
150 x 25mm	1.52	1.67	0.20	2.52	nr	**4.19**
150 x 38mm	1.60	1.76	0.22	2.78	nr	**4.54**
225 x 25mm	2.14	2.35	0.25	3.15	nr	**5.50**
225 x 38mm	2.20	2.42	0.25	3.15	nr	**5.57**
Stop End; Triple Compartment						
75 x 25mm	0.97	1.07	0.21	2.65	nr	**3.72**
75 x 38mm	1.03	1.13	0.20	2.52	nr	**3.65**
100 x 25mm	1.18	1.30	0.25	3.15	nr	**4.45**
100 x 38mm	1.18	1.30	0.25	3.15	nr	**4.45**
150 x 25mm	1.52	1.67	0.28	3.53	nr	**5.20**
150 x 38mm	1.60	1.76	0.30	3.79	nr	**5.55**
225 x 25mm	2.14	2.35	0.30	3.79	nr	**6.14**
225 x 38mm	2.20	2.42	0.33	4.16	nr	**6.58**

Y:MECHANICAL AND ELECTRICAL SERVICES

Item	Net Price £	Material £	Labour hours	Labour £	Unit	Total Rate £
Y60 : CONDUIT AND CABLE TRUNKING : METAL AND PLASTICS CONDUIT (contd)						
Zinc Coated Steel Underfloor Ducting Fittings; (cutting and jointing to fittings is included.) (contd)						
Flanged Connector; Double Compartment						
75 x 25mm	0.90	0.99	0.15	1.89	nr	2.88
75 x 38mm	0.90	0.99	0.15	1.89	nr	2.88
100 x 25mm	1.03	1.13	0.15	1.89	nr	3.02
100 x 38mm	1.11	1.22	0.15	1.89	nr	3.11
150 x 25mm	1.23	1.35	0.15	1.89	nr	3.24
150 x 38mm	1.32	1.45	0.15	1.89	nr	3.34
225 x 25mm	1.37	1.51	0.15	1.89	nr	3.40
225 x 38mm	1.46	1.61	0.15	1.89	nr	3.50
Flanged Connector; Triple Compartment						
75 x 25mm	0.90	0.99	0.20	2.52	nr	3.51
75 x 38mm	0.90	0.99	0.20	2.52	nr	3.51
100 x 25mm	1.03	1.13	0.20	2.52	nr	3.65
100 x 38mm	1.11	1.22	0.20	2.52	nr	3.74
150 x 25mm	1.23	1.35	0.20	2.52	nr	3.87
150 x 38mm	1.32	1.45	0.20	2.52	nr	3.97
225 x 25mm	1.37	1.51	0.20	2.52	nr	4.03
225 x 38mm	1.46	1.61	0.20	2.52	nr	4.13
Rising Bend; Double Compartment						
75 x 25mm	10.74	11.82	0.55	6.94	nr	18.76
75 x 38mm	10.89	11.98	0.55	6.94	nr	18.92
100 x 25mm	11.98	13.18	0.65	8.21	nr	21.39
100 x 38mm	12.07	13.28	0.70	8.84	nr	22.12
150 x 25mm	15.33	16.87	0.70	8.84	nr	25.71
150 x 38mm	18.31	20.14	0.75	9.47	nr	29.61
225 x 25mm	18.85	20.74	0.80	10.10	nr	30.84
225 x 38mm	19.12	21.03	0.80	10.10	nr	31.13
Rising Bend; Triple Compartment						
75 x 25mm	13.08	14.39	0.65	8.21	nr	22.60
75 x 38mm	13.26	14.59	0.65	8.21	nr	22.80
100 x 25mm	14.43	15.88	0.75	9.47	nr	25.35
100 x 38mm	14.70	16.17	0.75	9.47	nr	25.64
150 x 25mm	18.40	20.24	0.80	10.10	nr	30.34
150 x 38mm	18.58	20.44	0.90	11.36	nr	31.80
225 x 25mm	19.12	21.03	0.90	11.36	nr	32.39
225 x 38mm	19.66	21.63	0.95	12.00	nr	33.63

Y: MECHANICAL AND ELECTRICAL SERVICES

Item	Net Price £	Material £	Labour hours	Labour £	Unit	Total Rate £
Horizontal Bend; Double Compartment						
75 x 25mm	12.63	13.89	0.45	5.68	nr	**19.57**
75 x 38mm	12.90	14.19	0.45	5.68	nr	**19.87**
100 x 25mm	14.16	15.58	0.55	6.94	nr	**22.52**
100 x 38mm	14.34	15.78	0.55	6.94	nr	**22.72**
150 x 25mm	18.13	19.94	0.55	6.94	nr	**26.88**
150 x 38mm	18.31	20.14	0.65	8.21	nr	**28.35**
225 x 25mm	18.85	20.74	0.65	8.21	nr	**28.95**
225 x 38mm	19.21	21.13	0.65	8.21	nr	**29.34**
Horizontal Bend; Triple Compartment						
75 x 25mm	13.08	14.39	0.55	6.94	nr	**21.33**
75 x 38mm	13.17	14.49	0.55	6.94	nr	**21.43**
100 x 25mm	14.34	15.78	0.65	8.21	nr	**23.99**
100 x 38mm	14.70	16.17	0.65	8.21	nr	**24.38**
150 x 25mm	18.31	20.14	0.65	8.21	nr	**28.35**
150 x 38mm	18.58	20.44	0.75	9.47	nr	**29.91**
225 x 25mm	19.12	21.03	0.75	9.47	nr	**30.50**
225 x 38mm	19.66	21.63	0.75	9.47	nr	**31.10**
Junction or Service Outlet Boxes; Terminal; Double Compartment						
up to 150mm	26.62	29.28	0.85	10.73	nr	**40.01**
225mm	42.18	46.40	0.85	10.73	nr	**57.13**
Junction or Service Outlet Boxes; Terminal; Triple Compartment						
up to 150mm	29.51	32.46	1.25	15.77	nr	**48.23**
225mm	44.74	49.21	1.25	15.77	nr	**64.98**
Junction or Service Outlet Boxes; Through or Angle; Double Compartment						
up to 150mm	26.62	29.28	0.85	10.73	nr	**40.01**
225mm	44.74	49.21	0.85	10.73	nr	**59.94**
Junction or Service Outlet Boxes; Through or Angle; Triple Compartment						
up to 150mm	29.51	32.46	1.25	15.77	nr	**48.23**
225mm	44.74	49.21	1.25	15.77	nr	**64.98**
Junction or Service Outlet Boxes; Tee; Double Compartment						
up to 150mm	26.62	29.28	0.85	10.73	nr	**40.01**
225mm	42.18	46.40	0.85	10.73	nr	**57.13**
Junction or Service Outlet Boxes; Tee; Triple Compartment						
up to 150mm	29.51	32.46	1.25	15.77	nr	**48.23**
225mm	44.74	49.21	1.25	15.77	nr	**64.98**

Y:MECHANICAL AND ELECTRICAL SERVICES

Item	Net Price £	Material £	Labour hours	Labour £	Unit	Total Rate £
Y60 : CONDUIT AND CABLE TRUNKING : METAL AND PLASTICS CONDUIT (contd)						
Zinc Coated Steel Underfloor Ducting Fittings; (cutting and jointing to fittings is included.) (contd)						
Junction or Service Outlet Boxes; Cross; Double Compartment						
up to 150mm	26.62	29.28	0.85	10.73	nr	**40.01**
225mm	42.18	46.40	0.85	10.73	nr	**57.13**
Junction or Service Outlet Boxes; Cross; Triple Compartment						
up to 150mm	29.51	32.46	1.25	15.77	nr	**48.23**
225mm	44.74	49.21	1.25	15.77	nr	**64.98**
Service outlet box comprising flat lid with flanged carpet trim; twin 13 A outlet and drilled plate for mounting 2 telephone outlets and terminal blocks; terminal outlet box; double compartment						
150 x 25mm trunking	55.61	61.17	2.00	25.24	nr	**86.41**
Service outlet box comprising flat lid with flanged carpet trim; twin 13 A outlet and drilled plate for mounting 2 telephone outlets and terminal blocks; terminal outlet box; triple compartment						
225 x 25mm trunking	69.63	76.59	2.50	31.55	nr	**108.14**

Keep your figures up to date, free of charge

This section, and most of the other information in this Price Book, is brought up to date every three months, until the next annual edition, in the *Price Book Update*.

The *Update* is available free to all Price Book purchasers.

To ensure you receive your copy, simply complete the reply card from the centre of the book and return it to us.

Y:MECHANICAL AND ELECTRICAL SERVICES

Item	Net Price £	Material £	Labour hours	Labour £	Unit	Total Rate £
Y61 : HV/LV CABLES AND WIRING : FLEXIBLE CORDS						
PVC Insulated PVC Sheathed, Flexible cord; heat resistant; copper conductor; BS 6141 (loose laid).						
300/500 Volt Grade; Two Core						
0.50 mm2	0.69	0.80	0.07	0.88	m	**1.68**
0.75 mm2	0.84	0.97	0.07	0.88	m	**1.85**
300/500 Volt Grade; Three Core						
0.50 mm2	0.94	1.09	0.07	0.88	m	**1.97**
0.75 mm2	1.00	1.16	0.07	0.88	m	**2.04**
1.00 mm2	1.39	1.57	0.07	0.88	m	**2.45**
1.50 mm2	1.90	2.19	0.08	1.01	m	**3.20**
2.50 mm2	2.85	3.29	0.08	1.01	m	**4.30**
VR Insulated, Tough Rubber Sheathed Cable; copper conductor; BS 6500 Table 6 (loose laid).						
300/500 Volt Grade; Two Core						
0.50 mm2	1.13	1.31	0.07	0.88	m	**2.19**
0.75 mm2	0.79	0.91	0.07	0.88	m	**1.79**
1.00 mm2	0.98	1.13	0.07	0.88	m	**2.01**
1.50 mm2	1.36	1.57	0.07	0.88	m	**2.45**
2.50 mm2	1.99	2.30	0.07	0.88	m	**3.18**
300/500 Volt Grade; Three Core						
0.50 mm2	1.40	1.62	0.07	0.88	m	**2.50**
0.75 mm2	1.01	1.17	0.07	0.88	m	**2.05**
1.00 mm2	1.18	1.37	0.07	0.88	m	**2.25**
1.50 mm2	1.64	1.89	0.07	0.88	m	**2.77**
2.50 mm2	2.42	2.79	0.07	0.88	m	**3.67**
300/500 Volt Grade; Four Core						
0.75 mm2	1.47	1.70	0.08	1.01	m	**2.71**
1.00 mm2	1.74	2.01	0.08	1.01	m	**3.02**
1.50 mm2	2.19	2.53	0.08	1.01	m	**3.54**
2.50 mm2	3.26	3.76	0.08	1.01	m	**4.77**
Rubber Insulated H.O.F.R. Sheathed, Circular flexible Cords; copper conductor; BS 6500 Table 9 (loose laid).						
300/500 Volt Grade; Two Core						
0.50 mm2	1.68	1.95	0.07	0.88	m	**2.83**
0.75 mm2	1.40	1.62	0.07	0.88	m	**2.50**
1.00 mm2	1.98	2.28	0.07	0.88	m	**3.16**
1.50 mm2	2.20	2.54	0.07	0.88	m	**3.42**
2.50 mm2	3.38	3.90	0.08	1.01	m	**4.91**
4.00 mm2	4.60	5.31	0.09	1.14	m	**6.45**

Y:MECHANICAL AND ELECTRICAL SERVICES

Item	Net Price £	Material £	Labour hours	Labour £	Unit	Total Rate £
Y61 : HV/LV CABLES AND WIRING : FLEXIBLE CORDS (contd)						
Rubber Insulated H.O.F.R. Sheathed, Circular flexible Cords; copper conductor; BS 6500 Table 9 (loose laid) (cont'd)						
300/500 Volt Grade; Three Core						
0.50 mm2	1.48	1.71	0.07	0.88	m	2.59
0.75 mm2	1.62	1.88	0.07	0.88	m	2.76
1.00 mm2	1.93	2.22	0.07	0.88	m	3.10
1.50 mm2	2.38	2.75	0.08	1.01	m	3.76
2.50 mm2	3.40	3.93	0.08	1.01	m	4.94
4.00 mm2	5.98	6.91	0.10	1.26	m	8.17
300/500 Volt Grade; Four Core						
0.75 mm2	2.88	3.33	0.07	0.88	m	4.21
1.00 mm2	3.35	3.87	0.07	0.88	m	4.75
1.50 mm2	4.25	4.91	0.08	1.01	m	5.92
2.50 mm2	5.46	6.30	0.08	1.01	m	7.31
4.00 mm2	13.27	15.32	0.10	1.26	m	16.58
PVC Insulated PVC Sheathed, Light flexible Cords; copper conductor; BS 6500 Table 15 (loose laid).						
300/300 Volt Grade; Two Core						
0.50 mm2	0.25	0.29	0.07	0.88	m	1.17
0.75 mm2	0.44	0.51	0.07	0.88	m	1.39
300/500 Volt Grade; Three Core						
0.50 mm2	0.34	0.39	0.07	0.88	m	1.27
0.75 mm2	0.61	0.70	0.07	0.88	m	1.58
300/500 Volt Grade; Parallel Twin						
0.50 mm2	0.37	0.43	0.07	0.88	m	1.31
0.75 mm2	0.46	0.53	0.07	0.88	m	1.41
PVC Insulated, Parallel Twin Flexible Cord; figure 8 type; copper conductor; BS 6500 Table 9 (loose laid).						
300/500 Volt Grade; Two Core						
0.50 mm2	0.35	0.41	0.07	0.88	m	1.29
0.75 mm2	0.49	0.56	0.07	0.88	m	1.44

Y:MECHANICAL AND ELECTRICAL SERVICES

Item	Net Price £	Material £	Labour hours	Labour £	Unit	Total Rate £
PVC insulated, PVC Sheathed Circular flexible cord copper conductor; BS 6500 Table 16 (loose laid).						
300/500 Volt Grade; Two Core						
0.75 mm2	0.46	0.53	0.07	0.88	m	**1.41**
1.00 mm2	0.60	0.69	0.07	0.88	m	**1.57**
1.50 mm2	0.85	0.98	0.07	0.88	m	**1.86**
2.50 mm2	1.79	2.07	0.07	0.88	m	**2.95**
300/500 Volt Grade; Three Core						
0.75 mm2	0.44	0.51	0.07	0.88	m	**1.39**
1.00 mm2	0.54	0.62	0.07	0.88	m	**1.50**
1.50 mm2	0.75	0.86	0.07	0.88	m	**1.74**
2.50 mm2	1.54	1.78	0.07	0.88	m	**2.66**
300/500 Volt Grade; Four Core						
0.75 mm2	1.24	1.43	0.07	0.88	m	**2.31**
1.00 mm2	1.46	1.68	0.07	0.88	m	**2.56**
1.50 mm2	2.07	2.40	0.07	0.88	m	**3.28**
2.50 mm2	3.20	3.70	0.07	0.88	m	**4.58**
300/500 Volt Grade; Five Core						
0.75 mm2	2.75	3.17	0.08	1.01	m	**4.18**
1.00 mm2	2.74	3.17	0.08	1.01	m	**4.18**
1.50 mm2	3.84	4.43	0.08	1.01	m	**5.44**
2.50 mm2	6.05	6.99	0.08	1.01	m	**8.00**
300/500 Volt Grade; Parallel Twin						
0.75 mm2	0.49	0.56	0.07	0.88	m	**1.44**
PVC Insulated Flexible Cord; copper conductor; BS 6500 Table 19 and BS 6004 Table 1c (loose laid).						
300/500 Volt Grade; Single Core						
0.50 mm2	0.14	0.17	0.07	0.88	m	**1.05**
0.75 mm2	0.17	0.19	0.07	0.88	m	**1.07**
1.00 mm2	0.21	0.24	0.07	0.88	m	**1.12**
1.50 mm2	0.30	0.35	0.08	1.01	m	**1.36**
2.50 mm2	0.57	0.65	0.08	1.01	m	**1.66**
V.R. Insulated, Tough Rubber Sheathed Flexible cord cooper conductor; screened; PVC sheathed; BS 6500 Table 7 (loose laid).						
300/500 Volt Grade; Three Core						
0.75 mm2	7.86	9.08	0.07	0.88	m	**9.96**
1.00 mm2	10.91	12.60	0.07	0.88	m	**13.48**
1.50 mm2	9.59	11.08	0.08	1.01	m	**12.09**
2.50 mm2	12.78	14.76	0.08	1.01	m	**15.77**

Y:MECHANICAL AND ELECTRICAL SERVICES

Item	Net Price £	Material £	Labour hours	Labour £	Unit	Total Rate £
Y61 : HV/LV CABLES AND WIRING : FLEXIBLE CORDS (contd)						
V.R. Insulated, Tough Rubber Sheathed Flexible cord cooper conductor; screened; PVC sheathed; BS 6500 Table 7 (loose laid) (contd)						
300/600 Volt Grade; Four Core						
0.75 mm2	11.68	13.49	0.07	0.88	m	**14.37**
1.00 mm2	12.35	14.27	0.08	1.01	m	**15.28**
1.50 mm2	10.83	12.51	0.08	1.01	m	**13.52**
2.50 mm2	13.53	15.62	0.08	1.01	m	**16.63**

Y:MECHANICAL AND ELECTRICAL SERVICES

Item	Net Price £	Material £	Labour hours	Labour £	Unit	Total Rate £
Y61 : HV/LV CABLES AND WIRING : CABLE						
Cable; Lead Covered, Paper Insulated; Textile bedded; SWA; PVC sheathed copper; clipped to backgrounds; BS 6350 and BS 6480/69						
6350/11000 Volt Grade, including fixings and connecting tails; Single Core						
120 mm2	11.38	13.14	0.40	5.05	m	**18.19**
150 mm2	11.85	13.69	0.42	5.30	m	**18.99**
185 mm2	13.27	15.33	0.47	5.93	m	**21.26**
240 mm2	16.12	18.62	0.53	6.69	m	**25.31**
300 mm2	18.01	20.81	0.60	7.57	m	**28.38**
6350/11000 Volt Grade, including fixings and connecting tails; Triple Core						
95 mm2	20.86	24.09	0.45	5.68	m	**29.77**
120 mm2	24.65	28.47	0.48	6.06	m	**34.53**
150 mm2	27.50	31.76	0.52	6.56	m	**38.32**
185 mm2	31.29	36.14	0.53	6.69	m	**42.83**
240 mm2	35.08	40.52	0.60	7.57	m	**48.09**
300 mm2	45.51	52.56	0.70	8.84	m	**61.40**
Cable; Lead Covered, Paper Insulated; Textile bedded; SWA; PVC sheathed copper; laid in trench with marker tape; BS 6350 and BS 6480/69 (Cable tiles measured elsewhere).						
Single Core						
120 mm2	11.38	13.14	0.18	2.27	m	**15.41**
150 mm2	11.85	13.69	0.18	2.27	m	**15.96**
185 mm2	13.27	15.33	0.18	2.27	m	**17.60**
240 mm2	16.12	18.62	0.27	3.41	m	**22.03**
300 mm2	18.01	20.81	0.30	3.79	m	**24.60**
Triple Core						
95 mm2	20.86	24.09	0.20	2.52	m	**26.61**
120 mm2	24.65	28.47	0.20	2.52	m	**30.99**
150 mm2	27.50	31.76	0.20	2.52	m	**34.28**
185 mm2	31.29	36.14	0.20	2.52	m	**38.66**
240 mm2	35.08	40.52	0.30	3.79	m	**44.31**
300 mm2	45.51	52.56	0.40	5.05	m	**57.61**

Y:MECHANICAL AND ELECTRICAL SERVICES

Item	Net Price £	Material £	Labour hours	Labour £	Unit	Total Rate £
Y61 : HV/LV CABLES AND WIRING : CABLE (contd)						
Cable Termination; for lead covered, paper insulated; textile bedded; SWA; PVC sheathed copper cable (contd)						
Brass weatherproof; PVC shroud; brass lock nut; brass earthing ring; drilling and cutting mild steel gland plate;						
Single Core; Internal						
120 mm2	56.46	62.11	1.60	20.19	nr	**82.30**
150 mm2	59.49	65.44	1.60	20.19	nr	**85.63**
185 mm2	59.49	65.44	1.60	20.19	nr	**85.63**
240 mm2	59.49	65.44	1.60	20.19	nr	**85.63**
Brass weatherproof; PVC shroud; brass lock nut; brass earthing ring; drilling and cutting mild steel gland plate;						
Three Core; Internal						
95 mm2	59.93	65.92	1.70	21.46	nr	**87.38**
120 mm2	59.93	65.92	1.70	21.46	nr	**87.38**
185 mm2	62.97	69.27	1.70	21.46	nr	**90.73**
240 mm2	62.97	69.27	1.70	21.46	nr	**90.73**
Brass weatherproof; PVC shroud; brass lock nut; brass earthing ring; drilling and cutting mild steel gland plate;						
Single Core; External						
120 mm2	73.39	80.73	1.60	20.19	nr	**100.92**
150 mm2	77.31	85.04	1.60	20.19	nr	**105.23**
185 mm2	77.31	85.04	1.60	20.19	nr	**105.23**
240 mm2	77.31	85.04	1.60	20.19	nr	**105.23**
Brass weatherproof; PVC shroud; brass lock nut; brass earthing ring; drilling and cutting mild steel gland plate;						
Three Core; External						
95 mm2	83.82	92.21	1.70	21.46	nr	**113.67**
120 mm2	83.82	92.21	1.70	21.46	nr	**113.67**
185 mm2	88.17	96.98	1.70	21.46	nr	**118.44**
240 mm2	83.82	92.21	1.70	21.46	nr	**113.67**

Y:MECHANICAL AND ELECTRICAL SERVICES

Item	Net Price £	Material £	Labour hours	Labour £	Unit	Total Rate £
Cable; XLPE Insulated; SWA; PVC sheathed copper; BS 5467; clipped to backgrounds; (Supports, fixings and connecting tails are excluded).						
600/1000 Volt Grade; Two Core						
25 mm2	2.27	2.62	0.30	3.79	m	**6.41**
35 mm2	2.73	3.15	0.32	4.04	m	**7.19**
50 mm2	3.45	3.99	0.33	4.16	m	**8.15**
70 mm2	4.22	4.88	0.35	4.42	m	**9.30**
90 mm2	5.35	6.18	0.37	4.67	m	**10.85**
120 mm2	6.43	7.42	0.38	4.80	m	**12.22**
150 mm2	7.70	8.89	0.40	5.05	m	**13.94**
185 mm2	9.99	11.54	0.43	5.43	m	**16.97**
240 mm2	12.63	14.59	0.46	5.81	m	**20.40**
300 mm2	15.35	17.73	0.48	6.06	m	**23.79**
600/1000 Volt Grade; Three Core						
25 mm2	2.85	3.29	0.35	4.42	m	**7.71**
35 mm2	3.46	4.00	0.37	4.67	m	**8.67**
50 mm2	4.55	5.26	0.38	4.80	m	**10.06**
70 mm2	5.65	6.52	0.39	4.92	m	**11.44**
90 mm2	7.36	8.50	0.40	5.05	m	**13.55**
120 mm2	8.87	10.24	0.44	5.55	m	**15.79**
150 mm2	11.53	13.32	0.44	5.55	m	**18.87**
185 mm2	13.70	15.83	0.47	5.93	m	**21.76**
240 mm2	17.92	20.69	0.49	6.19	m	**26.88**
300 mm2	21.72	25.09	0.52	6.56	m	**31.65**
400 mm2	27.62	31.91	0.60	7.57	m	**39.48**
600/1000 Volt Grade; Four Core						
25 mm2	3.32	3.84	0.37	4.67	m	**8.51**
35 mm2	4.20	4.85	0.39	4.92	m	**9.77**
50 mm2	5.66	6.54	0.40	5.05	m	**11.59**
70 mm2	7.32	8.45	0.44	5.55	m	**14.00**
90 mm2	9.33	10.77	0.46	5.81	m	**16.58**
120 mm2	12.21	14.10	0.46	5.81	m	**19.91**
150 mm2	14.41	16.65	0.49	6.19	m	**22.84**
185 mm2	17.35	20.04	0.51	6.44	m	**26.48**
240 mm2	23.11	26.70	0.53	6.69	m	**33.39**
300 mm2	28.10	32.45	0.57	7.19	m	**39.64**
400 mm2	37.78	43.64	0.63	7.95	m	**51.59**
Cable; XLPE insulated; SWA; PVC sheathed copper; BS 5467; laid in trench including marker tape. (Cable tiles measured elsewhere.)						
600/1000 Volt Grade; Two Core						
25 mm2	2.27	2.62	0.14	1.77	m	**4.39**
35 mm2	2.73	3.15	0.14	1.77	m	**4.92**
50 mm2	3.45	3.99	0.16	2.02	m	**6.01**
70 mm2	4.22	4.88	0.16	2.02	m	**6.90**
90 mm2	5.35	6.18	0.18	2.27	m	**8.45**
120 mm2	6.43	7.42	0.18	2.27	m	**9.69**

Y:MECHANICAL AND ELECTRICAL SERVICES

Item	Net Price £	Material £	Labour hours	Labour £	Unit	Total Rate £
Y61 : HV/LV CABLES AND WIRING : CABLE (contd)						
Cable; XLPE insulated; SWA; PVC sheathed copper; BS 5467; laid in trench including marker tape. (Cable tiles measured elsewhere.) (contd)						
600/1000 Volt Grade; Two Core (contd)						
150 mm2	7.70	8.89	0.20	2.52	m	**11.41**
185 mm2	9.99	11.54	0.20	2.52	m	**14.06**
240 mm2	12.63	14.59	0.22	2.78	m	**17.37**
300 mm2	15.35	17.73	0.22	2.78	m	**20.51**
600/1000 Volt Grade; Three Core						
25 mm2	2.85	3.29	0.15	1.89	m	**5.18**
35 mm2	3.46	4.00	0.15	1.89	m	**5.89**
50 mm2	4.55	5.26	0.18	2.27	m	**7.53**
70 mm2	5.65	6.52	0.18	2.27	m	**8.79**
90 mm2	7.36	8.50	0.20	2.52	m	**11.02**
120 mm2	8.87	10.24	0.20	2.52	m	**12.76**
150 mm2	11.53	13.32	0.22	2.78	m	**16.10**
185 mm2	13.70	15.83	0.22	2.78	m	**18.61**
240 mm2	17.92	20.69	0.24	3.03	m	**23.72**
300 mm2	21.72	25.09	0.24	3.03	m	**28.12**
400 mm2	27.62	31.91	0.26	3.28	m	**35.19**
600/1000 Volt Grade; Four Core						
25 mm2	3.32	3.84	0.18	2.27	m	**6.11**
35 mm2	4.20	4.85	0.18	2.27	m	**7.12**
50 mm2	5.66	6.54	0.21	2.65	m	**9.19**
70 mm2	7.32	8.45	0.21	2.65	m	**11.10**
90 mm2	9.33	10.77	0.23	2.90	m	**13.67**
120 mm2	12.21	14.10	0.23	2.90	m	**17.00**
150 mm2	14.41	16.65	0.25	3.15	m	**19.80**
185 mm2	17.35	20.04	0.25	3.15	m	**23.19**
240 mm2	23.11	26.70	0.27	3.41	m	**30.11**
300 mm2	28.10	32.45	0.27	3.41	m	**35.86**
400 mm2	37.78	43.64	0.29	3.66	m	**47.30**
Cable Termination; XLPE or PVC insulated armoured cables (including drilling and cutting mild steel gland plate).						
Brass weatherproof gland with inner and outer seal; PVC shroud, brass locknut, brass earthing ring (Type E1W)						
Two Core						
25 mm2	8.79	9.67	0.98	12.37	nr	**22.04**
35 mm2	8.79	9.67	1.05	13.26	nr	**22.93**
50 mm2	8.79	9.67	1.10	13.88	nr	**23.55**
70 mm2	12.25	13.47	1.12	14.15	nr	**27.62**
90 mm2	12.25	13.47	1.15	14.52	nr	**27.99**
120 mm2	22.41	24.65	1.20	15.15	nr	**39.80**

Y:MECHANICAL AND ELECTRICAL SERVICES

Item	Net Price £	Material £	Labour hours	Labour £	Unit	Total Rate £
150 mm2	22.41	24.65	1.28	16.16	nr	40.81
185 mm2	31.66	34.83	1.72	21.72	nr	56.55
240 mm2	32.83	36.11	1.83	23.11	nr	59.22
300 mm2	55.09	60.60	1.93	24.36	nr	84.96
Brass weatherproof gland with inner and outer seal; PVC shroud, brass locknut, brass earthing ring, (type E1W)						
Three Core						
25 mm2	8.89	9.78	1.05	13.26	nr	23.04
35 mm2	12.25	13.47	1.15	14.52	nr	27.99
50 mm2	12.25	13.47	1.20	15.15	nr	28.62
70 mm2	12.25	13.47	1.25	15.77	nr	29.24
90 mm2	22.41	24.65	1.28	16.16	nr	40.81
120 mm2	31.66	34.83	1.37	17.31	nr	52.14
150 mm2	31.66	34.83	1.80	22.74	nr	57.57
185 mm2	32.83	36.11	1.80	22.74	nr	58.85
240 mm2	55.09	60.60	2.00	25.24	nr	85.84
300 mm2	56.26	61.88	2.17	27.43	nr	89.31
400 mm2	71.29	78.41	2.83	35.75	nr	114.16
Brass weatherproof gland with inner and outer seal; PVC shroud, brass locknut, brass earthing ring, (Type E1W)						
Four Core						
25 mm2	12.25	13.47	1.15	14.52	nr	27.99
35 mm2	12.25	13.47	1.30	16.41	nr	29.88
50 mm2	12.25	13.47	1.34	16.89	nr	30.36
70 mm2	22.41	24.65	1.45	18.32	nr	42.97
90 mm2	31.66	34.83	1.48	18.70	nr	53.53
120 mm2	32.83	36.11	1.62	20.45	nr	56.56
150 mm2	32.83	36.11	1.70	21.46	nr	57.57
185 mm2	55.09	60.60	1.87	23.63	nr	84.23
240 mm2	56.26	61.88	2.40	30.34	nr	92.22
300 mm2	71.29	78.41	2.50	31.55	nr	109.96
400 mm2	229.67	252.64	3.00	37.90	nr	290.54
Cable; XLPE Insulated; PVC Sheathed; SWA; PVC Insulated; copper; BS 5467 : 1989; fixed with clips to backgrounds. (Supports and fixings, connecting tails are included.)						
600/1000 Volt Grade (Single Strand)						
Two Core						
1.5 mm2	1.70	1.96	0.10	1.26	m	3.22
2.5 mm2	2.11	2.44	0.11	1.39	m	3.83
Three Core						
1.5 mm2	2.06	2.38	0.11	1.39	m	3.77
2.5 mm2	2.53	2.92	0.12	1.51	m	4.43

Y:MECHANICAL AND ELECTRICAL SERVICES

Item	Net Price £	Material £	Labour hours	Labour £	Unit	Total Rate £
Y61 : HV/LV CABLES AND WIRING : CABLE (contd)						
Cable; XLPE Insulated; PVC Sheathed; SWA; PVC Insulated; copper; BS 5467 : 1989; fixed with clips to backgrounds. (Supports and fixings, connecting tails are included.) (contd)						
600/1000 Volt Grade (Single Strand) (contd)						
Four Core						
1.5 mm2	2.21	2.55	0.12	1.51	m	**4.06**
2.5 mm2	2.81	3.25	0.13	1.64	m	**4.89**
Five Core						
1.5 mm2	2.77	3.20	0.13	1.64	m	**4.84**
2.5 mm2	3.64	4.21	0.14	1.77	m	**5.98**
Seven Core						
1.5 mm2	3.13	3.61	0.15	1.89	m	**5.50**
2.5 mm2	4.25	4.91	0.16	2.02	m	**6.93**
Ten Core						
1.5 mm2	5.13	5.93	0.17	2.15	m	**8.08**
2.5 mm2	6.34	7.33	0.17	2.15	m	**9.48**
Twelve Core						
1.5 mm2	5.10	5.89	0.18	2.27	m	**8.16**
2.5 mm2	6.67	7.71	0.19	2.40	m	**10.11**
Nineteen Core						
1.5 mm2	7.74	8.94	0.23	2.90	m	**11.84**
2.5 mm2	10.46	12.08	0.24	3.03	m	**15.11**
Twenty Seven Core						
1.5 mm2	11.37	13.13	0.26	3.28	m	**16.41**
2.5 mm2	14.16	16.35	0.27	3.38	m	**19.73**
Thirty Seven Core						
1.5 mm2	14.61	16.87	0.30	3.79	m	**20.66**
2.5 mm2	18.76	21.67	0.31	3.91	m	**25.58**
Forty Eight Core						
1.5 mm2	15.66	18.09	0.35	4.42	m	**22.51**
2.5 mm2	22.91	26.46	0.35	4.42	m	**30.88**

Y:MECHANICAL AND ELECTRICAL SERVICES

Item	Net Price £	Material £	Labour hours	Labour £	Unit	Total Rate £
600/1000 Volt Grade (Multi Strand)						
Two Core						
1.5 mm2	1.66	1.91	0.09	1.14	m	**3.05**
2.5 mm2	2.07	2.39	0.10	1.26	m	**3.65**
4.0 mm2	3.10	3.58	0.11	1.39	m	**4.97**
6.0 mm2	3.97	4.59	0.12	1.51	m	**6.10**
10.0 mm2	6.30	7.28	0.13	1.64	m	**8.92**
16.0 mm2	6.47	7.47	0.14	1.77	m	**9.24**
Three Core						
1.5 mm2	2.02	2.33	0.10	1.26	m	**3.59**
2.5 mm2	2.47	2.85	0.11	1.39	m	**4.24**
4.0 mm2	3.90	4.51	0.12	1.51	m	**6.02**
6.0 mm2	5.24	6.06	0.13	1.64	m	**7.70**
10.0 mm2	8.37	9.66	0.14	1.77	m	**11.43**
16.0 mm2	10.27	11.87	0.15	1.89	m	**13.76**
Four Core						
1.5 mm2	2.16	2.49	0.11	1.39	m	**3.88**
2.5 mm2	2.76	3.18	0.12	1.51	m	**4.69**
4.0 mm2	4.96	5.73	0.13	1.64	m	**7.37**
6.0 mm2	6.02	6.95	0.14	1.77	m	**8.72**
10.0 mm2	8.58	9.91	0.15	1.89	m	**11.80**
16.0 mm2	10.49	12.12	0.16	2.02	m	**14.14**
Seven Core						
1.5 mm2	3.06	3.54	0.14	1.77	m	**5.31**
2.5 mm2	4.17	4.81	0.15	1.89	m	**6.70**
4.0 mm2	5.98	6.91	0.16	2.02	m	**8.93**
Ten Core						
1.5 mm2	5.12	5.91	0.14	1.77	m	**7.68**
2.5 mm2	6.33	7.31	0.14	1.77	m	**9.08**
Twelve Core						
1.5 mm2	5.00	5.78	0.17	2.15	m	**7.93**
2.5 mm2	6.54	7.56	0.18	2.27	m	**9.83**
Nineteen Core						
1.5 mm2	7.59	8.77	0.22	2.78	m	**11.55**
2.5 mm2	10.25	11.84	0.23	2.90	m	**14.74**
Twenty Seven Core						
1.5 mm2	11.14	12.87	0.25	3.15	m	**16.02**
2.5 mm2	13.88	16.03	0.26	3.28	m	**19.31**
Thirty Seven Core						
1.5 mm2	14.32	16.54	0.29	3.66	m	**20.20**
2.5 mm2	18.40	21.25	0.30	3.79	m	**25.04**

Y:MECHANICAL AND ELECTRICAL SERVICES

Item	Net Price £	Material £	Labour hours	Labour £	Unit	Total Rate £
Y61 : HV/LV CABLES AND WIRING : CABLE (contd)						
Cable; XLPE Insulated; PVC Sheathed; SWA; PVC Insulated; copper; BS 5467 : 1989; fixed with clips to backgrounds. (Supports and fixings, connecting tails are included.) (contd)						
600/1000 Volt Grade (Multi Strand) (contd)						
Forty Eight Core						
1.5 mm2	15.08	17.42	0.29	3.66	m	**21.08**
2.5 mm2	22.06	25.48	0.30	3.79	m	**29.27**
Cable; XLPE PVC Sheathed; SWA; PVC insulated; copper; BS 5467: 1989; laid in trench; including marker tape. (Cable tiles measured eleswhere).						
600/1000 Volt Grade (Single Strand)						
Two Core						
1.5 mm2	1.70	1.96	0.03	0.38	m	**2.34**
2.5 mm2	2.11	2.44	0.04	0.50	m	**2.94**
Three Core						
1.5 mm2	2.06	2.38	0.05	0.63	m	**3.01**
2.5 mm2	2.53	2.92	0.06	0.76	m	**3.68**
Four Core						
1.5 mm2	2.21	2.55	0.07	0.88	m	**3.43**
2.5 mm2	2.81	3.25	0.07	0.88	m	**4.13**
Five Core						
1.5 mm2	2.77	3.20	0.08	1.01	m	**4.21**
2.5 mm2	3.64	4.21	0.08	1.01	m	**5.22**
Seven Core						
1.5 mm2	3.13	3.61	0.10	1.26	m	**4.87**
2.5 mm2	4.25	4.91	0.10	1.26	m	**6.17**
Ten Core						
1.5 mm2	5.13	5.93	0.11	1.39	m	**7.32**
2.5 mm2	6.34	7.33	0.11	1.39	m	**8.72**
Twelve Core						
1.5 mm2	5.10	5.89	0.12	1.51	m	**7.40**
2.5 mm2	6.67	7.71	0.12	1.51	m	**9.22**

Y:MECHANICAL AND ELECTRICAL SERVICES

Item	Net Price £	Material £	Labour hours	Labour £	Unit	Total Rate £
Nineteen Core						
1.5 mm2	7.74	8.94	0.15	1.89	m	10.83
2.5 mm2	10.46	12.08	0.15	1.89	m	13.97
Twenty Seven Core						
1.5 mm2	11.37	13.13	0.18	2.27	m	15.40
2.5 mm2	14.16	16.35	0.18	2.27	m	18.62
Thirty Seven Core						
1.5 mm2	14.61	16.87	0.21	2.65	m	19.52
2.5 mm2	18.76	21.67	0.21	2.65	m	24.32
Forty Eight Core						
1.5 mm2	15.66	18.09	0.24	3.03	m	21.12
2.5 mm2	22.91	26.46	0.24	3.03	m	29.49
600/1000 Volt Grade (Multi Strand);						
Two Core						
1.5 mm2	1.66	1.91	0.03	0.38	m	2.29
2.5 mm2	2.07	2.39	0.03	0.38	m	2.77
4 mm2	3.10	3.58	0.05	0.63	m	4.21
6 mm2	3.97	4.59	0.05	0.63	m	5.22
10 mm2	6.30	7.28	0.06	0.76	m	8.04
16 mm2	6.47	7.47	0.06	0.76	m	8.23
Three Core						
1.5 mm2	2.02	2.33	0.05	0.63	m	2.96
2.5 mm2	2.47	2.85	0.05	0.63	m	3.48
4 mm2	3.90	4.51	0.07	0.88	m	5.39
6 mm2	5.24	6.06	0.07	0.88	m	6.94
10 mm2	8.37	9.66	0.07	0.88	m	10.54
16 mm2	10.27	11.87	0.07	0.88	m	12.75
Four Core						
1.5 mm2	2.16	2.49	0.07	0.88	m	3.37
2.5 mm2	2.76	3.18	0.07	0.88	m	4.06
4 mm2	4.96	5.73	0.07	0.88	m	6.61
6 mm2	6.02	6.95	0.07	0.88	m	7.83
10 mm2	8.58	9.91	0.08	1.01	m	10.92
16 mm2	10.49	12.12	0.08	1.01	m	13.13
Seven Core						
1.5 mm2	3.06	3.54	0.08	1.01	m	4.55
2.5 mm2	4.17	4.81	0.08	1.01	m	5.82
4 mm2	5.98	6.91	0.08	1.01	m	7.92
Twelve Core						
1.5 mm2	5.00	5.78	0.09	1.14	m	6.92
2.5 mm2	6.54	7.56	0.09	1.14	m	8.70

Y:MECHANICAL AND ELECTRICAL SERVICES

Item	Net Price £	Material £	Labour hours	Labour £	Unit	Total Rate £
Y61 : HV/LV CABLES AND WIRING : CABLE (contd)						
Cable; XLPE PVC Sheathed; SWA; PVC insulated; copper; BS 5467: 1989; laid in trench; including marker tape. (Cable tiles measured eleswhere) (contd)						
600/1000 Volt Grade (Multi Strand) (contd)						
Nineteen Core						
1.5 mm2	7.59	8.77	0.11	1.39	m	**10.16**
2.5 mm2	10.25	11.84	0.11	1.39	m	**13.23**
Twenty Seven Core						
1.5 mm2	11.14	12.87	0.12	1.51	m	**14.38**
2.5 mm2	13.88	16.03	0.12	1.51	m	**17.54**
Thirty Seven Core						
1.5 mm2	14.32	16.54	0.13	1.64	m	**18.18**
2.5 mm2	18.40	21.25	0.13	1.64	m	**22.89**
Cable Terminations; for PVC Insulated SWA PVC sheathed copper (including drilling and cutting mild steel gland plate)						
Brass weatherproof gland with outer seal; brass locknut and earthing ring; PVC shroud (Type E1W);						
Two Core						
1.5 mm2	5.01	5.51	0.48	6.06	nr	**11.57**
2.5 mm2	5.05	5.56	0.48	6.06	nr	**11.62**
4 mm2	5.05	5.56	0.48	6.06	nr	**11.62**
6 mm2	5.92	6.52	0.50	6.31	nr	**12.83**
10 mm2	8.89	9.78	0.65	8.21	nr	**17.99**
16 mm2	8.89	9.78	0.77	9.72	nr	**19.50**
Three Core						
1.5 mm2	5.01	5.51	0.52	6.56	nr	**12.07**
2.5 mm2	5.05	5.56	0.52	6.56	nr	**12.12**
4 mm2	5.05	5.56	0.52	6.56	nr	**12.12**
6 mm2	5.92	6.52	0.53	6.69	nr	**13.21**
10 mm2	8.50	9.78	0.72	9.09	nr	**18.87**
16 mm2	8.89	9.78	0.83	10.48	nr	**20.26**
Four Core						
1.5 mm2	5.05	5.56	0.57	7.19	nr	**12.75**
2.5 mm2	5.05	5.56	0.57	7.19	nr	**12.75**
4 mm2	5.92	6.52	0.57	7.19	nr	**13.71**
6 mm2	5.92	6.52	0.58	7.32	nr	**13.84**
10 mm2	8.89	9.78	0.82	10.35	nr	**20.13**
16 mm2	8.89	9.78	0.93	11.74	nr	**21.52**

Y:MECHANICAL AND ELECTRICAL SERVICES

Item	Net Price £	Material £	Labour hours	Labour £	Unit	Total Rate £
Five Core						
1.5 mm2	5.05	5.56	0.63	7.95	nr	**13.51**
2.5 mm2	5.05	5.56	0.63	7.95	nr	**13.51**
Seven Core						
1.5 mm2	5.05	5.56	0.73	9.22	nr	**14.78**
2.5 mm2	5.92	6.52	0.73	9.22	nr	**15.74**
4 mm2	8.89	9.78	0.76	9.60	nr	**19.38**
Ten Core						
1.5 mm2	5.92	6.52	0.85	10.73	nr	**17.25**
Twelve Core						
1.5 mm2	5.92	6.52	0.85	10.73	nr	**17.25**
2.5 mm2	8.89	9.78	0.85	10.73	nr	**20.51**
Nineteen Core						
1.5 mm2	8.89	9.78	1.05	13.26	nr	**23.04**
2.5 mm2	8.89	9.78	1.05	13.26	nr	**23.04**
Twenty Seven Core						
1.5 mm2	12.25	13.47	1.28	16.16	nr	**29.63**
2.5 mm2	12.25	13.47	1.28	16.16	nr	**29.63**
Thirty Seven Core						
1.5 mm2	12.25	13.47	1.55	19.57	nr	**33.04**
2.5 mm2	22.41	24.65	1.55	19.57	nr	**44.22**
Forty Eight Core						
1.5 mm2	12.25	13.47	2.13	26.91	nr	**40.38**
2.5 mm2	22.41	24.65	2.13	26.91	nr	**51.56**
Cable, Mineral Insulated; copper sheathed with copper conductors; fixed with clips to backgrounds. BASEC approval to BS 6207 Part 1 1995; complies with BS 6387 Category CWZ						
Light duty 500 Volt grade; Bare						
2L 1.0	1.05	1.22	0.16	2.06	m	**3.28**
2L 1.5	1.26	1.46	0.16	2.07	m	**3.53**
2L 2.5	1.64	1.90	0.17	2.15	m	**4.05**
2L 4.0	2.43	2.80	0.18	2.21	m	**5.01**
3L 1.0	1.30	1.50	0.16	2.08	m	**3.58**
3L 1.5	1.65	1.91	0.17	2.13	m	**4.04**
3L 2.5	2.55	2.95	0.17	2.18	m	**5.13**
4L 1.0	1.56	1.80	0.17	2.12	m	**3.92**
4L 1.5	1.99	2.30	0.17	2.17	m	**4.47**
4L 2.5	3.10	3.58	0.18	2.26	m	**5.84**

Y:MECHANICAL AND ELECTRICAL SERVICES

Item	Net Price £	Material £	Labour hours	Labour £	Unit	Total Rate £
Y61 : HV/LV CABLES AND WIRING : CABLE (contd)						
Cable, Mineral Insulated; copper sheathed with copper conductors; fixed with clips to backgrounds. BASEC approval to BS 6207 Part 1 1995; complies with BS 6387 Category CWZ (contd)						
Light duty 500 Volt grade; Bare (contd)						
7L 1.0	2.30	2.66	0.17	2.20	m	4.86
7L 1.5	2.86	3.30	0.18	2.28	m	5.58
7L 2.5	3.75	4.33	0.16	2.04	m	6.37
Light duty 500 Volt grade; LSF sheathed						
2L 1.0	1.23	1.43	0.16	2.06	m	3.49
2L 1.5	1.46	1.68	0.16	2.07	m	3.75
2L 2.5	1.85	2.14	0.17	2.15	m	4.29
2L 4.0	2.58	2.98	0.18	2.21	m	5.19
3L 1.0	1.50	1.73	0.16	2.08	m	3.81
3L 1.5	1.83	2.11	0.17	2.12	m	4.23
3L 2.5	2.70	3.12	0.17	2.18	m	5.30
4L 1.0	1.77	2.04	0.17	2.12	m	4.16
4L 1.5	2.23	2.58	0.17	2.17	m	4.75
4L 2.5	3.23	3.73	0.18	2.26	m	5.99
7L 1.0	2.64	3.05	0.17	2.20	m	5.25
7L 1.5	3.26	3.77	0.18	2.28	m	6.05
7L 2.5	4.18	4.83	0.16	2.04	m	6.87
Cable, Mineral Insulated; copper sheathed with copper conductors; fixed with clips to backgrounds; BASEC approval to BS 6207 Part 1 1995; complies with complies with BS 6387 Category CWZ						
Heavy duty 750 Volt grade; Bare						
1H 10	2.36	2.73	0.18	2.30	m	5.03
1H 16	3.24	3.74	0.19	2.40	m	6.14
1H 25	4.52	5.22	0.17	2.18	m	7.40
1H 35	6.03	6.97	0.19	2.40	m	9.37
1H 50	7.66	8.85	0.22	2.75	m	11.60
1H 70	9.95	11.49	0.26	3.27	m	14.76
1H 95	13.07	15.10	0.28	3.56	m	18.66
1H 120	15.98	18.46	0.33	4.13	m	22.59
1H 150	19.81	22.88	0.39	4.97	m	27.85
1H 185	24.12	27.85	0.48	6.05	m	33.90
1H 240	31.31	36.16	0.68	8.57	m	44.73
2H 1.5	2.01	2.32	0.18	2.23	m	4.55
2H 2.5	2.47	2.85	0.19	2.33	m	5.18
2H 4	3.11	3.60	0.16	2.06	m	5.66
2H 6	4.15	4.79	0.18	2.25	m	7.04
2H 10	5.37	6.21	0.21	2.63	m	8.84
2H 16	7.73	8.93	0.25	3.18	m	12.11
2H 25	10.86	12.54	0.29	3.71	m	16.25

![BICC **Pyrotenax**]

The problem's the same.

The solution is simple...
"*<u>Pyrotenax</u>"

Pyrotenax is the ultimate cable
for all fire survival situations.
Together with cable on reels and the
Pyro Mate seal you have a cost effective and
versatile cable system.

**ENJOY REEL
BENEFITS**

**FIT A *PYRO MATE
IN SECONDS**

BICC PYROTENAX LIMITED, P.O.BOX 20, PRESCOT, MERSEYSIDE L34 5GB, UNITED KINGDOM.   TEL: (+44) 0151 430 4000   FAX: (+44) 0151 430 4004
Pyrotenax and BICC are trade marks of BICC plc.

Everything you need to know about Circuit Protection

In its 72 fully illustrated pages, the Crabtree Circuit Protection Technical Guideline provides complete technical information on the use of circuit breakers and residual current devices in domestic, commercial and industrial electrical installations. It also provides advice on protection against over current and earth faults, electric shock and fire risk protection.

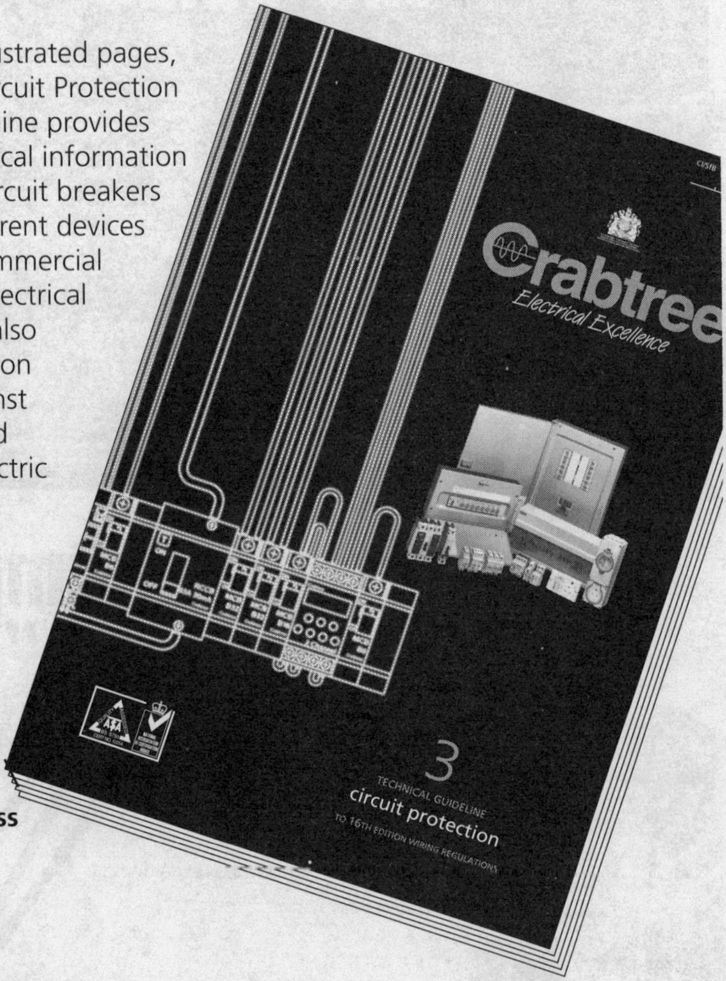

For your complimentary copy, contact us at the address below.

Crabtree Electrical Industries Limited
Head Office: Lincoln Works, Walsall, England WS1 2DN
Telephone: 01922 721202 Fax: 01922 721321
A Hanson Company

Y:MECHANICAL AND ELECTRICAL SERVICES

Item	Net Price £	Material £	Labour hours	Labour £	Unit	Total Rate £
3H 1.5	2.22	2.57	0.18	2.27	m	**4.84**
3H 2.5	2.80	3.23	0.16	1.98	m	**5.21**
3H 4	3.55	4.10	0.17	2.17	m	**6.27**
3H 6	6.63	7.66	0.19	2.37	m	**10.03**
3H 10	6.63	7.66	0.23	2.86	m	**10.52**
3H 16	9.31	10.75	0.25	3.15	m	**13.90**
3H 25	14.28	16.49	0.33	4.16	m	**20.65**
4H 1.5	2.77	3.20	0.15	1.96	m	**5.16**
4H 2.5	3.48	4.02	0.17	2.11	m	**6.13**
4H 4	4.35	5.02	0.19	2.35	m	**7.37**
4H 6	5.80	6.70	0.21	2.63	m	**9.33**
4H 10	8.24	9.51	0.26	3.22	m	**12.73**
4H 16	12.02	13.88	0.30	3.79	m	**17.67**
4H 25	17.46	20.16	0.40	5.01	m	**25.17**
7H 1.5	3.82	4.42	0.18	2.23	m	**6.65**
7H 2.5	5.20	6.01	0.20	2.50	m	**8.51**
12H 1.5	6.82	7.88	0.24	2.99	m	**10.87**
12H 2.5	9.07	10.48	0.25	3.15	m	**13.63**
19H 1.5	14.00	16.17	0.28	3.52	m	**19.69**
Heavy duty 750 Volt grade; LSF Sheathed						
1H 10	2.57	2.97	0.18	2.30	m	**5.27**
1H 16	3.52	4.06	0.19	2.40	m	**6.46**
1H 25	4.86	5.61	0.17	2.18	m	**7.79**
1H 35	6.37	7.36	0.19	2.40	m	**9.76**
1H 50	8.05	9.30	0.22	2.75	m	**12.05**
1H 70	10.46	12.08	0.26	3.27	m	**15.35**
1H 95	13.73	15.85	0.28	3.56	m	**19.41**
1H 120	16.75	19.34	0.33	4.13	m	**23.47**
1H 150	20.64	23.83	0.39	4.97	m	**28.80**
1H 185	25.33	29.25	0.48	6.05	m	**35.30**
1H 240	32.65	37.71	0.68	8.57	m	**46.28**
2H 1.5	2.26	2.61	0.18	2.23	m	**4.84**
2H 2.5	2.69	3.11	0.19	2.33	m	**5.44**
2H 4	3.39	3.92	0.16	2.06	m	**5.98**
2H 6	4.49	5.19	0.18	2.25	m	**7.44**
2H 10	5.80	6.70	0.21	2.63	m	**9.33**
2H 16	8.19	9.46	0.25	3.18	m	**12.64**
2H 25	10.86	12.54	0.29	3.71	m	**16.25**
3H 1.5	2.50	2.89	0.18	2.27	m	**5.16**
3H 2.5	3.07	3.55	0.16	1.98	m	**5.53**
3H 4	3.92	4.52	0.17	2.17	m	**6.69**
3H 6	4.92	5.68	0.19	2.37	m	**8.05**
3H 10	7.07	8.17	0.23	2.86	m	**11.03**
3H 16	9.99	11.54	0.25	3.15	m	**14.69**
3H 25	15.06	17.39	0.33	4.16	m	**21.55**
4H 1.5	3.02	3.49	0.15	1.96	m	**5.45**
4H 2.5	3.79	4.38	0.17	2.11	m	**6.49**
4H 4	4.70	5.43	0.19	2.35	m	**7.78**
4H 6	6.18	7.14	0.21	2.63	m	**9.77**
4H 10	8.72	10.08	0.26	3.22	m	**13.30**
4H 16	12.73	14.70	0.30	3.79	m	**18.49**
4H 25	18.55	21.43	0.40	5.01	m	**26.44**
7H 1.5	4.18	4.83	0.18	2.23	m	**7.06**
7H 2.5	5.61	6.48	0.20	2.50	m	**8.98**
12H 1.5	7.28	8.40	0.24	2.99	m	**11.39**
12H 2.5	9.69	11.20	0.25	3.15	m	**14.35**
19H 1.5	14.76	17.05	0.28	3.52	m	**20.57**

Y:MECHANICAL AND ELECTRICAL SERVICES

Item	Net Price £	Material £	Labour hours	Labour £	Unit	Total Rate £
Y61 : HV/LV CABLES AND WIRING : CABLE (contd)						
Cable terminations for M.I. Cable; Polymeric one piece moulding; containing grey sealing compound; testing; phase marking and connection						
Brass gland; polymeric one moulding containing grey sealing compound; coloured conductor sleeving; Earth tag; plastic gland shroud						
Light Duty 500 Volt grade						
2L 1.5	2.18	2.40	0.15	1.89	nr	4.29
2L 2.5	2.18	2.40	0.15	1.89	nr	4.29
3L 1.5	2.18	2.40	0.15	1.89	nr	4.29
4L 1.5	2.18	2.40	0.15	1.89	nr	4.29
Cable Terminations; for MI copper sheathed cable. Certified for installation in potentially explosive atmospheres; testing; phase marking and conNnection; BS 6207 Part 2 1995						
Brass gland; brass pot with earth tail; pot closure; sealing compound; conductor sleving; plastic gland shroud; identification markers						
Light Duty 500 Volt grade						
2L 1.0	1.98	2.18	0.30	3.79	nr	5.97
2L 1.5	1.98	2.18	0.30	3.79	nr	5.97
2L 2.5	1.98	2.18	0.30	3.79	nr	5.97
2L 4.0	1.99	2.19	0.30	3.79	nr	5.98
3L 1.0	1.99	2.19	0.30	3.79	nr	5.98
3L 1.5	1.99	2.19	0.30	3.79	nr	5.98
3L 2.5	1.99	2.19	0.30	3.79	nr	5.98
4L 1.0	1.99	2.19	0.33	4.16	nr	6.35
4L 1.5	1.99	2.19	0.33	4.16	nr	6.35
4L 2.5	1.99	2.19	0.33	4.16	nr	6.35
7L 1.0	5.33	5.87	0.53	6.69	nr	12.56
7L 1.5	5.33	5.87	0.53	6.69	nr	12.56
7L 2.5	5.33	5.87	0.53	6.69	nr	12.56
Brass gland; brass pot with earth tail; pot closure; sealing compound; conductor sleeving; plastic gland shroud; identification markers						
Heavy duty 750 Volt grade						
1H 10	5.34	5.88	0.22	2.78	nr	8.66
1H 16	5.34	5.88	0.22	2.78	nr	8.66
1H 25	8.95	9.84	0.30	3.79	nr	13.63
1H 35	8.95	9.84	0.30	3.79	nr	13.63
1H 50	16.00	17.59	0.30	3.79	nr	21.38

Y:MECHANICAL AND ELECTRICAL SERVICES

Item	Net Price £	Material £	Labour hours	Labour £	Unit	Total Rate £
1H 70	3.77	4.15	0.38	4.80	nr	8.95
1H 95	3.77	4.15	0.38	4.80	nr	8.95
1H 120	8.67	9.54	0.55	6.94	nr	16.48
1H 150	7.09	7.80	0.55	6.94	nr	14.74
1H 185	7.09	7.80	0.75	9.47	nr	17.27
1H 240	14.08	15.49	0.75	9.47	nr	24.96
2H 1.5	2.42	2.66	0.30	3.79	nr	6.45
2H 2.5	2.42	2.66	0.30	3.79	nr	6.45
2H 4	5.34	5.88	0.30	3.79	nr	9.67
2H 6	5.34	5.88	0.38	4.80	nr	10.68
2H 10	8.95	9.82	0.38	4.80	nr	14.62
2H 16	16.00	17.59	0.41	5.17	nr	22.76
2H 25	16.00	17.59	0.41	5.17	nr	22.76
3H 1.5	2.42	2.66	0.30	3.79	nr	6.45
3H 2.5	5.34	5.88	0.30	3.79	nr	9.67
3H 4	5.34	5.88	0.42	5.30	nr	11.18
3H 6	5.34	5.88	0.42	5.30	nr	11.18
3H 10	9.89	10.88	0.42	5.30	nr	16.18
3H 16	16.00	17.59	0.45	5.68	nr	23.27
3H 25	16.00	17.59	0.45	5.68	nr	23.27
4H 1.5	2.42	2.66	0.40	5.05	nr	7.71
4H 2.5	5.34	5.88	0.40	5.05	nr	10.93
4H 4	5.34	5.88	0.38	4.80	nr	10.68
4H 6	8.95	9.84	0.43	5.43	nr	15.27
4H 10	8.95	9.84	0.43	5.43	nr	15.27
4H 16	16.00	17.59	0.46	5.81	nr	23.40
4H 25	16.00	17.59	0.46	5.81	nr	23.40
7H 1.5	5.34	5.88	0.53	6.69	nr	12.57
7H 2.5	5.34	5.88	0.53	6.69	nr	12.57
12H 1.5	8.97	9.87	0.58	7.32	nr	17.19
12H 2.5	8.97	9.87	0.58	7.32	nr	17.19
19H 2.5	16.55	18.20	0.75	9.47	nr	27.67

PVC Insulated PVC Sheathed Copper Cable BS 6004 Tables 4 and 5; clipped to backgrounds; (Supports and fixings included.)

300/500 Volt Grade

Single Core

Item	Net Price £	Material £	Labour hours	Labour £	Unit	Total Rate £
1.0 mm2	0.34	0.39	0.03	0.38	m	0.77
1.5 mm2	0.44	0.51	0.03	0.38	m	0.89
2.5 mm2	0.75	0.86	0.03	0.38	m	1.24
4.0 mm2	1.21	1.40	0.04	0.50	m	1.90
6.0 mm2	1.61	1.86	0.04	0.50	m	2.36
10.0 mm2	2.59	2.99	0.05	0.63	m	3.62
16.0 mm2	3.37	3.90	0.05	0.63	m	4.53
25.0 mm2	6.37	7.36	0.06	0.76	m	8.12
35.0 mm2	9.79	11.30	0.06	0.76	m	12.06

Y:MECHANICAL AND ELECTRICAL SERVICES

Item	Net Price £	Material £	Labour hours	Labour £	Unit	Total Rate £
Y61 : HV/LV CABLES AND WIRING : CABLE (contd)						
PVC Insulated PVC Sheathed Copper Cable BS 6004 Tables 4 and 5; clipped to backgrounds; (Supports and fixings included.) (contd)						
300/500 Volt Grade (contd)						
Twin Core						
1.0 mm2	0.57	0.66	0.04	0.50	m	1.16
1.5 mm2	0.73	0.85	0.04	0.50	m	1.35
2.5 mm2	1.04	1.20	0.04	0.50	m	1.70
4.0 mm2	1.66	1.91	0.05	0.63	m	2.54
6.0 mm2	2.27	2.62	0.05	0.63	m	3.25
10.0 mm2	3.68	4.25	0.05	0.63	m	4.88
16.0 mm2	5.73	6.61	0.06	0.76	m	7.37
Twin Core with Earth Wire						
1.0 mm2	0.55	0.63	0.05	0.63	m	1.26
1.5 mm2	0.72	0.84	0.05	0.63	m	1.47
2.5 mm2	1.01	1.17	0.05	0.63	m	1.80
4.0 mm2	2.53	2.93	0.05	0.63	m	3.56
6.0 mm2	3.04	3.51	0.05	0.63	m	4.14
10.0 mm2	4.86	5.62	0.06	0.76	m	6.38
16.0 mm2	7.76	8.96	0.06	0.76	m	9.72
Three Core						
1.0 mm2	0.98	1.14	0.05	0.63	m	1.77
1.5 mm2	1.36	1.57	0.05	0.63	m	2.20
2.5 mm2	2.23	2.58	0.05	0.63	m	3.21
4.0 mm2	2.75	3.17	0.06	0.76	m	3.93
6.0 mm2	3.74	4.32	0.06	0.76	m	5.08
10.0 mm2	5.77	6.66	0.06	0.76	m	7.42
16.0 mm2	9.11	10.52	0.07	0.88	m	11.40
Three Core with Earth Wire						
1.0 mm2	0.83	0.96	0.06	0.76	m	1.72
1.5 mm2	1.36	1.57	0.06	0.76	m	2.33
2.5 mm2	2.35	2.72	0.06	0.76	m	3.48
4.0 mm2	3.86	4.46	0.06	0.76	m	5.22
Single Core, PVC Insulated Cable; non-sheathed copper; BS 6004 Table 1a						
450/700 Volt Grade						
Single Strand - laid in trunking						
1.0 mm2	0.19	0.22	0.03	0.38	m	0.60
1.5 mm2	0.31	0.36	0.03	0.38	m	0.74
2.5 mm2	0.44	0.51	0.03	0.38	m	0.89

Y:MECHANICAL AND ELECTRICAL SERVICES

Item	Net Price £	Material £	Labour hours	Labour £	Unit	Total Rate £
Multi Strand - laid in trunking						
1.5 mm2	0.31	0.36	0.03	0.38	m	**0.74**
2.5 mm2	0.44	0.51	0.03	0.38	m	**0.89**
4.0 mm2	0.75	0.87	0.03	0.38	m	**1.25**
6.0 mm2	1.09	1.26	0.04	0.50	m	**1.76**
10.0 mm2	2.18	2.52	0.04	0.50	m	**3.02**
16.0 mm2	3.33	3.85	0.04	0.50	m	**4.35**
25.0 mm2	4.44	4.88	0.05	0.63	m	**5.51**
35.0 mm2	5.57	6.43	0.05	0.63	m	**7.06**
50.0 mm2	8.11	9.37	0.05	0.63	m	**10.00**
70.0 mm2	11.16	12.89	0.05	0.63	m	**13.52**
95.0 mm2	17.20	19.86	0.06	0.76	m	**20.62**
120.0 mm2	23.64	27.31	0.07	0.88	m	**28.19**
150.0 mm2	29.81	34.43	0.08	1.01	m	**35.44**
Single Strand - drawn into conduit						
1.0 mm2	0.19	0.22	0.04	0.50	m	**0.72**
1.5 mm2	0.31	0.36	0.04	0.50	m	**0.86**
2.5 mm2	0.44	0.51	0.04	0.50	m	**1.01**
Multi Strand - drawn into conduit						
1.5 mm2	0.31	0.36	0.03	0.38	m	**0.74**
2.5 mm2	0.44	0.51	0.03	0.38	m	**0.89**
4.0 mm2	0.75	0.87	0.04	0.50	m	**1.37**
6.0 mm2	1.09	1.26	0.04	0.50	m	**1.76**
10.0 mm2	2.18	2.52	0.05	0.63	m	**3.15**
16.0 mm2	3.33	3.85	0.05	0.63	m	**4.48**
25.0 mm2	4.23	4.88	0.05	0.63	m	**5.51**
35.0 mm2	5.57	6.43	0.07	0.88	m	**7.31**

H.O.F.R. (Heat, Oil, Fire Retardent) instrument/Control cable; silicon rubber insulated copper conductor; aluminium/ hard grade L.S.O.H. laminate sheath; clipped to backgrounds; (Including supports fixings and connecting tails) BS 6387; IEC 331 Part 1.

250/440 Volt Grade; with drain wire

Two Core

Item	Net Price £	Material £	Labour hours	Labour £	Unit	Total Rate £
1.0 mm2	1.16	1.34	0.07	0.88	m	**2.22**
1.5 mm2	1.40	1.62	0.07	0.88	m	**2.50**
2.5 mm2	1.80	2.08	0.07	0.88	m	**2.96**
4.0 mm2	2.80	3.24	0.07	0.88	m	**4.12**
Three Core						
1.0 mm2	1.48	1.70	0.07	0.88	m	**2.58**
1.5 mm2	1.84	2.13	0.07	0.88	m	**3.01**
2.5 mm2	2.29	2.64	0.08	1.01	m	**3.65**
4.0 mm2	3.77	4.36	0.08	1.01	m	**5.37**

Y:MECHANICAL AND ELECTRICAL SERVICES

Item	Net Price £	Material £	Labour hours	Labour £	Unit	Total Rate £
Y61 : HV/LV CABLES AND WIRING : CABLE (contd)						
H.O.F.R. (Heat, Oil, Fire Retardent) instrument/Control cable; silicon rubber insulated coper conductor; aluminium/ hard grade L.S.O.H. laminate sheath; clipped to backgrounds; (Including supports fixings and connecting tails) BS 6387; IEC 331 Part 1. (contd)						
250/440 Volt Grade; with drain wire (contd)						
Four Core						
1.0 mm2	1.80	2.07	0.08	1.01	m	**3.08**
1.5 mm2	2.24	2.59	0.08	1.01	m	**3.60**
2.5 mm2	3.12	3.61	0.09	1.14	m	**4.75**
4.0 mm2	4.90	5.66	0.09	1.14	m	**6.80**
Seven Core						
1.0 mm2	2.60	3.00	0.09	1.14	m	**4.14**
1.5 mm2	3.21	3.71	0.09	1.14	m	**4.85**
2.5 mm2	4.21	4.87	0.09	1.14	m	**6.01**
Twelve Core						
1.0 mm2	7.75	8.95	0.10	1.26	m	**10.21**
1.5 mm2	10.18	11.75	0.10	1.26	m	**13.01**
Nineteen Core						
1.5 mm2	14.10	16.29	0.11	1.39	m	**17.68**
2.5 mm2	18.32	21.16	0.13	1.64	m	**22.80**
Cable Terminations for H.O.F.R. Cable (Including drilling and cutting mild steel gland plate)						
250 Volt grade; with drain wire; gland; nylon compression seal; locknut.						
Two Core						
1.0 mm2	0.31	0.35	0.13	1.64	nr	**1.99**
1.5 mm2	0.31	0.35	0.13	1.64	nr	**1.99**
2.5 mm2	0.42	0.47	0.15	1.89	nr	**2.36**
4.0 mm2	0.42	0.47	0.15	1.89	nr	**2.36**
Three Core						
1.0 mm2	0.31	0.35	0.16	2.02	nr	**2.37**
1.5 mm2	0.31	0.35	0.16	2.02	nr	**2.37**
2.5 mm2	0.42	0.47	0.18	2.27	nr	**2.74**
4.0 mm2	0.42	0.47	0.18	2.27	nr	**2.74**

Y:MECHANICAL AND ELECTRICAL SERVICES

Item	Net Price £	Material £	Labour hours	Labour £	Unit	Total Rate £
Four Core						
1.0 mm2	0.31	0.35	0.21	2.65	nr	**3.00**
1.5 mm2	0.42	0.47	0.23	2.90	nr	**3.37**
2.5 mm2	0.42	0.47	0.27	3.41	nr	**3.88**
4.0 mm2	0.73	0.80	0.30	3.79	nr	**4.59**
Seven Core						
1.0 mm2	0.42	0.47	0.35	4.42	nr	**4.89**
1.5 mm2	0.42	0.47	0.35	4.42	nr	**4.89**
2.5 mm2	0.73	0.80	0.40	5.05	nr	**5.85**
Twelve Core						
1.5 mm2	0.73	0.80	0.55	6.94	nr	**7.74**
2.5 mm2	1.00	1.10	0.60	7.57	nr	**8.67**
Nineteen Core						
1.5 mm2	1.00	1.10	0.80	10.10	nr	**11.20**
2.5 mm2	1.00	1.10	0.85	10.73	nr	**11.83**
250/440 Volt grade; with drain wire; gland; brass A1/A2 weatherproof rated I.P.66 PVC Shroud; brass locknut, identification ferrules						
Two Core						
1.0 mm2	3.03	3.33	0.13	1.64	nr	**4.97**
1.5 mm2	3.03	3.33	0.13	1.64	nr	**4.97**
Three Core						
1.0 mm2	3.03	3.33	0.16	2.02	nr	**5.35**
1.5 mm2	3.03	3.33	0.16	2.02	nr	**5.35**
Four Core						
1.0 mm2	3.03	3.33	0.23	2.90	nr	**6.23**
1.5 mm2	3.03	3.33	0.23	2.90	nr	**6.23**
Seven Core						
1.0 mm2	3.03	3.33	0.35	4.42	nr	**7.75**
1.5 mm2	3.03	3.33	0.35	4.42	nr	**7.75**
Twelve Core						
1.5 mm2	3.40	3.74	0.53	6.69	nr	**10.43**
Nineteen Core						
1.5 mm2	5.36	5.90	0.83	10.48	nr	**16.38**

Y:MECHANICAL AND ELECTRICAL SERVICES

Item	Net Price £	Material £	Labour hours	Labour £	Unit	Total Rate £
Y62 : BUSBAR TRUNKING : BUSBARS						
Rising Main Busbar System; fixed to backgrounds including supports, fixings and connections/jointing to equipment. (Provision and fixing of plates, discs etc, for identification is included.)						
315 Amp TP&N						
Copper busbar enclosed in metal trunking with insulated supports, earth continuity bar fixed to outside of trunking, single hole joints and hangers.	62.32	68.55	6.02	76.02	m	**144.57**
Extra for trunking fittings						
End feed unit	136.33	149.96	4.00	50.48	nr	**200.44**
Centre feed unit	186.96	205.66	6.02	76.02	nr	**281.68**
End cap	14.61	16.07	1.00	12.62	nr	**28.69**
Fire Resistant Barries	37.00	40.70	1.00	12.62	nr	**53.32**
Direct Tap-Off unit including HRC fuses						
32 Amp	63.29	69.62	3.00	37.90	nr	**107.52**
63 Amp	73.03	80.33	3.00	37.90	nr	**118.23**
100 Amp	134.38	147.82	3.00	37.90	nr	**185.72**
250 Amp	429.42	472.37	3.00	37.90	nr	**510.27**
Direct fused switch including HRC fuses						
32 Amp	92.51	101.76	3.00	37.90	nr	**139.66**
63 Amp	113.93	125.32	3.00	37.90	nr	**163.22**
100 Amp	158.72	174.59	3.00	37.90	nr	**212.49**
200 Amp	429.42	472.37	3.00	37.90	nr	**510.27**
400 Amp TP&N						
Copper busbar enclosed in metal trunking with insulated supports, earth continuity bar fixed to outside of trunking, single hole joints and hangers	80.82	88.90	6.02	76.02	m	**164.92**
Extra for trunking fittings						
End feed unit	166.51	183.16	4.00	50.48	nr	**233.64**
Centre feed unit	217.15	238.86	6.02	76.02	nr	**314.88**
End cap	15.58	17.14	1.00	12.62	nr	**29.76**
Fire Resistant Barrier	37.00	40.70	1.00	12.62	nr	**53.32**
Direct tap-off unit including HRC fuses						
32 Amp	63.29	69.62	3.00	37.90	nr	**107.52**
63 Amp	73.03	80.33	3.00	37.90	nr	**118.23**
100 Amp	134.38	147.82	3.00	37.90	nr	**185.72**
250 Amp	429.42	472.37	3.00	37.90	nr	**510.27**

Y:MECHANICAL AND ELECTRICAL SERVICES

Item	Net Price £	Material £	Labour hours	Labour £	Unit	Total Rate £
Direct fused switch including HRC fuses						
32 Amp	92.51	101.76	3.00	37.90	nr	**139.66**
60 Amp	113.93	125.32	3.00	37.90	nr	**163.22**
100 Amp	158.72	174.59	3.00	37.90	nr	**212.49**
200 Amp	429.42	472.37	3.00	37.90	nr	**510.27**
600 Amp TP&N						
Copper busbar enclosed in metal trunking with insulated supports, earth continuity bar fixed to outside of trunking, single hole joints and hangers	113.93	125.32	6.02	76.02	m	**201.34**
600 Amp TP&N; extra for trunking fittings						
End feed unit	237.60	261.35	4.00	50.48	nr	**311.83**
Centre feed unit	311.60	342.76	6.02	76.02	nr	**418.78**
End cap	19.48	21.42	1.00	12.62	nr	**34.04**
Fire Resistant Barrier	50.64	55.70	1.00	12.62	nr	**68.32**
600 Amp TP&N; direct tap-off unit including HRC fuses						
32 Amp	63.29	69.62	3.00	37.90	nr	**107.52**
63 Amp	73.03	80.33	3.00	37.90	nr	**118.23**
100 Amp	134.38	147.82	3.00	37.90	nr	**185.72**
250 Amp	429.42	472.37	3.00	37.90	nr	**510.27**
600 Amp TP&N; direct fused switch including HRC fuses						
32 Amp	92.51	101.76	3.00	37.90	nr	**139.66**
63 Amp	113.93	125.32	3.00	37.90	nr	**163.22**
100 Amp	158.72	174.59	3.00	37.90	nr	**212.49**
200 Amp	429.42	472.37	3.00	37.90	nr	**510.27**
800 Amp TP&N						
Copper busbar enclosed in metal trunking with insulated supports, earth continuity bar fixed to outside of trunking, single hole joints and hangers	137.30	151.03	7.04	88.87	m	**239.90**
800 Amp TP&N; extra for trunking fittings						
End feed unit	283.36	311.70	4.00	50.48	nr	**362.18**
Centre feed unit	372.95	410.24	6.02	76.02	nr	**486.26**
End cap	21.42	23.56	1.00	12.62	nr	**36.18**
Fire Resistant Barrier	56.48	62.13	1.00	12.62	nr	**74.75**
800 Amp TP&N; direct tap-off unit including HRC fuses						
32 Amp	63.29	69.62	3.00	37.90	nr	**107.52**
63 Amp	73.03	80.33	3.00	37.90	nr	**118.23**
100 Amp	134.38	147.82	3.00	37.90	nr	**185.72**
250 Amp	429.42	472.37	3.00	37.90	nr	**510.27**

Y:MECHANICAL AND ELECTRICAL SERVICES

Item	Net Price £	Material £	Labour hours	Labour £	Unit	Total Rate £
Y62 : BUSBAR TRUNKING : BUSBARS (contd)						
Rising Main Busbar System; fixed to backgrounds including supports, fixings and connections/jointing to equipment. (Provision and fixing of plates, discs etc, for identification is included.) (contd)						
800 Amp TP&N; direct fused switch including HRC fuses						
32 Amp	92.51	101.76	3.00	37.90	nr	139.66
63 Amp	113.93	125.32	3.00	37.90	nr	163.22
100 Amp	158.72	174.59	3.00	37.90	nr	212.49
200 Amp	429.42	472.37	3.00	37.90	nr	510.27
1000 Amp TP&N						
Copper busbar enclosed in metal trunking with insulated supports, earth continuity bar fixed to outside of trunking, single hole joints and hangers	160.67	176.74	8.00	100.96	m	277.70
1000 Amp TP&N; extra for trunking fittings						
End feed unit	345.68	380.25	4.00	50.48	nr	430.73
Centre feed unit	453.77	499.14	6.02	76.02	nr	575.16
End cap	24.34	26.78	1.00	12.62	nr	39.40
Fire Resistant Barrier	58.42	64.27	1.00	12.62	nr	76.89
1000 Amp TP&N; direct tap-off unit including HRC fuses						
32 Amp	63.29	69.62	3.00	37.90	nr	107.52
63 Amp	73.03	80.33	3.00	37.90	nr	118.23
100 Amp	134.38	147.82	3.00	37.90	nr	185.72
250 Amp	429.42	472.37	3.00	37.90	nr	510.27
1000 Amp TP&N; direct fused switch including HRC fuses						
32 Amp	92.51	101.76	3.00	37.90	nr	139.66
63 Amp	113.93	125.32	3.00	37.90	nr	163.22
100 Amp	158.72	174.59	3.00	37.90	nr	212.49
200 Amp	429.42	472.37	3.00	37.90	nr	510.27

Y:MECHANICAL AND ELECTRICAL SERVICES

Item	Net Price £	Material £	Labour hours	Labour £	Unit	Total Rate £
Busbar Chambers; fixed to background including all supports, fixings, connections/jointing to equipment. (Provision and fixing of plates, discs etc, for identification is included.)						
Sheet steel case enclosing 4 pole 550 Volt copper bars, detachable metal end plates; 600mm long						
200 Amp	199.18	219.09	3.51	44.28	nr	263.37
300 Amp	198.97	218.86	4.00	50.48	nr	269.34
500 Amp	291.13	320.24	5.00	63.10	nr	383.34
Sheet steel case enclosing 4 pole 550 Volt copper bars, detachable metal end plates; 900mm long						
200 Amp	286.70	315.37	3.76	47.44	nr	362.81
300 Amp	338.13	371.95	4.26	53.70	nr	425.65
500 Amp	391.51	430.67	5.26	66.42	nr	497.09
Sheet steel case enclosing 4 pole 550 Volt copper bars, detachable metal end plates; 1350mm long						
200 Amp	400.42	440.47	4.00	50.48	nr	490.95
300 Amp	461.25	507.38	4.50	56.85	nr	564.23
500 Amp	516.44	568.08	5.52	69.72	nr	637.80

Keep your figures up to date, free of charge

This section, and most of the other information in this Price Book, is brought up to date every three months, until the next annual edition, in the *Price Book Update*.

The *Update* is available free to all Price Book purchasers.

To ensure you receive your copy, simply complete the reply card from the centre of the book and return it to us.

Y:MECHANICAL AND ELECTRICAL SERVICES

Item	Net Price £	Material £	Labour hours	Labour £	Unit	Total Rate £
Y63 : SUPPORT COMPONENTS - CABLES : CABLE TRAY						
Galvanised Steel Cable Tray to BS 729; including standard coupling joints, fixings and earth continuity straps. (Supports and hangers are excluded.)						
Tray						
75 mm	4.46	4.90	0.22	2.78	m	7.68
100 mm	5.44	5.99	0.32	4.04	m	10.03
150 mm	6.54	7.19	0.32	4.04	m	11.23
225 mm	8.46	9.31	0.38	4.80	m	14.11
300 mm	12.54	13.79	0.50	6.31	m	20.10
450 mm	19.22	21.14	0.57	7.19	m	28.33
600 mm	24.26	26.69	0.80	10.10	m	36.79
Tray with return flange						
75 mm	8.64	9.51	0.35	4.42	m	13.93
100 mm	9.05	9.95	0.35	4.42	m	14.37
150 mm	10.98	12.08	0.35	4.42	m	16.50
225 mm	13.97	15.36	0.40	5.05	m	20.41
300 mm	16.02	17.63	0.55	6.94	m	24.57
450 mm	25.04	27.54	0.60	7.57	m	35.11
600 mm	31.48	34.62	0.85	10.73	m	45.35
750 mm	39.22	43.14	0.90	11.36	m	54.50
900 mm	46.04	50.65	1.00	12.62	m	63.27
Tray with return flange with epoxy bonding coat						
75 mm	16.23	17.85	0.37	4.67	m	22.52
100 mm	18.09	19.90	0.37	4.67	m	24.57
150 mm	21.71	23.88	0.38	4.80	m	28.68
225 mm	26.65	29.32	0.42	5.30	m	34.62
300 mm	32.09	35.30	0.57	7.19	m	42.49
450 mm	49.91	54.90	0.62	7.83	m	62.73
600 mm	61.54	67.69	0.88	11.11	m	78.80
750 mm	77.02	84.72	0.93	11.74	m	96.46
900 mm	92.62	101.88	1.03	13.01	m	114.89
Galvanised Steel Tray and Fittings to BS 729; (Cutting and jointing tray to fittings is included.)						
Bend						
75 mm	3.77	4.15	0.20	2.52	nr	6.67
100 mm	4.16	4.58	0.22	2.78	nr	7.36
150 mm	5.05	5.55	0.22	2.78	nr	8.33
225 mm	6.79	7.47	0.28	3.53	nr	11.00
300 mm	9.55	10.50	0.32	4.04	nr	14.54
450 mm	16.55	18.20	0.53	6.69	nr	24.89
600 mm	25.78	28.36	0.67	8.46	nr	36.82

Y:MECHANICAL AND ELECTRICAL SERVICES

Item	Net Price £	Material £	Labour hours	Labour £	Unit	Total Rate £
Bend; with return flange						
75 mm	21.75	23.93	0.22	2.78	nr	**26.71**
100 mm	24.72	27.20	0.22	2.78	nr	**29.98**
150 mm	26.31	28.94	0.25	3.15	nr	**32.09**
225 mm	31.29	34.42	0.33	4.16	nr	**38.58**
300 mm	37.98	41.78	0.35	4.42	nr	**46.20**
450 mm	58.88	64.77	0.55	6.94	nr	**71.71**
600 mm	82.75	91.02	0.70	8.84	nr	**99.86**
750 mm	88.82	97.70	1.00	12.62	nr	**110.32**
900 mm	134.73	148.20	1.00	12.62	nr	**160.82**
Bend; with return flange with epoxy bonding coat						
75 mm	41.31	45.44	0.24	3.03	nr	**48.47**
100 mm	42.24	46.47	0.24	3.03	nr	**49.50**
150 mm	47.15	51.86	0.27	3.41	nr	**55.27**
225 mm	60.21	66.23	0.32	4.04	nr	**70.27**
300 mm	71.31	78.44	0.37	4.67	nr	**83.11**
450 mm	117.40	129.14	0.57	7.19	nr	**136.33**
600 mm	155.89	171.49	0.70	8.84	nr	**180.33**
750 mm	194.89	214.38	0.85	10.73	nr	**225.11**
900 mm	264.32	290.76	1.02	12.88	nr	**303.64**
Equal Tee						
50 mm	4.88	5.37	0.30	3.79	nr	**9.16**
75 mm	4.99	5.49	0.30	3.79	nr	**9.28**
100 mm	5.30	5.83	0.32	4.04	nr	**9.87**
150 mm	6.15	6.77	0.32	4.04	nr	**10.81**
225 mm	9.97	4.39	1.00	12.62	nr	**17.01**
300 mm	15.28	16.80	0.50	6.31	nr	**23.11**
450 mm	26.42	29.06	0.77	9.72	nr	**38.78**
600 mm	26.55	29.20	1.10	13.88	nr	**43.08**
Equal Tee; with return flange						
75 mm	25.46	28.01	0.32	4.04	nr	**32.05**
100 mm	29.18	32.10	0.32	4.04	nr	**36.14**
150 mm	31.09	34.20	0.35	4.42	nr	**38.62**
225 mm	36.71	40.38	0.42	5.30	nr	**45.68**
300 mm	41.16	45.28	0.53	6.69	nr	**51.97**
450 mm	69.17	76.09	0.80	10.10	nr	**86.19**
600 mm	96.54	106.19	1.13	14.26	nr	**120.45**
750 mm	120.94	133.03	1.19	15.02	nr	**148.05**
900 mm	142.16	156.37	1.50	18.95	nr	**175.32**
Equal Tee; with return flange with epoxy bonding coat						
75 mm	25.46	28.01	0.34	4.29	nr	**32.30**
100 mm	55.54	61.10	0.34	4.29	nr	**65.39**
150 mm	61.61	67.78	0.37	4.66	nr	**72.44**
225 mm	79.94	87.94	0.47	5.93	nr	**93.87**
300 mm	92.07	101.28	0.55	6.94	nr	**108.22**
450 mm	149.26	164.19	0.88	11.11	nr	**175.30**
600 mm	210.28	231.31	1.15	14.52	nr	**245.83**
750 mm	262.92	289.21	1.21	15.28	nr	**304.49**
900 mm	317.07	348.78	1.52	19.21	nr	**367.99**

Y:MECHANICAL AND ELECTRICAL SERVICES

Item	Net Price £	Material £	Labour hours	Labour £	Unit	Total Rate £
Y63 : SUPPORT COMPONENTS - CABLES: CABLE TRAY (contd)						
Galvanised Steel Tray and Fittings to BS 729; (Cutting and jointing tray to fittings is included.) (contd)						
Reducer						
75 mm	6.90	7.59	0.15	1.89	nr	9.48
100 mm	7.11	7.82	0.15	1.89	nr	9.71
150 mm	10.18	11.20	0.15	1.89	nr	13.09
225 mm	13.79	15.17	0.23	2.90	nr	18.07
300 mm	17.72	19.49	0.23	2.90	nr	22.39
450 mm	28.22	31.04	0.37	4.67	nr	35.71
600 mm	36.07	39.68	0.37	4.66	nr	44.34
Reducer; with return flange						
100 mm	15.81	17.39	0.17	2.15	nr	19.54
150 mm	17.51	19.26	0.17	2.15	nr	21.41
225 mm	20.27	22.29	0.25	3.15	nr	25.44
300 mm	22.81	25.09	0.25	3.15	nr	28.24
450 mm	36.07	39.68	0.40	5.05	nr	44.73
600 mm	48.80	53.68	0.40	5.05	nr	58.73
750 mm	68.96	75.85	0.61	7.70	nr	83.55
900 mm	86.14	94.76	0.61	7.70	nr	102.46
Reducer; with return flange with epoxy bonding coat						
100 mm	35.37	38.91	0.19	2.40	nr	41.31
150 mm	37.81	41.59	0.19	2.40	nr	43.99
225 mm	45.98	50.58	0.27	3.41	nr	53.99
300 mm	55.78	61.36	0.27	3.41	nr	64.77
450 mm	83.43	91.77	0.42	5.30	nr	97.07
600 mm	87.18	95.89	0.42	5.30	nr	101.19
750 mm	91.61	100.77	0.63	7.95	nr	108.72
900 mm	95.92	105.51	0.63	7.95	nr	113.46
Rigid PVC Cable Tray; including standard coupling joints and fixings. (Excluding supports or hangers.)						
Light Duty Tray						
50 mm	3.36	3.88	0.40	5.05	m	8.93
75 mm	3.88	4.49	0.40	5.05	m	9.54
100 mm	4.23	4.88	0.40	5.05	m	9.93
150 mm	5.12	5.91	0.45	5.68	m	11.59
200 mm	8.97	10.36	0.45	5.68	m	16.04
300 mm	10.93	12.63	0.45	5.68	m	18.31
Medium Duty Tray						
50 mm	3.36	3.88	0.40	5.05	m	8.93
75 mm	3.80	4.39	0.40	5.05	m	9.44
100 mm	4.23	4.88	0.40	5.05	m	9.93
150 mm	5.12	5.91	0.40	5.05	m	10.96

Y:MECHANICAL AND ELECTRICAL SERVICES

Item	Net Price £	Material £	Labour hours	Labour £	Unit	Total Rate £
Medium Duty Cover						
50 mm	1.66	1.92	0.10	1.26	m	**3.18**
75 mm	1.99	2.30	0.10	1.26	m	**3.56**
100 mm	3.47	4.01	0.10	1.26	m	**5.27**
150 mm	3.99	4.60	0.10	1.26	m	**5.86**
Heavy Duty Tray						
100 mm	5.68	6.56	0.40	5.05	m	**11.61**
150 mm	5.86	6.77	0.40	5.05	m	**11.82**
200 mm	9.46	10.92	0.45	5.68	m	**16.60**
300 mm	10.06	11.62	0.45	5.68	m	**17.30**
400 mm	16.05	18.54	0.55	6.94	m	**25.48**
600 mm	37.61	43.44	0.55	6.94	m	**50.38**
Heavy Duty Cover						
100 mm	3.67	4.24	0.10	1.26	m	**5.50**
150 mm	4.15	4.79	0.10	1.26	m	**6.05**
200 mm	7.40	8.55	0.10	1.26	m	**9.81**
300 mm	10.06	11.62	0.10	1.26	m	**12.88**
400 mm	16.05	18.54	0.15	1.89	m	**20.43**
600 mm	23.55	27.20	0.15	1.89	m	**29.09**
Rigid PVC Tray and Fittings; (Cutting and jointing to fittings included.)						
Bend						
50 mm	16.47	18.11	0.40	5.05	nr	**23.16**
100 mm	16.47	18.11	0.40	5.05	nr	**23.16**
150 mm	21.94	24.13	0.40	5.05	nr	**29.18**
200 mm	25.54	28.10	0.45	5.68	nr	**33.78**
300 mm	34.99	38.49	0.45	5.68	nr	**44.17**
400 mm	73.44	80.78	0.55	6.94	nr	**87.72**
600 mm	95.75	105.32	0.55	6.94	nr	**112.26**
Bend Cover						
50 mm	7.29	8.02	0.10	1.26	nr	**9.28**
100 mm	7.29	8.02	0.10	1.26	nr	**9.28**
150 mm	10.18	11.20	0.10	1.26	nr	**12.46**
200 mm	12.58	13.84	0.10	1.26	nr	**15.10**
300 mm	19.62	21.58	0.10	1.26	nr	**22.84**
400 mm	34.39	37.83	0.15	1.89	nr	**39.72**
600 mm	43.76	48.14	0.15	1.89	nr	**50.03**
Reducer						
100 mm	20.58	22.64	0.40	5.05	nr	**27.69**
150 mm	20.58	22.64	0.40	5.05	nr	**27.69**
200 mm	20.58	22.64	0.45	5.68	nr	**28.32**
300 mm	58.98	64.87	0.45	5.68	nr	**70.55**
400 mm	72.86	80.14	0.55	6.94	nr	**87.08**
600 mm	72.86	80.14	0.55	6.94	nr	**87.08**

Y:MECHANICAL AND ELECTRICAL SERVICES

Item	Net Price £	Material £	Labour hours	Labour £	Unit	Total Rate £
Y63 : SUPPORT COMPONENTS - CABLES: CABLE TRAY (contd)						
GRP Cable Tray including standard coupling joints and fixings; (Supports and hangers excluded).						
Tray						
100 mm	8.76	10.12	0.40	5.05	m	**15.17**
200 mm	11.73	13.55	0.45	5.68	m	**19.23**
400 mm	26.37	30.46	0.55	6.94	m	**37.40**
Cover						
100 mm	6.62	7.65	0.10	1.26	m	**8.91**
200 mm	9.92	11.45	0.10	1.26	m	**12.71**
400 mm	15.20	17.56	0.15	1.89	m	**19.45**
GRP Tray and Fittings; (Cutting and jointing to fittings included).						
Bend						
100 mm	27.28	30.01	0.40	5.05	nr	**35.06**
200 mm	44.23	48.66	0.45	5.68	nr	**54.34**
400 mm	90.72	99.79	0.15	1.89	nr	**101.68**
Bend Cover						
100 mm	10.42	11.46	0.10	1.26	nr	**12.72**
200 mm	12.20	13.41	0.10	1.26	nr	**14.67**
400 mm	14.88	16.37	0.15	1.89	nr	**18.26**
Tee						
100 mm	31.45	34.59	0.40	5.05	nr	**39.64**
200 mm	50.87	55.95	0.45	5.68	nr	**61.63**
400 mm	104.33	114.76	0.55	6.94	nr	**121.70**
Tee Cover						
100 mm	15.06	16.56	0.40	5.05	nr	**21.61**
200 mm	16.10	17.71	0.45	5.68	nr	**23.39**
400 mm	17.83	19.62	0.55	6.94	nr	**26.56**
Reducer						
200 mm	50.87	55.95	0.10	1.26	nr	**57.21**
400 mm	104.33	114.76	0.15	1.89	nr	**116.65**
Reducer Cover						
200 mm	14.64	16.10	0.40	5.05	nr	**21.15**
400 mm	15.07	16.58	0.45	5.68	nr	**22.26**

Y:MECHANICAL AND ELECTRICAL SERVICES

Item	Net Price £	Material £	Labour hours	Labour £	Unit	Total Rate £
Y63 : SUPPORT COMPONENTS - CABLES : CABLE RACK						
Galvanised Steel Ladder Rack; fixed to backgrounds; including supports, fixings and brackets; earth continuity straps.						
Straight						
150 mm wide ladder	9.60	11.09	0.67	8.46	m	**19.55**
200 mm wide ladder	9.81	11.33	0.67	8.46	m	**19.79**
300 mm wide ladder	10.18	11.76	0.87	10.98	m	**22.74**
400 mm wide ladder	11.86	13.70	1.08	13.64	m	**27.34**
500 mm wide ladder	13.51	15.60	1.30	16.41	m	**32.01**
600 mm wide ladder	13.89	16.04	1.58	19.97	m	**36.01**
800 mm wide ladder	16.05	18.54	1.75	22.10	m	**40.64**
1000 mm wide ladder	16.74	19.33	1.90	23.99	m	**43.32**
Galvanised Steel Ladder Rack Fittings; (Cutting and jointing racking to fittings is included.)						
Internal Radius Bend						
150 mm wide ladder	23.06	25.37	0.27	3.41	nr	**28.78**
200 mm wide ladder	23.52	25.88	0.27	3.41	nr	**29.29**
300 mm wide ladder	26.66	29.33	0.50	6.31	nr	**35.64**
400 mm wide ladder	31.18	34.30	0.83	10.48	nr	**44.78**
500 mm wide ladder	33.12	36.43	0.92	11.61	nr	**48.04**
600 mm wide ladder	34.87	38.36	1.00	12.62	nr	**50.98**
800 mm wide ladder	43.63	48.00	1.02	12.88	nr	**60.88**
1000 mm wide ladder	63.93	70.32	1.03	13.01	nr	**83.33**
External Radius Bend						
200 mm wide ladder	36.81	40.49	0.27	3.41	nr	**43.90**
300 mm wide ladder	41.60	45.77	0.33	4.20	nr	**49.97**
400 mm wide ladder	44.74	49.22	0.83	10.48	nr	**59.70**
500 mm wide ladder	53.14	58.45	0.58	7.32	nr	**65.77**
600 mm wide ladder	58.21	64.03	0.83	10.48	nr	**74.51**
800 mm wide ladder	87.64	96.40	1.00	12.62	nr	**109.02**
T-Junction						
200 mm wide ladder	38.28	42.11	0.37	4.67	nr	**46.78**
300 mm wide ladder	41.14	45.26	0.57	7.19	nr	**52.45**
400 mm wide ladder	44.10	48.51	1.17	14.78	nr	**63.29**
500 mm wide ladder	46.68	51.35	1.17	14.78	nr	**66.13**
600 mm wide ladder	48.80	53.68	1.17	14.78	nr	**68.46**
800 mm wide ladder	83.76	92.14	1.25	15.77	nr	**107.91**
1000 mm wide ladder	92.25	101.48	1.33	16.80	nr	**118.28**

Y:MECHANICAL AND ELECTRICAL SERVICES

Item	Net Price £	Material £	Labour hours	Labour £	Unit	Total Rate £
Y63 : SUPPORT COMPONENTS - CABLES: CABLE RACK (contd)						
Galvanised Steel Ladder Rack Fittings; (Cutting and jointing racking to fittings is included.) (contd)						
X-Junction						
200 mm wide ladder	50.28	55.30	0.50	6.31	nr	**61.61**
300 mm wide ladder	54.61	60.07	0.67	8.46	nr	**68.53**
400 mm wide ladder	59.50	65.45	1.17	14.78	nr	**80.23**
500 mm wide ladder	66.97	73.67	1.27	16.04	nr	**89.71**
600 mm wide ladder	71.68	78.85	1.37	17.31	nr	**96.16**
800 mm wide ladder	81.73	89.91	1.44	18.18	nr	**108.09**
1000 mm wide ladder	92.25	101.48	1.50	18.95	nr	**120.43**
Riser						
150 mm wide ladder	32.29	35.52	0.33	4.20	nr	**39.72**
200 mm wide ladder	35.05	38.56	0.50	6.31	nr	**44.87**
300 mm wide ladder	38.93	42.82	0.67	8.44	nr	**51.26**
400 mm wide ladder	41.79	45.97	0.83	10.48	nr	**56.45**
500 mm wide ladder	42.71	46.98	0.92	11.61	nr	**58.59**
600 mm wide ladder	43.36	47.69	1.03	13.01	nr	**60.70**
800 mm wide ladder	46.13	50.74	1.10	13.88	nr	**64.62**
1000 mm wide ladder	57.19	62.91	1.17	14.78	nr	**77.69**
End Connector						
Any ladder width	2.31	2.54	0.27	3.41	nr	**5.95**
Universal Coupling						
Any ladder width	3.04	3.35	0.27	3.41	nr	**6.76**

Y:MECHANICAL AND ELECTRICAL SERVICES

Item	Net Price £	Material £	Labour hours	Labour £	Unit	Total Rate £
Y70 : HV SWITCHGEAR : **CIRCUIT BREAKERS**						
H.V. Circuit Breakers; installed on prepared foundations including all supports, fixings and inter panel connections (Excluding main and multi core cabling and heat shrink cable termination kits.)						
Three phase 11kV, 630 Amp circuit breaker panels; hand charged spring closing operation; prospective fault level up to 20 kA; feeders include ammeter with phase selector switch, 3 pole IDMTL, overcurrent and earth fault relays with necessary current relays with necessary current transformers; incomers include 3 phase voltage transformer, voltmeter and phase selector switch; vacuum or SF6 insulation						
Single panel with incoming and outgoing cable box	14600.00	16060.00	30.30	382.42	nr	**16442.42**
Three panel with one incomer and two feeders	44000.00	48400.00	50.00	631.00	nr	**49031.00**
Five panel with two incoming, two feeders and a bus section	75000.00	82500.00	62.50	788.75	nr	**83288.75**
Three phase 11kV, 630 Amp circuit breaker panels; hand charged spring closing operation; prospective fault level up to 20 kA; feeders include ammeter with phase selector switch, 3 pole IDMTL, overcurrent and earth fault relays with necessary current relays with necessary current transformers; incomers include 3 phase voltage transformer, voltmeter and phase selector switch; oil insulation						
Single cubicle panel	13700.00	15070.00	30.30	382.42	nr	**15452.42**
Three cubicle panel, two incoming and one outgoing	40500.00	44550.00	50.00	631.00	nr	**45181.00**
Five cubicle panel, two incoming, two outgoing and a bus section	70000.00	77000.00	62.50	788.75	nr	**77788.75**
Three phase; 11kV; non extensible ring main unit for outdoor duty; comprising 2 x 630 Amp fault make load breaker ring switches and 1 x 250 Amp Tee off circuit breaker with C.T. trips and time fuses						
SF6 Insulation	4500.00	4950.00	30.30	382.42	nr	**5332.42**
Note: Prices are exclusive of metering section by electricity authorities.						

Y:MECHANICAL AND ELECTRICAL SERVICES

Item	Net Price £	Material £	Labour hours	Labour £	Unit	Total Rate £
Y71 : LV SWITCHGEAR AND DISTRIBUTION BOARDS : SWITCHES						
Switches; in sheet steel case; surface fixed to backgrounds including supports, fixings, connections/jointing to equipment. (Provision and fixing ofplates, discs, for identification is included.) NOTE: Prices are exclusive of fuse links.						
20 Amp/500V; Isolating Switches; Sheet Steel Case						
D.P.	40.90	44.99	1.00	12.62	nr	**57.61**
T.P.&N.	46.60	51.26	1.25	15.77	nr	**67.03**
20 Amp/500V; Switch Fuses; Sheet Steel Case						
S.P.&N.	46.22	50.84	1.00	12.62	nr	**63.46**
D.P.	49.36	54.30	1.00	12.62	nr	**66.92**
T.P.&N.	57.99	63.79	1.75	22.10	nr	**85.89**
32 Amp/500V; Isolating Switches; Sheet Steel Case						
D.P.	49.95	54.95	1.00	12.62	nr	**67.57**
T.P.&N.	57.46	63.20	1.25	15.77	nr	**78.97**
32 Amp/500V; Switch Fuses; Sheet Steel Case						
S.P.&N.	56.91	62.61	1.25	15.77	nr	**78.38**
D.P.	60.74	66.82	1.25	15.77	nr	**82.59**
T.P.&N	74.56	82.01	2.00	25.24	nr	**107.25**
63 Amp/500V; Isolating Switches; Sheet Steel Case						
D.P.	72.88	80.17	1.50	18.95	nr	**99.12**
T.P.&N	93.63	103.00	1.75	22.10	nr	**125.10**
63 Amp/500V; Switch Fuses; Sheet Steel Case						
S.P.&N	90.21	99.23	0.67	8.41	nr	**107.64**
D.P.	99.22	109.14	1.50	18.95	nr	**128.09**
T.P.&N	128.13	140.94	2.25	28.42	nr	**169.36**
100 Amp; Isolating Switches; Sheet Steel Case						
D.P.	113.87	125.25	1.50	18.95	nr	**144.20**
T.P.&N.	145.82	160.41	1.75	22.10	nr	**182.51**
100 Amp; Switch Fuses; Sheet Steel Case						
S.P.&N	142.52	156.77	1.75	22.10	nr	**178.87**
D.P.	166.30	182.93	1.75	22.10	nr	**205.03**
T.P.&N	216.60	238.26	2.50	31.55	nr	**269.81**

Y:MECHANICAL AND ELECTRICAL SERVICES

Item	Net Price £	Material £	Labour hours	Labour £	Unit	Total Rate £
63 Amp; Fuse Switches; Sheet Steel Case H.R.C.						
S.P.&N.	90.21	99.23	1.50	18.95	nr	**118.18**
D.P.	98.02	107.82	1.50	18.95	nr	**126.77**
T.P.&N.	123.19	135.51	1.75	22.10	nr	**157.61**
100 Amp; Fuse Switches; Sheet Steel Case H.R.C.						
S.P.&N.	142.49	156.74	1.75	22.10	nr	**178.84**
D.P.	163.49	179.83	1.75	22.10	nr	**201.93**
T.P.&N.	214.79	236.26	2.25	28.42	nr	**264.68**
T.P.&S.N	260.43	286.48	2.50	31.55	nr	**318.03**
200 Amp; Fuse Switches; Sheet Steel Case H.R.C.						
S.P.& N.	283.86	312.25	2.25	28.42	nr	**340.67**
D.P.	302.29	332.52	2.25	28.42	nr	**360.94**
T.P.&N.	329.97	362.97	2.50	31.55	nr	**394.52**
T.P.&S.N.	386.35	424.98	2.75	34.77	nr	**459.75**
315 Amp; Fuse Switches; Sheet Steel Case; H.R.C.						
T.P.&N.	590.28	649.31	2.75	34.77	nr	**684.08**
T.P.&S.N.	630.65	693.71	3.26	41.11	nr	**734.82**
400 Amp; Fuse Switches; Sheet Steel Case; H.R.C.						
T.P.&N.	644.99	709.49	3.00	37.90	nr	**747.39**
T.P.&S.N.	749.72	824.69	4.50	56.85	nr	**881.54**
630 Amp; Fuse Switches; Sheet Steel Case; H.R.C.						
T.P.&N.	1247.82	1372.61	3.76	47.44	nr	**1420.05**
T.P.&S.N.	1451.80	1596.98	4.50	56.85	nr	**1653.83**
800 Amp; Fuse Switches; Sheet Steel Case; H.R.C.						
T.P.&N.	1464.45	1610.90	4.50	56.85	nr	**1667.75**
T.P.&S.N.	1609.52	1770.47	4.76	60.10	nr	**1830.57**

Y:MECHANICAL AND ELECTRICAL SERVICES

Item	Net Price £	Material £	Labour hours	Labour £	Unit	Total Rate £
Y71 : LV SWITCHGEAR AND DISTRIBUTION BOARDS : CIRCUIT BREAKERS						
L.V. Circuit Breakers; cassette type withdrawable breakers fitted with CTZ overcurrent device for purpose made switchboard; 600V to 50 Kva; including all necessary connections to equipment. (Provision and fixing of plates, discs etc, for identification is included.)						
3 second; Independant Manual Operation; Triple Pole						
800 Amp	1768.00	1944.80	2.00	25.24	nr	1970.04
1250 Amp	1842.00	2026.20	2.00	25.24	nr	2051.44
1600 Amp	2043.00	2247.30	2.00	25.24	nr	2272.54
2000 Amp	2576.00	2833.60	3.00	37.90	nr	2871.50
2500 Amp	3274.00	3601.40	3.00	37.90	nr	3639.30
3150 Amp	4403.00	4843.30	3.00	37.90	nr	4881.20
3500 Amp	5702.00	6272.20	4.00	50.48	nr	6322.68
4000 Amp	7002.00	7702.20	4.00	50.48	nr	7752.68
3 second Independant Manual Operation; Four Pole						
800 Amp	2170.00	2387.00	3.00	37.90	nr	2424.90
1250 Amp	2237.00	2460.70	3.00	37.90	nr	2498.60
1600 Amp	2631.00	2894.10	3.00	37.90	nr	2932.00
2000 Amp	3256.00	3581.60	4.00	50.48	nr	3632.08
2500 Amp	3992.00	4391.20	4.00	50.48	nr	4441.68
3150 Amp	4732.00	5205.20	4.00	50.48	nr	5255.68
3500 Amp	6032.00	6635.20	5.00	63.10	nr	6698.30
4000 Amp	9092.00	10001.20	5.00	63.10	nr	10064.30
Solenoid Operation; Triple Pole						
800 Amp	1932.00	2125.20	3.00	37.90	nr	2163.10
1250 Amp	2006.00	2206.60	3.00	37.90	nr	2244.50
1600 Amp	2208.00	2428.80	3.00	37.90	nr	2466.70
2000 Amp	2740.00	3014.00	4.00	50.48	nr	3064.48
2500 Amp	3437.00	3780.70	4.00	50.48	nr	3831.18
3150 Amp	4566.00	5022.60	4.00	50.48	nr	5073.08
3500 Amp	5866.00	6452.60	5.00	63.10	nr	6515.70
4000 Amp	7166.00	7882.60	5.00	63.10	nr	7945.70
Solenoid Operation; Four Pole						
800 Amp	2334.00	2567.40	4.00	50.48	nr	2617.88
1250 Amp	2400.00	2640.00	4.00	50.48	nr	2690.48
1600 Amp	2796.00	3075.60	4.00	50.48	nr	3126.08
2000 Amp	3420.00	3762.00	5.00	63.10	nr	3825.10
2500 Amp	4156.00	4571.60	5.00	63.10	nr	4634.70
3150 Amp	4732.00	5205.20	5.00	63.10	nr	5268.30
3500 Amp	7470.00	8217.00	6.02	76.02	nr	8293.02
4000 Amp	9257.00	10182.70	6.02	76.02	nr	10258.72

Y:MECHANICAL AND ELECTRICAL SERVICES

Item	Net Price £	Material £	Labour hours	Labour £	Unit	Total Rate £
Motor Spring Operation; Triple Pole						
800 Amp	2098.00	2307.80	5.00	63.10	nr	2370.90
1250 Amp	2098.00	2307.80	5.00	63.10	nr	2370.90
1600 Amp	2372.00	2609.20	5.00	63.10	nr	2672.30
2000 Amp	2905.00	3195.50	6.02	76.02	nr	3271.52
2500 Amp	3604.00	3964.40	6.02	76.02	nr	4040.42
3150 Amp	4732.00	5205.20	7.04	88.87	nr	5294.07
3500 Amp	6032.00	6635.20	8.00	100.96	nr	6736.16
4000 Amp	7331.00	8064.10	8.00	100.96	nr	8165.06
Motor Spring Operation; Four Pole						
800 Amp	2500.00	2750.00	6.02	76.02	nr	2826.02
1250 Amp	2566.00	2822.60	6.02	76.02	nr	2898.62
1600 Amp	2960.00	3256.00	6.02	76.02	nr	3332.02
2000 Amp	3585.00	3943.50	7.04	88.87	nr	4032.37
2500 Amp	4320.00	4752.00	7.04	88.87	nr	4840.87
3150 Amp	5847.00	6431.70	7.04	88.87	nr	6520.57
3500 Amp	5566.00	6122.60	8.00	100.96	nr	6223.56
4000 Amp	9423.00	10365.30	8.00	100.96	nr	10466.26
Hand Charged Spring Operation; Triple Pole						
800 Amp	2015.00	2216.50	2.00	25.24	nr	2241.74
1250 Amp	2087.00	2295.70	2.00	25.24	nr	2320.94
1600 Amp	2289.00	2517.90	2.00	25.24	nr	2543.14
2000 Amp	2822.00	3104.20	3.00	37.90	nr	3142.10
2500 Amp	3520.00	3872.00	3.00	37.90	nr	3909.90
3150 Amp	4649.00	5113.90	3.00	37.90	nr	5151.80
3500 Amp	5949.00	6543.90	4.00	50.48	nr	6594.38
4000 Amp	7249.00	7973.90	4.00	50.48	nr	8024.38
Hand Charged Spring Operation; Four Pole						
800 Amp	2417.00	2658.70	3.00	37.90	nr	2696.60
1250 Amp	2563.00	2819.30	3.00	37.90	nr	2857.20
1600 Amp	2877.00	3164.70	3.00	37.90	nr	3202.60
2000 Amp	3502.00	3852.20	4.00	50.48	nr	3902.68
2500 Amp	4238.00	4661.80	4.00	50.48	nr	4712.28
3150 Amp	5764.00	6340.40	4.00	50.48	nr	6390.88
3500 Amp	7551.00	8306.10	5.00	63.10	nr	8369.20
4000 Amp	9338.00	10271.80	5.00	63.10	nr	10334.90
Extra For						
Shunt trip release device	40.00	44.00	2.00	25.24	nr	69.24
Under voltage release device - instant	87.00	95.70	2.00	25.24	nr	120.94
Under voltage release device - time delay	116.00	127.60	2.00	25.24	nr	152.84
Solid state breaker mounted supply unit for solenoid operated breakers	122.00	134.20	3.00	37.90	nr	172.10
Loose figure interlock for key control	11.00	12.10	3.00	37.90	nr	50.00
Change over interlock	318.00	349.80	3.00	37.90	nr	387.70

Y:MECHANICAL AND ELECTRICAL SERVICES

Item	Net Price £	Material £	Labour hours	Labour £	Unit	Total Rate £
Y71 : LV SWITCHGEAR AND DISTRIBUTION BOARDS : CIRCUIT BREAKERS (contd)						
L.V. Circuit Breakers; cassette type withdrawable breakers fitted with CTZ overcurrent device for purpose made switchboard; 600V to 50 Kva; including all necessary connections to equipment. (Provision and fixing of plates, discs etc, for identification is included.) (contd)						
Busbar Earthing Device						
3 Pole 800A - 1600 Amp	270.00	297.00	2.00	25.24	nr	322.24
3 Pole 2000 - 2500 Amp	291.00	320.10	2.00	25.24	nr	345.34
3 Pole 3150 Amp	372.00	409.20	3.00	37.90	nr	447.10
3 Pole 3500 - 4000 Amp	478.00	525.80	4.00	50.48	nr	576.28
4 Pole 800A - 1600 Amp	348.00	382.80	2.00	25.24	nr	408.04
4 Pole 2000 - 2500 Amp	374.00	411.40	4.00	50.48	nr	461.88
4 Pole 3150 Amp	478.00	525.80	3.00	37.90	nr	563.70
4 Pole 3500 - 4000 Amp	615.00	676.50	4.00	50.48	nr	726.98
Overcurrent or Manual Reset Device						
Any amp	43.00	47.30	2.00	25.24	nr	72.54
Moulded Case Circuit Breakers; in sheet steel enclosure with thermal/magnetic trips; including supports connections jointing to equipment; BS 4572/IEC947; (provision and fixing of plates, discs etc, for identification is included.)						
7.5-200 Amp range in nominal ratings indicated; T.P.&N.						
25-50 Amp	129.10	142.01	2.75	34.77	nr	176.78
63 Amp	133.06	146.36	2.75	34.77	nr	181.13
80-100 Amp	135.89	149.48	2.75	34.77	nr	184.25
125 Amp	192.51	211.76	3.00	37.90	nr	249.66
160 Amp	203.84	224.23	3.51	44.28	nr	268.51
200 Amp	249.14	274.05	3.51	44.28	nr	318.33
225 Amp	297.83	327.61	3.76	47.44	nr	375.05
250-400 Amp range in nominal ratings indicated; T.P.&S.N.						
250-315 Amp	433.35	476.69	3.76	47.44	nr	524.13
400 Amp	459.62	505.58	3.76	47.44	nr	553.02
315-800 Amp range in nominal ratings indicated; T.P.&S.N.						
315-500 Amp	669.74	736.71	4.00	50.48	nr	787.19
500-630 Amp	669.74	736.71	4.50	56.85	nr	793.56
800 Amp	764.30	840.73	4.76	60.10	nr	900.83

Y:MECHANICAL AND ELECTRICAL SERVICES

Item	Net Price £	Material £	Labour hours	Labour £	Unit	Total Rate £
Miniature Circuit Breakers for Distribution Boards; BS 3871 DIN rail mounting; including connecting to circuit						
Extra for miniature circuit breakers installation and connecting of wiring;						
Single pole						
6 Amp	5.47	6.02	0.10	1.26	nr	**7.28**
10 - 40 Amp	5.11	5.62	0.10	1.26	nr	**6.88**
50 - 60 Amp	5.47	6.02	0.15	1.89	nr	**7.91**
Double pole						
6 Amp	15.88	17.47	0.20	2.52	nr	**19.99**
10 - 40 Amp	14.84	16.32	0.20	2.52	nr	**18.84**
50 - 60 Amp	15.88	17.47	0.30	3.79	nr	**21.26**
Triple pole						
6 Amp	23.41	25.75	0.30	3.79	nr	**29.54**
10 - 40 Amp	22.02	24.22	0.45	5.68	nr	**29.90**
50 - 60 Amp	23.41	25.75	0.45	5.68	nr	**31.43**

Y:MECHANICAL AND ELECTRICAL SERVICES

Item	Net Price £	Material £	Labour hours	Labour £	Unit	Total Rate £
Y71 : LV SWITCHGEAR AND DISTRIBUTION BOARDS : DISTRIBUTION BOARDS						
Distribution Boards; sheet steel case; fully shrouded; fixed to backgrounds; including supports, fixings/connections and equipment. (Provision and fixing of plates, discs etc, for identification is included.) NOTE: prices exclude fuse links.						
500V suitable for 2-20 Amp H.R.C. fuses; S.P.&N.						
4 way	75.01	82.51	1.00	12.62	nr	95.13
6 way	90.55	99.61	1.25	15.77	nr	115.38
8 way	106.13	116.75	1.50	18.92	nr	135.67
12 way	137.31	151.04	2.00	25.24	nr	176.28
500V suitable for 2-20 Amp H.R.C. fuses; T.P.&N.						
4 way	134.39	147.83	1.50	18.92	nr	166.75
6 way	165.63	182.20	1.90	23.99	nr	206.19
8 way	213.07	234.38	2.30	29.08	nr	263.46
12 way	271.60	298.77	3.40	42.93	nr	341.70
500V suitable for 2-30 Amp H.R.C. fuses; S.P.&N.						
4 way	89.28	98.21	1.50	18.92	nr	117.13
6 way	117.72	129.50	1.40	17.68	nr	147.18
8 way	138.66	152.53	1.65	20.83	nr	173.36
12 way	180.44	198.49	2.15	27.14	nr	225.63
500V suitable for 2-30 Amp H.R.C. fuses; T.P.&N.						
2 way	108.25	119.08	1.00	12.62	nr	131.70
4 way	156.66	172.33	1.90	23.99	nr	196.32
6 way	208.58	229.44	2.40	30.34	nr	259.78
8 way	252.45	277.70	2.80	35.35	nr	313.05
10 way	302.56	332.81	3.21	40.45	nr	373.26
12 way	348.49	383.34	3.80	47.98	nr	431.32
500V suitable for 35-63 Amp H.R.C. fuses; T.P.&N.						
2 way	250.43	275.47	1.60	20.19	nr	295.66
4 way	336.34	369.97	2.50	31.55	nr	401.52
6 way	426.61	469.27	3.00	37.90	nr	507.17
8 way	509.25	560.17	3.51	44.28	nr	604.45
10 way	611.61	672.77	3.75	47.27	nr	720.04

Y:MECHANICAL AND ELECTRICAL SERVICES

Item	Net Price £	Material £	Labour hours	Labour £	Unit	Total Rate £
500V suitable for 35-100 Amp H.R.C. fuses; T.P.&N.						
2 way	334.79	368.27	1.90	23.99	nr	**392.26**
3 way	397.31	437.04	2.40	30.34	nr	**467.38**
4 way	509.95	560.95	2.75	34.77	nr	**595.72**
6 way	657.65	723.42	3.26	41.11	nr	**764.53**
8 way	790.75	869.83	3.51	44.28	nr	**914.11**
Distribution Boards; Sheet steel case; fully shrouded; fixed to backgrounds; including supports, fixings/connections to equipment. (Provision and fixings of plates, discs etc, for identification is included.) NOTE: prices include MCB's and doors.						
500V suitable for 6-63 Amp M.C.B's; S.P.&N.						
12 way	122.69	111.54	3.00	37.90	nr	**160.59**
16 way	140.41	154.45	4.00	50.48	nr	**204.93**
500V suitable for 6-63 Amp M.C.B's; T.P.&N.						
4 way	113.83	103.48	3.00	37.90	nr	**151.73**
8 way	244.48	222.53	3.00	37.90	nr	**282.38**
12 way	300.61	330.67	4.00	50.48	nr	**381.15**

Y:MECHANICAL AND ELECTRICAL SERVICES

Item	Net Price £	Material £	Labour hours	Labour £	Unit	Total Rate £
Y71 : LV SWITCHGEAR AND DISTRIBUTION BOARDS : CONSUMER UNITS						
Consumer units; sheet steel case; fully shrouded; fixed to backgrounds; including supports, fixings/connections to equipment. (Provision and fixings of plates, discs etc, for identification is included.) NOTE: prices include MCB's and doors.						
500V suitable for 6-40 Amp M.C.B's; S.P.&N.						
7 way	68.12	74.94	3.00	37.90	nr	**112.84**
10 way	89.76	98.74	4.00	50.48	nr	**149.22**
13 way	105.98	116.58	4.00	50.48	nr	**167.06**
Consumer Units; fixed to backgrounds; including supports, fixings, connections/jointing to equipment. (Provision and fixing of plates, discs etc, for identification is included.)						
Switched and insulated; moulded plastic case, 60 Amp 240 Volt D.P./A.C						
fitted rewireable fuses						
2 way	12.82	15.97	1.25	15.77	nr	**31.74**
3 way	19.59	21.55	1.50	18.95	nr	**40.50**
4 way	24.28	26.71	1.50	18.95	nr	**45.66**
fitted cartridge fuses						
2 way	15.34	16.87	1.25	15.77	nr	**32.64**
3 way	20.82	22.91	1.50	18.95	nr	**41.86**
4 way	23.82	26.20	1.50	18.95	nr	**45.15**
fitted MCB's						
2 way	23.19	25.51	1.25	15.77	nr	**41.28**
3 way	32.62	35.88	1.50	18.95	nr	**54.83**
4 way	41.63	45.79	1.50	18.95	nr	**64.74**
Switched and insulated; moulded plastic case, 100 Amp 240 Volt D.P./A.C.						
fitted rewireable fuses						
6 way	31.74	34.92	1.83	23.11	nr	**58.03**
8 way	40.65	44.41	2.00	25.24	nr	**69.65**
fitted cartridge fuses						
6 way	34.21	37.63	1.83	23.11	nr	**60.74**
8 way	43.64	48.01	2.00	25.24	nr	**73.25**

Y:MECHANICAL AND ELECTRICAL SERVICES

Item	Net Price £	Material £	Labour hours	Labour £	Unit	Total Rate £
fitted MCBs						
6 way	62.75	63.66	1.83	23.11	nr	**86.77**
8 way	75.07	82.58	2.00	25.24	nr	**107.82**
Switched and insulated; enamelled steel case; 60 Amp 240 Volt D.P./A.C.						
fitted rewireable fuses						
2 way	17.28	19.01	1.25	15.77	nr	**34.78**
3 way	24.41	26.85	1.50	18.95	nr	**45.80**
4 way	28.10	30.91	1.50	18.95	nr	**49.86**
fitted cartridge fuses						
2 way	18.11	19.92	1.25	15.77	nr	**35.69**
3 way	25.63	28.20	1.50	18.95	nr	**47.15**
4 way	29.74	32.72	1.50	18.95	nr	**51.67**
fitted MCB's						
2 way	25.96	28.56	1.25	15.77	nr	**44.33**
3 way	37.43	41.18	1.50	18.95	nr	**60.13**
4 way	45.45	50.00	1.50	18.95	nr	**68.95**
Switched and insulated; enamelled steel case; 100 Amp 240 Volt D.P./A.C.;						
fitted rewireable fuses						
6 way	35.91	39.51	1.83	23.11	nr	**62.62**
8 way	43.75	48.13	1.83	23.11	nr	**71.24**
12 way	78.56	86.43	1.83	23.11	nr	**109.54**
fitted cartridge fuses						
6 way	38.33	42.22	1.83	23.11	nr	**65.33**
8 way	47.03	51.74	2.00	25.24	nr	**76.98**
12 way	83.50	91.86	3.00	37.90	nr	**129.76**
fitted MCB's						
6 way	62.04	68.25	1.83	23.11	nr	**91.36**
8 way	78.44	86.30	2.00	25.24	nr	**111.54**
12 way	130.82	143.92	3.00	37.90	nr	**181.82**
Switched and insulated; moulded plastic case; D.P. RCCB control on MCB ways; 240 Volt A.C.						
80 Amp; 30 mA						
3 way	63.62	69.98	1.20	15.15	nr	**85.13**
5 way	66.84	73.52	1.20	15.15	nr	**88.67**
7 way	70.46	77.50	1.50	18.95	nr	**96.45**
11 way	76.83	84.52	1.90	23.99	nr	**108.51**
17 way	98.48	108.33	2.10	26.51	nr	**134.84**

Y:MECHANICAL AND ELECTRICAL SERVICES

Item	Net Price £	Material £	Labour hours	Labour £	Unit	Total Rate £
Y71 : LV SWITCHGEAR AND DISTRIBUTION BOARDS : CONSUMER UNITS (contd)						
Consumer Units; fixed to backgrounds; including supports, fixings, connections/jointing to equipment. (Provision and fixing of plates, discs etc, for identification is included.) (contd)						
Switched and insulated; moulded plastic case; D.P. RCCB control for all MCB ways; 240 Volt A.C.						
80 Amp; 100mA						
3 way	61.32	67.45	1.20	15.15	nr	82.60
5 way	64.12	70.54	1.40	17.68	nr	88.22
7 way	67.74	74.52	1.50	18.95	nr	93.47
9 way	74.12	81.53	1.90	23.99	nr	105.52
11 way	95.77	105.34	2.10	26.51	nr	131.85
Switched and insulated; enamelled steel case; D.P. RCCB control for MCB ways; 240 Volt A.C.						
80 Amps; 30mA						
5 way	68.80	75.68	1.40	17.68	nr	93.36
7 way	72.83	80.11	1.75	22.10	nr	102.21
11 way	80.17	88.18	2.10	26.51	nr	114.69
80 Amps; 100mA						
5 way	66.08	72.69	1.40	17.68	nr	90.37
7 way	70.11	77.12	1.50	18.95	nr	96.07
11 way	77.45	85.19	1.90	23.99	nr	109.18
Switched and insulated; moulded plastic case; split load range; 100 Amp D.P. main switch with 63 Amp D.P. RCCB on selected ways; 240 Volt AC.						
9 MCB way; 6 RCCB Way; 100mA	84.31	92.74	1.75	22.10	nr	114.84
9 MCB way; 6 RCCB Way; 30mA	88.87	97.75	1.75	22.10	nr	119.85
9 MCB way; 5 RCCB Way; 100mA	84.31	92.74	1.75	22.10	nr	114.84
9 MCB way; 5 RCCB Way; 30mA	88.87	97.75	1.75	22.10	nr	119.85
9 MCB way; 4 RCCB Way; 100mA	84.31	92.74	1.75	22.10	nr	114.84
9 MCB way; 4 RCCB Way; 30mA	88.87	97.75	1.75	22.10	nr	119.85
9 MCB way; 3 RCCB Way; 100mA	84.31	92.74	1.75	22.10	nr	114.84
9 MCB way; 3 RCCB Way; 30mA	88.87	97.75	1.75	22.10	nr	119.85

Y:MECHANICAL AND ELECTRICAL SERVICES

Item	Net Price £	Material £	Labour hours	Labour £	Unit	Total Rate £
Switched and insulated; enamelled steel case; split load range; 100 Amp D.P. main switch with 63 Amp D.P. RCCB on selected ways; 240 Volt AC.						
5 MCCB way; 3 RCCB ways; 100mA	80.72	88.79	1.60	20.19	nr	**108.98**
5 MCCB way; 3 RCCB ways; 30mA	85.28	93.81	1.60	20.19	nr	**114.00**
5 MCCB way; 2 RCCB ways; 100mA	80.72	88.79	1.60	20.19	nr	**108.98**
5 MCCB way; 2 RCCB ways; 30mA	85.28	93.81	1.60	20.19	nr	**114.00**
9 MCCB way; 6 RCCB ways; 100mA	87.64	96.40	1.60	20.19	nr	**116.59**
9 MCCB way; 6 RCCB ways; 30mA	92.20	101.42	1.75	22.10	nr	**123.52**
9 MCCB way; 5 RCCB ways; 100mA	87.64	96.40	1.75	22.10	nr	**118.50**
9 MCCB way; 5 RCCB ways; 30mA	92.20	101.42	1.75	22.10	nr	**123.52**
9 MCCB way; 4 RCCB ways; 100mA	87.64	96.40	1.75	22.10	nr	**118.50**
9 MCCB way; 4 RCCB ways; 30mA	92.20	101.42	1.75	22.10	nr	**123.52**
9 MCCB way; 3 RCCB ways; 100mA	87.64	96.40	1.75	22.10	nr	**118.50**
9 MCCB way; 3 RCCB ways; 30mA	92.20	101.42	1.75	22.10	nr	**123.52**

Y:MECHANICAL AND ELECTRICAL SERVICES

Item	Net Price £	Material £	Labour hours	Labour £	Unit	Total Rate £
Y71 : LV SWITCHGEAR AND DISTRIBUTION BOARDS : VOLTAGE STABLISERS/LINE CONDITIONERS						
Voltage Stabilisers; solid state circuitry; sheet steel enclosure; maintains output +/- 3% over zero to full load with input voltage swings of +/- 15%; positioned on site, final connections and commissioning. (Include delivery up to 75 miles.)						
Electronic; Single Phase; 220/230/240 Volt AC, 50 Hz						
0.3 kVA	318.00	349.80	2.00	25.24	nr	375.04
0.6 kVA	334.00	367.40	2.00	25.24	nr	392.64
1 kVA	364.00	400.40	2.00	25.24	nr	425.64
2 kVA	474.00	521.40	2.00	25.24	nr	546.64
3 kVA	558.00	613.80	3.00	37.90	nr	651.70
4.5 kVA	776.00	853.60	4.00	50.48	nr	904.08
6 kVA	1039.00	1142.90	4.00	50.48	nr	1193.38
9 kVA	1310.00	1441.00	6.02	76.02	nr	1517.02
12 kVA	1462.00	1608.20	7.04	88.87	nr	1697.07
15 kVA	1656.00	1821.60	7.52	94.89	nr	1916.49
20 kVA	1936.00	2129.60	8.00	100.96	nr	2230.56
25 kVA	2249.00	2473.90	8.00	100.96	nr	2574.86
30 kVA	2600.00	2860.00	10.00	126.20	nr	2986.20
40 kVA	2959.00	3254.90	12.05	152.05	nr	3406.95
50 kVA	3592.00	3951.20	14.08	177.75	nr	4128.95
Three phase, 380/400/415 Volt AC, 50 Hz						
3.0 kVA	940.00	1034.00	4.00	50.48	nr	1084.48
6.0 kVA	1150.00	1265.00	4.00	50.48	nr	1315.48
9.0 kVA	1450.00	1595.00	6.02	76.02	nr	1671.02
13.0 kVA	1959.00	2154.90	7.04	88.87	nr	2243.77
18.0 kVA	2519.00	2770.90	8.00	100.96	nr	2871.86
27.0 kVA	3012.00	3313.20	8.00	100.96	nr	3414.16
36.0 kVA	3373.00	3710.30	10.00	126.20	nr	3836.50
50.0 kVA	4229.00	4651.90	14.08	177.75	nr	4829.65
60.0 kVA	6948.00	7642.80	16.13	203.55	nr	7846.35
75.0 kVA	8015.00	8816.50	16.13	203.55	nr	9020.05
100.0 kVA	7573.00	8330.30	16.13	203.55	nr	8533.85
125.0 kVA	8118.00	8929.80	18.18	229.45	nr	9159.25
150.0 kVA	8516.00	9367.60	24.39	307.80	nr	9675.40

Y:MECHANICAL AND ELECTRICAL SERVICES

Item	Net Price £	Material £	Labour hours	Labour £	Unit	Total Rate £
Line conditioners; solid state circuitry; sheet steel enclosure; maintains output +/- 2% over zero to full load with input voltage swings of +/- 15%; discrimination against voltage spikes and r.f. interference; positioned on site, final connections and commissioning. (Includes delivery up to 75 miles.)						
Single phase, 240 Volt AC, 50 Hz						
4.5 kVA	1098.00	1207.80	4.00	50.48	nr	**1258.28**
6.0 kVA	1432.00	1575.20	4.00	50.48	nr	**1625.68**
9.0 kVA	1692.00	1861.20	6.02	76.02	nr	**1937.22**
12.0 kVA	1840.00	2024.00	7.04	88.87	nr	**2112.87**
Three phase, 415 Volt AC, 50Hz						
18.0 kVA	3195.00	3514.50	8.00	100.96	nr	**3615.46**
27.0 kVA	4130.00	4543.00	8.00	100.96	nr	**4643.96**
60.0 kVA	6948.00	7642.80	16.13	203.55	nr	**7846.35**
100.0 kVA	10006.00	11006.60	16.13	203.55	nr	**11210.15**
150.0 kVA	11760.00	12936.00	24.39	307.80	nr	**13243.80**

Keep your figures up to date, free of charge

This section, and most of the other information in this Price Book, is brought up to date every three months, until the next annual edition, in the *Price Book Update.*

The *Update* is available free to all Price Book purchasers.

To ensure you receive your copy, simply complete the reply card from the centre of the book and return it to us.

Y:MECHANICAL AND ELECTRICAL SERVICES

Item	Net Price £	Material £	Labour hours	Labour £	Unit	Total Rate £
Y71 : LV SWITCHGEAR AND DISTRIBUTION BOARDS : FUSIBLE BOARDS						
Fusible Links for Distribution Boards						
Extra for fuse links; Q1 fusing factor; BS 88 Part 2; fitting and connecting						
6 Amp	4.25	4.68	0.10	1.26	nr	**5.94**
10 Amp	4.25	4.68	0.10	1.26	nr	**5.94**
16 Amp	4.25	4.68	0.10	1.26	nr	**5.94**
20 Amp	4.25	4.68	0.10	1.26	nr	**5.94**
32 Amp	4.60	5.06	0.10	1.26	nr	**6.32**
50 Amp	4.85	5.34	0.10	1.26	nr	**6.60**
63 Amp	4.85	5.34	0.10	1.26	nr	**6.60**
80 Amp	7.96	8.76	0.10	1.26	nr	**10.02**
100 Amp	7.96	8.76	0.10	1.26	nr	**10.02**
125 Amp	13.25	14.58	0.10	1.26	nr	**15.84**
160 Amp	14.31	15.74	0.10	1.26	nr	**17.00**
200 Amp	15.12	16.64	0.10	1.26	nr	**17.90**
250 Amp	18.56	20.41	0.10	1.26	nr	**21.67**
315 Amp	22.45	24.70	0.10	1.26	nr	**25.96**
355 Amp	28.03	30.83	0.10	1.26	nr	**32.09**
400 Amp	30.77	33.85	0.10	1.26	nr	**35.11**
560 Amp	47.89	52.68	0.10	1.26	nr	**53.94**
630 Amp	56.89	62.58	0.10	1.26	nr	**63.84**
710 Amp	71.93	79.12	0.10	1.26	nr	**80.38**
750 Amp	78.32	86.15	0.10	1.26	nr	**87.41**
800 Amp	91.77	100.95	0.10	1.26	nr	**102.21**
Extra for fuse links; extended motor range, fitting and connecting motor circuit						
20 M25 (3HP)	2.03	2.23	0.10	1.26	nr	**3.49**
20 M32 (4HP)	2.03	2.23	0.10	1.26	nr	**3.49**
32 M35 (5.5HP)	2.77	3.05	0.10	1.26	nr	**4.31**
32 M50 (7.5HP)	2.94	3.24	0.10	1.26	nr	**4.50**
32 M63 (10HP)	3.23	3.56	0.10	1.26	nr	**4.82**
63 M80 (15HP)	6.09	6.70	0.10	1.26	nr	**7.96**
63 M100 (20HP)	6.33	6.97	0.10	1.26	nr	**8.23**
100 M125 (25HP)	9.69	10.66	0.10	1.26	nr	**11.92**
100 M160 (30HP)	9.83	10.81	0.10	1.26	nr	**12.07**
100 M200 (50HP)	10.96	12.06	0.10	1.26	nr	**13.32**
200 M250 (60HP)	16.99	18.69	0.10	1.26	nr	**19.95**
200 M315 (70HP)	19.52	21.47	0.10	1.26	nr	**22.73**
315 M355 (100HP)	27.97	30.76	0.10	1.26	nr	**32.02**
400 M450 (150HP)	38.98	42.88	0.10	1.26	nr	**44.14**

Y:MECHANICAL AND ELECTRICAL SERVICES

Item	Net Price £	Material £	Labour hours	Labour £	Unit	Total Rate £
Y72 : CONTACTORS AND STARTERS : CONTRACTORS/PUSH BUTTONS						
Contactor Relays; pressed steel enclosure; fixed to backgrounds including supports, fixings, connections/jointing to equipment. (Provision and fixing of plates, discs, etc, for identification is included.)						
Relays						
6 Amp, 415/240 Volt, 4 pole N/O	23.69	26.06	0.50	6.31	nr	**32.37**
6 Amp, 415/240 Volt, 8 pole N/O	28.81	31.69	0.75	9.47	nr	**41.16**
Push Button Stations; Heavy gauge pressed steel enclosure; polycarbonate cover; IP65; fixed to backgrounds including supports, fixings, connections/joinging to equipment. (plates, discs, etc, for identification is included)						
Standard Units						
One button (start or stop)	8.90	9.79	0.33	4.16	nr	**13.95**
Two button (start or stop)	14.02	15.42	0.42	5.30	nr	**20.72**
Three button (forward-reverse-stop)	19.39	21.32	0.50	6.31	nr	**27.63**

Y:MECHANICAL AND ELECTRICAL SERVICES

Item	Net Price £	Material £	Labour hours	Labour £	Unit	Total Rate £
Y73 : LUMINAIRES AND LAMPS : **LUMINAIRES**						
Fluorescent Luminaires; surface fixed to backgrounds. (Including supports and fixings.)						
Fluorescent light fittings; batten type; tubes fitted; surface mounted						
600 mm Single - 18 W	11.18	12.30	0.50	6.31	nr	**18.61**
600 mm Twin - 18 W	19.25	21.17	0.50	6.31	nr	**27.48**
1200 mm Single - 36 W	14.37	15.80	0.65	8.21	nr	**24.01**
1200 mm Twin - 36 W	27.55	30.31	0.65	8.21	nr	**38.52**
1500 mm Single - 58 W	16.74	18.41	0.75	9.47	nr	**27.88**
1500 mm Twin - 58 W	32.73	36.00	0.75	9.47	nr	**45.47**
1800 mm Single - 70 W	20.21	22.23	1.00	12.62	nr	**34.85**
1800 mm Twin - 70 W	35.88	39.47	1.00	12.62	nr	**52.09**
2400 mm Single - 100 W	27.65	30.42	1.25	15.77	nr	**46.19**
2400 mm Twin - 100 W	47.01	51.71	1.25	15.77	nr	**67.48**
Surface mounted, reeded diffuser; tubes fitted						
600 mm Single - 18 W	6.89	7.58	0.50	6.31	nr	**13.89**
600 mm Twin - 18 W	7.68	8.45	0.50	6.31	nr	**14.76**
1200 mm Single - 36 W	8.79	9.67	0.65	8.21	nr	**17.88**
1200 mm Twin - 36 W	17.16	18.88	0.65	8.21	nr	**27.09**
1500 mm Single - 58 W	9.91	10.91	0.75	9.47	nr	**20.38**
1500 mm Twin - 58 W	21.35	23.49	0.75	9.47	nr	**32.96**
1800 mm Single - 70 W	11.39	12.53	1.00	12.62	nr	**25.15**
1800 mm Twin - 70 W	24.80	27.28	1.00	12.62	nr	**39.90**
2400 mm Single - 100 W	16.82	18.50	1.25	15.77	nr	**34.27**
2400 mm Twin - 100 W	32.78	36.06	1.25	15.77	nr	**51.83**
Modular type; recessed mounting; opal diffuser; switchstart; tubes fitted						
300 x 1200 mm - 36 W, 1 lamp	58.80	64.68	1.50	18.95	nr	**83.63**
300 x 1200 mm - 36 W, 2 lamp	69.44	76.38	1.50	18.95	nr	**95.33**
300 x 1800 mm - 70 W, 1 lamp	67.99	74.79	1.75	22.10	nr	**96.89**
600 x 600 mm - 40 W, 2 lamp	79.95	87.94	1.50	18.95	nr	**106.89**
600 x 600 mm - 40 W, 3 lamp	99.68	109.65	1.50	18.95	nr	**128.60**
600 x 1200 mm - 36 W, 3 lamp	98.85	108.74	1.75	22.10	nr	**130.84**
600 x 1200 mm - 36 W, 4 lamp	107.78	118.56	1.75	22.10	nr	**140.66**
600 x 1800 mm - 70 W, 2 lamp	104.51	114.96	2.00	25.24	nr	**140.20**
600 x 1800 mm - 70 W, 3 lamp	119.45	131.39	2.15	27.14	nr	**158.53**
600 x 1800 mm - 70 W, 4 lamp	119.45	131.39	2.25	28.42	nr	**159.81**
Corrosion resistant GRP body; gasket sealed; acrylic diffuser; tubes fitted						
600 mm Single - 18 W	23.05	25.36	0.50	6.31	nr	**31.67**
600 mm Twin - 18 W	31.96	35.16	0.50	6.31	nr	**41.47**
1200 mm Single - 36 W	28.80	31.68	0.65	8.21	nr	**39.89**
1200 mm Twin - 36 W	38.74	42.61	0.65	8.21	nr	**50.82**
1500 mm Single - 58 W	31.72	34.89	0.75	9.47	nr	**44.36**
1500 mm Twin - 58 W	42.97	47.26	0.75	9.47	nr	**56.73**
1800 mm Single - 70 W	37.10	40.81	1.00	12.62	nr	**53.43**
1800 mm Twin - 70 W	52.53	57.78	1.00	12.62	nr	**70.40**

Y:MECHANICAL AND ELECTRICAL SERVICES

Item	Net Price £	Material £	Labour hours	Labour £	Unit	Total Rate £
Flameproof to IIA/IIB,I.P. 64; Aluminium Body; BS 229 and 899; Tubes fitted						
600 mm Single - 18 W	254.75	280.22	1.00	12.62	nr	**292.84**
1200 mm Single - 36 W	279.34	307.27	1.25	15.77	nr	**323.04**
1200 mm Twin - 36 W	342.41	376.65	1.00	12.62	nr	**389.27**
1500 mm Single - 58 W	298.97	328.87	1.50	18.95	nr	**347.82**
1500 mm Twin - 58 W	358.48	394.33	1.50	18.95	nr	**413.28**
1800 mm Single - 70 W	328.38	361.21	1.75	22.10	nr	**383.31**
1800 mm Twin - 70 W	375.60	413.16	1.75	22.14	nr	**435.30**
Emergency Lighting Fittings; surface fixed to backgrounds. (Including supports and fixings.)						
Self contained non-maintained 3 hour duration; aluminium body; glass diffuser; vandel resistant fixings; IP65;						
4 W	55.69	61.25	1.00	12.62	nr	**73.87**
Self contained non-maintained 3 hour duration; polycarbonate body; prismatic diffuser; vandel resistant fixings; IP65;						
4 W	48.38	53.22	1.00	12.62	nr	**65.84**

Y:MECHANICAL AND ELECTRICAL SERVICES

Item	Net Price £	Material £	Labour hours	Labour £	Unit	Total Rate £
Y74 : ACCESSORIES FOR ELECTRICAL SERVICES : SWITCHES						
Local Switches; fixed to backgrounds (Including fixings.)						
6 Amp metal clad surface mounted switch, gridswitch; one way						
1 Gang	2.61	2.87	0.50	6.31	nr	9.18
2 Gang	2.61	2.87	0.70	8.84	nr	11.71
3 Gang	4.48	4.93	1.00	12.62	nr	17.55
4 Gang	4.48	4.93	1.17	14.78	nr	19.71
6 Gang	8.32	9.16	1.35	17.05	nr	26.21
8 Gang	8.32	9.16	1.50	18.95	nr	28.11
12 Gang	12.53	13.78	1.75	22.10	nr	35.88
Extra for; Blank inset						
6 Amp – Two way switch	1.78	1.96	0.03	0.38	nr	2.34
20 Amp – Two way switch	2.30	2.53	0.04	0.50	nr	3.03
20 Amp – Intermediate	4.13	4.55	0.09	1.14	nr	5.69
20 Amp – One way SP switch	2.60	2.86	0.09	1.14	nr	4.00
Steel blank plate; 1 Gang	0.89	0.98	0.03	0.38	nr	1.36
Steel blank plate; 2 Gang	1.33	1.46	0.03	0.38	nr	1.84
6 Amp modular type switch; galvanised steel box, bronze or satin chrome coverplate; metalclad switches; flush mounting; one way						
1 Gang	10.41	11.46	0.50	6.31	nr	17.77
2 Gang	15.06	16.56	0.70	8.84	nr	25.40
3 Gang	23.59	25.94	1.00	12.62	nr	38.56
4 Gang	28.23	31.05	1.17	14.78	nr	45.83
6 Gang	44.08	48.48	1.50	18.95	nr	67.43
8 Gang	53.26	58.58	2.20	27.80	nr	86.38
9 Gang	62.54	68.79	2.50	31.55	nr	100.34
12 Gang	76.47	84.11	3.00	37.90	nr	122.01
6 Amp modular type swtich; galvanised steel box; bronze or satin chrome coverplate; flush mounting; two way						
1 Gang	10.93	12.02	0.50	6.31	nr	18.33
2 Gang	16.08	17.69	0.70	8.84	nr	26.53
3 Gang	25.12	27.64	1.00	12.62	nr	40.26
4 Gang	30.28	33.31	1.17	14.78	nr	48.09
6 Gang	40.59	44.65	1.50	18.95	nr	63.60
8 Gang	57.36	63.09	2.20	27.80	nr	90.89
9 Gang	67.15	73.86	2.50	31.55	nr	105.41
12 Gang	82.61	90.88	3.00	37.90	nr	128.78

Y:MECHANICAL AND ELECTRICAL SERVICES

Item	Net Price £	Material £	Labour hours	Labour £	Unit	Total Rate £
Plate switches; 6 Amp flush mounted, white plastic fronted; 16mm metal box; fitted brass earth terminal						
1 Gang 1 Way, Single Pole	2.09	2.30	0.25	3.15	nr	**5.45**
1 Gang 2 Way, Single Pole	2.33	2.57	0.33	4.16	nr	**6.73**
2 Gang 2 Way, Single Pole	3.61	3.97	0.49	6.19	nr	**10.16**
3 Gang 2 Way, Single Pole	5.17	5.68	0.60	7.57	nr	**13.25**
1 Gang Intermediate	4.94	5.43	0.50	6.31	nr	**11.74**
1 Gang 1 Way, Double Pole	4.43	4.87	0.33	4.16	nr	**9.03**
1 Gang Single Pole with bell symbol	3.62	3.99	0.25	3.15	nr	**7.14**
1 Gang Single Pole marked "PRESS"	3.36	3.70	0.25	3.15	nr	**6.85**
Time delay switch, suppressed	28.12	30.93	0.44	5.55	nr	**36.48**
Plate switches; 6 Amp flush mounted white plastic fronted; 25mm metal box; fitted brass earth terminal						
4 Gang 2 Way, Single Pole	8.70	9.57	0.25	3.15	nr	**12.72**
6 Gang 2 Way, Single Way	16.34	17.97	0.25	3.15	nr	**21.12**
Architrave plate switches; 6 Amp flush mounted, white plastic fronted; 27mm metal box; brass earth terminal						
1 Gang 2 Way, Single Pole	2.40	2.65	0.25	3.15	nr	**5.80**
2 Gang 2 Way, Single Pole	4.97	5.46	0.25	3.15	nr	**8.61**
Ceiling switches, white moulded plastic, pull cord; standard unit						
6 Amp, 1 Way, Single Pole	3.28	3.61	0.25	3.15	nr	**6.76**
6 Amp, 2 Way, Single Pole	3.76	4.14	0.30	3.79	nr	**7.93**
16 Amp, 1 Way, Double Pole	5.04	5.55	0.35	4.42	nr	**9.97**
45 Amp, 1 Way, Double Pole with neon indicator	9.72	10.70	0.40	5.05	nr	**15.75**
10 Amp splash proof moulded switch with plain, threaded or PVC entry						
1 Gang, 1 Way Single Pole	6.61	7.82	0.35	4.42	nr	**12.24**
1 Gang, 1 Way Double Pole	8.26	9.77	0.40	5.05	nr	**14.82**
1 Gang, 2 Way Single Pole	7.63	9.02	0.40	5.05	nr	**14.07**
6 Amp Watertight switch; metalclad; BS 3676; ingress protected to IP65 surface mounted						
1 Gang, 2 Way; terminal entry	15.18	17.95	0.38	4.80	nr	**22.75**
1 Gang, 2 Way; through entry	16.22	19.17	0.41	5.17	nr	**24.34**
2 Gang, 2 Way; terminal entry	24.76	29.28	0.54	6.82	nr	**36.10**
2 Gang, 2 Way; through entry	25.79	30.50	0.57	7.19	nr	**37.69**
2 Way replacement switch	2.78	3.28	0.06	0.76	nr	**4.04**

Y:MECHANICAL AND ELECTRICAL SERVICES

Item	Net Price £	Material £	Labour hours	Labour £	Unit	Total Rate £
Y74 : ACCESSORIES FOR ELECTRICAL SERVICES : SWITCHES (contd)						
Local Switches; fixed to backgrounds (Including fixings.) (contd)						
15 Amp Watertight switch; metalclad; BS 3676;ingress protected to IP65; surface mounted						
1 Gang 2 Way, terminal entry	16.05	18.98	0.40	5.05	nr	**24.03**
1 Gang 2 Way, through entry	17.18	20.31	0.43	5.43	nr	**25.74**
2 Gang 2 Way, terminal entry	26.50	31.33	0.56	7.07	nr	**38.40**
2 Gang 2 Way, through entry	27.54	32.57	0.59	7.45	nr	**40.02**
Intermediate interior only	4.82	5.70	0.08	1.01	nr	**6.71**
2 way interior only	3.58	4.23	0.08	1.01	nr	**5.24**
Double pole interior only	4.32	5.10	0.08	1.01	nr	**6.11**
Electrical Accessories; fixed to backgrounds (Including fixings.)						
Dimmer switches; Rotary action; for individual lights; moulded plastic case; metal backbox; flush mounted						
1 Gang, 1 Way; 250 Watt	9.03	9.93	0.35	4.42	nr	**14.35**
1 Gang, 1 Way; 400 Watt	13.83	15.21	0.35	4.42	nr	**19.63**
Dimmer switches; Push on/off action; for individual lights; moulded plastic case; metal backbox; flush mounted						
1 Gang, 2 Way; 250 Watt	13.83	15.21	0.38	4.80	nr	**20.01**
3 Gang, 2 Way; 250 Watt	48.30	53.13	0.53	6.69	nr	**59.82**
4 Gang, 2 Way; 250 Watt	66.54	73.19	0.62	7.83	nr	**81.02**
Dimmer switches; Rotary action; metal cald; metal backbox; BS 5518 and BS 800; flush mounted						
1 Gang, 1 Way; 400 Watt	26.94	29.64	0.35	4.42	nr	**34.06**

Y:MECHANICAL AND ELECTRICAL SERVICES

Item	Net Price £	Material £	Labour hours	Labour £	Unit	Total Rate £
Y74 : ACCESSORIES FOR ELECTRICAL SERVICES : SOCKET OUTLETS						
Electrical Accessories; fixed to backgrounds. (Including fixings.)						
Socket outlet: unswitched; 13 Amp metal clad; BS 1363; galvanised steel box and coverplate with white plastic inserts; fixed surface mounted						
1 Gang	4.90	5.39	0.45	5.68	nr	**11.07**
2 Gang	8.68	9.55	0.45	5.68	nr	**15.23**
Socket outlet: switched; 13 Amp metal clad; BS 1363; galvanised steel box and coverplate with white plastic inserts; fixed surface mounted						
1 Gang	6.70	7.37	0.45	5.68	nr	**13.05**
2 Gang	11.90	13.09	0.45	5.68	nr	**18.77**
Socket outlet: switched with neon indicator; 13 Amp metal clad; BS 1363; galvanised steel box and coverplate with white plastic inserts; fixed surface mounted						
1 Gang	9.00	9.90	0.45	5.68	nr	**15.58**
2 Gang	15.85	17.44	0.45	5.68	nr	**23.12**
Socket outlet: unswitched; 13 Amp; BS 1363; white moulded plastic box and coverplate; fixed surface mounted						
1 Gang	4.04	4.45	0.45	5.68	nr	**10.13**
2 Gang	7.74	8.51	0.45	5.68	nr	**14.19**
Socket outlet; switched; 13 Amp; BS 1363; white moulded plastic box and coverplate; fixed surface mounted						
1 Gang	4.64	5.11	0.45	5.68	nr	**10.79**
2 Gang	9.07	9.98	0.45	5.68	nr	**15.66**
Socket outlet: switched with neon indicator; 13 Amp; BS 1363; white moulded plastic box and coverplate; fixed surface mounted						
1 Gang	6.80	7.48	0.45	5.68	nr	**13.16**
2 Gang	12.89	14.18	0.45	5.68	nr	**19.86**
Socket outlet: switched; 13 Amp; BS 1363; galvanised steel box, white moulded coverplate; flush fitted						
1 Gang	3.86	4.25	0.45	5.68	nr	**9.93**
2 Gang	6.80	7.48	0.45	5.68	nr	**13.16**

UNIVERSITY COLLEGE LONDON LIBRARY

Y:MECHANICAL AND ELECTRICAL SERVICES

Item	Net Price £	Material £	Labour hours	Labour £	Unit	Total Rate £
Y74 : ACCESSORIES FOR ELECTRICAL SERVICES : SOCKET OUTLETS (contd)						
Electrical Accessories; fixed to backgrounds. (Including fixings.) (contd)						
Socket outlet: switched with neon indicator; 13 Amp; BS 1363; galvanised steel box, white moulded coverplate; flush fixed						
1 Gang	6.01	6.61	0.45	5.68	nr	12.29
2 Gang	10.62	11.68	0.45	5.68	nr	17.36
Socket outlet: switched; 13 Amp; BS 1363; galvanised steel box, satin chrome coverplate; BS 4662; flush fixed						
1 Gang	10.34	11.38	0.45	5.68	nr	17.06
2 Gang	12.99	14.29	0.45	5.68	nr	19.97
Socket outlet: switched with neon indicator; 13 Amp; BS 1363; steel backbox, satin chrome coverplate; BS 4662; flush fixed						
1 Gang	12.24	13.47	0.45	5.68	nr	19.15
2 Gang	22.90	25.19	0.45	5.68	nr	30.87
Weatherproof socket outlet: 13 Amp; switched; single gang; RCD protected; water and dust protected to I.P.66; surface mounted						
40A 30mA tripping current protecting 1 socket	60.23	66.25	0.75	9.47	nr	75.72
40A 30mA tripping current protecting 2 sockets	74.13	81.55	0.75	9.47	nr	91.02
Plug for weatherproof socket outlet: protected to I.P.66						
13Amp plug	10.65	11.71	0.20	2.52	nr	14.23

Y:MECHANICAL AND ELECTRICAL SERVICES

Item	Net Price £	Material £	Labour hours	Labour £	Unit	Total Rate £
Y74 : ACCESSORIES FOR ELECTRICAL SERVICES : CONNECTOR UNITS						
Electrical Accessories; fixed to backgrounds (Including fixings.)						
Connection units: Moulded pattern; BS5733; moulded plastic box; white coverplate; knockout for flex outlet; surface mounted – standard fused						
DP Switched	6.87	7.55	0.45	5.68	nr	**13.23**
Unswitched	6.33	6.97	0.45	5.68	nr	**12.65**
DP Switched with neon indicator	8.85	9.74	0.45	5.68	nr	**15.42**
Connection units: moulded pattern; BS 5733; galvanised steel box; white coverplate; knockout for flex outlet; surface mounted						
DP Switched	6.85	7.53	0.45	5.68	nr	**13.21**
Unswitched	6.32	6.95	0.45	5.68	nr	**12.63**
DP Switched with neon indicator	8.83	9.72	0.45	5.68	nr	**15.40**
Connection units: galvanised pressed steel pattern; galvanised steel box; satin chrome or satin brass finish; white moulded plastic inserts; flush mounted – standard fused						
DP Switched	8.82	9.70	0.45	5.68	nr	**15.38**
Unswitched	7.77	8.55	0.45	5.68	nr	**14.23**
DP Switched with neon indicator	10.71	11.78	0.45	5.68	nr	**17.46**
Connection units: galvanised steel box; satin chrome or satin brass finish; white moulded plastic inserts; flex outlet; flush mounted – standard fused						
Switched	9.98	10.98	0.45	5.68	nr	**16.66**
Unswitched	9.41	10.35	0.45	5.68	nr	**16.03**
Switched with neon indicator	11.95	13.15	0.45	5.68	nr	**18.83**
Connector Unit: moulded white plastic cover and block; galvanised steel box Immersion Heaters						
3 Kw up to 915mm Long; fitted thermostat	15.44	16.99	0.75	9.47	nr	**26.46**

Y:MECHANICAL AND ELECTRICAL SERVICES

Item	Net Price £	Material £	Labour hours	Labour £	Unit	Total Rate £
Y74 : ACCESSORIES FOR ELECTRICAL SERVICES : LAMP HOLDERS, OUTLETS AND CONTROL UNITS						
Electrical Accessories; fixed to backgrounds. (Including fixings.)						
Telephone outlet: moulded plastic plate with box; fitted and connected; flush or surface mounted						
Single Master outlet	5.20	5.72	0.30	3.79	nr	**9.51**
Single Secondary outlet	3.63	4.00	0.30	3.79	nr	**7.79**
Telephone outlet: bronze or satin chrome plate; with box; fitted and connected; flush or surface mounted						
Single Master outlet	9.28	10.21	0.30	3.79	nr	**14.00**
Single Secondary outlet	7.71	8.48	0.30	3.79	nr	**12.27**
TV Co-Axial socket outlet: moulded plastic box: flush or surface mounted						
One way Direct Connection	4.03	4.43	0.30	3.79	nr	**8.22**
Two way Direct Connection	5.52	6.07	0.30	3.79	nr	**9.86**
One way Isolated UHF/VHF	6.96	7.66	0.30	3.79	nr	**11.45**
Two way Isolated UHF/VHF	9.48	10.43	0.30	3.79	nr	**14.22**
Ceiling Rose: white moulded plastic; flush fixed to conduit box						
Plug in type; ceiling socket with 2 terminals , loop-in and ceiling plug with 3 terminals and cover	3.30	3.63	0.33	4.16	nr	**7.79**
BC Lampholder; white moulded plastic; Heat resistent PVC insulated and sheathed cable; flush fixed						
2 Core; 0.75mm2	1.38	1.52	0.30	3.79	nr	**5.31**
Battern Holder: white moulded plastic; 3 terminals; BS 5042; fixed to conduit						
Straight pattern; 2 terminals with loop-in and Earth	2.66	2.92	0.25	3.15	nr	**6.07**
Angled pattern ; looped in terminal	3.31	3.64	0.25	3.15	nr	**6.79**
Shaver Unit: self setting overload device; 200/250 voltage supply; white moulded plastic faceplate; unswitched						
Surface type with moulded plastic box	12.85	14.14	0.42	5.30	nr	**19.44**
Flush type with galvanised steel box	12.74	14.01	0.45	5.68	nr	**19.69**

Y:MECHANICAL AND ELECTRICAL SERVICES

Item	Net Price £	Material £	Labour hours	Labour £	Unit	Total Rate £
Shaver Unit: dual voltage supply unit; white moulded plastic faceplate; unswitched						
Surface type with moulded plastic box	30.11	33.12	0.55	6.94	nr	**40.06**
Flush type with galvanised steel box	29.41	32.35	0.60	7.57	nr	**39.92**
Cooker Control Unit: BS 4177; 45 amp D.P. main switch; 13 Amp switched socket outlet; metal coverplate; plastic inserts; neon indicators						
Surface mounted with mounting box	23.10	25.41	0.45	5.68	nr	**31.09**
Flush mounted with galvanised steel box	20.71	22.78	0.45	5.68	nr	**28.46**
Cooker Control Unit: BS 4177; 45 Amp D.P. main switch; 13 Amp switched socket outlet; moulded plastic box and coverplate; surface mounted						
Standard	13.05	14.36	0.45	5.68	nr	**20.04**
With neon indicators	17.15	18.86	0.45	5.68	nr	**24.54**

Y:MECHANICAL AND ELECTRICAL SERVICES

Item	Net Price £	Material £	Labour hours	Labour £	Unit	Total Rate £
Y74 : ACCESSORIES FOR ELECTRICAL SERVICES : CABLE TILES						
Cable Tiles; single width; laid in trench above cables on prepared sand bed. (Costs of excavation and filling sand excluded.)						
Reinforced concrete covers; concave/ convex ends						
914 x 152 x 63/38 mm	2.15	2.37	0.10	1.26	m	3.63
914 x 229 x 63/38 mm	2.71	2.98	0.10	1.26	m	4.24
914 x 305 x 63/38 mm	3.32	3.65	0.10	1.26	m	4.91

Keep your figures up to date, free of charge

This section, and most of the other information in this Price Book, is brought up to date every three months, until the next annual edition, in the *Price Book Update*.

The *Update* is available free to all Price Book purchasers.

To ensure you receive your copy, simply complete the reply card from the centre of the book and return it to us.

Y:MECHANICAL AND ELECTRICAL SERVICES

Item	Net Price £	Material £	Labour hours	Labour £	Unit	Total Rate £
Y89 : SUNDRY COMMON ELECTRICAL ITEMS : JUNCTION BOXES						
Weatherproof Junction Boxes; enclosures with rail mounted terminal blocks; side hung door to receive padlock; fixed to backgrounds, including all supports and fixings. (Suitable for cable up to 2.5mm2; including glandplates and gaskets.)						
Sheet steel with zinc spray finish enclosure						
Overall Size 229 x 152; suitable to receive 3 x 20(A) glands per gland plate	59.41	65.35	1.65	20.83	nr	86.18
Overall Size 306 x 306; suitable to receive 14 x 20(A) glands per gland plate	82.27	90.50	2.50	31.55	nr	122.05
Overall Size 458 x 382; suitable to receive 18 x 20(A) glands per gland plate	112.41	123.65	4.26	53.70	nr	177.35
Overall Size 762 x 508; suitable to receive 26 x 20(A) glands per gland plate	198.56	218.41	6.02	76.02	nr	294.43
Overall Size 914 x 610; suitable to receive 45 x 20(A) glands per gland plate	256.45	282.10	9.01	113.69	nr	395.79
Weatherproof Junction Boxes; enclosures with rail mounted terminal blocks; screw fixed lid; fixed to backgrounds, including all supports and fixings. (Suitable for cable up to 2.5mm2; including glandplates and gaskets.)						
Glassfibre reinforced polycarbonate enclosure						
Overall Size 190 x 190 x 130	18.58	20.44	1.65	20.83	nr	41.27
Overall Size 190 x 190 x 180	18.58	20.44	1.80	22.70	nr	43.14
Overall Size 280 x 190 x 130	24.47	26.92	2.50	31.55	nr	58.47
Overall Size 280 x 190 x 180	27.15	29.86	2.80	35.35	nr	65.21
Overall Size 380 x 190 x 130	28.98	31.88	4.00	50.48	nr	82.36
Overall Size 380 x 190 x 180	33.03	36.34	3.95	49.88	nr	86.22
Overall Size 380 x 280 x 130	38.02	41.82	5.99	75.57	nr	117.39
Overall Size 380 x 280 x 180	43.43	47.77	7.04	88.87	nr	136.64
Overall Size 560 x 280 x 130	54.87	60.36	9.01	113.69	nr	174.05
Overall Size 560 x 380 x 180	77.13	84.84	10.00	126.20	nr	211.04

Reliable
Easy-to-use
Spon's
Price data
On disc
Now!
So get
Estimating with

Construction cost estimating with Windows™

Prepare estimates with 100% confidence using data you know you can rely on

Save time, money and stress with this simple to use new estimating system

reSPONse
it's never been easier

Davis Langdon & Everest,
Chartered Quantity Surveyors

Olympic Computer Solutions

Choose from the latest data from the 1996 edition of either **Spon's Architects' & Builders' Price Book** or **Spon's Mechanical & Electrical Services Price Book**

Windows™ based software available on 3·5" discs

E & F N SPON
An imprint of Chapman & Hall

All this for less than £400!

Telephone us for an information pack on

0171 522 9966

or write to:
The Marketing Department, E & F N Spon, 2 - 6 Boundary Row, London SE1 8HN

PART FOUR

Daywork

Heating and Ventilating Industry, *page 411*
Electrical Industry, *page 416*
Building Industry Plant Hire Costs, *page 422*

When work is carried out in connection with a contract which cannot be valued in any other way, it is usual to assess the value on a cost basis with suitable allowances to cover overheads and profit. The basis of costing is a matter for agreement between the parties concerned but definitions of prime cost for the Heating and Ventilating and Electrical Industries have been prepared and published jointly by the Royal Institution of Chartered Surveyors and the appropriate bodies of the industries concerned for the convenience of those who wish to use them together with a schedule of basic plant charges published by the Royal Institution of Chartered Surveyors. These documents are reproduced by kind permission of the publishers.

Building Services Engineering

2nd Edition

D V Chadderton, Southampton Institute of Higher Education and also Chairman of CIBSE Southern Region, UK

This book is a compilation of information regarding all sections of building services. The coverage is comprehensive and engages the reader in the design of calculations carried out by the building services engineer. It explains the working principles of the major services, supported by line diagrams and copious calculations and self-assessment exercises. Each chapter is introduced with a statement of the learning objectives and key terms employed and ends with a glossary of technical terms used.

Contents: Preface. Acknowledgements. Units and constants. Symbols. The Built environment. Energy economics. Heat loss calculations. Hot and cold water supplies. Soil and waste systems. Surface-water drainage. Below-ground drainage. Condensation in buildings. Lighting. Electrical installations. Mechanical transport. Refuse disposal. Plant and service areas. Solving problems with programmable calculators and microcomputers. References. Index.

August 1994: 246x189: c.352pp, 55 line illus
Paperback: 0-419-19530-0: £18.99

For further information please contact: The Marketing Dept,. E & F N Spon, 2-6 Boundary Row, London SE1 8HN Tel: 0171 865 0066 Fax: 0171 522 9623

E & F N Spon

An imprint of Chapman & Hall

HEATING AND VENTILATING INDUSTRY

DEFINITION OF PRIME COST OF DAYWORK CARRIED OUT UNDER A HEATING, VENTILATING, AIR CONDITIONING, REFRIGERATION, PIPEWORK AND/OR DOMESTIC ENGINEERING CONTRACT (JULY 1980 EDITION)

This Definition of Prime Cost is published by the Royal Institution of Chartered Surveyors and the Heating and Ventilating Contractors Association for convenience, and for use by people who choose to use it. Members of the Heating and Ventilating Contractors Association are not in any way debarred from defining Prime Cost and rendering accounts for work carried out on that basis in any way they choose. Building owners are advised to reach agreement with contractors on the Definition of Prime Cost to be used prior to entering into a contract or sub-contract.

SECTION 1: APPLICATION

    1.1   This Definition provides a basis for the valuation of daywork executed under such heating, ventilating, air conditioning, refrigeration, pipework and or domestic engineering contracts as provide for its use.

    1.2   It is not applicable in any other circumstances, such as jobbing or other work carried out as a separate or main contract nor in the case of daywork executed after a date of practical completion.

    1.3   The terms 'contract' and 'contractor' herein shall be read as 'sub-contract' and 'sub-contractor' as applicable.

SECTION 2: COMPOSITION OF TOTAL CHARGES

    2.1   The Prime Cost of daywork comprises the sum of the following costs:
        (a)   Labour as defined in Section 3.
        (b)   Materials and goods as defined in Section 4.
        (c)   Plant as defined in Section 5.

    2.2   Incidental costs, overheads and profit as defined in Section 6, as provided in the contract and expressed therein as percentage adjustments, are applicable to each of 2.1 (a)-(c).

SECTION 3: LABOUR

    3.1   The standard wage rates, emoluments and expenses referred to below and the standard working hours referred to in 3.2 are those laid down for the time being in the rules or decisions or agreements of the Joint Conciliation Committee of the Heating, Ventilating and Domestic Engineering Industry applicable to the works (or those of such other body as may be appropriate) and to the grade of operative concerned at the time when and the area where the daywork is executed.

    3.2   Hourly base rates for labour are computed by dividing the annual prime cost of labour, based upon the standard working hours and as defined in 3.4, by the number of standard working hours per annum. See example.

    3.3   The hourly rates computed in accordance with 3.2 shall be applied in respect of the time spent by operatives directly engaged on daywork, including those operating mechanical plant and transport and erecting and dismantling other plant (unless otherwise expressly provided in the contract) and handling and distributing the materials and goods used in the daywork.

    3.4   The annual prime cost of labour comprises the following:
        (a)   Standard weekly earnings (i.e. the standard working week as determined at the appropriate rate for the operative concerned).
        (b)   Any supplemental payments.
        (c)   Any guaranteed minimum payments (unless included in Section 6.1(a)-(p)).
        (d)   Merit money.
        (e)   Differentials or extra payments in respect of skill, responsibility, discomfort, inconvenience or risk (excluding those in respect of supervisory responsibility - see 3.5)
        (f)   Payments in respect of public holidays.
        (g)   Any amounts which may become payable by the contractor to or in respect of operatives arising from the rules etc. referred to in 3.1 which are not provided for in 3.4 (a)-(f) nor in Section 6.1 (a)-(p).

HEATING AND VENTILATING INDUSTRY

(h)   Employers contributions to the Annual Holiday with Pay and Welfare Benefits Scheme or payments in lieu thereof.

(i)   Employers National Insurance contributions as applicable to 3.4 (a)-(h).

(j)   Any contribution, levy or tax imposed by Statute, payable by the contractor in his capacity as an employer.

3.5   Differentials or extra payments in respect of supervisory responsibility are excluded from the annual prime cost (see Section 6). The time of principals, staff, foremen, chargehands and the like when working manually is admissible under this Section at the rates for the appropriate grades.

SECTION 4: MATERIALS AND GOODS

4.1   The prime cost of materials and goods obtained specifically for the daywork is the invoice cost after deducting all trade discounts and any portion of cash discounts in excess of 5%.

4.2   The prime cost of all other materials and goods used in the daywork is based upon the current market prices plus any appropriate handling charges.

4.3   The prime cost referred to in 4.1 and 4.2 includes the cost of delivery to site.

4.4   Any Value Added Tax which is treated, or is capable of being treated, as input tax (as defined by the Finance Act 1972, or any re-enactment or amendment thereof or substitution therefore) by the contractor is excluded.

SECTION 5: PLANT

5.1   Unless otherwise stated in the contract, the prime cost of plant comprises the cost of the following:

(a)   use or hire of mechanically-operated plant and transport for the time employed on and/or provided or retained for the daywork;

(b)   use of non-mechanical plant (excluding non-mechanical hand tools) for the time employed on and/or provided or retained for the daywork;

(c)   transport to and from the site and erection and dismantling where applicable.

5.2   The use of non-mechanical hand tools and of erected scaffolding, staging, trestles or the like is excluded (see Section 6), unless specifically retained for the daywork.

SECTION 6: INCIDENTAL COSTS, OVERHEADS AND PROFIT

6.1   The percentage adjustments provided in the contract which are applicable to each of the totals of Sections 3, 4 and 5 comprise the following:

(a)   Head office charges.

(b)   Site staff including site supervision.

(c)   The additional cost of overtime (other than that referred to in 6.2).

(d)   Time lost due to inclement weather.

(e)   The additional cost of bonuses and all other incentive payments in excess of any included in 3.4.

(f)   Apprentices' study time.

(g)   Fares and travelling allowances.

(h)   Country, lodging and periodic allowances.

(i)   Sick pay or insurances in respect thereof, other than as included in 3.4.

(j)   Third party and employers' liability insurance.

(k)   Liability in respect of redundancy payments to employees.

(l)   Employer's National Insurance contributions not included in 3.4.

(m)   Use and maintenance of non-mechanical hand tools.

(n)   Use of erected scaffolding, staging, trestles or the like (but see 5.2).

(o)   Use of tarpaulins, protective clothing, artificial lighting, safety and welfare facilities, storage and the like that may be available on site.

(p)   Any variation to basic rates required by the contractor in cases where the contract provides for the use of a specified schedule of basic plant charges (to the extent that no other provision is made for such variation - see 5.1).

HEATING AND VENTILATING INDUSTRY

    (q)   In the case of a sub-contract which provides that the sub-contractor shall allow a cash discount, such provision as is necessary for the allowance of the prescribed rate of discount.

    (r)   All other liabilities and obligations whatsoever not specifically referred to in this Section nor chargeable under any other Section.

    (s)   Profit.

6.2   The additional cost of overtime where specifically ordered by the Architect/Supervising Officer shall only be chargeable in the terms of a prior written agreement between the parties.

HEATING AND VENTILATING INDUSTRY

Example of calculation of a typical standard hourly base rate (as defined in Section 3) for Advanced Fitter (qualified gas and arc) and Mate employed in the Heating, Ventilating, Air Conditioning, Piping and Domestic Engineering Industry under NJIC Rules based upon rates ruling at 4 February 1980.

	Adv Fitter (qual gas/arc)		Mate	
	Rate/hr £	Rate/annum £	Rate/hr £	Rate/annum £
Standard Weekly Earnings - 48 weeks x 38 hours	2.26	4122.24	1.64	2991.36
Welding Supplement (gas & arc) - 48 weeks x 38 hours	0.20	364.80		
Merit Money and other variables as applicable [see Note (a)]		*		*
		4487.04		2991.36
Employer's National Insurance Contribution at 13.5%		605.75		403.83
		5092.79		3395.19
CITB Annual Levy		60.00		6.00
Weekly Holiday Credit and Welfare Stamp- 48 weeks	9.83	471.84	8.08	387.84
Annual Labour Cost as defined in Section 3		£5624.63		£3789.03

Hourly base rates as defined in Section 3, clause 3.2.	$\dfrac{£5624.63}{1763.2}$ = **£3.19**	$\dfrac{£3789.03}{1763.2}$ = **£2.15**

Standard working hours per annum calculated as follows:

52 weeks at 38 hours			1976.0	
Less:				
4 weeks holiday at 38 hrs	=	152.0		
8 days Public Holidays at av 7.6 hrs per day	=	60.8		212.8
				1763.2

HEATING AND VENTILATING INDUSTRY

Note: For the convenience of readers the example which appears previously has been up-dated by the Editors as at March 1994.

	Adv Fitter (qual gas/arc)		Mate	
	Rate/hr £	Rate/annum £	Rate/hr £	Rate/annum £
Standard Weekly Earnings - 46.8 wks x 37.5 hrs	5.56	9761.14	4.04	7092.62
Welding Supplement (gas and arc) - 46.8 wks x 37.5 hrs	0.51	895.36		
Merit Money and other variables as applicable [See note (a)]		*		*
		10656.50		7092.62
Employer's National Insurance Contribution	10.2%	1086.96	7.6%	539.04
		11743.46		7631.66
Weekly Holiday Credit and Welfare Stamp - 52 weeks	30.29	1575.08	24.30	1263.60
Annual Labour Cost as defined in Section 3		£13318.54		£8895.26

Hourly base rates as defined in Section 3, Clause 3.2

$$\frac{£13318.54}{1755.6} = £7.59 \qquad \frac{£8895.26}{1755.6} = £5.07$$

Standard working hours per annum are calculated as follows:

52 weeks at 38 hours		1976.0
Less:		
4.2 weeks holiday at 38 hours	=	159.6
8 days Public Holidays at (av.) 7.6 hrs per day	=	60.8 220.4
		1755.6

Notes

(a) Where applicable, Merit Money and other variables (e.g. Daily Abnormal Conditions Money), which attract Employer's National Insurance Contribution, should be included at *. It should be noted that all labour costs incurred by the Contractor in his capacity as an Employer, other than those contained in the hourly base rate, as defined under Section 3, are to be taken into account under Section 6.

(b) The above example is for the convenience of users only and does not form part of the Definition; all the basic costs are subject to re-examination according to the time when and where the daywork is executed.

ELECTRICAL INDUSTRY

DEFINITION OF PRIME COST OF DAYWORK CARRIED OUT UNDER AN ELECTRICAL CONTRACT (MARCH 1981 EDITION)

This Definition of Prime Cost is published by The Royal Institution of Chartered Surveyors and The Electrical Contractors' Associations for convenience and for use by people who choose to use it. Members of The Electrical Contractors' Association are not in any way debarred from defining Prime Cost and rendering accounts for work carried out on that basis in any way they choose. Building owners are advised to reach agreement with contractors on the Definition of Prime Cost to be used prior to entering into a contract or sub-contract.

SECTION 1: APPLICATION

1.1 This Definition provides a basis for the valuation of daywork executed under such electrical contracts as provide for its use.

1.2 It is not applicable in any other circumstances, such as jobbing, or other work carried out as a separate or main contract, nor in the case of daywork executed after the date of practical completion.

1.3 The terms 'contract' and 'contractor' herein shall be read as 'sub-contract' and 'sub-contractor' as the context may require.

SECTION 2: COMPOSITION OF TOTAL CHARGES

2.1 The Prime Cost of daywork comprises the sum of the following costs:
    (a) Labour as defined in Section 3.
    (b) Materials and goods as defined in Section 4.
    (c) Plant as defined in Section 5.

2.2 Incidental costs, overheads and profit as defined in Section 6, as provided in the contract and expressed therein as percentage adjustments, are applicable to each of 2.1 (a)-(c).

SECTION 3: LABOUR

3.1 The standard wage rates, emoluments and expenses referred to below and the standard working hours referred to in 3.2 are those laid down for the time being in the rules and determinations or decisions of the Joint Industry Board or the Scottish Joint Industry Board for the Electrical Contracting Industry (or those of such other body as may be appropriate) applicable to the works and relating to the grade of operative concerned at the time when and in the area where daywork is executed.

3.2 Hourly base rates for labour are computed by dividing the annual prime cost of labour, based upon the standard working hours and as defined in 3.4 by the number of standard working hours per annum. See examples.

3.3 The hourly rates computed in accordance with 3.2 shall be applied in respect of the time spent by operatives directly engaged on daywork, including those operating mechanical plant and transport and erecting and dismantling other plant (unless otherwise expressly provided in the contract) and handling and distributing the materials and goods used in the daywork.

3.4 The annual prime cost of labour comprises the following:
    (a) Standard weekly earnings (ie the standard working week as determined at the appropriate rate for the operative concerned).
    (b) Payments in respect of public holidays.
    (c) Any amounts which may become payable by the Contractor to or in respect of operatives arising from operation of the rules etc. referred to in 3.1 which are not provided for in 3.4(a) and (b) nor in Section 6.
    (d) Employer's National Insurance Contributions as applicable to 3.4 (a)-(c).
    (e) Employer's contributions to the Joint Industry Board Combined Benefits Scheme or Scottish Joint Industry Board Holiday and Welfare Stamp Scheme, and holiday payments made to apprentices in compliance with the Joint Industry Board National Working Rules and Industrial Determinations.
    (f) Any contribution, levy or tax imposed by Statute, payable by the Contractor in his capacity as an employer.

ELECTRICAL INDUSTRY

3.5 Differentials or extra payments in respect of supervisory responsibility are excluded from the annual prime cost (see Section 6). The time of principals and similar categories, when working manually, is admissible under this Section at the rates for the appropriate grades.

SECTION 4: MATERIALS AND GOODS

4.1 The prime cost of materials and goods obtained specifically for the daywork is the invoice cost after deducting all trade discounts and any portion of cash discounts in excess of 5%.

4.2 The prime cost of all other materials and goods used in the daywork is based upon the current market prices plus any appropriate handling charges.

4.3 The prime cost referred to in 4.1 and 4.2 includes the cost of delivery to site.

4.4 Any Value Added Tax which is treated, or is capable of being treated, as input tax (as defined by the Finance Act 1972, or any re-enactment or amendment thereof or substitution therefore) by the Contractor is excluded.

SECTION 5: PLANT

5.1 Unless otherwise stated in the contract, the prime cost of plant comprises the cost of the following:
   (a) Use or hire of mechanically-operated plant and transport for the time employed on and/or provided or retained for the daywork;
   (b) Use of non-mechanical plant (excluding non-mechanical hand tools) for the time employed on and/or provided or retained for the daywork;
   (c) Transport to and from the site and erection and dismantling where applicable.

5.2 The use of non-mechanical hand tools and of erected scaffolding, staging, trestles or the likes is excluded (see Section 6), unless specifically retained for daywork.

5.3 Note: Where hired or other plant is operated by the Electrical Contractor's operatives, such time is to be included under Section 3 unless otherwise provided in the contract.

SECTION 6: INCIDENTAL COSTS, OVERHEADS AND PROFIT

6.1 The percentage adjustments provided in the contract which are applicable to each of the totals of Sections 3, 4 and 5, compromise the following:
   (a) Head Office charges.
   (b) Site staff including site supervision.
   (c) The additional cost of overtime (other than that referred to in 6.2).
   (d) Time lost due to inclement weather.
   (e) The additional cost of bonuses and other incentive payments.
   (f) Apprentices' study time.
   (g) Travelling time and fares.
   (h) Country and lodging allowances.
   (i) Sick pay or insurance in lieu thereof, in respect of apprentices.
   (j) Third party and employers' liability insurance.
   (k) Liability in respect of redundancy payments to employees.
   (l) Employers' National Insurance Contributions not included in 3.4.
   (m) Use and maintenance of non-mechanical hand tools.
   (n) Use of erected scaffolding, staging, trestles or the like (but see 5.2.).
   (o) Use of tarpaulins, protective clothing, artificial lighting, safety and welfare facilities, storage and the like that may be available on site.
   (p) Any variation to basic rates required by the Contractor in cases where the contract provides for the use of a specified schedule of basic plant charges (to the extent that no other provision is made for such variation - see 5.1).
   (q) All other liabilities and obligations whatsoever not specifically referred to in this Section nor chargeable under any other Section.
   (r) Profit.
   (s) In the case of a sub-contract which provides that the sub-contractor shall allow a cash discount, such provision as is necessary for the allowance of the prescribed rate of discount.

6.2 The additional cost of overtime where specifically ordered by the Architect/Supervising Officer shall only be chargeable in the terms of a prior written agreement between the parties.

ELECTRICAL INDUSTRY

Example of calculation of typical standard hourly base rate (as defined in Section 3) for Approved Electrician and Apprentice employed in the Electrical Contracting Industry under JIB Rules applicable to England, Wales, Northern Ireland, Isle of Man and the Channel Islands based upon rates ruling at 14 September 1980.

	Approved Elect.		Apprentice age 18	
	Rate/hr £	*Rate/annum* £	*Rate/hr* £	*Rate/annum* £
Standard Weekly Earnings				
Approved Electrician				
48 weeks x 38 hrs = 1824 hrs	2.70	4924.80		
Apprentice				
52 weeks x 38 hrs = 1976 hrs				
Less Study Time				
40 weeks @ 8 hrs = <u>320</u> hrs				
1656 hrs			1.24	2053.44
[See note (c)]				
Employer's National Insurance				
Contributions @ 13.7%		674.70		281.32
Employer's Contributions to:				
CITB Levy, per annum		43.20		
JIB Benefits Scheme 48 wks	9.20	441.60		
Annual labour costs as defined				
in Section 3		£6084.30		£2334.76

Hourly base rate defined	$\dfrac{£6084.30}{1760.0}$ =	**£3.46**	$\dfrac{£2334.76}{1440.0}$ =	**£1.62**
in Section 3, Clause 3.2				

Standard working hours per annum calculated as follows:

52 weeks @ 38 hours			1976.0	1976.0
Less				
4 weeks holiday @ 38 hrs =		152.0		
8 days public holiday @ 8 hrs =		64.0	<u>216.0</u>	<u>216.0</u>
			1760.0	1760.0
Study time 40 weeks @ 8 hrs =				<u>320.0</u>
				1440.0

Notes

(a) It should be noted that all labour costs incurred by the Contractor in his capacity as an Employer, other than those contained in the hourly rate above, must be taken into account under Section 6.

(b) The above example is for the convenience of users only and does not form part of the rules laid down in the Definition; all the basic costs are subject to re-examination according to the time when and in the area where the daywork is executed.

(c) Public Holidays are included in annual earnings.

ELECTRICAL INDUSTRY

Example of calculation of typical standard hourly base rate (as defined in Section 3) for Approved Electrician and Electrician employed in the Electrical Contracting Industry under SJIB Rules applicable to Scotland based upon rates ruling at 14 September 1980.

	Approved Elect.		Electrician	
	Rate/hr £	Rate/annum £	Rate/hr £	Rate/annum £
Standard Weekly Earnings				
47 weeks x 38 hrs = 1786 hrs	2.70	4822.20	2.48	4429.28
Employer's National Insurance				
Contribution @ 13.7%		660.64		606.81
Employer's Contribution to:				
CITB Levy, per annum		43.20		43.20
SJIB Holiday and Welfare				
Scheme, 52 weeks	11.15	579.80	10.36	538.72
Annual labour cost as defined		————		————
in Section 3		£6105.84		£5618.01
		————		————

Hourly base rate defined	$\dfrac{£6105.84}{1760.0}$	=	**£3.47**	$\dfrac{£5618.01}{1760.0}$ =	**£3.19**
in Section 3, Clause 3.2					

Standard working hours per annum calculated as follows:

52 weeks @ 38 hours			1976.0
Less			
4 weeks holiday @ 38 hours	=	152.0	
8 days public holiday @ 8 hours	=	<u>64.0</u>	<u>216.0</u>
			<u>1760.0</u>

Notes

(a)   It should be noted that all labour costs incurred by the Contractor in his capacity as an Employer, other than those contained in the hourly rate above, must be taken into account under Section 6.

(b)   The above example is for the convenience of users only and does not form part of the rules laid down in the Definition; all the basic costs are subject to re-examination according to the time when and in the area where the daywork is executed.

ELECTRICAL INDUSTRY

Note: For the convenience of readers the example for England, Wales, Northern Ireland, Isle of Man and the Channel Islands which appears previously has been up-dated by the Editors as at May 1994. With the revision of the Apprentice Training Scheme, this example includes figures based on a Senior Apprentice (Stage 2).

	Approved Elect.		*Senior Apprentice (Stage 2)*	
	Rate/hr £	*Rate/annum* £	*Rate/hr* £ £	*Rate/annum* £
Standard Weekly Earnings Approved Electrician 47.8 wks x 37.5 hrs = 1792.5 hrs	6.00	10755.00		
Apprentice 52 wks x 37.5 hrs = 1950.0 hrs Less Study Time 40 wks @ 7.5 hrs = <u>300.0</u> hrs 1650.0 hrs [See Note (c)]			3.53	5824.50
Employer's National Insurance Contributions @	10.4%	1118.52	6.6%	384.42
Employer's Contributions to: JIB Benefits Scheme 47 wks	24.77	1164.19		
Annual labour costs as defined in Section 3.		£13037.71		£6208.92
Hourly base rate as defined in Section 3, Clause 3.2	<u>£13037.71</u> = 1732.5	**£7.53**	<u>£6208.92</u> = 1432.5	**£4.33**

Standard working hours per annum calculated as follows:

52 weeks @ 37.5 hours Less			1950.0	1950.0
4.2 weeks holiday @ 37.5 hours = 157.5 8 days public holiday @ 8 hours = 60.0			<u>217.5</u> <u>1732.5</u>	<u>217.5</u> <u>1732.5</u>
Study time 40 wks @ 7.5 hrs				300.0
				<u>1432.5</u>

Notes

(a)   It should be noted that all labour costs incurred by the Contractor in his capacity as an Employer, other than those contained in the hourly rate above, must be taken into account under Section 6.

(b)   The above example is for the convenience of users only and does not form part of the rules laid down in the Definition; all the basic costs are subject to re-examination according to the time when and in the area where the daywork is executed.

(c)   Public Holidays are included in annual earnings

ELECTRICAL INDUSTRY

Note: For the convenience of readers the example for Scotland which appears previously has been up-dated by the Editors as at May 1994.

	Approved Elect. *Rate/hr* £	*Rate/annum* £	*Electrician* *Rate/hr* £	*Rate/annum* £
Standard Weekly Earnings 46.2 wks x 37.5 hrs = 1732.5 hrs	6.40	11088.00	5.94	10291.05
Employer's National Insurance Contribution @	10.2%	1130.98	10.2%	1049.69
Employer's Contribution to: SJIB Holiday and Welfare Scheme, 52 weeks	23.57	1225.64	22.20	1154.40
Annual labour cost as defined in Section 3, Clause 3.2		£13444.62		£12495.14

Hourly base rate as defined in Section 3, Clause 3.2	$\dfrac{£13444.62}{1732.5}$ =	**£7.76**	$\dfrac{£12495.14}{1732.5}$ =	**£7.21**

Standard working hours per annum calculated as follows:

			Approved Electrician *& Electrician*
52 wks @ 37.5 hrs		1950.00	
Less			
4.2 wks holiday @ 37.5 hrs	=	157.50	
8 days public holiday @ 7.5 hrs	=	60.00	217.50
			1732.50

Notes

(a)   It should be noted that all labour costs incurred by the Contractor in his capacity as an Employer, other than those contained in the hourly rate above, must be taken into account under Section 6.

(b)   The above example is for the convenience of users only and does not form part of the rules laid down in the Definition; all the basic costs are subject to re-examination according to the time when and in the area where the daywork is executed.

BUILDING INDUSTRY PLANT HIRE COSTS

SCHEDULE OF BASIC PLANT CHARGES (JANUARY 1990)

This Schedule is published by the Royal Institution of Chartered Surveyors and is for use in connection with Dayworks under a Building Contract.

EXPLANATORY NOTES

1   The rates in the Schedule are intended to apply solely to daywork carried out under and incidental to a Building Contract. They are NOT intended to apply
to:
(i)   jobbing or any other work carried out as a main or separate contract; or
(ii) work carried out after the date of commencement of the Defects Liability Period.

2   The rates in the Schedule are basic and may be subject to an overall adjustment to be quoted by the Contractor prior to the placing of the Contract.

3   The rates apply to plant and machinery already on site, whether hired or owned by the Contractor.

4   The rates, unless otherwise stated, include the cost of fuel and power of every description, lubricating oils, grease, maintenance, sharpening of tools, replacement of spare parts, all consumable stores and for licences and insurances applicable to items of plant. They do not include the costs of drivers and attendants (unless otherwise stated).

5   The rates should be applied to the time during which the plant is actually engaged in daywork.

6   Whether or not plant is chargeable on daywork depends on the daywork agreement in use and the inclusion of an item of plant in this schedule does not necessarily indicate that item is chargeable.

7   Rates for plant not included in the Schedule or which is not on site and is specifically hired for daywork shall be settled at prices which are reasonably related to the rates in the Schedule having regard to any overall adjustment quoted by the Contractor in the Conditions of Contract.

NOTE: All rates in the schedule are expressed per hour and were calculated during the first quarter of 1989.

BUILDING INDUSTRY PLANT HIRE COSTS

MECHANICAL PLANT AND TOOLS

Item of plant	Description		Unit	Rate per hr £
Bar-bending and Shearing Machines	Power Driven			
Bar bending machine	Up to 2 in (51 mm) dia. rods		each	2.01
Bar cropper machine	Up to 2 in (51 mm) dia. rods		each	1.51
Bar shearing machine	Up to 1½ in (38 mm) dia. rods		each	2.00
	Up to 2 in (51 mm) dia. rods		each	2.00
Block & stone splitter	Hydraulic		each	0.83
Brick Saws				
Brick saw (use of abrasive disc to be charged net & credited)	Power driven (bench type clipper or similar or portable)		each	1.83
Compressors				
Portable compressors (machine only)	*Nominal delivery of free air per min at 100 lb/sq. in. (7 kg/sq.m.) pressure*			
	(cfm.)	(m3/min)		
	80/85	2.41	each	1.30
	125-140	3.50- 3.92	each	1.58
	160-175	4.50- 4.95	each	2.37
	250	7.08	each	3.25
	380	10.75	each	4.80
	600-630	16.98-17.84	each	6.00
Lorry Mounted Compressors (Machine plus lorry only)	101-150	2.86-4.24	each	4.12
Tractor Mounted Compressor (Machine plus rubber tyred tractor only)	101-120	2.86-3.40	each	3.63
Compressed Air Equipment (with & including up to 50 ft (15.24 m) of air hose)				
Breakers with six steels (Light/Medium/Heavy)			each	0.47
Light pneumatic pick with six steels			each	0.37
Pneumatic clay spade and one blade			each	0.22
Chipping hammer plus six steels			each	0.65
Hand-held rock drill and rod	15 lb (7.0 kg) Class		each	0.33
(bits to be paid for at a net	35 lb (16.0 kg) Class		each	0.37
cost and credited)	45 lb (20.0 kg) Class		each	0.37
	55 lb (25.0 kg) Class		each	0.39
Drill, rotary	Up to ¾ in (19 mm)		each	0.37
(bits to be paid for at net cost and credited)	Up to 1¼ in (32mm)		each	0.73
Sander/Grinder			each	0.39
Scrabbler (heads extra)	Single head		each	0.48
	Triple head		each	0.65

BUILDING INDUSTRY PLANT HIRE COSTS

MECHANICAL PLANT AND TOOLS - *continued*

Item of plant	*Description*		*Unit*	*Rate per hr* £
Compressor Equipment				
Accessories				
Additional hoses	Per 50 ft (15.0 m)		each	0.09
Muffler, tool silencer			each	0.08
Concrete Breaker				
Concrete breaker, portable hydraulic				
complete with power pack			each	1.33
Concrete/Mortar				
Mixers				
Concrete mixer	Diesel, electric or petrol			
	cu ft	(cu m)		
Open drum without hopper	3/2	(0.09/0.06)	each	0.40
	4/3	(0.12/0.09)	each	0.44
	5/3.5	(0.15/0.10)	each	0.54
Open drum with hopper	7/5	(0.20/0.15)	each	0.67
Closed drum	8/6.5	(0.25/0.18)	each	0.90
Reversing drum with				
hopper weigher & feed	10/7	(0.28/0.20)	each	2.44
shovel	21/14	(0.60/0.40)	each	4.17
Concrete Pump				
Lorry mounted concrete pump				
(meterage charge to be added net)			each	13.70
Concrete Equipment				
Vibrator, poker type	Petrol, diesel or electric			
	Up to 3" diameter		each	0.87
	Air, excluding compressor and hose			
	Up to 3" diameter		each	0.65
	Extra heads		each	0.50
Vibrator-tamper	With tamping board		each	0.87
	Double beam screeder		each	1.19
Power float	29"-36"		each	1.13
Conveyor Belts				
	Power operated up to 25 ft (7.62 m)			
	long, 16 in (400 mm) wide		each	3.15
Cranes	*Maximum capacity*			
Mobile, rubber tyred	Up to 15 cwt (762 kg)		each	5.15
Lorry mounted, telescopic	6 tons (tonnes)		each	10.47
jib, 2 wheel (All rates	7 tons (tonnes)		each	11.22
inclusive of driver)	8 tons (tonnes)		each	13.67
	10 tons (tonnes)		each	14.99
	12 tons (tonnes)		each	16.91
	15 tons (tonnes)		each	19.21
	18 tons (tonnes)		each	20.34
	20 tons (tonnes)		each	22.04
	25 tons (tonnes)		each	24.86

BUILDING INDUSTRY PLANT HIRE COSTS

MECHANICAL PLANT AND TOOLS - *continued*

		Unit	Rate Per hr £
Item of plant	*Description*		

Cranes *continued*

Item of plant	Description	Unit	Rate £
Lorry mounted, telescopic jib, 4 wheel (All rates inclusive of driver)	10 tons (tonnes)	each	15.30
	12 tons (tonnes)	each	17.26
	15 tons (tonnes)	each	19.59
	20 tons (tonnes)	each	22.48
	25 tons (tonnes)	each	25.36
	30 tons (tonnes)	each	28.82
	45 tons (tonnes)	each	31.12
	50 tons (tonnes)	each	33.64

Item of plant	Capacity metre/tonnes	Height under hook above ground (m)	Unit	Rate £
Track/mounted tower crane (electric)(Capacity = max lift in tons x max rad at which can be lifted) (All rates inclusive of driver)	10	17	each	7.40
	15	17	each	7.95
	20	18	each	8.50
	25	20	each	10.70
	30	22	each	12.76
	40	22	each	16.75
	50	22	each	21.70
	60	22	each	22.52
	70	22	each	23.07
	80	22	each	23.99
	110	22	each	24.49
	125	30	each	27.20
	150	30	each	29.95

Static tower cranes	To be charged at a percentage of the above rates	90% of the above

Crane Equipment

Item of plant	cu yd	(cu m)	Unit	Rate £
Tipping bucket (circular)	Up to ¼	(0.19)	each	0.25
	¾	(0.57)	each	0.28
Skip, muck	Up to ½	(0.38)	each	0.25
	¾	(0.57)	each	0.25
	1½	(0.76)	each	0.30
Skip, concrete	Up to ½	(0.38)	each	0.44
	¾	(0.57)	each	0.53
	1	(0.76)	each	0.76
	1½	(1.15)	each	0.61
Skip, concrete lay down or roll over	½	(0.38)	each	0.44

Dehumidifiers

Item of plant	Description	Unit	Rate £
110/240V, Water extraction per 24 hrs	15 galls (68 litres)	each	0.84
	20 galls (90 litres)	each	0.97

BUILDING INDUSTRY PLANT HIRE COSTS

MECHANICAL PLANT AND TOOLS - *continued*

Item of plant	*Description*		*Unit*	*Rate per hr £*
Diamond Drilling and Chasing				
Chasing machine drilling	6" (152mm)		each	1.35
(Diamond core)	Mini rig up to 35mm		each	3.16
	3"-8" (76mm-203mm) electric 110v		each	3.33
	3"-8" (76mm-203mm) Air		each	4.23
Dumpers				
Dumper (site use only excl. Tax, Insurance & extra cost of DERV, etc., when operating on highway)	*Makers capacity*			
2 wheel drive				
Gravity tip	15 cwt	(762 kg)	each	1.04
Hydraulic tip	20 cwt	(1016 kg)	each	1.25
Hydraulic tip	23 cwt	(1168 kg)	each	1.31
4 wheel drive				
Gravity tip	23 cwt	(1168 kg)	each	1.42
Hydraulic tip	25 cwt	(1270 kg)	each	1.61
Hydraulic tip	30 cwt	(1524 kg)	each	1.83
Hydraulic tip	35 cwt	(1778 kg)	each	2.08
Hydraulic tip	40 cwt	(2032 kg)	each	2.33
Hydraulic tip	50 cwt	(2540 kg)	each	2.96
Hydraulic tip	60 cwt	(3048 kg)	each	3.68
Hydraulic tip	80 cwt	(4064 kg)	each	4.72
Hydraulic tip	100 cwt	(5080 kg)	each	6.38
Hydraulic tip	120 cwt	(6096 kg)	each	7.54
Electric Hand Tools				
Breakers	Heavy:	Kango 1800	each	0.87
		Kango 2500	each	0.90
	Medium:	Hilti TP800	each	0.73
Rotary Hammer	Heavy:	Kango 950	each	0.55
	Medium:	Kango 627/637	each	0.47
	Light:	Hilti TE12/TE17	each	0.39
Pipe drilling tackle	Ordinary type		set	0.51
	Under pressure type		set	0.83
Excavators				
Hydraulic full circle slew	cu yd	(cu m)		
Crawler mounted, backactor	5/8	(0.50)	each	5.01
	3/4	(0.60)	each	5.92
	7/8	(0.70)	each	8.24
	1	(0.75)	each	13.25
Wheeled tractor type	hydraulic excavator, JCB type 3C or similar		each	9.23

BUILDING INDUSTRY PLANT HIRE COSTS

MECHANICAL PLANT AND TOOLS - *continued*

Item of plant	*Description*		*Unit*	*Rate per hr £*
Mini Excavators				
Kubota mini excavators	360 tracked - 1 tonne		each	5.38
	360 tracked - 3.5 tonnes		each	7.23
Forklifts	*Payload*	*Max lift*		
2 wheel drive	20 cwt (1016 kg)	21 ft 4 in (6.50 m)	each	4.38
"Rough terrain"	20 cwt (1016 kg)	26 ft 0 in (7.92 m)	each	4.38
	30 cwt (1524 kg)	20 ft 0 in (6.09 m)	each	4.50
	36 cwt (1829 kg)	18 ft 0 in (5.48 m)	each	4.50
	50 cwt (2540 kg)	12 ft 0 in (3.66 m)	each	4.73
4 wheel drive	30 cwt (1524 kg)	20 ft 0 in (6.09 m)	each	5.60
"Rough terrain"	40 cwt (2032 kg)	20 ft 0 in (6.09 m)	each	5.78
	50 cwt (2540 kg)	12 ft 0 in (3.66 m)	each	7.79
Hammers				
Cartridge (excluding	Hammer DX450		each	0.40
cartridges and studs)	Hammer DX600		each	0.48
	Hammer spit TS		each	0.48
Heaters, Space				
	Paraffin/electric Btu/hr			
	50 000- 75 000		each	0.73
	80 000-100 000		each	0.77
	150 000		each	0.88
	320 000		each	1.20
Hoists				
Scaffold	Up to 5 cwt (254 kg)		each	0.90
Mobile (goods only)	Up to 10 cwt (508 kg)		each	1.43
Static (goods only)	10 to 15 cwt (508 kg - 762 kg)		each	1.67
Rack and pinion (goods				
only)	16 cwt (813 kg)		each	2.75
	20 cwt (1016 kg)		each	3.00
	25 cwt (1270 kg)		each	3.30
Rack and pinion (goods				
& passenger)	8 person, 1433 lbs (650 kg)		each	4.50
	12 person, 2205 lbs (1000 kg)		each	4.95
Lorries	*Plated gross vehicle*			
	Weight, ton/tonne			
Fixed body	Up to 5.50		each	9.60
	Up to 7.50		each	10.20
Tipper	Up to 16.00		each	11.73
	Up to 24.00		each	14.98
	Up to 30.00		each	17.73

 Daywork

BUILDING INDUSTRY PLANT HIRE COSTS

MECHANICAL PLANT AND TOOLS - *continued*

Item of plant	Description	Unit	Rate per hr £
Pipe Work Equipment			
Pipe bender	Power driven, 50-150 mm dia	each	0.75
	Hydraulic, 2" capacity	each	0.96
Pipe cutter	Hydraulic	each	1.02
Pipe defrosting equipment	Electrical	set	1.83
Pipe testing equipment	Compressed air	set	1.04
	Hydraulic	set	0.68
Pipe threading equipment	Up to 4"	set	1.65
Pumps			
Including 20 ft (6 m) length of suction and/or delivery hose, couplings, valves and strainers			
Diaphragm:			
'Simplite' 2 in (50 mm)	Petrol or electric	each	0.79
'Wickham' 3 in (76 mm)	Diesel	each	1.22
Submersible 2 in (50 mm)	Electric	each	0.85
Induced flow:			
'Spate' 3 in (76 mm)	Diesel	each	1.42
'Spate' 4 in (102 mm)	Diesel	each	1.88
Centrifugal, self-priming:			
'Univac' 2 in (50 mm)	Diesel	each	1.87
'Univac' 4 in (120 mm)	Diesel	each	2.40
'Univac' 6 in (152 mm)	Diesel	each	3.69

Pumping Equipment	in	(mm)	Unit	Rate per hr £
Pump hoses, per 20 ft (6 m)	2	(51)	each	0.15
flexible, suction or delivery,	3	(76)	each	0.17
inc coupling valve & strainer	4	(102)	each	0.19
	6	(152)	each	0.32

Rammers and Compactors			Unit	Rate per hr £
Power rammer, Pegson or similar			each	0.88
Soil compactor, plate type				
Plate size				
12 x 13 in (305 x 330 mm)	172 lb (78 kg)		each	0.78
20 x 18 in (508 x 457 mm)	264 lb (120 kg)		each	0.93

Rollers	cwt	(kg)	Unit	Rate per hr £
Vibrating rollers	7½ - 8¼	(368-420)	each	1.00
Single roller	10½	(533)	each	1.53
Twin roller	13¾	(698)	each	1.75
	16¾	(851)	each	2.01
Twin roller with seat	21	(1067)	each	2.75
and steering wheel	27½	(1397)	each	2.88
Pavement rollers	ton	tonne		
dead weight	3	4	each	2.89
	Over 4	6	each	3.75
	Over 6	10	each	4.40

BUILDING INDUSTRY PLANT HIRE COSTS

MECHANICAL PLANT AND TOOLS - *continued*

Item of plant	*Description*		*Unit*	*Rate per hr* £
Saws, Mechanical	in	(m)		
Chain saw	21	(0.53)	each	0.78
	30	(0.76)	each	1.12
Bench saw	Up to 20	(0.51) blade	each	0.80
	Up to 24	(0.61) blade	each	1.20
Screed Pump	*Maximum delivery*			
Wkg vol 7 cu ft	Vertical 300 ft (91 m)			
(200 ltrs.)	Horizontal 600 ft (182 m)		each	8.33
Screed pump hose	50/65 mm dia.			
	13.3 m long		each	1.83
Screwing Machines				
	13-50 mm dia.		each	0.43
	25-100 mm dia.		each	0.87
Tractors	cu yd	(cu m)		
Shovel, tractor	Up to ¾	(0.57)	each	4.70
(crawler) any type	1	(0.76)	each	5.70
of bucket	1¼	(0.96)	each	6.62
	1½	(1.15)	each	8.15
	1¾	(1.34)	each	8.98
	2	(1.53)	each	10.69
	2¾-3½	(2.10-2.70)	each	17.27
Shovel, tractor	Up to ½	(0.38)	each	3.36
(wheeled)	¾	(0.57)	each	4.14
	1	(0.76)	each	5.02
	2½	(1.91)	each	7.30
Tractor (crawler) with	*Maker's rated flywheel horsepower*			
dozer	75		each	9.31
	140		each	12.52
Tractor-wheeled (rubber tyred)		Light 48 hp		
each	4.23			
Agricultural type	Heavy 65 hp		each	4.68
Traffic Lights				
	Mains/generator 2-way		set	2.23
	Mains/generator 3-way		set	4.40
	Mains/generator 4-way		set	5.45
	Trailer mounted 2-way		set	2.21
Welding and Cutting and Burning Sets				
Welding and cutting set (inc. oxygen and acetylene, excl. underwater equip. and thermic boring)			each	4.33
Welding set, diesel	300 amp, single operator		each	1.84
(excluding electrodes)	600 amp, double operator		each	1.89
	600 amp, double operator		each	2.35

BUILDING INDUSTRY PLANT HIRE COSTS

NON-MECHANICAL PLANT

Item of plant	Description	Unit	Rate per hr £
Bar-bending and Shearing Machines			
Bar bending machine manual	Up to 1 in (25 mm) dia. rods	each	0.26
Shearing machine, manual	Up to ⅝ in (16 mm) dia. rods	each	0.15
Brother or Sling Chains			
	Not exceeding 2 ton/tonne	set	0.28
	Exceeding 2 ton/tonne, not exceeding 5 ton/tonne	set	0.40
	Exceeding 5 ton/tonne, not exceeding 10 ton/tonne	set	0.48
Drain Testing Equipment		set	0.42
Lifting and Jacking Gear	ton/tonne		
Pipe winch - inc. shear legs	½	set	0.35
Pipe winch - inc. gantry	2	set	0.72
	3	set	0.84
Chain blocks up to 20 ft (6.10 m) lift	ton/tonne		
	1	each	0.35
	2	each	0.44
	3	each	0.51
	4	each	0.66
Pull lift (Tirfor type)	¾ tons/tonnes	each	0.35
	1½ tons/tonnes	each	0.47
	3 tons/tonnes	each	0.61
Pipe Benders			
	13-75 mm dia.	each	0.32
	50-100 mm dia.	each	0.52
Plumber's Furnace	Calor gas or similar	each	1.20
Road Works, Equipment			
Barrier trestles or similar		ten	0.50
Crossing plates (steel sheets)		each	0.32
Danger lamp, inc. oil		each	0.03
Warning signs		each	0.07
Road cones		ten	0.11
Flasher unit (battery to be charged at cost)		each	0.06
Flashing bollard (battery to be charged at cost)		each	0.10

BUILDING INDUSTRY PLANT HIRE COSTS

NON-MECHANICAL PLANT - continued

Item of plant	Description	Unit	Rate per hr £
Rubbish Chutes			
Length of sections 154cm	Standard section	each	0.15
giving working length of	Brand section	each	0.20
130cm	Hopper	each	0.16
	Fixing frame	each	0.20
Winch Type A	20m for erection	each	0.29
Winch Type B	40m for erection	each	0.67
Scaffolding			
Boards		100ft (30.48m)	0.06
Castor wheels	Steel or rubber tyred	100	0.87
Fall ropes	Up to 200 ft (61 m)	each	0.44
Fittings (inc. couplers,			
and base plates)	Steel or alloy	100	0.07
Ladders, pole	20 rung	each	0.11
	30 rung	each	0.15
	40 rung	each	0.26
Ladder, extension	Extended length		
	ft (m)		
	20 (6.10)	each	0.31
	26 (7.92)	each	0.43
	35 (10.67)	each	0.64
Putlogs	Steel or alloy	100	0.06
Splithead	Small	10	0.12
	Medium	10	0.15
	Large	10	0.16
Staging, lightweight		100ft (30.48 m)	1.00
Tube	Steel	100ft (30.48 m)	0.02
	Alloy		0.04
Wheeled tower	Working platform up to 20 ft (6m)		
7 x 7 ft (2.13 x 2.13 m)	high including castors and boards	each	0.58
10 x 10 ft (3.05 x 3.05 m)		each	0.59
Tarpaulins		10m2	0.04
Trench Struts and Sheets			
Adjustable steel trench			
strut	All sizes from 1 ft (305 mm)		
	closed to 5 ft 6 in (1.68m)		
	extended	10	0.06
Steel trench sheet	5-14 ft (1.52-4.27 m) lengths	100ft (30.48 m)	0.16

The *only* source of information on cost data for Europe...

Spon's European Construction Costs Handbook

2nd Edition

Edited by **Davis Langdon & Everest**, Chartered Quantity Surveyors, UK

"This is a book that many of you have been waiting for. This work is currently the only one of its kind that we know of that is available on the public market... an absolute must for anybody doing international construction work...[it] will pay for itself." - **Cost Engineering**

✦ **Countries covered:** Austria, Belgium, Bulgaria, Cyprus, Czech Republic, Denmark, Finland, France, Germany, Greece, Hungary, Ireland, Italy, Japan, Luxembourg, Malta, The Netherlands, Norway, Poland, Portugal, Romania, Russia, Slovak Republic, Spain, Sweden, Switzerland, Turkey, UK, Ukraine, USA

Spon's European Construction Costs Handbook is the only book of its kind - a unique compilation of cost data on the single most important construction market in the world. This updated edition expands its coverage of countries and once again gives details of the difficult-to-research markets of Eastern Europe as well as Western Europe, North and South.

October 1995: 234x156: c.600pp, 25 line illus
Hardback: 0-419-19650-1: £75.00

For further information and to order please contact: The Marketing Dept., E & F N Spon, 2-6 Boundary Row, London SE1 8HN
Tel: 0171 865 0066 Fax: 0171 522 9623

E & F N Spon

An imprint of Chapman & Hall

PART FIVE

Fees for Professional Services

Quantity Surveyors' Fees, *page 435*

Standards for Thermal Comfort

Indoor air temperature standards for the 21st century

Fergus Nicol, Michael Humphreys, Oliver Sykes and **Susan Roaf**, Oxford Brookes University, UK

✦ Definitve new reference on the setting of standards

Standards for Thermal Comfort is a comprehensive review of indoor air temperature standards, and is the result of the first international seminar on this topic since the 1970s.

Current international standards are inappropriate for many regions of the world and the result of laboratory tests rather than practical, on-site investigations. It brings together expert contributions from around the world recommending new approaches to climatic and cultural variables to ensure standards appropriate for the region.

Contents: List of participants. Preface. Introduction -*Dr Ken C Parsons*. Thermal comfort temperatures and the habits of Hobbits. Towards new indoor comfort temperatures for Pakistani buildings Temperature standards for the tropics? New thermal comfort standard of the Czech Republic. Comfort, preferences or design data? Pale green, simple and user friendly: occupant perceptions of thermal comfort in office buildings. Designing for the individual: a radical reading of ISO 7730. An empirical model for predicting air movement preferred in warm office environments. ISO standards and thermal comfort: recent developments. Comfort and air movement in a naturally ventilated room. Thermal comfort in Thai air-conditioned and naturally ventilated offices. Thermal comfort in air-conditioned buildings in the tropics. Deliberate design. Discussions on human thermal comfort in Vietnam. Thermal comfort and temperature standards in Pakistan. Comfort conditions in PASCOOL surveys. Comfort standards from field surveys in the leisure industry. What is thermal comfort in a naturally ventilated building? The energy implications of a climate-based indoor air temperature standard. An adaptive guideline for UK office temperatures. Design parameters of a non-air-conditioned passive solar house for cold climate of Srinager, India. Higher PMV causes higher energy consumption in air-conditioned buildings: a case study in Jakarta, Indonesia. Thermal comfort of factory workers in Northern India. Warm and sweaty: thermal comfort in two naturally ventilated offices in Sydney, NSW.

April 1995: 234x156: 264pp, 85 line illus
Hardback: 0-419-20420-2:£45.00

For further information and to order please contact: The Marketing Dept.,
E & F N Spon, 2-6 Boundary Row, London SE1 8HN
Tel: 0171 865 0066 Fax: 0171 522 9623

E & F N Spon

An imprint of Chapman & Hall

QUANTITY SURVEYORS' FEES

In recent years there has been an increasing tendency for Bills of Quantities to be prepared for mechanical and electrical services. Where this is done by Chartered Quantity Surveyors the fees will be in accordance with the appropriate scale of fees issued by The Royal Institution of Chartered Surveyors, 12 Great George Street, London SW1P 3AD.

Reproduced below, by kind permission of the R.I.C.S., are the additional fees where the Chartered Quantity Surveyor executes these services in association with building and civil engineering projects. Services and fees referred to in paragraphs outside the range of those reproduced are for the main building project.

SCALE 36. INCLUSIVE PROFESSIONAL CHARGES FOR QUANTITY SURVEYING SERVICES FOR BUILDING WORKS

PRE AND POST CONTRACT SERVICES BASED ON BILLS OF QUANTITIES

2.2 **Air Conditioning, heating, ventilating and electrical services**

(a)   When the service outlined in para. 1.3. are provided by the quantity surveyor for the air conditioning, heating, ventilating and electrical services there shall be a fee for these services in addition to the fee calculated in accordance with para. 2.1. as follows:

Value of Work £			Additional fee £			£
Up to		120,000		5.0%		
120,000	–	240,000	6,000 +	4.7%	on balance over	120,000
240,000	–	480,000	11,640 +	4.0%	on balance over	240,000
480,000	–	750,000	21,240 +	3.6%	on balance over	480,000
750,000	–	1,000,000	30,960 +	3.0%	on balance over	750,000
1,000,000	–	4,000,000	38,460 +	2.7%	on balance over	1,000,000
over		4,000,000	119,460 +	2.4%	on balance over	4,000,000

(b)   The value of such services, whether the subject of separate tenders or not, shall be aggregated and the total value of the work so obtained used for the purpose of calculating the additional fee chargeable in accordance with para. (a). (Except that when, more than one firm of consulting engineers is engaged on the design of these services, the separate values for which each such firm is responsible shall be aggregated and the additional fees charged shall be calculated independently on each such total value so obtained).

(c)   Fees shall be calculated upon the basis of the account for the whole of the air conditioning, heating, ventilating and electrical services for which bills of quantities and final accounts have been prepared by the quantity surveyor.

Fees for Professional Services

QUANTITY SURVEYORS' FEES

SCALE 37. ITEMISED PROFESSIONAL CHARGES FOR QUANTITY SURVEYING SERVICES FOR BUILDING WORK

PRE-CONTRACT SERVICES BASED ON BILLS OF QUANTITIES

2.2   Air Conditioning, heating, ventilating and electrical services

(a)   Where bills of quantities are prepared by the quantity surveyor for the air conditioning, heating, ventilating and electrical services there shall be a fee for these services (which shall include examining tenders received and reported thereon), in addition to the fee calculated in accordance with para. 2.1., as follows:

Value of Work		Additional fee		
	£	£		£
Up to	120,000	2.5%		
120,000 -	240,000	3,000 + 2.25%	on balance over	120,000
240,000 -	480,000	5,700 + 2.00%	on balance over	240,000
480,000 -	750,000	10,500 + 1.75%	on balance over	480,000
750,000 -	1,000,000	15,225 + 1.25%	on balance over	750,000
over	1,000,000	18,350 + 1.15%	on balance over	1,000,000

(b)   The value of such services, whether the subject of separate tenders or not, shall be aggregated and the total value of the work so obtained used for the purpose of calculating the additional fee chargeable in accordance with para. (a).
(Except that when, more than one firm of consulting engineers is engaged on the design of these services, the separate values for which each such firm is responsible shall be aggregated and the additional fees charged shall be calculated independently on each such total value so obtained).

(c)   Fees shall be calculated upon the accepted tender for the whole of the air conditioning, heating, ventilating and electrical services for which bills of quantities have been prepared by the quantity surveyor. In the event of no tender being accepted, fees shall be calculated upon the basis of the lowest original bona fide tender received. In the event of no such tender being received, the fees shall be calculated upon a reasonable valuation of the services based upon the original bills of quantities.

NOTE: In the foregoing context `bona fide tender' shall be deemed to mean a tender submitted in good faith without major errors of computation and not subsequently withdrawn by the tenderer.

(d)   When cost planning services are provided by the quantity surveyor for air conditioning, heating, ventilating and electrical services (or for any part of such services) there shall be an additional fee based on the time involved (see paras. 19.1. and 19.2.) Alternatively the fee may be on a lump sum or percentage basis agreed between the employer and the quantity surveyor.

NOTE: The incorporation of figures for air conditioning, heating, ventilating and electrical services provided by the consulting engineer is deemed to be included in the quantity surveyor's services under para. 2.1.

QUANTITY SURVEYORS' FEES

POST-CONTRACT SERVICES BASED ON BILLS OF QUANTITIES

(Where the Quantity Surveyor also provided the pre-contract services)

5.3   **Air conditioning, heating, ventilating and electrical services**

(a)   Where final accounts are prepared by the quantity surveyor for the air conditioning, heating, ventilating and electrical services there shall be a fee for these services, in addition to the fee calculated in accordance with para. 5.2., as follows:

Value of Work			Additional fee		
£		£			£
Up to	120,000	2.00%			
120,000 -	240,000	2,400 + 1.60%	on balance over	120,000	
240,000 -	1,000,000	4,320 + 1.25%	on balance over	240,000	
1,000,000 -	4,000,000	13,820 + 1.00%	on balance over	1,000,000	
over	4,000,000	43,820 + 0.90%	on balance over	4,000,000	

(b)   The value of such services, whether the subject of separate tenders or not, shall be aggregated and the total value of the work so obtained used for the purpose of calculating the additional fee chargeable in accordance with para. (a).
(Except that when, more than one firm of consulting engineers is engaged on the design of these services, the separate values for which each such firm is responsible shall be aggregated and the additional fees charged shall be calculated independently on each such total value so obtained).

(c)   The scope of the scale of the services to be provided by the quantity surveyor under para. (a) above shall be deemed to be equivalent to those described for the basic scale for post-contract services.

(d)   When the quantity surveyor is required to prepare periodic valuations of materials or goods off site, an additional fee shall be charged based on the time involved (see paras. 19.1 and 19.2).

(e)   The basic scale for post-contract services included for a simple routine of periodically estimating final costs. When the employer specifically requests a cost monitoring service which involves the quantity surveyor in additional or abortive measurement an additional fee shall be based on the time involved (see paras. 19.1 and 19.2), or alternatively on a lump sum or percentage basis agreed between the employer and the quantity surveyor.

(f)   Fees shall be calculated upon the basis of the account for the whole of the air conditioning, heating, ventilating and electrical services for which final accounts have been prepared by the quantity surveyor.

QUANTITY SURVEYORS' FEES

BILLS OF APPROXIMATE QUANTITIES, INTERIM CERTIFICATES AND FINAL ACCOUNTS

7.2   **Air conditioning, heating, ventilating and electrical services**

(a)   Where bills of approximate quantities and final accounts are prepared by the quantity surveyor for the air conditioning, heating, ventilating and electrical services there shall be a fee for these services, in addition to the fee calculated in accordance with para. 7.1., as follows:

Value of Work		Additional fee			
£		£			£
Up to	120,000		4.50%		
120,000 –	240,000	5,400 +	3.85%	on balance over	120,000
240,000 –	480,000	10,020 +	3.25%	on balance over	240,000
480,000 –	750,000	17,820 +	3.00%	on balance over	480,000
750,000 –	1,000,000	25,920 +	2.50%	on balance over	750,000
1,000,000 –	4,000,000	32,170 +	2.15%	on balance over	1,000,000
over	4,000,000	96,670 +	2.05%	on balance over	4,000,000

(b)   The value of such services, whether the subject of separate tenders or not, shall be aggregated and the value of the work so obtained used for the purpose of calculating the additional fee chargeable in accordance with para. (a). (Except that when, more than one firm of consulting engineers is engaged on the design of these services, the separate values for which each such firm is responsible shall be aggregated and the additional fees charged shall be calculated independently on each such total value so obtained).

(c)   The scope of the services to be provided by the quantity surveyor under para. (a) above shall be deemed to be equivalent to those described for the basic scale for pre-contract and post-contract services.

(d)   When the quantity surveyor is required to prepare valuations of materials or goods off site, an additional fee shall be charged based on the time involved (see paras. 19.1 and 19.2).

(e)   The basic scale for post-contract services includes for a simple routine of periodically estimating final costs. When the employer specifically requests a cost monitoring services which involves the quantity surveyor in additional or abortive measurement, an additional fee shall be based on the time involved (see paras. 19.1 and 19.2), or alternatively on a lump sum or percentage basis agreed between the employer and the quantity surveyor.

(f)   Fees shall be calculated upon the basis of the account for the whole of the air conditioning, heating, ventilating and electrical services for which final accounts have been prepared by the quantity surveyor.

(g)   When cost planning services are provided by the quantity surveyor for air conditioning, heating, ventilating and electrical services (or for any part of such services) there shall be an additional fee based on the time involved (see paras. 19.1 and 19.2), or alternatively on a lump sum or percentage basis agreed between the employer and quantity surveyor.

      NOTE: The incorporation of figures for air conditioning, heating, ventilating and electrical services provided by the consulting engineer is deemed to be included in the quantity surveyor's services under para 7.1.

QUANTITY SURVEYORS' FEES

SCALE 38. ITEMISED CHARGES FOR QUANTITY SURVEYING SERVICES FOR CIVIL ENGINEERING WORKS

PRE-CONTRACT SERVICES BASED ON BILLS OF QUANTITIES

2.2   **Mechanical and electrical installations**

(a)   When bills of quantities giving estimated quantities of the work required are prepared by the quantity surveyor for the mechanical and electrical services installed in structures for domestic or environmental purposes common to buildings generally (i.e. purposes other than industrial processes) there shall be a fee for these services (which shall include examining tenders received and reporting thereon) in addition to the fee calculated in accordance with paragraph 2.1 as follows:

Value of Work £		Additional fee £			£
Up to	120,000		1.80%		
120,000 –	240,000	2,160 +	1.55%	on balance over	120,000
240,000 –	480,000	4,020 +	1.30%	on balance over	240,000
480,000 –	750,000	7,140 +	1.20%	on balance over	480,000
750,000 –	1,000,000	10,380 +	1.00%	on balance over	750,000
1,000,000 –	4,000,000	12,880 +	0.85%	on balance over	1,000,000
over	4,000,000	38,380 +	0.82%	on balance over	4,000,000

(b)   When bills of quantities giving estimated quantities of the work required are prepared for machinery and other permanent mechanical and electrical installation or plant is connection with industrial processes other than those included in paragraph 2.1 then a fee in addition to that envisaged under paragraph 2.1 shall be negotiated.

(c)   Fees shall be calculated upon the accepted tender for each individual contract.

POST CONTRACT SERVICES BASED ON BILLS OF QUANTITIES

4.2   **Mechanical and electrical installations**

(a)   When final accounts are prepared by the quantity surveyor for mechanical and electrical services installed in structures for domestic or environmental purposes common to buildings generally (i.e. purpose other than industrial processes) there shall be a fee for these services in addition to the fee calculated in accordance with paragraph 4.1 as follows:

Value of Work £		Category II fee £			£
Up to	120,000		2.70%		
120,000 –	240,000	3,240 +	2.30%	on balance over	120,000
240,000 –	480,000	6,000 +	1.95%	on balance over	240,000
480,000 –	750,000	10,680 +	1.80%	on balance over	480,000
750,000 –	1,000,000	15,540 +	1.50%	on balance over	750,000
1,000,000 –	4,000,000	19,290 +	1.30%	on balance over	1,000,000
over	4,000,000	58,290 +	1.23%	on balance over	4,000,000

(b)   When final accounts are required to be prepared by the quantity surveyor for machinery and other permanent mechanical and electrical installations or plant in connection with industrial processes other than those included in paragraph 4.1 then a fee in addition to that envisaged under paragraph 4.1. shall be negotiated.

(c)   Fees shall be calculated upon the basis of the accounts for each individual contract for which the final account has been prepared by the quantity surveyor.

NEW EDITION

Building Regulations Explained

1995 Revision
5th Edition

John Stephenson

(from the 3rd and 4th editions)
"...more than a guide in the usual sense, it sets the Regulations in context"
- AJ Focus

"This is a must for every Technical Library or builder's bookshelf, and will be of immense value to architects, surveyors and others in interpreting and applying these complex regulations." *- Geoff Hatton, ASI Journal*

✦ Contains important new material

✦ Comprehensive and wide-ranging

Thoroughly revised and updated, the latest edition of **Building Regulations Explained** incorporates the most recent amendments to the Regulations which came into effect during 1995.

It starts with the development of building control and then describes procedural aspects before dealing with all 14 Approved Documents in detail, showing their relevance and method of operation, with the help of numerous tables and diagrams.

An excellent reference source you can't afford to be without!

August 1995: 297x210: 512pp, 627 line illus
Hardback: 0-419-19690-0: £49.50

For further information and to order please contact: The Marketing Dept.,
E & F N Spon, 2-6 Boundary Row, London SE1 8HN
Tel: 0171 865 0066 Fax: 0171 522 9623

E & F N Spon

An imprint of Chapman & Hall

Tables and Memoranda

Spon's Asia Pacific Construction Costs Handbook

Edited by **Davis Langdon & Seah**, Quantity Surveyors, Singapore

- covers fastest growing economies of the world
- DLS are the leading quantity surveyors in region with access to greatest amount of cost data
- UK and USA data included for purposes of comparison
- cost data from: Australia, Brunei, Canada, China, Hong Kong, Indonesia, Japan, Malaysia, New Zealand, Philippines, Singapore, South Korea, Thailand, Vietnam

The economies of the Pacific rim are among the fastest growing of the world. Construction is therefore booming. This unique reference covers construction costs and procedures in the main countries of the area. It will be required reading for construction professionals in the Far East, for those contemplating work or investment there and for multinational companies, development agencies and others needing a world overview.

Contents: Introduction. The South East Asian construction industries. Country sections: Australia, Brunei, Canada, China, Hong Kong, Indonesia, Japan, Malaysia, New Zealand, Philippines, Singapore, South Korea, Thailand, UK, USA. Comparative data. Index.

November 1993: 234x156: 352pp, 17 line illus
Hardback: 0-419-17570-9: £65.00

For further information please contact: The Marketing Dept,. E & F N Spon, 2-6 Boundary Row, London SE1 8HN Tel: 0171 865 0066 Fax: 0171 522 9623

E & F N Spon

An imprint of Chapman & Hall

CONVERSION TABLES

Example : 1mm = 0.039in - 1in = 25.4mm

1 LINEAR

0.039	in	← 1 →	mm	25.4
0.281	ft	← 1 →	metre	0.305
1.094	yd	← 1 →	metre	0.914

2 WEIGHT

0.020	cwt	← 1 →	kg	50.802
0.984	ton	← 1 →	tonne	1.016
2.205	lb	← 1 →	kg	0.454

3 CAPACITY

1.76	pt	← 1 →	litre	0.568
0.220	gal	← 1 →	litre	4.546

4 AREA

0.002	in²	← 1 →	mm²	645.16
10.764	ft²	← 1 →	m²	0.093
1.196	yd²	← 1 →	m²	0.836
2.471	acre	← 1 →	ha	0.405
0.386	mile²	← 1 →	km²	2.59

5 VOLUME

0.061	in³	← 1 →	cm³	16.387
35.315	ft³	← 1 →	m³	0.028
1.308	yd³	← 1 →	m³	0.765

6 POWER

1.310	HP	← 1 →	kW	0.746

CONVERSION TABLES

LENGTH   Conversion Factors

Centimetre	(cm)	1 in	=	2.54	cm	:	1 cm	=	0.3937	in
Metre	(m)	1 ft	=	0.3048	m	:	1 m	=	3.2808	ft
Kilometre	(km)	1 yd	=	0.9144	m	:	1 m	=	1.0936	yd
Kilometre	(km)	1 mile	=	1.6093	km	:	1 km	=	0.6214	mile

NOTE :	1 cm	=	10	mm	1 ft	=	12	in
	1 m	=	100	cm	1 yd	=	3	ft
	1 km	=	1000	m	1 mile	=	1760	yd

AREA

Square Millimetre	(mm²)	1 in²	=	645.2	mm²	:	1 mm²	=	0.0016	in²
Square Centimetre	(cm²)	1 in²	=	6.4516	cm²	:	1 cm²	=	0.1550	in²
Square Metre	(m²)	1 ft²	=	0.0929	m²	:	1 m²	=	10.764	ft²
Square Metre	(m²)	1 yd²	=	0.8361	m²	:	1 m²	=	1.1960	yd²
Square Kilometre	(km²)	1 mile²	=	2.590	km²	:	1 km²	=	0.3861	mile²

NOTE :	1 cm²	=	100	mm²	1 ft²	=	144	in²
	1 m²	=	10000	cm²	1 yd²	=	9	ft²
	1 km²	=	100	ha	1 mile²	=	640	acre
					1 acre	=	4840	yd²

VOLUME

Cubic Centimetre	(cm³)	1 cm³	=	0.0610	in³	:	1 in³	=	16.387	cm³
Cubic Decimetre	(dm³)	1 dm³	=	0.0353	ft³	:	1 ft³	=	28.329	dm³
Cubic Metre	(m³)	1 m³	=	35.315	ft³	:	1 ft³	=	0.0283	m³
Cubic Metre	(m³)	1 m³	=	1.3080	yd³	:	1 yd³	=	0.7646	m³
Litre	(L)	1 L	=	1.76	pint	:	1 pint	=	0.5683	L
Litre	(L)	1 L	=	2.113	US pt	:	1 pint	=	0.4733	US L

NOTE :	1 dm³	=	1000	cm³	1 ft³	=	1728	in³
	1 m³	=	1000	dm³	1 yd³	=	27	ft³
	1 L	=	1	dm³	1 pint	=	20	fl oz
	1 HL	=	100	L	1 gal	=	8	pints

MASS

Milligram	(mg)	1 mg	=	0.0154	grain	:	1 grain	=	64.935	mg
Gram	(g)	1 g	=	0.0353	oz	:	1 oz	=	28.35	g
Kilogram	(kg)	1 kg	=	2.2046	lb	:	1 lb	=	0.4536	kg
Tonne	(t)	1 t	=	0.9842	ton	:	1 ton	=	1.016	t

NOTE :	1 g	=	1000	mg	1 oz	=	437.5	grains
	1 kg	=	1000	g	1 lb	=	16	oz
	1 t	=	1000	kg	1 stone	=	14	lb
					1 cwt	=	112	lb
					1 ton	=	20	cwt

CONVERSION TABLES

FORCE   Conversion Factors

Newton	(N)	1 lb f	=	4.448	N	: 1 kg f	=	9.807	N
Kilonewton	(kN)	1 lb f	=	0.004448	kN	: 1 ton f	=	9.964	kN
Meganewton	(mN)	100 ton f =		0.9964	mN				

PRESSURE AND STRESS

Kilonewton	1 lb f/in²	=	6.895	kN/m²	:		
per square metre (kN/m²)	1 bar	=	100	kN/m²	:		
Meganewton	1 ton f/ft²	=	107.3	kN/m²		=	0.1073 mN/m²
per square metre (mN/m²)	1 kg f/cm² =		98.07	kN/m²	:		
	1 lb f/ft²	=	0.0479	kN/m²	:		

TEMPERATURE

Degree Celcius (°C)   °C   =   5 ÷ 9 (°F - 32°)   °F   =   9 ÷ 5 (°C + 32°)

FORMULAE

Two dimensional figures

Figure	Area
Triangle	0.5 x base x height, or N(s(s - a)(s - b)(s - c)) where s = 0.5 x the sum of the three sides and a,b and c are the lengths of the three sides, or $a^2 = b^2 + c^2 - 2 \times bc \times COS\ A$ where A is the angle opposite side a
Hexagon	2.6 x (side)^2
Octagon	4.83 x (side)^2
Trapezoid	height x 0.5 (base + top)
Circle	3.142 x radius^2 or 0.7854 x diameter^2 (circumference = 2 x 3.142 x radius or 3.142 x diameter)
Sector of a circle	0.5 x length of arc x radius
Segment of a circle	area of sector - area of triangle
Ellipse of a circle	3.142 x AB (where A = 0.5 x height and B = 0.5 x length)
Spandrel	3/14 x radius^2

Three dimensional figure

Figure	VolumeSurface	Area
Prism x height	Area of base x height	circumference of base
Cube	(length of side) cubed	6 x (length of side)^2
Cylinder	3.142 x radius^2 x height	2 x 3.142 x radius x (height - radius)
Sphere	4/3 x 3.142 x radius^3	4 x 3.142 x radius^2
Segment of a sphere	[(3.142 x h)/6] x (3 x r^2 + h^2)	(2 x 3.142 x r x h)
Pyramid	1/3 x (area of base x height)	0.5 x circumference of base x slant height

FRACTIONS, DECIMALS AND MILLIMETRE EQUIVALENTS

Fractions	Decimals	mm	Fractions	Decimals	mm
1/64	0.015625	0.396875	33/64	0.515625	13.096875
1/32	0.03125	0.79375	17/32	0.53125	13.49375
3/64	0.046875	1.190625	35/64	0.546875	13.890625
1/16	0.0625	1.5875	9/16	0.5625	14.2875
5/64	0.078125	1.984375	37/64	0.578125	14.684375
3/32	0.09375	2.38125	19/32	0.59375	15.08125
7/64	0.109375	2.778125	39/64	0.609375	15.478125
1/8	0.125	3.175	5/8	0.625	15.875
9/64	0.140625	3.571875	41/64	0.640625	16.271875
5/32	0.15625	3.96875	21/32	0.65625	16.66875
11/64	0.171875	4.365625	43/64	0.671875	17.065625
3/16	0.1875	4.7625	11/16	0.6875	17.4625
13/64	0.203125	5.159375	45/64	0.703125	17.859375
7/32	0.21875	5.55625	23/32	0.71875	18.25625
15/64	0.234375	5.953125	47/64	0.734375	18.653125
1/4	0.25	6.35	3/4	0.75	19.05
17/64	0.265625	6.746875	49/64	0.765625	19.446875
9/32	0.28125	7.14375	25/32	0.78125	19.84375
19/64	0.296875	7.540625	51/64	0.796875	20.240625
5/16	0.3125	7.9375	13/16	0.8125	20.6375
21/64	0.328125	8.334375	53/64	0.828125	21.034375
11/32	0.34375	8.73125	27/32	0.84375	21.43125
23/64	0.359375	9.128125	55/64	0.859375	21.828125
3/8	0.375	9.525	7/8	0.875	22.225
25/64	0.390625	9.921875	57/64	0.890625	22.621875
13/32	0.40625	10.31875	29/32	0.90625	23.01875
27/64	0.421875	10.71563	59/64	0.921875	23.415625
7/16	0.4375	11.1125	15/16	0.9375	23.8125
29/64	0.453125	11.50938	61/64	0.953125	24.209375
15/32	0.46875	11.90625	31/32	0.96875	24.60625
31/64	0.484375	12.30313	63/64	0.984375	25.003125
1/2	0.5	12.7	1.0	1	25.4

Tables and Memoranda

IMPERIAL STANDARD WIRE GAUGE (SWG)

SWG No	Diameter in	SWG mm	No	Diameter in	mm
7/0	0.5	12.7	23	0.024	0.61
6/0	0.464	11.79	24	0.022	0.559
5/0	0.432	10.97	25	0.02	0.508
4/0	0.4	10.16	26	0.018	0.457
3/0	0.372	9.45	27	0.0164	0.417
2/0	0.348	8.84	28	0.0148	0.376
1/0	0.324	8.23	29	0.0136	0.345
1	0.3	7.62	30	0.0124	0.315
2	0.276	7.01	31	0.0116	0.295
3	0.252	6.4	32	0.0108	0.274
4	0.232	5.89	33	0.01	0.254
5	0.212	5.38	34	0.009	0.234
6	0.192	4.88	35	0.008	0.213
7	0.176	4.47	36	0.008	0.193
8	0.16	4.06	37	0.007	0.173
9	0.144	3.66	38	0.006	0.152
10	0.128	3.25	39	0.005	0.132
11	0.116	2.95	40	0.005	0.122
12	0.104	2.64	41	0.004	0.112
13	0.092	2.34	42	0.004	0.102
14	0.08	2.03	43	0.004	0.091
15	0.072	1.83	44	0.003	0.081
16	0.064	1.63	45	0.003	0.071
17	0.056	1.42	46	0.002	0.061
18	0.048	1.22	47	0.002	0.051
19	0.04	1.016	48	0.002	0.041
20	0.036	0.914	49	0.001	0.031
21	0.032	0.813	50	0.001	0.025
22	0.028	0.711			

WATER PRESSURE DUE TO HEIGHT

Imperial

Head Feet	Pressure lb/in²	Head Feet	Pressure lb/in²
1	0.43	70	30.35
5	2.17	75	32.51
10	4.34	80	34.68
15	6.5	85	36.85
20	8.67	90	39.02
25	10.84	95	41.18
30	13.01	100	43.35
35	15.17	105	45.52
40	17.34	110	47.69
45	19.51	120	52.02
50	21.68	130	56.36
55	23.84	140	60.69
60	26.01	150	65.03
65	28.18		

Metric

Head m	Pressure bar	Head m	Pressure bar
0.5	0.049	18.0	1.766
1.0	0.098	19.0	1.864
1.5	0.147	20.0	1.962
2.0	0.196	21.0	2.06
3.0	0.294	22.0	2.158
4.0	0.392	23.0	2.256
5.0	0.491	24.0	2.354
6.0	0.589	25.0	2.453
7.0	0.687	26.0	2.551
8.0	0.785	27.0	2.649
9.0	0.883	28.0	2.747
10.0	0.981	29.0	2.845
11.0	1.079	30.0	2.943
12.0	1.177	32.5	3.188
13.0	1.275	35.0	3.434
14.0	1.373	37.5	3.679
15.0	1.472	40.0	3.924
16.0	1.57	42.5	4.169
17.0	1.668	45.0	4.415

1 bar	=	14.5038 lbf/in²
1 metre	=	3.2808 ft or 39.3701 in
1 foot	=	0.3048 metres
1 lbf/in²	=	0.06895 bar
1 in wg	=	2.5 mbar (249.1 N/m²)

TABLE OF WEIGHTS FOR STEELWORK

Mild Steel Bar

Diameter (mm)	Weight (kg/m)	Diameter (mm)	Weight (kg/m)
6	0.22	20	2.47
10	0.62	25	3.85
12	0.89	30	5.55
16	1.58	32	6.31

Mild Steel Flat

Size (mm)	Weight (kg/m)	Size (mm)	Weight (kg/m)
15 x 3	0.36	15 x 5	0.59
20 x 3	0.47	20 x 5	0.79
25 x 3	0.59	25 x 5	0.98
30 x 3	0.71	30 x 5	1.18
40 x 3	0.94	40 x 5	1.57
45 x 3	1.06	45 x 5	1.77
50 x 3	1.18	50 x 5	1.96
20 x 6	0.94	20 x 8	1.26
25 x 6	1.18	25 x 8	1.57
30 x 6	1.41	30 x 8	1.88
40 x 6	1.88	40 x 8	2.51
45 x 6	2.12	45 x 8	2.83
50 x 6	2.36	50 x 8	3.14
55 x 6	2.60	55 x 8	3.45
60 x 6	2.83	60 x 8	3.77
65 x 6	3.06	65 x 8	4.08
70 x 6	3.30	70 x 8	4.40
75 x 6	3.53	75 x 8	4.71
100 x 6	4.71	100 x 8	6.28
20 x 10	1.57	20 x 12	1.88
25 x 10	1.96	25 x 12	2.36
30 x 10	2.36	30 x 12	2.83
40 x 10	3.14	40 x 12	3.77
45 x 10	3.53	45 x 12	4.24
50 x 10	3.93	50 x 12	4.71
55 x 10	4.32	55 x 12	5.12
60 x 10	4.71	60 x 12	5.65
65 x 10	5.10	65 x 12	6.12
70 x 10	5.50	70 x 12	6.59
75 x 10	5.89	75 x 12	7.07
100 x 10	7.85	100 x 12	9.42

TABLE OF WEIGHTS FOR STEELWORK

Mild Steel Equal Angle

Size (mm)	Weight (kg/m)	Size (mm)	Weight (kg/m)
13 x 13 x 3	0.56	60 x 60 x 10	8.69
20 x 20 x 3	0.88	70 x 70 x 10	10.30
25 x 25 x 3	1.11	75 x 75 x 10	11.05
30 x 30 x 3	1.36	80 x 80 x 10	11.90
40 x 40 x 3	1.82	90 x 90 x 10	13.40
45 x 45 x 3	2.06	100 x 100 x 10	15.00
50 x 50 x 3	2.30	120 x 120 x 10	18.20
		150 x 156 x 10	23.00
30 x 30 x 6	2.56		
40 x 40 x 6	3.52	75 x 75 x 12	13.07
45 x 45 x 6	4.00	80 x 80 x 12	14.00
50 x 50 x 6	4.47	90 x 90 x 12	15.90
60 x 60 x 6	5.42	100 x 120 x 12	21.90
70 x 70 x 6	6.38	120 x 120 x 12	21.60
75 x 75 x 6	6.82	150 x 150 x 12	27.30
80 x 80 x 6	7.34	200 x 200 x 12	36.74
90 x 90 x 6	8.30		
40 x 40 x 8	4.55		
50 x 50 x 8	5.82		
60 x 60 x 8	7.09		
70 x 70 x 8	8.36		
75 x 75 x 8	8.96		
80 x 80 x 8	9.63		
90 x 90 x 8	10.90		
100 x 100 x 8	12.20		
120 x 120 x 8	14.70		

Mild Steel Unequal Angle

Size (mm)	Weight (kg/m)	Size (mm)	Weight (kg/m)
40 x 25 x 6	2.79	100 x 65 x 10	12.30
50 x 40 x 6	4.24	100 x 75 x 10	13.00
60 x 30 x 6	3.99	125 x 75 x 10	15.00
65 x 50 x 6	5.16	150 x 75 x 10	17.00
75 x 50 x 6	5.65	150 x 90 x 10	18.20
80 x 60 x 6	6.37	200 x 100 x 10	23.00
125 x 75 x 6	9.18		
75 x 50 x 8	7.39	100 x 75 x 12	15.40
80 x 60 x 8	8.34	125 x 75 x 12	17.80
100 x 65 x 8	9.94	150 x 75 x 12	20.20
100 x 75 x 8	10.60	150 x 90 x 12	21.60
125 x 75 x 8	12.20	200 x 100 x 12	27.30
137 x 102 x 8	14.88	200 x 150 x 12	32.00

Tables and Memoranda

TABLE OF WEIGHTS FOR STEELWORK

Rolled Steel Channels

Size (mm)	Weight (kg/m)	Size (mm)	Weight (kg/m)
32 x 27	2.80	178 x 76	20.84
38 x 19	2.49	178 x 79	26.81
51 x 25	4.46	203 x 76	23.82
51 x 38	5.80	203 x 89	29.78
64 x 25	6.70	229 x 76	26.06
76 x 38	7.46	229 x 89	32.76
76 x 51	9.45	254 x 76	28.29
102 x 51	10.42	254 x 89	35.74
127 x 64	14.90	305 x 89	41.67
152 x 76	17.88	305 x 102	46.18
152 x 89	23.84	381 x 102	55.10

Rolled Steel Joists

Size (mm)	Weight (kg/m)	Size (mm)	Weight (kg/m)
76 x 38	6.25	152 x 76	17.86
76 x 76	12.65	152 x 89	17.09
102 x 44	7.44	152 x 127	37.20
102 x 64	9.65	178 x 102	21.54
102 x 102	23.06	203 x 102	25.33
127 x 76	13.36	203 x 152	52.03
127 x 114	26.78	254 x 114	37.20
127 x 114	29.76	254 x 203	81.84
		305 x 203	96.72

Universal Columns

Size (mm)	Weight (kg/m)	Size (mm)	Weight (kg/m)
152 x 152	23.00	254 x 254	89.00
152 x 152	30.00	254 x 254	107.00
152 x 152	37.00	254 x 254	132.00
203 x 203	46.00	254 x 254	167.00
203 x 203	52.00	305 x 305	97.00
203 x 203	60.00	305 x 305	118.00
203 x 203	71.00	305 x 305	137.00
203 x 203	86.00	305 x 305	158.00
254 x 254	73.00	305 x 305	198.00

TABLE OF WEIGHTS FOR STEELWORK

Universal Beams

Size (mm)	Weight (kg/m)	Size (mm)	Weight (kg/m)
203 x 133	25.00	305 x 127	48.00
203 x 133	30.00	305 x 165	40.00
254 x 102	22.00	305 x 165	46.00
254 x 102	25.00	305 x 165	54.00
254 x 102	28.00	356 x 127	33.00
254 x 146	31.00	356 x 127	39.00
254 x 146	37.00	356 x 171	45.00
254 x 146	43.00	356 x 171	51.00
305 x 102	25.00	356 x 171	57.00
305 x 102	28.00	356 x 171	67.00
305 x 102	33.00	381 x 152	52.00
305 x 127	37.00	381 x 152	60.00
305 x 127	42.00	381 x 152	67.00

Circular Hollow Sections

Size (mm)	Weight (kg/m)	Size (mm)	Weight (kg/m)
21.3 x 3.2	1.43	76.1 x 3.2	5.75
26.9 x 3.2	1.87	76.1 x 4.0	7.11
33.7 x 2.6	1.99	76.1 x 5.0	8.77
33.7 x 3.2	2.41	88.9 x 3.2	6.76
33.7 x 4.0	2.93	88.9 x 4.0	8.36
42.4 x 2.6	2.55	88.9 x 5.0	10.30
42.4 x 3.2	3.09	114.3 x 3.6	9.83
42.4 x 4.0	3.79	114.3 x 5.0	13.50
48.3 x 3.2	3.56	114.3 x 6.3	16.80
48.3 x 4.0	4.37	139.7 x 5.0	16.60
48.3 x 5.0	5.34	139.7 x 6.3	20.70
60.3 x 3.2	4.51	139.7 x 8.0	26.00
60.3 x 4.0	5.55	139.7 x 10.0	32.00
60.3 x 5.0	6.82	168.3 x 5.0	20.10

TABLE OF WEIGHTS FOR STEELWORK

Square Hollow Sections

Size (mm)	Weight (kg/m)	Size (mm)	Weight (kg/m)
20 x 20 x 2.0	1.12	90 x 90 x 3.6	9.72
20 x 20 x 2.6	1.39	90 x 90 x 5.0	13.30
30 x 30 x 2.6	2.21	90 x 90 x 6.3	16.40
30 x 30 x 3.2	2.65	100 x 100 x 4.0	12.00
40 x 40 x 2.6	3.03	100 x 100 x 5.0	14.80
40 x 40 x 3.2	3.66	100 x 100 x 6.3	18.40
40 x 40 x 4.0	4.46	100 x 100 x 8.0	22.90
50 x 50 x 3.2	4.66	100 x 100 x 10.0	27.90
50 x 50 x 4.0	5.72	120 x 120 x 5.0	18.00
50 x 50 x 5.0	6.97	120 x 120 x 6.3	22.30
60 x 60 x 3.2	5.67	120 x 120 x 8.0	27.90
60 x 60 x 4.0	6.97	120 x 120 x 10.0	34.20
60 x 60 x 5.0	8.54	150 x 150 x 5.0	22.70
70 x 70 x 3.2	7.46	150 x 150 x 6.3	28.30
70 x 70 x 5.0	10.10	150 x 150 x 8.0	35.40
80 x 80 x 3.6	8.59	150 x 150 x 10.0	43.60
80 x 80 x 5.0	11.70		
80 x 80 x 6.3	14.40		

Rectangular Hollow Sections

Size (mm)	Weight (kg/m)	Size (mm)	Weight (kg/m)
50 x 30 x 2.6	3.03	120 x 80 x 5.0	14.80
50 x 30 x 3.2	3.66	120 x 80 x 6.3	18.40
60 x 40 x 3.2	4.66	120 x 80 x 8.0	22.90
60 x 40 x 4.0	5.72	120 x 80 x 10.0	27.90
80 x 40 x 3.2	5.67	150 x 100 x 5.0	18.70
80 x 40 x 4.0	6.97	150 x 100 x 6.3	23.30
90 x 50 x 3.6	7.46	150 x 100 x 8.0	29.10
90 x 50 x 5.0	10.10	150 x 100 x 10.0	35.70
100 x 50 x 3.2	7.18	160 x 80 x 5.0	18.00
100 x 50 x 4.0	8.86	160 x 80 x 6.3	22.30
100 x 50 x 5.0	10.90	160 x 80 x 8.0	27.90
100 x 60 x 3.6	8.59	160 x 80 x 10.0	34.20
100 x 60 x 5.0	11.70	200 x 100 x 5.0	22.70
100 x 60 x 6.3	14.40	200 x 100 x 6.3	28.30
120 x 60 x 3.6	9.72	200 x 100 x 8.0	35.40
120 x 60 x 5.0	13.30	200 x 100 x 10.0	43.60
120 x 60 x 6.3	16.40		

DIMENSIONS AND WEIGHTS OF COPPER PIPES TO BS 2871 PART 1

Outside Diameter (mm)	Internal Diameter (mm)	Weight per Metre (kg)	Internal Diameter (mm)	Weight per Metre (kg)	Internal Diameter (mm)	Weight per Metre (kg)
	Table X		Table Y		Table Z	
6	4.80	0.0911	4.40	0.1170	5.00	0.0774
8	6.80	0.1246	6.40	0.1617	7.00	0.1054
10	8.80	0.1580	8.40	0.2064	9.00	0.1334
12	10.80	0.1914	10.40	0.2511	11.00	0.1612
15	13.60	0.2796	13.00	0.3923	14.00	0.2031
18	16.40	0.3852	16.00	0.4760	16.80	0.2918
22	20.22	0.5308	19.62	0.6974	20.82	0.3589
28	26.22	0.6814	25.62	0.8985	26.82	0.4594
35	32.63	1.1334	32.03	1.4085	33.63	0.6701
42	39.63	1.3675	39.03	1.6996	40.43	0.9216
54	51.63	1.7691	50.03	2.9052	52.23	1.3343
76.1	73.22	3.1287	72.22	4.1437	73.82	2.5131
108	105.12	4.4666	103.12	7.3745	105.72	3.5834
133	130.38	5.5151	--	--	130.38	5.5151
159	155.38	8.7795	--	--	156.38	6.6056

DIMENSIONS OF STAINLESS STEEL PIPES TO BS 4127

Outside Diameter (mm)	Maximun Outside Diameter (mm)	Minimum Outside Diameter (mm)	Wall Thickness (mm)	Working Pressure (bar)
6	6.045	5.94	0.6	330
8	8.045	7.94	0.6	260
10	10.045	9.94	0.6	210
12	12.045	11.94	0.6	170
15	15.045	14.94	0.6	140
18	18.045	17.94	0.7	135
22	22.055	21.95	0.7	110
28	28.055	27.95	0.8	121
35	35.070	34.96	1.0	100
42	42.070	41.96	1.1	91
54	54.090	53.94	1.2	77

Tables and Memoranda

DIMENSIONS OF STEEL PIPES TO BS 1387

Nominal Size	Approx. Outside Diameter	Outside Diameter				Thickness		
		Light		Medium and Heavy		Light	Medium	Heavy
		Maximum	Minimum	Maximum	Minimum			
mm	mm	mm	mm	mm	mm	mm	mm	mm
6	10.20	10.10	9.70	10.40	9.80	1.80	2.00	2.65
8	13.50	13.60	13.20	13.90	13.30	1.80	2.35	2.90
10	17.20	17.10	16.70	17.40	16.80	1.80	2.35	2.90
15	21.30	21.40	21.00	21.70	21.10	2.00	2.65	3.25
20	26.90	26.90	26.40	27.20	26.60	2.35	2.65	3.25
25	33.70	33.80	33.20	34.20	33.40	2.65	3.25	4.05
32	42.40	42.50	41.90	42.90	42.10	2.65	3.25	4.05
40	48.30	48.40	47.80	48.80	48.00	2.90	3.25	4.05
50	60.30	60.20	59.60	60.80	59.80	2.90	3.65	4.50
65	76.10	76.00	75.20	76.60	75.40	3.25	3.65	4.50
80	88.90	88.70	87.90	89.50	88.10	3.25	4.05	4.85
100	114.30	113.90	113.00	114.90	113.30	3.65	4.50	5.40
125	139.70	--	--	140.60	138.70	--	4.85	5.40
150	165.1*	--	--	166.10	164.10	--	4.85	5.40

* 165.1mm (6.5in) outside diameter is not generally recommended except where screwing to BS 21 is necessary.

All dimensions are in accordance with ISO R65 except approximate outside diameters which are in accordance with ISO R64.

Light quality is equivalent to ISO R65 Light Series II.

APPROXIMATE METRES PER TONNE OF TUBES TO BS 1387

Nom. Size mm	BLACK						GALVANISED					
	Plain/screwed ends			Screwed & socketed			Plain/screwed ends			Screwed & socketed		
	Light m	Medium m	Heavy m	Light m	Medium m	Heavy m	Light m	Medium m	Heavy m	Light m	Medium m	Heavy m
6	2765	2461	2030	2743	2443	2018	2604	2333	1948	2584	2317	1937
8	1936	1538	1300	1920	1527	1292	1826	1467	1254	1811	1458	1247
10	1483	1173	979	1471	1165	974	1400	1120	944	1386	1113	939
15	1050	817	688	1040	811	684	996	785	665	987	779	661
20	712	634	529	704	628	525	679	609	512	673	603	508
25	498	410	336	494	407	334	478	396	327	474	394	325
32	388	319	260	384	316	259	373	308	254	369	305	252
40	307	277	226	303	273	223	296	268	220	292	264	217
50	244	196	162	239	194	160	235	191	158	231	188	157
65	172	153	127	169	151	125	167	149	124	163	146	122
80	147	118	99	143	116	98	142	115	97	139	113	96
100	101	82	69	98	81	68	98	81	68	95	79	67
125	--	62	56	--	60	55	--	60	55	--	59	54
150	--	52	47	--	50	46	--	51	46	--	49	45

The figures for `plain or screwed ends' apply also to tubes to BS 1775 HFW of equivalent size and thickness.

    Tables and Memoranda

FLANGE DIMENSION CHART TO BS 4504 & BS 10

Normal Pressure Rating (N.P.6) 6 Bar

Nom. Size	Flange Outside Diam.	Table 6/2 Forged Welding Neck	Table 6/3 Plate Slip	Table 6/4 Forged Bossed Screwed	Table 6/5 Forged Bossed Slip on	Table 6/8 Plate Blank	Raised Face Diam.	Raised Face T'ness	Pitch Circle Diam.	Nr. Bolt Holes	Size of Bolt
15	80	12	12	12	12	12	40	2	55	4	M10 x 40
20	90	14	14	14	14	14	50	2	65	4	M10 x 45
25	100	14	14	14	14	14	60	2	75	4	M10 x 45
32	120	14	16	14	14	14	70	2	90	4	M12 x 45
40	130	14	16	14	14	14	80	3	100	4	M12 x 45
50	140	14	16	14	14	14	90	3	110	4	M12 x 45
65	160	14	16	14	14	14	110	3	130	4	M12 x 45
80	190	16	18	16	16	16	128	3	150	4	M16 x 55
100	210	16	18	16	16	16	148	3	170	4	M16 x 55
125	240	18	20	18	18	18	178	3	200	8	M16 x 60
150	265	18	20	18	18	18	202	3	225	8	M16 x 60
200	320	20	22	--	20	20	258	3	280	8	M16 x 60
250	375	22	24	--	22	22	312	3	335	12	M16 x 65
300	440	22	24	--	22	22	365	4	395	12	M20 x 70

Normal Pressure Rating (N.P.16) 16 Bar

Nom. Size	Flange Outside Diam.	Table 6/2 Forged Welding Neck	Table 6/3 Plate Slip	Table 6/4 Forged Bossed Screwed	Table 6/5 Forged Bossed Slip on	Table 6/8 Plate Blank	Raised Face Diam.	Raised Face T'ness	Pitch Circle Diam.	Nr. Bolt Holes	Size of Bolt
15	95	14	14	14	14	14	45	2	55	4	M12 x 45
20	105	16	16	16	16	16	58	2	65	4	M12 x 50
25	115	16	16	16	16	16	68	2	75	4	M12 x 50
32	140	16	16	16	16	16	78	2	90	4	M16 x 55
40	150	16	16	16	16	16	88	3	100	4	M16 x 55
50	165	18	18	18	18	18	102	3	110	4	M16 x 60
65	185	18	18	18	18	18	122	3	130	4	M16 x 60
80	200	20	20	20	20	20	138	3	150	8	M16 x 60
100	220	20	20	20	20	20	158	3	170	8	M16 x 65
125	250	22	22	22	22	22	188	3	200	8	M16 x 70
150	285	22	22	22	22	22	212	3	225	8	M20 x 70
200	340	24	24	--	24	24	268	3	280	12	M20 x 75
250	405	26	26	--	26	26	320	3	335	12	M24 x 90
300	460	28	28	--	28	28	378	4	395	12	M24 x 90

MINIMUM DISTANCES BETWEEN SUPPORTS/FIXINGS

Material	BS Nominal Pipe Size		Pipes - Vertical	Pipes - Horizontal on to low gradients
	inch	mm	support distance in metres	support distance in metres
Copper	0.50	15.00	1.90	1.30
	0.75	22.00	2.50	1.90
	1.00	28.00	2.50	1.90
	1.25	35.00	2.80	2.50
	1.50	42.00	2.80	2.50
	2.00	54.00	3.90	2.50
	2.50	67.00	3.90	2.80
	3.00	76.10	3.90	2.80
	4.00	108.00	3.90	2.80
	5.00	133.00	3.90	2.80
	6.00	159.00	3.90	2.80
muPVC	1.25	32.00	1.20	0.50
	1.50	40.00	1.20	0.50
	2.00	50.00	1.20	0.60
Polypropylene	1.25	32.00	1.20	0.50
	1.50	40.00	1.20	0.50
uPVC	--	82.40	1.20	0.50
	--	110.00	1.80	0.90
	--	160.00	1.80	1.20

LITRES OF WATER STORAGE REQUIRED PER PERSON PER BUILDING TYPE

Type of Building	Storage per person litres
Houses and flats	90
Hostels	90
Hotels	135
Nurses home and medical quarters	115
Offices with canteen	45
Offices without canteen	35
Restaurants, per meal served	7
Boarding schools	90
Day schools	30

COLD WATER PLUMBING - INSULATION THICKNESS REQUIRED AGAINST FROST

Bore of Tube		Pipework within Buildings - Declared Thermal Conductivity (W/m degrees C)		
inch	mm	Up to 0.040	0.041 to 0.055	0.056 to 0.070
0.50	15	32	50	75
0.75	20	32	50	75
1.00	25	32	50	75
1.25	32	32	50	75
1.50	40	32	50	75
2.00	50	25	32	50
2.50	65	25	32	50
3.00	80	25	32	50
4.00	100	19	25	38

CAPACITY AND DIMENSIONS OF GALVANISED MILD STEEL CISTERNS - BS 417

Capacity (litres)	BS type (SCM)	Dimensions		
		Length (mm)	Width (mm)	Depth (mm)
18	45	457	305	305
36	70	610	305	371
54	90	610	406	371
68	110	610	432	432
86	135	610	457	482
114	180	686	508	508
159	230	736	559	559
191	270	762	584	610
227	320	914	610	584
264	360	914	660	610
327	450/1	1220	610	610
336	450/2	965	686	686
423	570	965	762	787
491	680	1090	864	736
709	910	1070	889	889

CAPACITY OF COLD WATER POLYPROPYLENE STORAGE CISTERNS - BS 4213

Capacity (litres)	BS type (PC)	Maximum Height mm
18	4	310
36	8	380
68	15	430
91	20	510
114	25	530
182	40	610
227	50	660
273	60	660
318	70	660
455	100	760

STORAGE CAPACITY AND RECOMMENDED POWER OF HOT WATER STORAGE BOILERS

Type of Building		Storage at 65°C (litres / person)	Boiler power to 65°C (kW / person)
Flats and Dwellings			
(a)	Low Rent Properties	25	0.5
(b)	Medium Rent Properties	30	0.7
(c)	High Rent Properties	45	1.2
Nurses Homes		45	0.9
Hostels		0.7	0.7
Hotels			
(a)	Top Quality - Up Market	1.2	1.2
(b)	Average Quality - Low Market	0.9	0.9
Colleges and Schools			
(a)	Live-in Accomodation	0.7	0.7
(b)	Public Comprehensive	0.1	0.1
Factories		0.1	0.1
Hospitals			
(a)	General	1.5	1.5
(b)	Infectious	1.5	1.5
(c)	Infirmaries	0.6	0.6
(d)	Infirmaries with Laundary	0.9	0.9
(e)	Maternity	2.1	2.1
(f)	Mental	0.7	0.7
Offices		0.1	0.1
Sports Pavilions		0.3	0.3

THICKNESS OF THERMAL INSULATION FOR HEATING INSTALLATIONS

Size of Tube mm	Declared Thermal Conductivity			
	Up to 0.025	0.026 to 0.040	0.041 to 0.055	0.056 to 0.070
LTHW Systems	Minimum thickness of insulation			
15	25	25	38	38
20	25	32	38	38
25	25	38	38	38
32	32	38	38	50
40	32	38	38	50
50	38	38	50	50
65	38	50	50	50
80	38	50	50	50
100	38	50	50	63
125	38	50	50	63
150	50	50	63	63
200	50	50	63	63
250	50	63	63	63
300	50	63	63	63
Flat Surfaces	50	63	63	63

MTHW Systems and Condensate	Minimum thickness of insulation			
15	25	38	38	38
20	32	38	38	50
25	38	38	38	50
32	38	50	50	50
40	38	50	50	50
50	38	50	50	50
65	38	50	50	50
80	50	50	50	63
100	50	63	63	63
125	50	63	63	63
150	50	63	63	63
200	50	63	63	63
250	50	63	63	75
300	63	63	63	75
Flat Surfaces	63	63	63	75

THICKNESS OF THERMAL INSULATION FOR HEATING INSTALLATIONS

Size of Tube mm	Declared Thermal Conductivity			
	Up to 0.025	0.026 to 0.040	0.041 to 0.055	0.056 to 0.070
HTHW Systems and Steam	Minimum thickness of insulation			
15	38	50	50	50
20	38	50	50	50
25	38	50	50	50
32	50	50	50	63
40	50	50	50	63
50	50	50	75	75
65	50	63	75	75
80	50	63	75	75
100	63	63	75	100
125	63	63	100	100
150	63	63	100	100
200	63	63	100	100
250	63	75	100	100
300	63	75	100	100
Flat Surfaces	63	75	100	100

CAPACITIES AND DIMENSIONS OF GALVANISED MILD STEEL INDIRECT CYLINDERS TO BS 1565

Approximate Capacity (litres)	BS Size No.	Internal Diameter (mm)	External height over dome (mm)
109	BSG.1M	457	762
136	BSG.2M	457	914
159	BSG.3M	457	1067
227	BSG.4M	508	1270
273	BSG.5M	508	1473
364	BSG.6M	610	1372
455	BSG.7M	610	1753
123	BSG.8M	457	838

CAPACITIES AND DIMENSIONS OF COPPER INDIRECT CYLINDERS (COIL TYPE) TO BS 1566

Approximate Capacity (litres)	BS Type	External Diameter (mm)	External height over dome (mm)
96	0	300	1600
72	1	350	900
96	2	400	900
114	3	400	1050
84	4	450	675
95	5	450	750
106	6	450	825
117	7	450	900
140	8	450	1050
162	9	450	1200
190	10	500	1200
245	11	500	1500
280	12	600	1200
360	13	600	1500
440	14	600	1800

RECOMMENDED AIR CONDITIONING DESIGN LOADS

Building Type	Design Loading
Computer rooms	10 m² of floor area per ton
Restaurants	16 m² of floor area per ton
Banks (main area)	22 m² of floor area per ton
Large Office Buildings (exterior zone)	25 m² of floor area per ton
Supermarkets	30 m² of floor area per ton
Large Office Block (interior zone)	32 m² of floor area per ton
Small Office Block (interior zone)	35 m² of floor area per ton

Index

Index to Advertisers

Advertisement Manager:
Karen Donnison
E. & F.N. Spon
2–6 Boundary Row
London SE1 8HN